THE INTERSTELLAR DISK-HALO CONNECTION IN GALAXIES

INTERNATIONAL ASTRONOMICAL UNION

UNION ASTRONOMIQUE INTERNATIONALE

THE INTERSTELLAR DISK-HALO CONNECTION IN GALAXIES

PROCEEDINGS OF THE 144TH SYMPOSIUM OF THE
INTERNATIONAL ASTRONOMICAL UNION,
HELD IN LEIDEN, THE NETHERLANDS,
JUNE 18–22, 1990

EDITED BY

HANS BLOEMEN

SRON, Leiden, The Netherlands

KLUWER ACADEMIC PUBLISHERS

DORDRECHT / BOSTON / LONDON

Library of Congress Cataloging-in-Publication Data

```
International Astronomical Union. Symposium (144th : 1990 : Leiden,
  Netherlands)
    The interstellar disk-halo connection in galaxies : proceedings of
  the 144th Symposium of the International Astronomical Union, held in
  Leiden, The Netherlands, June 18-22, 1990 / edited by Hans Bloemen.
      p.    cm.
    Includes index.
    ISBN 0-7923-1255-4 (HB : acid free paper)
    1. Galaxies--Congresses.  2. Milky Way--Congresses.  3. Disks
  (Astrophysics)--Congresses.  4. Galactic halos--Congresses.
  5. Interstellar matter--Congresses.   I. Bloemen, Johannes
  Bernhardus Gerhardus Maria, 1957-      II. Title.
  QB856.I58   1990a
  523.1'12--dc20                                               91-15239
```
ISBN 0-7923-1255-4

Published on behalf of
the International Astronomical Union
by
Kluwer Academic Publishers, P.O. Box 17, 3300 AA Dordrecht, The Netherlands.

Kluwer Academic Publishers incorporates
the publishing programmes of
D. Reidel, Martinus Nijhoff, Dr W. Junk and MTP Press.

Sold and distributed in the U.S.A. and Canada
by Kluwer Academic Publishers,
101 Philip Drive, Norwell, MA 02061, U.S.A.

In all other countries, sold and distributed
by Kluwer Academic Publishers Group,
P.O. Box 322, 3300 AH Dordrecht, The Netherlands.

Printed on acid-free paper

All Rights Reserved
© 1991 International Astronomical Union

No part of the material protected by this copyright notice may be reproduced or utilized in any form or by any means, electronic or mechanical including photocopying, recording or by any information storage and retrieval system, without written permission from the publisher.

Printed in the Netherlands

TABLE OF CONTENTS

Preface and editorial notes — xi

Committees and supporting organizations — xii

Participants — xiii

Opening address by *Professor J.H. Oort* — xvii

F.D. Kahn — 1
    Introduction to Fountains, Wind, Infall (and Magnetic Fields and Dynamo Mechanisms)

I. THE DISK-HALO INTERFACE IN OUR GALAXY

F.J. Lockman — 15
    The Neutral Halo in the Inner Galaxy

B.P. Wakker — 27
    High-velocity Clouds

L. Blitz — 41
    Molecular Clouds at High z

L. Danly — 53
    Optical and Ultraviolet Absorption Studies of Cool Gas in the Milky Way Halo

R.J. Reynolds — 67
    Ionized Disk/Halo Gas: Insight from Optical Emission Lines and Pulsar Dispersion Measures

D.W. Sciama — 77
    Dark Matter Decay and the Heating and Ionisation of HI Regions

M. Tosi   79
    Constraints on Galactic Infall from Studies of Chemical Evolution

I.F. Mirabel   89
    Infall of HVC's and the Origin of HI Supershells

G.L. Verschuur   93
    Wave Structure within HI Filaments at High Galactic Latitude and the Nature of "Clouds" in Interstellar Space

R.F.G. Wyse and G. Gilmore   97
    The Stellar Disk-Halo Connection

W. Tobin   109
    Star Formation at Large Galactic z?

J. Brand and J.G.A. Wouterloot   121
    Molecular Clouds and Star Formation at Large R

B.D. Savage   131
    UV Absorption and Emission Lines from Highly Ionized Gas in the Galactic Halo

D.P. Cox   143
    Hot Gas in the Disk, Halo, and Disk-Halo Interaction

F.X. Désert   149
    The High-Latitude Sky at IR, Optical, and UV Wavelengths

K.S. de Boer, U. Herbstmeier, and U. Mebold   161
    Metals and Molecules in Halo Clouds

B.-C. Koo, C. Heiles, and W.T. Reach   165
    Galactic Worms

Y. Sofue   169
    Magnetic Fields in the Disk-Halo Interface

V.A. Dogiel   175
    The Cosmic-ray Halo: Insight from Gamma Rays and Cosmic-ray Observations

W. Reich   187
    Radio Studies of Cosmic Rays in the Galaxy

X. Chi and A.W. Wolfendale 197
    Ionization in the Interstellar HI Region by Low-energy Cosmic-ray Electrons (poster)

X. Chi and A.W. Wolfendale 198
    Optical Radiation Field in the Disk and Halo (poster)

II. THE DISK-HALO INTERFACE IN OTHER GALAXIES

J.M. van der Hulst and J. Kamphuis 201
    Large-scale Structure of HI in other Galaxies

R. Braun 213
    Small-scale Properties of HI in Nearby Galaxies

R.A.M. Walterbos 223
    Diffuse Ionized Gas in Nearby Galaxies

E.M. Berkhuijsen, G. Golla, and R. Beck 233
    Is there Evidence for Disk-Halo Connections in M31?

A.C. Fabian 237
    X-ray Haloes and Cooling Flows

E.A. Valentijn 245
    Opaque Spiral Disks: Some Empirical Facts and Consequences

E. Hummel 257
    Radio Studies of Cosmic Rays in Nearby Galaxies

R. Beck 267
    Magnetic Fields in Disks and Halos of Spiral Galaxies

E.R. Seaquist and N. Odegard 281
    Synchrotron Emission as a Tracer of the Outflow in M82

R.J. Allen and S. Sukumar 287
    NGC 891: A Summary of Observations

R.-J. Dettmar, J.W. Keppel, M.S. Roberts, and J.S. Gallagher 295
    The Diffuse Ionized Gas Perpendicular to the Plane of NGC 891

S. Garcia-Burillo, M. Dahlem, and M. Guélin 299
    A CO survey of the Halo of NGC 891

F.P. Israel   303
    Cool Ionized Gas in Galaxy Thick Disks

Y. Sofue, N. Nakai, T. Handa, G. Golla, H.-P. Reuter, and R. Wielebinski   307
    Slow Rotation of Gas in the Halos of Edge-on Galaxies M82 and NGC 4631 (poster)

Y. Sofue, K. Wakamatsu, and D.F. Malin   309
    Boiling-Steaming Galactic Disk: Vertical Dust Jets in the Disk-Halo Interface (poster)

III. THEORY AND MODELLING

M. Crézé   313
    The Galactic Gravitational Potential

S.A. Stephens   323
    Hydrostatic Equilibrium of the Gas-Field System in the Galaxy and its Stability

H. de Boer   333
    Some Problems for Galactic Hydrostatic Equilibria

C.A. Norman   337
    The Global Mass, Energy, and Photoionization Balance of the Disk-Halo Interaction

H.J. Völk   345
    Cosmic-ray–powered Fountains and Winds

D.F. Cioffi   355
    The X-ray Appearance of Supernova Remnants in Tenuous Media

F.C. Jones   359
    Static versus Dynamical Cosmic-ray Halos

M. Pohl   369
    The Influence of Extended Source Distributions on Cosmic-ray Spectral Index Variations in the Galactic Wind Model

D. Breitschwerdt, J.F. McKenzie, and H.J. Völk   373
    Dynamical Implications of Diffusive and Convective Cosmic-ray Propagation in Galactic Halos

R. Schlickeiser 377
    Particle Acceleration in the Disk-Halo System

J.N. Bregman and G.A. Ashe 387
    The Structure of the Interstellar Medium

F. Ferrini 397
    Grain Evolution in the Framework of Disk-Halo Interactions

K. Tomisaka 407
    Numerical Simulations of Galactic Outflow and Inflow Phenomena

P.R. Shapiro 417
    Large-scale Gas Dynamical Processes Affecting the Origin and Evolution of Gaseous Galactic Halos

R. Matsumoto and K. Shibata 429
    Nonlinear Evolution of the Parker Instability

RAPPORTEUR PAPERS

C. Heiles 433
    The Interstellar Disk-Halo Connection in Galaxies: Review of Observational Aspects

Contents poster book 449

Author index 453

Subject index 455

PREFACE AND EDITORIAL NOTES

IAU Symposium No. 144 was held in Leiden in the summer of 1990, hosted by the Leiden Observatory and the Laboratory for Space Research Leiden. The meeting was devoted to the interstellar medium in the broadest sense, with emphasis on the interface between disks and halos of galaxies. This disk-halo connection is of major importance for studies of interstellar processes and galactic structure and evolution. Because of the wealth of new observational information, ranging from radio frequencies to the gamma-ray regime, and the considerable progress in theoretical studies in recent years, it was felt that this was the appropriate time to review the state of the art in an IAU symposium.

One of the main goals of the symposium was to bring together scientists working on widely different aspects of the field. The Scientific Organizing Committee, representing a broad range of expertise and nationalities, invited 34 speakers and selected 17 contributed oral presentations. Ample time was reserved for more than 60 poster presentations, which could all be displayed during the entire meeting. There were about 140 participants, coming from as many as 21 countries. Their long and lively discussions contributed significantly to the great success of the symposium.

This book contains the proceedings of the oral contributions, presented here in chronological order. A poster book, containing the substance of the poster presentations, was published by the Leiden Observatory prior to the meeting. Apparently, having this poster book available during the symposium was greatly appreciated by the participants. A few poster papers are included in the present volume, because they were accidentally excluded from the poster book. Copies of the poster proceedings can be obtained from the editor.

The non-scientific highlight of the meeting was a trip in a fleet of six classical flat-bottom sailing boats, followed by the conference banquet at the "Kager Plassen" lakes and a priceless after-dinner talk by Prof. H.C. van de Hulst. Photographs of this happening can be found scattered throughout the book.

The organizers express their gratitude to the institutions and companies that have financially supported the symposium. They are also very indebted to the Mayor of Leiden and the University of Leiden for the warm reception offered to the participants in the City Hall. I am grateful to every member of the Scientific and Local Organizing committees, especially to Janet Soulsby and Ernestine Spendel for their invaluable help in the organization of the meeting and the editing of the proceedings. Jan Melisse presented smoothly all slides, five days long, from a dark and stuffy projection room.

Hans Bloemen

SPONSORING IAU COMMISSIONS

33 Structure and Dynamics of the Galactic System
28 Galaxies
34 Interstellar Matter
48 High-energy Astrophysics

SCIENTIFIC ORGANIZING COMMITTEE

J.B.G.M. Bloemen (The Netherlands, Chair)
W.B. Burton (The Netherlands)
C.J. Cesarsky (France)
L.L. Cowie (U.S.A.)
V.A. Dogiel (U.S.S.R.)
T.W. Hartquist (F.R.G.)
C. Heiles (U.S.A.)
S. Ikeuchi (Japan)
J. Lequeux (France)
S.A. Stephens (India)
P.C. van der Kruit (The Netherlands)

LOCAL ORGANIZING COMMITTEE

J.B.G.M. Bloemen
W.B. Burton
F.P. Israel (Chair)
J.R. Soulsby
E.G.A. Spendel
in collaboration with the *Leids Congres Bureau*

SUPPORTING ORGANIZATIONS

International Astronomical Union
Kluwer Academic Publishers
Laboratory for Space Research Leiden
Leiden Observatory / Leiden University
Leids Kerkhoven-Bosscha Fonds
Philips
Royal Dutch Academy of Arts and Sciences (KNAW)

PARTICIPANTS

P. ABRAHAM	Konkoly Observatory Budapest, Hungary
R.J. ALLEN	STScI Baltimore, USA
S. AMES	Radioastronomisches Inst. Universität Bonn, FRG
V.A. ANTONOV	Inst. Theoretical Astronomy Leningrad, USSR
B. BARSELLA	Dipartimento di Fisica Pisa, Italy
A. BARTELDREES	MPIfR Bonn, FRG
R. BECK	MPIfR Bonn, FRG
R.A. BENJAMIN	University of Texas Austin, USA
E.M. BERKHUIJSEN	MPIfR Bonn, FRG
O. BIENAYME	Besançon Observatory, France
R.C. BISHOP	University of Texas Austin, USA
L. BLITZ	University of Maryland College Park, USA
J.B.G.M. BLOEMEN	Sterrewacht Leiden/Space Research Leiden, The Netherlands
N. BOCK CROSBY	Niels Bohr Institute Copenhagen, Denmark
J.C.J. DE BOER	Space Research Leiden, The Netherlands
K.S. DE BOER	Sternwarte Bonn, FRG
D.J. BOMANS	Sternwarte Bonn, FRG
D. BOWEN	Institute of Astronomy Cambridge, England
J. BRAND	Oss. Arcetri Florence, Italy
R. BRAUN	NFRA Dwingeloo, The Netherlands
J.N. BREGMAN	University of Michigan Ann Arbor, USA
D. BREITSCHWERDT	MPI für Kernphysik Heidelberg, FRG
L. BRETT	Manchester University, England
G. BURBIDGE	UCSD La Jolla, USA
W.B. BURTON	Sterrewacht Leiden, The Netherlands
A.M. BYKOV	Polytechnical Institute Leningrad, USSR
D. CAPTYN	Niels Bohr Institute Copenhagen, Denmark
C.L. CARILLI	CfA Cambridge, USA
C. CESARSKY	Service d'Astrophysique Saclay, France
S.K. CHAKRABARTI	Tata Institute Bombay, India
X. CHI	Physics Dept. University of Durham, England
P. CINZANO	Dipartimento di Astronomia Padova, Italy
D.F. CIOFFI	North Carolina State University Raleigh, USA
D. COX	University of Wisconsin Madison, USA
M. CREZE	Observatoire de Strasbourg, France
L. DANLY	STScI Baltimore, USA
F.X. DESERT	DEMIRM Meudon, France
R.-J. DETTMAR	Radioastronomisches Inst. Universität Bonn, FRG
V.A. DOGIEL	Lebedev Physical Institute Moscow, USSR
V.L. DORMAN	Lebedev Physical Institute Moscow, USSR
M.K. DOUGHERTY	MPI für Kernphysik Heidelberg, FRG
A. FABIAN	Institute of Astronomy Cambridge, England
A. FERRARA	Ist. di Astronomia Univ. di Firenze, Italy

F. FERRINI	Dipartimento di Fisica Pisa, Italy
S. GARCIA-BURILLO	IRAM Grenoble, France
G. GILMORE	Institute of Astronomy Cambridge, England
G. GOLLA	MPIfR Bonn, FRG
I. GRENIER	C.E.N. Saclay / University of Paris, FRANCE
A. HABE	Hokkaido University Sapporo, Japan
T. HANDA	Institute of Astronomy, University of Tokyo, Japan
L. HARTMANN	Sterrewacht Leiden, The Netherlands
M. HATTORI	RIKEN Wako Saitama, Japan
U. HAUD	Tartu Astrophysical Observatory Toravere, Estonia - USSR
C. HEILES	University of Colorado JILA Boulder/UC Berkeley, USA
M. HERNANZ	Centre d'Etudes Avancats de Blanes (CSIC), Spain
J.J. HESTER	IPAC/Caltech Pasadena, USA
H.C. VAN DE HULST	Sterrewacht Leiden, The Netherlands
J.M. VAN DER HULST	Kapteyn Institute Groningen, The Netherlands
E. HUMMEL	Jodrell Bank Macclesfield, Cheshire, England
J.A. IRWIN	Herzberg Institute of Astrophysics Ottawa, Canada
F.P. ISRAEL	Sterrewacht Leiden, The Netherlands
F. JANSEN	Space Research Institute Berlin, BRD
P.M.M. JENNISKENS	Sterrewacht Leiden, The Netherlands
J.L. JONAS	Rhodes University Grahamstown, South Africa
F.C. JONES	NASA/GSFC Greenbelt, USA
F.D. KAHN	University of Manchester, England
J. KAMPHUIS	Kapteyn Institute Groningen, The Netherlands
E. KASAK	Tartu Astrophysical Institute Toravere, Estonia - USSR
D.-W. KIM	CfA Cambridge, USA
C.-M. KO	MPI für Kernphysik Heidelberg, FRG
B.-C. KOO	University of California Berkeley, USA
B. KORIBALSKI	MPIfR Bonn, FRG
P.C. VAN DER KRUIT	Kapteyn Institute Groningen, The Netherlands
K.D. KUNTZ	STScI Baltimore, USA
S.E. LABOV	Lawrence Livermore National Laboratory, USA
K. LANZETTA	Institute of Astronomy Cambridge, England
F. LI	National Astronomical Observatory Mitaka, Japan
F.J. LOCKMAN	NRAO Charlottesville, USA
R. MATSUMOTO	Chiba University, Japan
J.P.M. MELISSE	Sterrewacht Leiden, The Netherlands
H. MEYERDIERKS	Radioastronomisches Inst. Universität Bonn, FRG
J. MILOGRADOV-TURIN	Inst. of Astronomy, University of Beograd, Yugoslavia
I.F. MIRABEL	Service d'Astrophysique Saclay, France
L. NETO	Observatoire de Paris Meudon, France
E.M. NEZHINSKIJ	Inst. Theoretical Astronomy Leningrad, USSR
C.A. NORMAN	STScI / JHU Baltimore, USA
R. OLLING	Columbia University New York, USA
A. ONCICA	Bucharest Observatory, Romania

J.H. OORT	Sterrewacht Leiden, The Netherlands
J. PALOUS	Astronomical Institute Prague, Czechoslovakia
C. PARDI	University of Milan, Italy
K. PEDERSEN	University of Copenhagen, Denmark
QIU-HE PENG	Nanjing University, China
M. POHL	MPIfR Bonn, FRG
W. REICH	MPIfR Bonn, FRG
H.-P. REUTER	MPIfR Bonn, FRG
R. J. REYNOLDS	University of Wisconsin Madison, USA
N. ROOS	Sterrewacht Leiden, The Netherlands
D.J. SAIKIA	Tata Institute Pune, India
S. SAKAMOTO	Faculty of Science, University of Tokyo, Japan
R. SANCISI	Kapteyn Institute Groningen, The Netherlands
W. SARGENT	Palomar Observatory Pasadena, USA
B.D. SAVAGE	University of Wisconsin Madison, USA
R. SCHLICKEISER	MPIfR Bonn, FRG
H. SCHULZ	Ruhr-University Bochum, FRG
U.J. SCHWARZ	Kapteyn Institute Groningen, The Netherlands
D.W. SCIAMA	Sissa Trieste, Italy
E.R. SEAQUIST	University of Toronto, Canada
P.R. SHAPIRO	University of Texas Austin, USA
M. SHAW	University of Manchester, England
M. SHULL	University of Colorado JILA Boulder, USA
V. SHUTENKOV	Lebedev Physical Institute Moscow, USSR
Y. SOFUE	Institute of Astronomy, University of Tokyo, Japan
I. SOUVATZIS	Radioastronomisches Inst. Universität Bonn, FRG
S.A. STEPHENS	Tata Institute Bombay, India
G. TENORIO-TAGLE	MPI für Physik & Astrophysik München, FRG
W. TOBIN	University of Canterbury Christchurch, New Zealand
K. TOMISAKA	Niigata University, Japan
M. TOSI	Osservatorio Astronomico di Bologna, Italy
L.V. TOTH	Eötvös University Budapest, Hungary
M. URBANIK	Astronomical Observatory Krakow, Poland
E.A. VALENTIJN	ESO Garching-bei-München, FRG
G.L. VERSCHUUR	4802 Brookstone Terrace, Bowie MD, USA
H.J. VOLK	MPI für Kernphysik Heidelberg, FRG
B.P. WAKKER	University of Illinois Urbana, USA
D.J. VAN DER WALT	Potchefstroom University, South Africa
R.A.M. WALTERBOS	University of California Berkeley, USA
P.P. VAN DER WERF	Kapteyn Institute Groningen, The Netherlands
H. VAN WOERDEN	Kapteyn Institute Groningen, The Netherlands
B. WOERMANN	Rhodes University Grahamstown, South Africa
G. YOUSSEFI	MPE Garching-bei-München, FRG
K.P. ZYBIN	Lebedev Physical Institute Moscow, USSR

OPENING ADDRESS BY PROFESSOR J.H. OORT

On behalf of the Organizing Committee I express a warm welcome to all participants. I hope you will find a favourable sphere to promote your discussions. I also hope the atmosphere will favour you on your sailing trip to the lakes.

As regards the sphere The Netherlands have a long and fruitful tradition in galactic research. As regards the atmosphere the tradition is more uncertain. We just hope for a rainless day with a fresh breeze.

Today's inauguration makes thoughts go back to the time when the International Astronomical Union for the first time organized a symposium. This was in 1953. Like our present symposium it took place in The Netherlands, on an estate near Groningen, and like the present its topic was the structure of the Galaxy, in fact on "Co-ordination of Galactic Research". That meeting was much cosier, almost a family gathering, with no more than 28, invited, participants. But it initiated an important new development in the I.A.U.

The second symposium, organized in the same year, this time in combination with the Union of Theoretical and Applied Dynamics, was on "Gas Dynamics of Cosmic Clouds" and thus came still closer to our present meeting. Among the participants were the physicist J.M. Burgers, from Delft, von Weizsäcker, Ludwig Biermann, Hoyle, Gold and such dynamic celebrities as von Kármán. The published account, edited by Van de Hulst, has been widely read and consulted.

Among the many topics that will be discussed there is one to which I would like to draw special attention, namely the ionization of hydrogen in dark clouds. This may lead to an unexpected new insight as will be explained by Dr. Sciama.

I wish that the present synposium may become equally inspiring as these two predecessors.

18 june 1990

INTRODUCTION TO FOUNTAINS, WIND, INFALL
(AND MAGNETIC FIELDS AND DYNAMO MECHANISMS)

F. D. KAHN
Department of Astronomy
The University
Manchester M13 9PL
England

ABSTRACT. The galactic fountain is formed by hot gas rising from the galactic disk. The lines of force of the interstellar magnetic field are dragged along in the flow. This lecture deals with the geometry and topology of the configuration of the field that is set up. The fountain flow acts as a dynamo, and the presence of the magnetic field plays an important part in the ionization balance of the gas as it returns towards the disk.

1. INTRODUCTION

The galactic fountain is essentially a dynamic phenomenon. That is the key to understanding its properties. The motion is driven by the hot inter-cloud medium which itself derives its energy from supernova remnants. Models of this kind have been discussed before (Bregman, 1980; Kahn, 1981, 1989), but here the treatment is extended to include effects due to the galactic magnetic field. The lines of force in such a flow eventually detach themselves from the material that they are linked to in the disk. The excursion into the fountain eventually causes the field to be amplified and therefore constitutes a dynamo mechanism. The presence of the field also has a dominant effect on the pressure balance in gas clouds falling back towards the disk, and needs to be allowed for in the interpretation of relevant observations. In particular an infalling cloud will only be observed as neutral gas at 21cm if it has a sufficiently large surface density.

2. SUPERNOVAE, THE SOURCES OF THE FOUNTAIN FLOW

A galactic fountain must be supported dynamically: it is very hard to argue for any alternative model, in which flow speeds are subsonic and support is essentially hydrostatic. The discussion here deals with dynamic fountains and will be based on a specification of the underlying galactic structure, as follows (Sellwood and Sanders, 1988):

50% by mass of the Galaxy belongs to Population II, with its contents distributed so that the mass contained within radial distance r of the centre is

$$M(r) = \mu r ; \tag{1}$$

50% by mass belongs to Population I, with its contents distributed in a thin disk, so that the mass contained within distance ϖ of the axis is

$$M(\varpi) = \mu \varpi ; \tag{2}$$

the rotation speed in the Galaxy is

$$V = 250 \, \text{km s}^{-1} , \tag{3}$$

the same at all axial distances, and then

$$\mu = 4.7 \times 10^{21} \text{gm cm}^{-1} . \tag{4}$$

Adopting a value of 10 kpc for the distance from the Sun to the galactic centre, the half-surface density of the disk becomes

$$\Sigma = \frac{\mu}{4\pi\varpi} = 1.24 \times 10^{-2} \text{gm cm}^{-2} . \tag{5}$$

Just above the galactic disk the gravitational acceleration towards the disk, due almost entirely to Population I, is

$$g_z = 4\pi G\Sigma = 10^{-8} \text{cm s}^{-2} . \tag{6}$$

The fountain rises to a height $Z = 3$ kpc above the galactic plane, according to the generally accepted view. The galactic disk has a half-thickness much smaller than Z; in a quasi-static fountain the scale height of the distribution would have to be

$$\frac{kT}{\overline{m}g_z} \sim Z , \tag{7}$$

where T is the temperature and $\overline{m}$ ($\sim 10^{-24}$ gm) the mean particle mass, and therefore

$$T \sim 7 \times 10^5 K . \tag{8}$$

The dynamical relaxation time of such a structure is

$$t_{\text{dyn}} = \left(\frac{kT}{\overline{m}}\right)^{\frac{1}{2}} / g_z = 10^{15} \text{s} = 30 my; \tag{9}$$

if the fountain can be set up by subsonic motions, t_{dyn} must be short compared with

$$t_{\text{therm}} = 3kT/\Lambda(T)n , \tag{10}$$

the thermal relaxation time. Here

$$\Lambda(T) \sim \lambda T^{-\frac{1}{2}} \tag{11}$$

is the usual cooling function, and $\lambda = 10^{-19}$ in CGS units, while n is the atom/ion density (Kahn, 1976). The condition implies that the column density in the halo is limited by the inequality

$$nZ \ll 3\left(\frac{\overline{m}^3}{k}\right)^{\frac{1}{2}} \frac{g^2 Z^2}{\lambda} = 2.5 \times 10^{19} \text{cm}^{-2}. \qquad (12)$$

Only a very tenuous fountain would obey this restriction. So a quasi-static model of the fountain proves to be unsatisfactory, and it clearly is necessary to consider dynamical models.

Three distinct possibilities of this kind have been proposed:

(i) The fountain flow consists of gas coming from the innermost parts of many supernova remnants. The essential point here is that the inner parts of a remnant are heated very strongly by the blast wave from the explosion, and take a very long time to lose their thermal energy by radiation. The flow into the fountain is therefore thought to consist of the amalgamated hot remains of many neighbouring SNR's.

(ii) A popular alternative view is that the major contribution to the flow comes from the combined remnants of many supernovae that have occurred close together, in space and time, at the sites of evolving OB associations. There is enough energy in such a combined explosion to create a blast that breaks right through the galactic disk, and ejects matter ballistically into the space above. The driving force is still provided by hot gas (Norman, 1990).

(iii) The third view stresses the possible importance of cosmic rays in driving the flow. Once more the fountain is spread more or less homogeneously over the galactic disk. The internal energy that sets up the motion is provided by the cosmic rays; they are coupled to the thermal plasma by Alfvén waves, and drag it along as they escape from the galactic disk (Völk, 1990).

Both (ii) and (iii) are dealt with in later papers at this Symposium. Here I shall explore the possibilities offered and the restrictions imposed by the first mechanism.

The first question to ask concerns the spacing of supernova remnants. If one waits long enough, then neighbouring supernova remnants will merge. By the time this happens a particular remnant will be well into Phase III. The Sedov solution predicts that the radius of a remnant is

$$r = at^{\frac{2}{5}} \equiv \left(\frac{2E}{\overline{\rho}}\right)^{\frac{1}{5}} t^{\frac{2}{5}} \qquad (13)$$

during Phase II; here the energy of the supernova explosion is $E (= 3 \times 10^{51}$ erg$)$ and the density of the surrounding medium is $\overline{\rho} (= 2 \times 10^{-24}$ gm cm$^{-3})$. Phase II comes to an end when radiative energy loss becomes important, at time

$$t = t_* = 1.1 \frac{E^{\frac{3}{14}}}{q^{\frac{5}{14}} \overline{\rho}^{\frac{4}{7}}}. \qquad (14)$$

The term q here is a coefficient introduced in a simplified cooling law suitable for interstellar gas dynamics, in the temperature range $3 \times 10^5 K < T < 3 \times 10^7 K$, see Kahn (1976). Its numerical value is 4×10^{32}, in CGS units, for a gas mixture with solar abundances.

Before time t_* the expansion of the SNR is driven mainly by gas pressure, and after time t_* it becomes momentum driven; the radius at time t is then given by

$$r = bt^{\frac{1}{4}}. \tag{15}$$

The coefficient here can be expressed in terms of E, q and $\bar{\rho}$ by the relation

$$b = 0.92 \frac{E^{\frac{13}{56}}}{q^{\frac{3}{56}}\bar{\rho}^{\frac{2}{7}}}, \tag{16}$$

with a very weak dependence on q, and therefore on the precise form of the cooling law.

Supernova explosions of type II seem to occur about three times per century per galaxy, effectively in a volume

$$\begin{aligned} \mathcal{V} &= \pi \times (10\text{kpc})^2 \times 200\text{pc} \\ &= 1.7 \times 10^{66}\text{cm}^3 \end{aligned}$$

and their rate of occurrence, per unit volume and time, is

$$\nu = 6 \times 10^{-76}\,\text{cm}^{-3}\,\text{s}^{-1}.$$

Now imagine that the Galaxy had started at time zero entirely free of supernova remnants, and that SNR's were subsequently created at rate ν and allowed to expand without interference from each other. The available space in the disk would be filled when

$$\frac{4\pi}{3}b^3 t^{\frac{3}{4}}\nu t = 1$$

or

$$t = \left(\frac{3}{4\pi}\right)^{\frac{4}{7}} b^{-\frac{12}{7}}\nu^{-\frac{4}{7}} = 3.9 \times 10^{13}\text{s} = 1.2 \text{ million years};$$

the radius of the typical remnant would be

$$r = 2.2 \times 10^{20}\text{cm} \sim 70\text{ pc}$$

then. These estimates compare well with the linear dimensions and inferred ages of the known loop structures in the neighbourhood of the Sun.

Even at this late stage a SNR still contains shock-heated gas which has not yet lost its thermal energy. The effect of radiative losses is conveniently described by the equation (Kahn 1976)

$$\frac{D}{Dt}\kappa^{\frac{3}{2}} = -q \tag{17}$$

where D/Dt denotes rate of change following the motion of a fluid element, and

$$\kappa \equiv \frac{P}{\rho^{\frac{5}{3}}} \tag{18}$$

is the adiabatic parameter. Equation (17) applies only while the gas still retains some heat; a particular fluid element has cooled off completely, and occupies only a small volume, when

κ reaches zero. If a fluid element is to stay hot for time $t_h = 10^6$ years, then it must acquire so much thermal energy from the blast wave that

$$\kappa^{\frac{3}{2}} = \kappa_s^{\frac{3}{2}} > qt_h = 1.2 \times 10^{46} \tag{19}$$

at the time that the blast wave passes across it. Each supernova explosion produces about 200 $M_\odot$ of interstellar gas which are shock-heated to this extent. On the present model this gas is still hot when neighbouring SNR's merge with one another, and is available for flow into the fountain. A supernova rate of three per century produces a flow of gas into the fountain at the rate of 6 $M_\odot$ per year. This estimate agrees well with the mass infall rate from the halo deduced by Wakker (1990a). The total mass of interstellar gas in the disk is

$$\begin{aligned} M_{is} &= 2 \times 10^{-24} \times 1.7 \times 10^{66} \,\text{gm} \\ &= 1.7 \times 10^9 \, M_\odot, \end{aligned}$$

with the present figures, so that the typical interstellar atom will typically make an excursion into the fountain once every 300 million years.

The flow takes place on each side of the galactic disk over an area some 10 kpc in radius; the mass flux is therefore

$$\begin{aligned} \Phi &= \frac{6}{6 \times 10^8} \, M_\odot \, \text{pc}^{-2} \, \text{yr}^{-1} \\ &= 7 \times 10^{-20} \, \text{gm cm}^{-2} \, \text{s}^{-1} \, . \end{aligned}$$

The fountain has to rise to a maximum height $Z = 3$ kpc, and so requires enthalpy

$$\mathcal{E} = g_z Z = 10^{14} \, \text{erg gm}^{-1} \, .$$

In terms of P, ρ and κ, then,

$$\frac{2}{3}\frac{P}{\rho} = \mathcal{E} \tag{20}$$

and if the adiabatic parameter given by

$$\kappa^{\frac{3}{2}} \sim 10^{46} \, (\text{CGS units}) \tag{21}$$

is taken as typical, then

$$\begin{aligned} P &= \left(\frac{2}{3}\mathcal{E}\right)^{\frac{5}{2}} \kappa^{-\frac{3}{2}} \\ &= 3.6 \times 10^{-12} \, \text{dyne cm}^{-2} \, . \end{aligned} \tag{22}$$

This value is somewhat larger than the interstellar pressure. A more careful discussion is needed for a more accurate determination of the parameter κ.

The upward flow leaves the disk with speed u_0, and produces a recoil force, per unit area,

$$\Phi u_0 = 10^{-12} \, \text{dyne cm}^{-2},$$

which agrees well with the usual value for the interstellar pressure.

If the fountain were purely one-dimensional there would be no need to consider any possible sideways drift parallel to the galactic plane. But the maximum height is comparable to typical distances within the disk, say that to the galactic centre. It is therefore relevant to note that the component of the gravitational acceleration in the direction towards the axis of symmetry is

$$g_\varpi = \frac{V^2}{\varpi_0} - \frac{V^2}{\varpi_0^2}\left(\Pi + \frac{|z|}{2}\right) + \ldots, \tag{23}$$

appropriate to the mass distribution assumed earlier. The symbols here mean that the fluid element has been displaced a distance z vertically and a distance Π axially from its original position in the plane, at axial distance ϖ_0. The material undergoes a Lindblad oscillation, whether or not it is in the galactic disk. The displacement $|z|$ away from the plane reduces the centripetal acceleration and causes the gas to drift outwards, and therefore also to lag in its rotation about the galactic axis. There is not space here to discuss the full implications of this effect, except to remark that the gas returning from the fountain will not, in general, fall back onto the galactic disk at exactly the same axial distance as that where it originated. Mixing is therefore promoted by the fountain flow, as well as transport of angular momentum, and cross-linkages in the galactic magnetic field.

3. AMPLIFICATION OF THE MAGNETIC FIELD.

The flow into the fountain consists of the combined parts of remnants from many supernova explosions, closely packed together like a set of bubbles in a foam. It can be described neither in terms of a single large bubble expelled by itself from the disk, nor in terms of a smooth continuum flow, since the individual bubbles themselves have diameters comparable with the thickness of the disk.

An important consequence of this model is that it naturally leads to the amplification of the galactic magnetic field. The 200 $M_\odot$ of hot gas which fill the bubble when it amalgamates with its neighbours were originally contained in a spherical volume of radius $r_0 = 12$ pc, and then expanded to a radius $r_f = 70$ pc, an increase in linear dimensions by a factor of about 6. The corresponding density decrease is by a factor 200 or so, and the flow upwards starts with a density $\rho_f \sim 10^{-26}$ gm cm^{-3}. The magnetic field linking the original distribution of material has strength $B_0 \sim 3 \times 10^{-6}$ gauss. The magnetic flux linked to the bubble is

$$\Phi_{\mathrm{mag}} = \pi B_0 r_0^2 = 1.22 \times 10^{36} \,\mathrm{gauss\,cm}^2$$

and remains conserved during the three-dimensional expansion. When the bubble reaches radius r_f, the horizontal field strength has therefore dropped to $B_f \sim 10^{-7}$ gauss, and the ratio B/ρ has increased by a factor 6, from 1.5 $\times 10^{18}$ gauss cm^3 gm^{-1} to 9 $\times 10^{18}$ gauss cm^3 gm^{-1}. Even after the expansion each bubble stays magnetically linked to the same elements of the gas as before the supernova explosion, all of them lying further out in the SNR. Consequently these elements will have been heated to a much smaller extent by the blast wave, and can cool off again by the time that the bubble drifts off into the fountain. So each rising bubble still has tubes of force trailing from it, connecting it to interstellar matter that remains behind in the disk. A complex magnetic pattern results in the fountain, with a systematic field component parallel to the original lines of force in the disk, as well as a vertical component, perpendicular to the disk, whose net flux is zero, but whose average field strength becomes dominant in the lower part of the flow.

If there were no reconnection of field lines the vertical component would eventually become so strong as to dominate flow entirely. Let the typical bubble have linear dimensions ℓ and contain horizontal magnetic field $B_\parallel$, linked to the galactic disk below. The associated vertical field $B_\perp$ then has total flux $B_\parallel \ell$ per unit width going up on one side of the bubble, and equal flux going down on the other. When the fountain has been flowing for some time, and there are n layers of bubbles stacked on top of one another, then tubes of force with an upward flux, per unit width, of $nB_\parallel \ell$, as well as an equal downward flux, have to be fitted into a length ℓ. The resulting vertical field averages to zero, and has typical magnitude

$$|B_\perp| = 2nB_\parallel = \frac{2nB_0}{f^2}. \qquad (24)$$

The spacing between vertical lines of force with opposite directions is typically ℓ/n. The Petschek mechanism (Petschek, 1964; Baum and Bratenahl, 1977) will cause reconnection in this field at the typical rate

$$\sigma \sim \frac{0.1\, B_\perp n}{\sqrt{4\pi\rho}\, \ell}. \qquad (25)$$

New vertical field is created by the flow at rate

$$\left(\frac{dB_\perp}{dt}\right)_{new} = 2B_\parallel \frac{v}{\ell} = \frac{B_\perp v}{n\ell}; \qquad (26)$$

reconnection halts the increase when

$$\sigma \equiv \frac{0.1\, B_\perp n}{\sqrt{4\pi\rho}\, \ell} = \frac{v}{n\ell}$$

or

$$n^2 = \frac{10\sqrt{4\pi\rho}\, v}{B_\perp}$$

or

$$n^3 = \frac{5\sqrt{4\pi\rho}\, v}{B_\parallel}. \qquad (27)$$

With the values adopted here, it is found that $n \sim 6$, and then setting $\ell = 140$ pc, the typical diameter of a bubble, it follows that the vertical magnetic field should extend to about 840 pc height; $|B_\perp|$ typically equals 1.2×10^{-6} gauss.

The vertical field will disappear at heights greater than about 1 kpc, and only a connected $B_\parallel$ field will remain. The later upward flow is essentially one-dimensional, and so is the final collapse back to the galactic plane. The gas that returns from the fountain has a value of $B_\parallel/\rho$ that is some six times larger than it was before it was expanded by the supernova remnant. The interstellar gas makes excursions into the fountain typically at intervals of 300 million years. On the face of it the magnetic field strength in the disk should increase by a factor 6 once every 300 million years, and so should double every 120 million years, but such a rate of increase is impossible to sustain.

In fact there are significant restrictions which limit the possible amplification rate. The enhanced magnetic field $B_\parallel$ can only be set up if each bubble in the fountain reconnects magnetically with other bubbles on either side. The linkage will be broken in places where there is no fountain flow. The amplification process will thus only produce extended magnetic loops lying above regions of the galactic disk with an active fountain flow.

A second limit is set by the conditions that the magnetic pressure term $B_\perp^2/8\pi$ must be significantly smaller than $\rho v^2 (\sim P_{is})$, the momentum flow into the fountain. Formally the restriction is that

$$\frac{B_\perp^2}{8\pi} = \frac{n^2 B_0^2}{2\pi f^4} \ll P_{is}, \tag{28}$$

the interstellar pressure, and writing relation (27) in the form

$$\frac{n^3 B_0}{f^2} = 5\sqrt{4\pi P_{is}} \tag{29}$$

one finds, after a simple calculation, that

$$\frac{B_0^2}{4\pi P_{is}} \ll \frac{f^4}{5 \times 2^{3/2}} \tag{30}$$

so that B_0 must be considerably smaller than 3×10^{-5} gauss, with $f = 6$. The amplification begins to fail when the galactic magnetic field is rather larger than 10 μG.

A lower limit is set on B_0 by the requirement that reconnection must occur fast enough, or else the entire galactic field might be swept into the fountain, leaving nothing behind to be amplified. The flux per unit width in the disk is $B_0 D$ and this must always be rather larger than $B_\perp \ell/2$, which is the flux still connected to bubbles in the fountain. So

$$B_0 D \gg \frac{1}{2} B_\perp \ell = n B_\parallel \ell = \frac{n B_0 \ell}{f^2}$$

or

$$\frac{n}{f^2} \ll \frac{D}{\ell} \tag{31}$$

and then with the help of relation (27) and a little manipulation

$$\frac{B_0^2}{4\pi P_i} \gg \frac{25 \ell^6}{f^8 D^6}. \tag{32}$$

With our values the implied minimum field strength is 1.3×10^{-8} gauss.

To sum up, this process will rearrange the magnetic field and create closed lines of force above regions in the galactic disk that have a fountain flow. All the interstellar gas is cycled through the fountain at intervals of a few hundred million years. The amplification process described here systematically increases the ratio $B_0 : \rho_0$; it operates provided that

$$10^{-8} \text{ gauss} < B_0 < 10^{-5} \text{ gauss}.$$

4. DESCENDING GAS IN THE FOUNTAIN: INFLUENCE OF THE MAGNETIC FIELD

The fountain flow plays an important role in maintaining the magnetic field of the Galaxy, as was shown in the previous section. Now I shall discuss the influence of the magnetic field on the structure and ionization balance of the gas as it descends towards the galactic plane. The flow is essentially one-dimensional, so that the horizontal field strength is related to the density by the formula

$$B_\| = K\rho \tag{33}$$

where typically $K = 10^{19}$ gauss cm^3 gm^{-1}. There is also the vertical component $B_\perp$ which has greater field strength at low altitudes but contributes no net flux, and can be ignored above about 1 kpc height. For this simple calculation it seems adequate to treat the motion as one-dimensional. Let the upward speed of flow be u at height z (greater than 1 kpc). Let there be a cooled sheet of gas at this height, with surface density Σ, descending at speed w. Two conditions specify its motion and its evolution: they are

conservation of mass

$$\frac{d}{dt}\Sigma w = \frac{\Phi}{u}(u+w) \tag{34}$$

and conservation of momentum

$$\frac{d}{dt}(\Sigma w) = \Sigma g - \Phi(u+w) . \tag{35}$$

The cooled sheet sets itself internally like an atmosphere under an effective gravitational acceleration

$$g_{\text{eff}} = g - \frac{dw}{dt} = \frac{\Phi}{\Sigma u}(u+w)^2 ; \tag{36}$$

in the simplest case it consists only of matter that has reached maximum height in the fountain, and is now falling back. In that case

$$w = \frac{u}{2} \tag{37}$$

and

$$\Sigma = 3\Phi\frac{u}{g} \tag{38}$$

and so

$$g_{\text{eff}} = \frac{3g}{4} . \tag{39}$$

But the falling sheet may be pushed from above by extragalactic matter that has entered from outside. In that case w will differ from the value given in (37), say

$$w = Wu/2 , \tag{40}$$

and the effective gravitational acceleration in the layer changes to

$$g_{\text{eff}} = \frac{g}{3}\left(1+\frac{W}{2}\right)^2 . \tag{41}$$

Suppose now that P_{eff} is the effective pressure (including magnetic effects) at the base of the descending layer, and let ρ_b be the gas density there, then

$$P_{\text{eff}} = \frac{B^2}{8\pi} + \rho_b c_i^2$$

$$= \frac{K^2\rho_b^2}{8\pi} + \rho_b c_i^2 , \tag{42}$$

and here c_i ($\equiv 10$ km s^{-1}) is the isothermal sound speed in the gas, assumed to be ionized. The momentum balance at the base of the sheet requires that

$$P_{\text{eff}} = \Phi u \left(1 + \frac{W}{2}\right)^2. \tag{43}$$

Magnetic pressure dominates whenever

$$\rho_b > \frac{8\pi c_i^2}{K^2} = 2.5 \times 10^{-25} \text{gm cm}^{-3},$$

with the values of the parameters being used here, and therefore also if

$$P_{\text{eff}} > \frac{16\pi c_i^4}{K^2} = 5 \times 10^{-13} \text{ dyne cm}^{-2},$$

or

$$u > \frac{P_{\text{eff}}}{\Phi\left(1 + \frac{W}{2}\right)^2} = \frac{7 \times 10^6}{\left(1 + \frac{W}{2}\right)^2} \text{ cm s}^{-1}. \tag{44}$$

W has to exceed unity, and so the least favourable value of the limiting speed on the right hand side is 30 km s^{-1}. The description from now on is confined to heights in the fountain greater than 1 kpc, so that the $B_\perp$ component of the field is absent, and more than 150 pc below the highest level in the flow, so that u exceeds 30 km s^{-1}, and magnetic pressure dominates. The condition of hydrostatic equilibrium in the descending sheet then states that

$$\frac{dP_{\text{eff}}}{dz} = \frac{K^2 \rho}{4\pi} \frac{d\rho}{dz} = -g_{\text{eff}} \rho$$

or

$$dz = -\frac{K^2}{4\pi g_{\text{eff}}} d\rho. \tag{45}$$

Let F_{ion} be the flux of Lyman continuum photons coming up into the fountain from unshielded early-type stars in the galactic disk. If F_{ion} is large enough, or the surface density Σ small enough, then the sheet will be turned entirely into an HII region. In such a case no high (or intermediate) velocity cloud will be detected at 21cm wavelength.

The value of F_{ion} defines a minimum surface density required before neutral hydrogen will be found in the sheet. The necessary condition is that

$$F_{\text{ion}} < \int b \, n_e^2 \, dz \tag{46}$$

where b is the usual recombination coefficient for hydrogen, n_e is the electron density and the integration extends through the sheet. The electron density in the ionized gas is related to the mass density by

$$n_e = \rho/m_a \tag{47}$$

where m_a ($= 2 \times 10^{-24}$ gm) equals the mass of atoms/ions present per free electron. With the help of (45) and (47) the inequality becomes

$$F_{\text{ion}} < \frac{b K^2 \rho_b^3}{12\pi g_{\text{eff}} m_a^2}$$

$$= \frac{1.92 \, b \, \Sigma^{3/2} g^{1/2} (1 + W/2)}{m_a^2 K} \tag{48}$$

after some reduction. The table below summarises the result. The calculation assumes that $W = 1$, in other words that the descending sheet is not pushed from behind by the infall of extragalactic material. But relation (48) shows that it is quite easy to allow for such an infall, and that the correction will not make a qualitative difference to the conclusions.

Mass per unit area Σ	10^{-5}	10^{-4}	10^{-3}	gm/cm^2
Upper limit to the ionizing flux from below $F_{\text{ion,max}}$	4.6×10^4	1.4×10^6	4.6×10^7	photon/cm^{2}s

The most likely suitable value for $F_{\text{ion,max}}$ is of order 10^6 photon/cm^{2}s (Bregman and Harrington, 1986; Reynolds, 1989). The results quoted in the table show that a Lyman c flux as strong as this will maintain full ionization in an infalling cloud unless the mass per unit area is at least of order 10^{-4} gm cm^{-2}, or the surface density of particles at least of order 10^{20} cm^{-2}. It is relevant therefore that the typical surface density of H atoms in an HVC is also of this order (Wakker, 1990b). Any interpretation of the HVC's must therefore make allowance for the inflow of the ionized part of a cloud that necessarily accompanies it.

REFERENCES

Baum, P. J., Bratenahl, A. (1977) *J. Plasma Phys.* **18**, 257.
Bregman, J. N. (1980) *Ap. J.* **236**, 577.
Bregman, J. N., Harrington, J. P. (1986) *Ap. J.* **309**, 833.
Kahn, F. D. (1976) *Astron. Astrophys.* **50**, 145.
Kahn, F. D. (1981) "Dynamics of the Galactic Fountain" in *Investigating the Universe*, ed. F. D.Kahn, Reidel, Dordrecht, p. 1.
Kahn, F. D. (1989) "Galactic Fountains" in *Structure and Dynamics of the Interstellar Medium*, IAU Colloquium 120, eds. G. Tenorio-Tagle, M. Moles, J. Melnick, Springer, p. 474.
Norman, C. (1990) Invited lecture at this Symposium.
Petschek, H. E. (1964) NASA SP-50, p. 425.
Reynolds, R. J. (1989) *Ap. J.* **339**, L29.
Völk, H. J. (1990) Invited lecture at this Symposium.
Wakker, B. P. (1990a) Ph.D.thesis, Rijksuniversiteit Groningen.
Wakker, B. P. (1990b) Invited lecture at this Symposium.

I

The Disk-Halo Interface in our Galaxy

THE NEUTRAL HALO IN THE INNER GALAXY

FELIX J. LOCKMAN
National Radio Astronomy Observatory [1]
Edgemont Rd., Charlottesville, VA USA

ABSTRACT. Neutral gas extends out of the disk to form a modest halo over most of the inner Galaxy except for the area at R $\lesssim 2.5$ kpc. The HI halo probably consists of several populations including a relatively quiescent layer and a cloudy component that has peculiar velocities. This latter component often appears between 150 and 400 pc from the plane. Most of the neutral gas at $|z| \gtrsim 500$ pc has the same kinematics as the gas below it.

1. INTRODUCTION

"... *clouds unquestionably exist at heights of the order of 1 kpc.*"
— Munch and Zirin (1961)

"... *well outside the real disk one still finds neutral hydrogen with an average density of between 5 and 10 per cent of the density in the plane.*" — Oort (1962)

Neutral gas that lies far from the galactic plane has been studied for at least thirty years, and has been the subject of debate for every bit of that time. It is fascinating to read the papers and discussion in the *Proceedings of the Third Symposium on Cosmical Gas Dynamics* held in June 1957, where the participants wrestled with issues that have remained puzzling into our time, in large part because the neutral halo is often difficult to detect and doubly difficult to disentangle from the disk. In the last decade, though, there have been major advances in the field including Albert's (1983) study of the halo's distribution and kinematics using optical absorption lines; measurement of halo hydrogen at 21 cm and in the Lyman-α ultraviolet absorption line (Lockman 1984; Kulkarni and Fich 1985; Savage and Massa 1987); and a growing number of studies that combine radio, optical and ultraviolet techniques (Hobbs *et al.* 1982; Lockman, Hobbs and Shull 1986; Albert

---

[1] The National Radio Astronomy Observatory is operated by Associated Universities, Inc., under agreement with the National Science Foundation.

et al. 1989). It is an active field. Recent reviews that discuss the topic of HI in the galactic halo include those by Kulkarni and Heiles (1987) and by Dickey and Lockman (1990; hereafter DL90), as well as contributions in this volume by Burton, Danly, Mirabel, Reynolds, Savage, and Wakker.

For this review of the neutral halo in the inner Galaxy I will first summarize some of the observations, especially the 21 cm studies, and then consider the more specific problems of halo kinematics and support. It is useful to begin with a list of the questions that, in my mind, form the background for this review. Many of these are likely to have interesting answers:

1. How pervasive is high-altitude HI? Is it, for example, entirely in clouds that have a small filling factor?
2. Is the amount of HI far from the galactic plane determined by the support available against gravity, or by ionization of the hydrogen? Does the temperature and ionization state of neutral gas change with z?
3. Are there several populations of neutral gas above the disk; is "halo gas" also found in the galactic plane?
4. How is the gas supported in the galactic potential?
5. Why does the HI halo change so dramatically at R=2.5 kpc but so little at larger radii?
6. What is the kinematics of halo gas; how tightly coupled is it to galactic rotation?
7. Is halo HI related to high-velocity clouds, the Magellanic Stream, or infall of extragalactic gas?

2. AVERAGE HI DENSITY ABOVE THE PLANE.

Figure 1 shows three recent experimental estimates of $n_H(z)$. The Dickey and Lockman (DL90) and the Lockman (L84) curves describe the distribution of 21 cm HI emission averaged over the range 3.4 − 6.8 kpc from the galactic center; the Savage and Massa curve is an exponential fit to Lyman-α column densities for about 60 stars within $\lesssim$10 kpc of the Sun, concentrated towards the inner Galaxy. The DL90 curve differs from L84 because it includes a correction for cool gas at low $|z|$. Quantitative values for these functions are given in DL90.

The DL90 curve is approximately Gaussian within a few hundred parsecs of the plane, where it is very similar to the functions found by previous 21 cm observers (e.g., Schmidt 1957; Baker and Burton 1975). For this review I am defining the "HI halo" as the gas well beyond the Gaussian core, typically at $|z| \gtrsim 400 - 500$ pc, where the observations, both radio and UV, show that there is a significant amount of neutral gas. Current estimates are in general agreement with Oort's (1962) values: at $R \sim R_0 = 8.5$ kpc between 5% and 10% of all HI lies at $|z| \gtrsim 500$ pc (L84; DL90).

It is interesting that optical and UV absorption line studies of the ISM have always derived a larger scale-height than 21 cm studies (cf., e.g., Bohlin, Savage

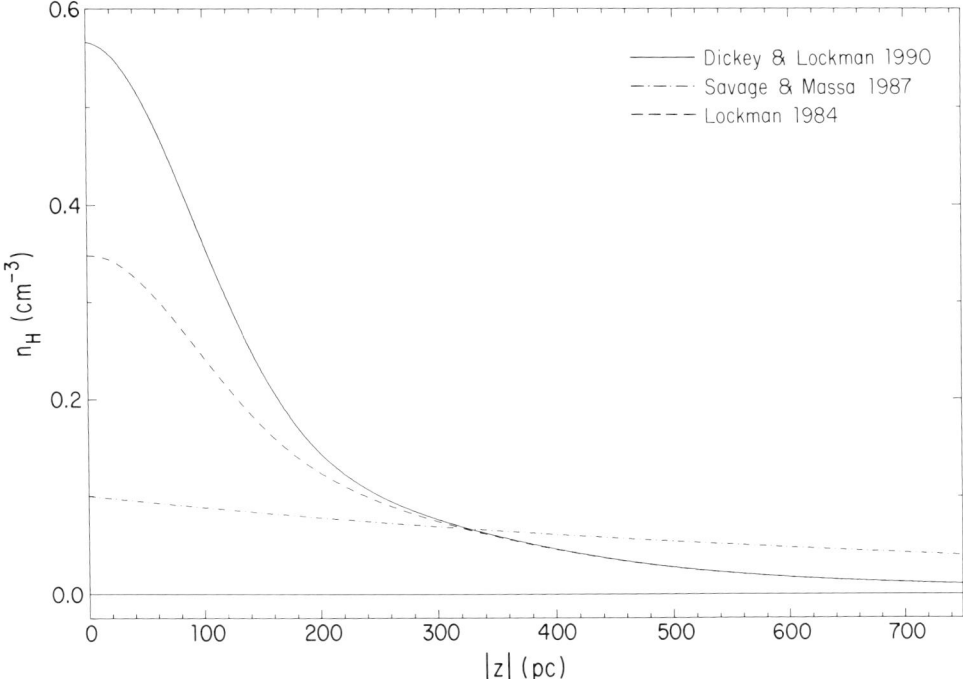

Figure 1. Recent estimates of the neutral hydrogen density above the galactic plane in the inner Galaxy. The Savage and Massa (1987) curve is derived from Lyman-α observations in the ultraviolet; the others come from 21 cm observations.

and Drake 1978; Celnik, Rohlfs and Braunsfurth 1979). Of course there are strong selection effects: the UV data are biased toward low reddening sitelines and positively avoid directions through dense clouds. This tendency artificially enhances the halo at the expense of the disk because the densest clouds are at low $|z|$. Also, the radio and UV data sample different parts of the Galaxy that might not have the same amount of halo gas. So all in all, within the range of uncertainty, and given the limited data, it is possible that the UV and radio estimates of the halo at $|z| \gtrsim 400$ pc do not actually disagree. But we cannot completely dismiss the possibility that the radio analyses are missing some gas at the higher altitudes, or more likely assigning it to the wrong location, because it has slightly different kinematics than disk gas. This is discussed further in section 5. Many more cooperative radio/optical/UV studies will be needed to establish the reality of the apparent differences.

There is some evidence that there is little neutral hydrogen at $|z| > 1$ kpc, at least in the part of the Galaxy near the Sun that has been probed by Lyman-α and optical absorption studies (Albert 1983; Albert et al. 1989). In this region the 21 cm column densities toward high-z stars are identical to the Lyman-α column densities, which give only the HI below the star, for all stars at $|z| \gtrsim 1500$ pc (Lockman, Hobbs and Shull 1986; DL90). On the other hand, there are places in the Galaxy where discrete HI structures are found more than 1 kpc from the plane (Smith 1963; L84; also see section 3). Most likely the upper boundary of the neutral halo varies from place to place.

Beyond the question of the accuracy of the derived $n_H(z)$ distributions is the question of how they should be interpreted. Do they supply evidence for a pervasive "intercloud" atmosphere around the galactic disk? Probably not. The 21 cm data are a smooth function of $|z|$ only when averaged over an enormous volume; individual positions sometimes show considerable clumpiness in the high-z gas. Likewise, the Lyman-α observations scatter about the simple exponential function to such an extent that a statistical analysis formally rejects the fit, unless it is assumed that there are line of sight variations of a factor of two in density (Danly et al. 1991; see also Lockman, Hobbs and Shull 1986). There are signs of structure in at least one halo cloud on a scale $\gtrsim 60$ pc, but the observations are so coarse that they are not very restrictive (Albert et al. 1989). In sum, we do not know how much of the gas at, say, z=500 pc, is in discrete structures with a low filling factor, and how much is so widely distributed that it would be useful to describe it as an atmosphere.

3. A NOTE ON SHELLS AND SUPERSHELLS

An all-sky map of integrated HI column densities (e.g. DL90) shows a number of arching filaments rising up out of the galactic plane to moderate and high latitudes. Many of these correspond to radio loops (e.g. the North Polar Spur; radio Loop II), others are parts of shells identified by Heiles (1979, 1984), others do not fit any clear pattern. Relatively dense HI associated with catalogued supershells (Heiles 1979) sometimes extends > 1 kpc from the plane. It can be difficult to detect old shells far from the Sun, but these data suggest that throughout the Galaxy there will be HI in the halo that has come from shells or their fragments, in addition to any diffuse component of halo HI.

4. VARIATION OF THE HI HALO IN THE INNER GALAXY

Figure 2 shows a cross-section of the distribution of HI in the Galaxy interior to the Sun (adapted from L84) with curves that enclose 90%, 75%, and 50% of the HI emission. It illustrates, first, that the halo undergoes an abrupt change around $R = 3$ kpc, and second, that the halo, and the entire HI layer, appears nearly constant between 3.5 and 7 kpc from the center. The absence of an HI halo in the innermost Galaxy is further illustrated in Figure 3, which shows normalized density profiles at locations 2.2 and 3.6 kpc from the galactic center. The vertical structure of the disk at the smaller radius approximates a single Gaussian, whereas at the larger radius it has wings and a second component in addition to the Gaussian core. The change in the halo at $R \sim 3$ kpc is a striking feature of our Galaxy. It

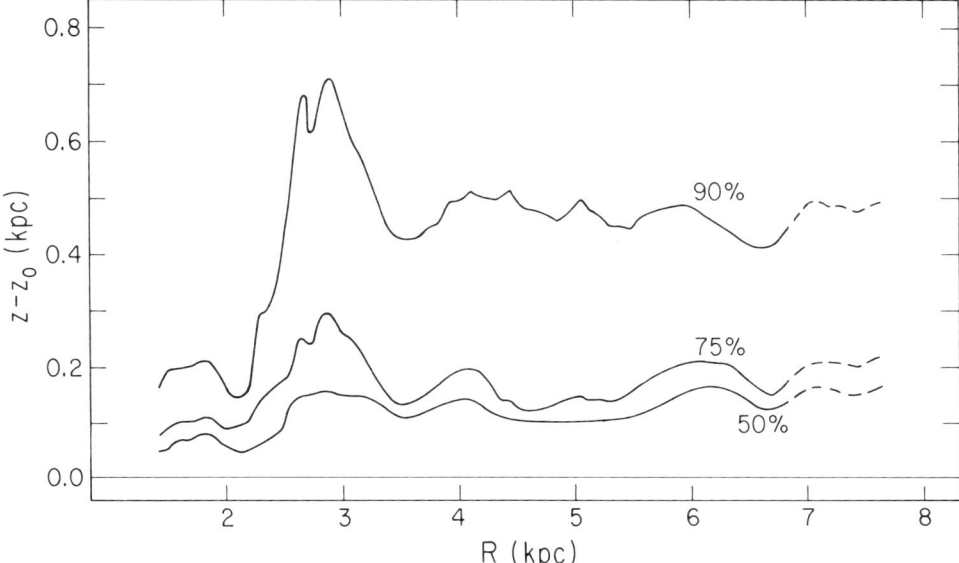

Figure 2. A cross-section of the galactic disk showing the extended HI halo. The contours enclose 50%, 75%, and 90% of the integrated HI column density (i.e., at R = 4 kpc 90% of galactic HI emission lies below 500 pc). Vertical distances are given from the mean location of the plane, which may differ from z= 0 by up to 100 pc. This figure is adapted from Lockman (1984) with the Sun at $R_0 = 8.5$ kpc. Note that there is no HI halo at R $\lesssim$2.5 kpc.

may be caused by a wind from the galactic bulge. Within the bulge the wind will sweep gas from the halo and outer parts of the disk, leaving behind only a narrow layer (Bregman 1980a). The observed break in the halo comes at approximately the radius of the galactic bulge.

The HI layer does not change much, on average, at R$\gtrsim$3 kpc, and this is puzzling because the gravitational force should vary by at least a factor of two over this range (Oort 1962). The HI thickness increases slightly with R beyond 3 kpc, but not nearly as much an one would expect assuming constant support. While the relative constancy of the HI scale-height over the inner Galaxy has recently been challenged by Knapp (1987), I do not believe that that analysis of the data is inconsistent with previous work given that there is a break in the distribution at R $\sim$ 3 kpc.

Beyond the Sun the situation is different. The HI layer thickens monotonically to the edge of the Galaxy even as it bends into the warp (Lozinskaya and Kardashev 1963). This is discussed by Burton elsewhere in this volume.

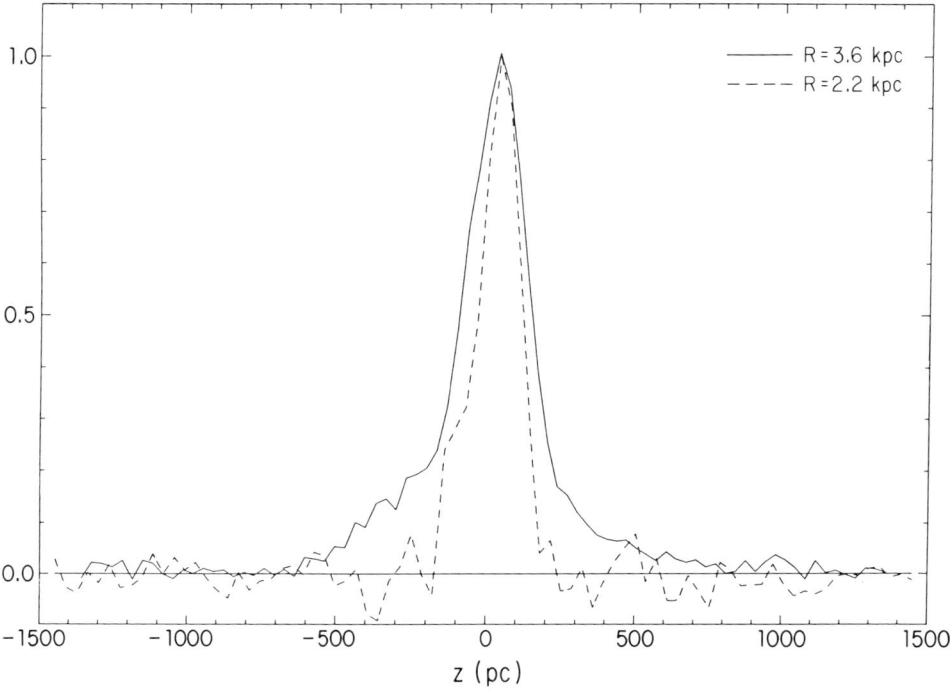

Figure 3. The normalized volume density of HI perpendicular to the galactic plane at two tangent points corresponding to R = 3.6 kpc and R = 2.2 kpc from the galactic center. The data are from the Weaver and Williams (1973) HI survey. This further illustrates the absence of a broad HI halo near 2 kpc and its presence beyond about 3 kpc.

5. KINEMATICS

It is now generally recognized that most galactic HI (at least in the inner Galaxy) is in cylindrical rotation (or "corotation"), i.e., its velocity is overwhelmingly rotational and the rotational velocity does not change with z. The 21 cm results shown in Figures 1, 2 and 3 refer to HI that is strictly corotating (L84). (Note that most of these results were derived from observations at $|b| \leq 10°$ so they are not greatly influenced by typical vertical motions). But in addition, there are interstellar components that are probably not corotating, an example being the classic high-velocity clouds (Giovanelli 1980; Kaelble, de Boer and Grewing 1985; Wakker this volume). There are also neutral clouds that are known to have large vertical velocities (e.g., Wessalius and Fejes 1973).

In a classic paper on the halo, Albert (1983) reported the comparison of optical absorption spectra (in lines of Ca, Na and Ti) toward pairs of high latitude stars that are nearly aligned on the sky but are at significantly different distances. From studying the difference between the spectra of foreground and background stars Albert concluded that the neutral halo has two principle components: a corotating thick disk that extends from the plane to well beyond most OB stars and lies within ± 10 km s^{-1} of zero velocity, and a high velocity gas observed only far from the plane. The two components differ in kinematics, distribution and abundances. This arrangement might result from a galactic fountain (Bregman 1980b).

These findings make it tempting to imagine that the ISM is fairly calm close to the plane, then becomes quite turbulent with large peculiar motions somewhat further out. This image, however, is not quite correct, for the low velocity corotating gas must also extend far from the plane. Furthermore, when I look at Albert's data it seems that the velocity extent of absorption below the foreground star is significantly smaller than that below the background star only when the foreground star is at $|z| < 100$ pc. If the foreground star is at $|z| > 200$ pc, its absorption velocities are similar to those seen against the background star. This implies that Albert's "high-velocity" layer starts at $|z| \sim 150$ pc.

5.1 The Halo in Three Interesting Directions

Figure 4 shows spectra toward three stars that are especially revealing about kinematics in the halo. The solid lines show the 21 cm spectra corrected for stray radiation; this is the total galactic HI in each direction. The stars were chosen because they have $N_{21} > N_{L\alpha}$ by a factor of at least two, i.e., there is at least as much HI above each star as below. Thus the neutral halo is especially pronounced in these directions. Finally, there are high velocity-resolution optical or UV interstellar absorption spectra available so that the spectral components below each star can be identified. The dashed lines show the absorption spectra inverted and drawn with an arbitrary vertical scale. Above each spectrum is a horizontal line (ending in an arrow) showing the expected relationship between velocity and distance for galactic rotation in the direction of the star. Tic marks are every kpc.

These stars are the only ones I know of that have $N_{21} \gtrsim 2\, N_{L\alpha}$, good absorption spectra, and lie at an interesting distance from the plane. The point of this display is to answer the question: what is the velocity of the large amounts of HI that lie above these stars?

Toward μ Col and ρ Leo the answer is simple: absorption covers virtually every velocity that has 21 cm emission. Thus the gas beyond these stars (from half to two-thirds of the total) must have the same kinematics as at least some of the gas below the stars. Only in the direction of the star closest to the plane, HD 28497, is there substantial HI emission at a velocity that shows no absorption. In this case the gas at $V \sim 10$ km s^{-1} must be in the halo beyond the star. For circular rotation this velocity corresponds to a distance of about 2 kpc and a $z = -1.2$ kpc. There is nothing astonishing about finding HI at this location. The inescapable conclusion from these data is that halo gas has the same kinematics as some (or all) of the disk gas below it. More precisely: *the large amounts of HI above each of these stars lies either at a velocity permitted by galactic rotation, or is in the range of velocities seen below each star.*

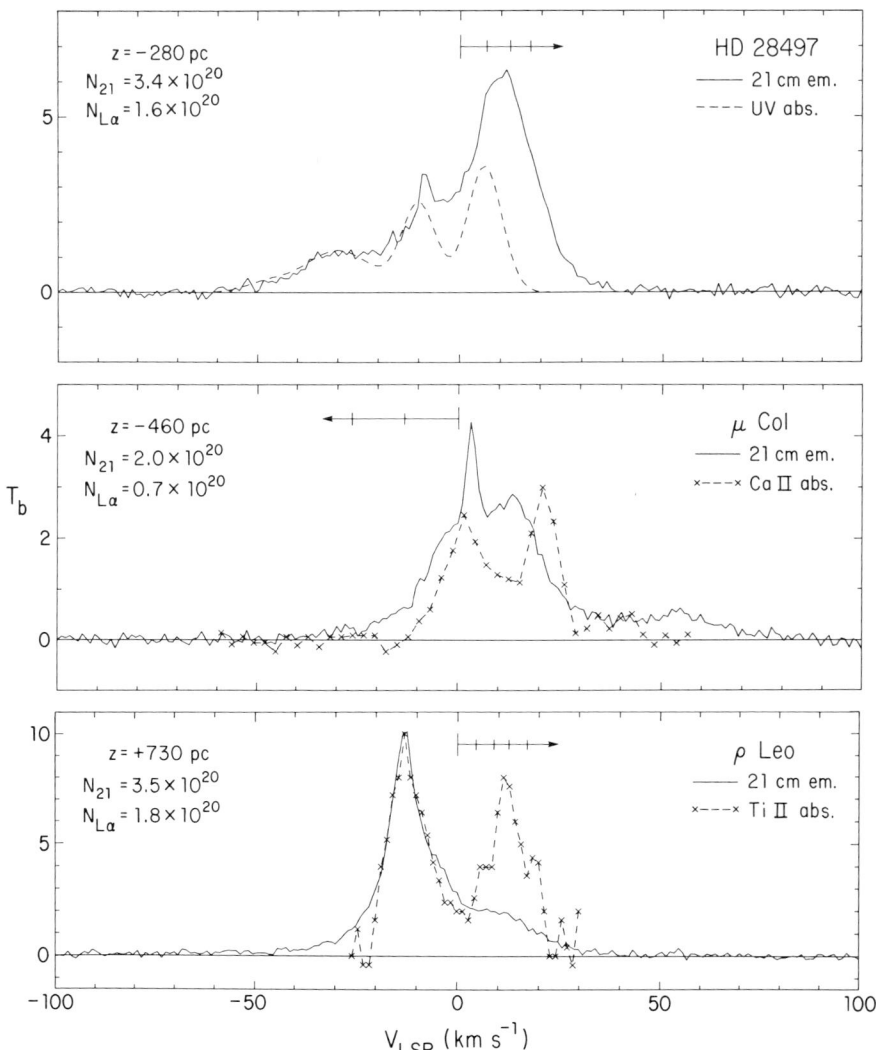

Figure 4. The kinematics of halo gas. In the direction of these stars more HI lies above the star than below ($N_{21} \gtrsim 2\, N_{L\alpha}$). The solid lines show the HI spectra. The other curves are inverted and arbitrarily scaled absorption spectra. The line that ends with an arrow in each panel shows the relationship between distance (tics are every kpc) and velocity for galactic rotation. The 21 cm data are from Lockman, Hobbs and Shull (1986). Absorption spectra are from Giovanelli et al. 1978 (an amalgam of low-ion stages observed by Shull and York, 1977, using the Copernicus satellite), the Ca II observations of Hobbs (1984) and the Ti II observations of Stokes (1978).

The region of "high-velocity" halo gas seems to be fully established by $|z| = 280$ pc as judged by the spectrum towards HD 28497. This is entirely consistent with the conclusion from Albert's data that the layer begins at $|z| \sim 150$ pc. But the situation cannot be as simple as this. The HI at $+50$ km s^{-1} toward μ Col, although only a small fraction of the total, does not appear in the Ca II absorption spectrum and so must lie at $|z| > 460$ pc. And in at least one other direction peculiar velocities do not appear until $|z| > 420$ pc (toward HD 119608; Albert *et al.* 1989). Also, it is a fact that many high- latitude directions show little gas at peculiar velocities. For example, virtually all of the HI is within 20 km s^{-1} of zero velocity toward ζ Lib and 53 Ari (Lockman, Hobbs and Shull 1986). These directions also have $N_{21} \sim N_{L\alpha}$, even though the stars are only 150 and -280 pc, respectively, from the plane.

The high-velocity component of the halo is clearly irregular. But it seems worthwhile italicizing a second conclusion at this point: *A layer of high-velocity halo gas (i.e. gas either not obeying galactic rotation or having a large V_z) is often present ~ 150 pc to ~ 500 pc from the plane.* In those cases where there is a significant amount of HI above 500 pc, it lies in the velocity range of the gas at lower z. I neglect, of course, the classic high-velocity clouds which occupy their own region of the halo at $|z| > 1$ kpc (see Wakker, this volume).

It is curious, and perhaps significant, that the directions discussed in Figure 4, where a large fraction of the gas is in the halo, also have a large fraction at velocities forbidden by galactic rotation. Most of the HI toward ρ Leo and μ Col has an anomalous velocity, and in the direction of HD 28497 only half of the HI obeys galactic rotation. To add to the mystery, the anomalous velocity gas is at relatively low $|z|$, i.e., it is below, not above, each star. A plausible explanation for these facts is that the neutral halo is most pronounced in regions where the HI layer near the plane has been disturbed. Much more data are needed to evaluate this suggestion, and careful consideration must be given to the role of the "local bubble" on the structure of the high-latitude sky (e.g., Cox and Reynolds 1987).

6. SUPPORT

Kulkarni and Fich (1985) pointed out that many 21 cm emission profiles have extended wings, implying that there is a population of "fast" HI clouds in the Galaxy (as first suggested by Radhakrishnan and Srinivasan, 1980). The energy in this component is sufficient to carry gas many hundreds of parsecs above the galactic plane, and it might produce a halo of sorts. [A similar discussion is in Albert (1983)]. Expanding on this suggestion, Lockman and Gehman (1991) have analyzed entire 21 cm emission profiles toward the galactic poles to see what sort of equilibrium layer might be established in the local gravitational field by the available kinetic energy. They calculated the density profile $n_H(z)$ that would give the observed $N_H(v)$ in the gravitational potential $\psi(z)$ believed to be appropriate for the solar neighborhood (e.g., Kuijken and Gilmore 1989).

Lockman and Gehman find that density functions like the L84 curve in Figure 1 can be maintained by the kinetic energy in observed 21 cm profiles, with no need for extra pressure support from cosmic rays or the magnetic field. Therefore, in the solar neighborhood there is enough energy in turbulent motions (for that is what determines an HI profile's width) to support a layer like the one we observe.

At the heart of the analysis is the assumption that galactic HI is composed of distinct components whose vertical distribution, rms velocity and density are independent. The components must not interact very much (as discussed by Kulkarni and Fich 1985). This requires a rather porous ISM and may imply that random motions of HI clouds are not isotropic. There is much work to be done in this area, but these initial results are quite suggestive.

7. CONCLUSIONS

Several questions were posed in the Introduction that can be answered to some degree:

1. **Pervasiveness**: The filling factor of high altitude HI is still not known. In some directions the halo seems to be populated by large clouds with peculiar kinematics but in others there is no discernable break with the disk. I would guess that much of the neutral mass in the halo comes from fragments of supershells, although the line of sight towards HD 28497 looks as if it intersects a smooth corotating layer in addition to clouds with peculiar velcities.

2. **Ionization**: The electron ("Reynolds") layer seems to have a greater scale-height than the neutral layer shown in Figure 1, but whether this is because the ionization fraction approaches unity as $|z| \rightarrow 1.5$ kpc, or because the electrons reside in a separate component, is not yet known.

3. **Populations**: There are probably several neutral components above the plane, distinguished by their kinematics, distribution and abundance (depletion). In some directions a "high-velocity" layer of gas appears between 150 and 400 pc from the plane.

4. **Support**: Locally, the kinetic energy in turbulence is sufficient to support the HI layer against gravity. Elsewhere in the Galaxy this may not be true.

5. **Large-scale Structure**: The lack of a halo at R < 2.5 kpc may signify that there is a wind flowing from the galactic bulge. Other factors that influence the form of the halo at different parts of the Galaxy are not known.

6. **Kinematics**: Most galactic HI corotates. A lot of halo HI also corotates, but some clearly does not. The halo seems to have a mix of kinematics. This may be a sign that a galactic fountain is at work.

7. **Other Halo Constituents**: A task for observers in the coming years is to untangle the relationship between the various populations of gas far above the disk, and also seek their connection with (or independence from) events in the disk itself. This work will be greatly helped by the new Green Bank Telescope, currently under construction, which will be a powerful tool for measuring faint 21 cm emission. The next decade will be an exciting time for studies of the HI halo.

REFERENCES

Albert, C.E. 1983, *Ap.J.*, 272: 509.
Albert, C.E., Blades, J.C., Morton, D.C., Proulx, M. and Lockman, F.J. 1989, in *Structure and Dynamics of the Interstellar Medium*, IAU Colloquium No. 120, ed: G. Tenorio-Tagle, M. Moles and J. Melnick, Springer-Verlag: Berlin, p. 442.
Baker, P.L., and Burton, W.B. 1975, *Ap.J.*, **198**, 281.
Bohlin, R.C., Savage, B.D., and Drake, J.F. 1978, *Ap.J.*, **224**, 132.
Bregman, J.N. 1980a, *Ap.J.*, **237**, 280.
Bregman, J.N. 1980b, *Ap.J.*, **236**, 577.
Celnik, W., Rohlfs, D., and Braunsfurth, E. 1979, *Astron. Ap.*, **76**, 24.
Cox, D.P., and Reynolds, R.J. 1987, *Ann. Rev. Astr. Ap.*, **25**, 303.
Danly, L., Lockman, F.J., Meade, M.R., and Savage, B.D. 1991, *Ap.J.*, in press.
Dickey, J.M. and Lockman, F.J. 1990, *Ann. Rev. Astron. Ap.*, **28**, 215. (**DL90**)
Giovanelli, R. 1980, *Astron. J.*, **85**, 1155.
Giovanelli, R., Haynes, M.P., York, D.G. and Shull, J.M. 1978, *Ap.J.*, **219**, 60.
Heiles, C. 1979, *Ap.J.*, **229**, 533.
Heiles, C. 1984, *Ap.J. Suppl.*, **55**, 585.
Hobbs, L.M. 1984, *Ap.J. Suppl.*, **56**, 315.
Hobbs, L.M., Morgan, W.W., Albert, C.E. and Lockman, F.J. 1982, *Ap.J.*, **263**, 690.
Kaelble, A., de Boer, K.S., and Grewing, M. 1985, *Astron. Ap.*, **143**, 408.
Knapp, G.R. 1987, *Pub. Astron. Soc. Pacific*, **90**, 1134.
Kuijken, K., and Gilmore, G. 1989, *MNRAS*, **239**, 605.
Kulkarni, S.R., and Fich, M. 1985, *Ap.J.*, **289**, 792.
Kulkarni, S.R., and Heiles, C. 1987, in *Interstellar Processess*, ed. D.J. Hollenbach and H.A. Thronson, Jr., Dordrecht: Reidel, p. 87.
Lockman, F.J. 1984, *Ap.J.*, **283**, 90. (**L84**)
Lockman, F.J., Hobbs, L.M. amd Shull, J.M. 1986, *Ap.J.*, **301**, 380.
Lockman, F.J., and Gehman, C. 1991, *Ap.J.*, in press.
Lozinskaya, T.A., Kardashev, N.S. 1963, *Sov. Astron. AJ*, **7**, 161.
Munch, G., and Zirin, H. 1961, *Ap.J.*, **133**, 11.
Oort, J.H. 1962, in *The Distribution and Motion of Interstellar Matter in Galaxies*, ed. L. Woltjer, (New York: Benjamin), pp. 3-21 and 71-77.
Radhakrishnan, V., and Srinivasan, G. 1980, *J. Astrop. Astr.*, **1**, 47.
Savage, B.D., and Massa, D. 1987, *Ap.J.*, **314**, 380.
Schmidt, M. 1957, *Bull. Astr. Inst. Netherlands*, **13**, 247.
Shull, J.M. and York, D.G. 1977, *Ap.J.*, **211**, 803.
Smith, G.P. 1963, *B.A.N.*, **17**, 203.
Stokes, G.M. 1978, *Ap.J. Suppl.*, **36**, 115.
Weaver, H., Williams, D.R.W. 1973, *Astron. Astrop. Suppl.*, **8**, 1.
Wesselius, P.R., Fejes, I. 1973, *Astron. Astrop.*, **24**, 15.

HIGH-VELOCITY CLOUDS

B.P. WAKKER
University of Illinois, Department of Astronomy, Urbana USA

INTRODUCTION

This contribution describes high-velocity clouds (HVCs), neutral hydrogen moving with velocities inexplicable by differential galactic rotation. They have been invoked as evidence for infall of gas to the Galaxy, as manifestations of a galactic fountain, as energy source for the formation of supershells, etc. It is becoming clear that a single model will not suffice to explain all HVCs. A better understanding is mainly hampered by the fact that the distance remains unknown. Many aspects to the study of HVCs will be discussed here.
1) Section 2 gives a historical overview of surveys and then describes some of the results of the latest surveys.
2) Section 3 discusses some attempts at distance determination.
3) Studies at high angular and velocity resolution are described in Sect. 4.
4) The appearance of HVCs at other wavelengths should give a better handle on their structure and environment. Unfortunately, only a very limited number of such detections exists (Sect. 5).
5) After knowing the observational facts, an interpretation is in order.
   In Sect. 6 an overview of various propositions is given.
   In Sect. 7 summarizes the conclusions and outlines outstanding problems.

SURVEYS FOR HIGH-VELOCITY NEUTRAL HYDROGEN

High-velocity gas was originally found by Muller *et al.* (1963). Surveys following this detection were at first fairly coarse and insensitive. Several of the brighter clouds were found: HVC 40-15+100 (Smith 1963), complexes A and C (Hulsbosch & Raimond, 1966) the "South Pole complex" (Dieter 1964), clouds M, AC I, AC II, AC III (Mathewson *et al.* 1966). All these detections revealed only gas at negative velocities relative to the LSR, which gave rise to the notion that HVCs have negative velocities and therefore must be falling into the Galaxy. After these initial surveys, ever more complete surveys were done: van Kuilenburg (1970, 1972) found the first few HVCs at positive velocity; the large and relatively sensitive survey by Meng

& Kraus (1970) showed many negative-velocity HVCs; and Wannier et al. (1972) detected a large number of faint positive-velocity clouds.

Another paradigm that formed after the initial surveys was that HVCs are at high galactic latitude. However, Hulsbosch (1975) already described a cloud in the plane, HVC 131+1−200, which deviates by at least 70 km s^{-1} from galactic rotation and is one of the brightest HVCs. It was ignored in most further studies.

The "South Pole Complex" was eventually shown to be part of the Magellanic Stream (mapped by Mathewson et al. 1974). Most probably this is a tidal tail between the Magellanic Clouds and the Galaxy (Murai & Fujimoto 1980).

The first era of HVC surveys culminated in the study of Giovanelli et al. (1973) who mapped the three northern complexes A, M and C on a fully sampled grid with 10′ and 1.37 km s^{-1} resolution and 0.5 K rms. Even now this survey is unsurpassed in completeness for these three complexes. Structural features ranged in size from 10′ to 60°.

In the following years many clouds were observed with higher resolution or improved coverage. Wright (1974) discovered a very-high-velocity cloud (VHVC, having $v_{lsr} < -200$ km s^{-1}). Many others were subsequently found (most notably HVC 110−7−465 and HVC 114−10−440, Hulsbosch 1978) and it became clear that many small VHVCs exist at southern galactic latitudes.

Giovanelli (1980) presented maps of AC I, II and III and of a stream from $(l,b)=(140°,-5°)$ to $(l,b)=(180°,-20°)$. Cohen (1981) studied HVC 160−50−110, which is a 10° long filament, coinciding with a similarly structured filament at -10 km s^{-1}. He also showed that a small cloud near $(l,b)=(165°,-45°)$ was the brightest spot in a complex with velocities up to -340 km s^{-1} (Cohen 1982). Mirabel & Morras (1984) made a sensitive survey around the galactic center. They found many small clouds at high negative velocities scattered on the sky. Together with the Anticenter clouds these clouds may provide evidence for an inflow of material toward the Galaxy.

Improvements in the sensitivity of radio receivers led to a survey by Giovanelli (1980) with the Green Bank 300-ft telescope. He found 800 HVC detections in 6000 spectra and thus estimated that they cover about 10% of the sky. His data allowed to make scatter diagrams of longitude against velocity, from which he showed that the distributions with respect to position, velocity and structure are very asymmetric: large and bright negative-velocity clouds occur mainly in the first two quadrants, while smaller and fainter positive-velocity clouds are seen in the third and fourth quadrants. Further, in the first two quadrants and the Anticenter region there is a population of clouds at very high negative velocities, well-separated in velocity from other HVCs. The separation suggests that these VHVCs have a different origin. The major drawbacks of this survey were the incomplete sky coverage and the fact that the beam was small compared to the grid.

The latest surveys repair these shortcomings. Hulsbosch & Wakker (1988) surveyed the whole sky north of declination $-18°$ on a $1°\times1°$ grid in galactic coordinates with the 25-m Dwingeloo telescope (35′.2 beam). The detection limit was 0.05 K (about $2 \cdot 10^{18}$ cm^{-2}). Bajaja et al. (1985) used the 25-m telescope at Villa Elisa (Argentina) to make a survey on a $2°\times2°$ grid with detection limit

0.08 K south of $-10°$. In these surveys a large number of small clouds was discovered and previously-known objects were shown to be much more extended. Also, HVC 165−45−280 and the AC clouds were shown to be connected.

An analysis of the statistical properties of the combined new surveys is given by Wakker (1990a). The latitude distribution of the HVCs shows that they are not limited to high b, but that there is a concentration toward low b. Further, negative velocities are not as predominant as was apparent from the first surveys. This impression was created by their lower sensitivity and the fact that most of the, generally fainter, positive-velocity clouds are in the southern sky. However, the total amount and the sky coverage of high-velocity gas at negative velocities still is greater than that at positive velocities.

At the survey limit of 0.05 K, 11% of the sky is covered by gas having $|v_{lsr}|$ < 100 km s^{-1}, an increase of a factor two over earlier surveys (detection limits ~0.2 K) (Wakker 1990a). Using ultra-violet absorption lines and assuming an element abundance one can probe much lower hydrogen column densities than in the 21-cm line. For instance, the Si II λ1260 line allows to detect material where $N_{HI} = 2 \cdot 10^{17}$ cm^{-2}. An extrapolation of the relation between sky coverage and detection limit down to this limit leads to the prediction that at that level between 30 and 60% of the sky should b covered by gas having $v_{lsr} < -100$ km s^{-1}. However, the sample of Danly (1989) contains 19 stars with D>1.5 kpc, in none of which absorption at $v_{lsr} < -100$ km s^{-1} is detected. Unless the HVCs have low heavy element abundances this implies that, statistically, they are farther away than 1.5 kpc.

Wakker & van Woerden (1990) used the surveys to construct the first homogeneous catalogue of HVCs. This includes all known clouds and lists their properties. Ten different populations (groups of clouds occupying certain regions in l-b-v_{lsr} space) are defined, each of which may have different physical properties, origin and relation to the Galaxy. Figure 1 shows the l-v_{lsr} distribution of the catalogued clouds, with different symbols indicating different populations.

The catalogue was also used to construct the $\log N(>S)$ and $\log N(>\Omega)$ distributions of HVCs. The slopes of these relations can be explained if the HVCs have a mass spectrum $n(M)dM = N_\circ M^{-3/2} dM$, are related to the Galaxy and visible throughout it (Wakker & van Woerden 1990). The slope of this mass spectrum is similar to that found for H I in the disk (−1.87, Dickey & Garwood 1989) and molecular clouds (−1.6, Scoville & Sanders 1987).

DISTANCE DETERMINATIONS

The main problem in the study of high-velocity gas is the impossibility to use the velocity to determine a distance. The only practical method known is to search for interstellar absorption lines at the velocity of the HVC in spectra of probe stars with known distance (e.g. van Woerden et al. 1989).

Ever since HVCs were discoverd, probe stars have been observed but no convincing detection exists. In recent years much progress has been made, including the proof of the existence of heavy elements in at least a few HVCs (Songaila 1981,

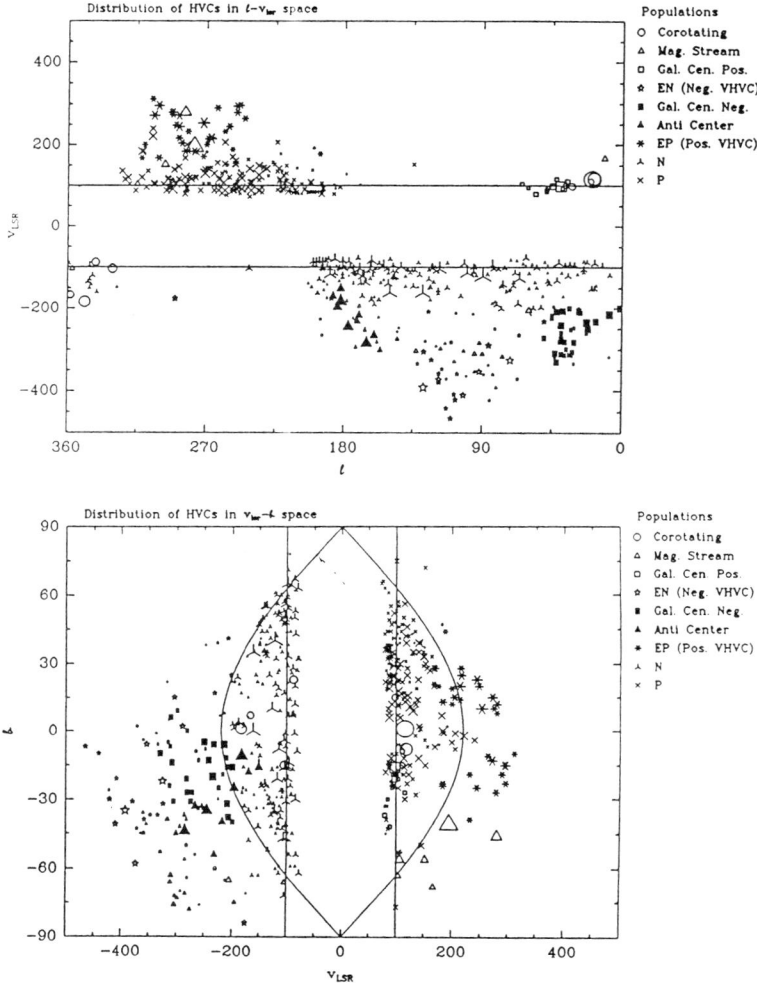

Fig. 1. Longitude-velocity and velocity-latitude distribution of HVCs in the catalogue of Wakker & van Woerden (1990).

West et al. 1985, Robertson et al. 1990). Songaila et al. (1988) claimed the detection of a line due to complex C in the spectrum of the star BT Dra, but Lilienthal et al. (1990) have shown that this line is likely to be stellar. In complex A two horizontal-branch stars have been observed and one in complex C, all lacking interstellar absorption at the HVC velocity (Pettini priv. comm., Schwarz et al. 1990).

As an extragalactic object projected against complex A does show a (faint) absorption (Schwarz et al. 1990), this would imply a distance larger than 5 kpc, but alternative explanations for the non-detections cannot be excluded at present.

FINE STRUCTURE

Up to the study of Giovanelli et al. (1973), HVCs were always observed with telescopes with half-degree beams. The cloud parameters they found at 10' resolution were different from those previously known. Column densities went up, estimated sizes decreased. Linewidths, which used to be of the order of 20 km s^{-1} appeared to be around 8 km s^{-1}.

Cram & Giovanelli (1976) analyzed 300-ft spectra of positions in complexes A, C and M and found a two-component structure; one component with mean widths of 23 km s^{-1} and another with mean widths 7 km s^{-1}. The presence of two components correlates with cloud morphology: narrower lines are seen in the bright concentrations. The same two-component structure was subsequently also found in positive-velocity HVCs (Giovanelli & Haynes 1976, Cohen & Ruelas Mayorga 1980, Morras & Bajaja 1983). These observations were interpreted as evidence for two thermally stable phases in the clouds. The warm component has a temperature of order 10^4 K, giving rise to the broad lines. The cool component was supposed to have a temperature around 100 K, though the linewidths corresponded to a temperature of order 1000 K. Therefore the existence of fine structure in concentrations on scales smaller than the resolution of the telescopes used was predicted.

Aperture synthesis observations of high-velocity clouds were first carried out with the Westerbork telescope by Schwarz et al. (1976), who studied part of A 0 at 2' resolution. Much fine structure was found, with sizes of order 5' and peak brightness temperatures rising from the 2.2 K observed with a 35' beam to 25 K. A breakthrough was the Westerbork study by Schwarz & Oort (1981), who mapped A I at 1' and 2 km s^{-1} resolution. Their analysis confirmed the prediction that the brighter cores seen at 10' resolution consist of many smaller concentrations, with sizes down to a few arcmin. The distribution of linewidths was very broad, with a modus at 5 km s^{-1}. Column densities were up to a few 10^{20} cm^{-2} and estimated volume densities varied between $25/D_{kpc}$ and $100/D_{kpc}$ cm^{-3} (with D_{kpc} the cloud distance in kpc). A remarkable feature was the fact that only 20% of the flux observed with a single-dish telescope is recovered in the WSRT maps. This indicates that the broad component indeed consists of warm gas and is not completely due to velocity crowding and beam smearing.

Since this study a large number of fields has been observed at Westerbork, i.a. five in the classical complexes A, M and HVC 131+1−200, and two VHVCs. A full description of these data is presented by Wakker (1990b) and Wakker & Schwarz (1990). Generally, the properties found by Schwarz & Oort are also found for these fields. In three fields 25% of the single-dish flux is recovered, while in the other three (including the two VHVCs) this is 40%. Estimates of the pressure show that it is likely that the hotter gaseous envelope is in equilibrium with the small cores. A

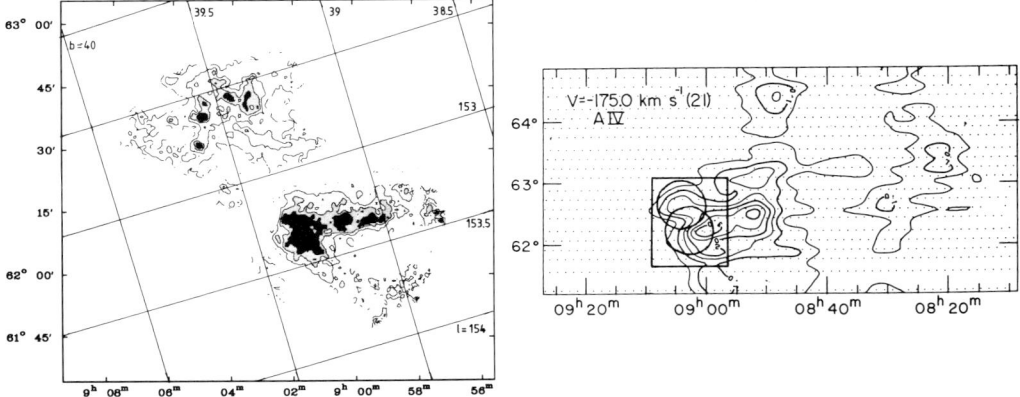

Fig. 2. Cloud A IV as observed with the Westerbork telescope.

comparison of timescales reveals that the fine structure within a cloud must evolve much faster than the whole cloud, so that different dynamical processes play a role to determine structure on different scales.

Density contrasts can be as high as 4 on scales of $4'$. For absorption-line studies this implies there is considerable uncertainty in the derivation of abundances. A small feature in the line of sight to a halo star may provide a strong absorption component, but go undetected in a single-dish beam due to beam-dilution effects. Alternatively, a cloud seen in H I but not in absorption could in fact lie in front of the probe star, but be patchy so that most of the gas lies just off the line of sight to the star

At each position in the WSRT data cubes a gaussian fit was made to the spectrum. Usually only one component is present, but profiles with two or three components are common. From the fits one obtains the velocity field of the cloud and the linewidths of its profiles. There is much velocity structure, with both smooth gradients and erratic jumps occurring often, just as is observed on large scales in the survey data (see figures in Wakker 1990a and 1990b). The origin of this structure is not understood, but it probably is a mixture of a number of different causes, like projection effects, macro-turbulent motions, and motions guided by an embedding medium or magnetic fields.

The linewidths are generally rather small and have a broad distribution with modus at 3–6 km s^{-1} FWHM. The kinetic temperature of the gas is likely below a few 100 K. The measured peak brightness temperatures provide a lower limit of 30 K.

STUDIES OTHER THAN IN 21-CM EMISSION

To determine the gas temperature directly emission-absorption observations have often been tried. These studies are difficult because few bright continuum sources are projected against HVC emission bright enough to expect a detectable line. Further, such observations require a small beam, so that most single-dish telescopes are not useful. Payne et al. (1980) failed to find absorptions using the Arecibo telescope. An exhaustive study of all bright continuum sources visible from Arecibo by Colgan et al. (1990) also gave negative results only. However, the limits that they can derive for the temperature are not very stringent. Wakker et al. (1990) describe the only known detection, obtained with the Westerbork telescope HVC 131+1−200. The spin temperature of this object is shown to be ~ 50 K.

Observations of HVCs at other wavelengths have often been attempted, with very little success. On the POSS plates no enhanced emission or extinction can be discerned. Kutyrev (1985), Reynolds (1987) and Kutyrev & Reynolds (1989) have searched for Hα and Hβ emission from some of the brighter clouds. No clear-cut result was obtained in these studies, although there is one possible detection. Because of the extremely low signal-to-noise ratio and because the method of observation is beset with difficulties, progress is slow.

A search for 100 μm emission, using IRAS data (Wakker & Boulanger 1985) showed that the amount of far-infrared radiation from clouds M I and A III is less than expected. The non-detection is interpreted as: 1) the amount of dust in these two HVCs is at least a factor three below the minimum amount found in low-velocity neutral hydrogen or 2) the dust is very cool, which implies a minimum cloud distance of 10 kpc.

CO observations have often been tried, although the negative results were never published. Some integrations of many hours still showed no emission at the HVC velocity. However, assuming a "normal" ratio of H I to H_2 and of H_2 to CO, no strong lines are to be expected.

INTERPRETATIONS

The paper by Oort (1966) already brings forward most of the interpretations for HVCs. He discusses the following set:
a) they are parts of nearby supernova shells,
b) they are condensations formed in a gaseous corona of high temperature,
c) they have been ejected from the galactic nucleus,
d) they have been ejected as cool clouds from the galactic disk,
e) they are intergalactic gas accreted by the Galaxy,
f) they are gas clouds in the Local group (i.e. independent galaxies or small satellites of the Milky Way).

In later work a few other alternatives have been proposed:
g) they are high-z spiral arms,

h) they are connected with distant globular clusters and nearby dwarf galaxies in the Local Group,

i) they were drawn out from the Magellanic Clouds on a previous passage of these galaxies, just as the Magellanic Stream was drawn out during the last passage.

To assess the implications of the survey data for these explanations, modeling is necessary. Ballistic models have been constructed (Wakker & Bregman 1990) that are based on the scheme described below.

1) Choose a galactic potential.
2) Inject clouds into this potential according to some prescription. A prescription consists of specifying position, velocity and mass at the time of formation.
3) Follow the clouds in their orbit until they either disperse or are destroyed.
4) Determine at the end of the calculation the observables longitude, latitude, velocity relative to the LSR, flux, area and brightness temperature.

Cloud evolution is modelled in a very simple way: they are assumed to be pressure-confined throughout their lifetime and therefore always keep the same size. Only when clouds hit the gaseous disk are they assumed to be destroyed. At formation a mass is given to each cloud, chosen from a power-law spectrum ($n(M)dM = N_\circ M^\alpha dM$, $\alpha = -1.5$ from the survey data). Assuming the same average density for all clouds makes it possible to calculate the cloud radius. Together with the distance this gives the cloud flux, brightness temperature and area. Fitting the predicted $\log N(>S)$ and $\log N(>\Omega)$ relations to the data sets the scale of the model and allows to determine the total mass in high-velocity gas.

The main problem with these ballistic models is that the clouds are modeled as particles and that gas dynamics (e.g. drag forces) is ignored. However, the consistency of the predicted and observed sky and velocity distributions can be checked.

Below, each of the possible explanations is discussed in more detail.

a) The hypothesis that HVCs are nearby supernova remnants was already rejected by Oort because of their sky distribution, the fact that if they are shells only one side is visible, and the indication of lower distance limits of at least 500 pc. Over the years these arguments have only been strengthened.

b) The idea that hot gas in a galactic corona condenses to form cool H I clouds that fall back to the disk has gained popularity over the years. The circulation of gas implied in this model is known as the "galactic fountain" (Shapiro & Field 1976). Oort (1966) originally argued that if there is pressure equilibrium between the hot and the cool gas, friction would quickly slow down the clouds. However, later he argued (Oort 1978) that a strong argument in favor of the fountain model is the observed fine structure. The random velocity jumps combined with the presence of a warm and a cold gas phase are easy to understand within this model. Nevertheless, more theoretical work is needed, aiming at predicting the ratio between ionized and neutral material, and the kind of structure that can be expected.

Bregman (1980) found that the fountain model gives a good approximation when applied to predict the sky and velocity distributions of the HVCs with $|v_{lsr}|$ < 200 km s^{-1}. The ballistic models of Wakker & Bregman (1990) fairly well reproduce the observed distributions (Fig. 3a). However, it is very difficult to produce

HVCs with $b > 45°$. Also, the VHVCs and the Anticenter complexes remain unexplained. The flow implied is $5\,M_\odot\,yr^{-1}$.

c) Ejection from the galactic nucleus does not seem to be a viable theory. The distribution of velocities on the sky is incompatible with this model (Oort 1966).

d) Oort (1966) calculated that it is possible that gas which was expelled from the galactic disk 70 Myr ago now appears near the Sun at a high negative velocity. Such gas could have been accelerated by the cumulative effect of many supernovae. However, to achieve the observed amount an excessive number is needed. Wakker & Bregman (1990) calculated a similar model under the name "cannonball" model. They show that the predicted velocities and latitudes are generally too low. Only some HVCs near the plane could originate this way.

e) If intergalactic gas clouds are falling into the Galaxy, they would be accelerated to the escape velocity (of the order of $450\,km\,s^{-1}$) of the Galaxy if there is no braking. When Oort (1966) first discussed this possibility, VHVCs were undiscovered and the highest known velocity was $200\,km\,s^{-1}$. He therefore argued that considerable deceleration must have occurred. With the discovery of VHVCs there is no need to assume deceleration and the hypothesis becomes more likely.

f) An explanation in terms of gas clouds moving around in the Local Group was also suggested by Oort (1966). He rejected this on the basis of 1) the large angular size of some clouds, 2) the structural similarity between HVCs and IVCs, some of which were shown to be nearby, 3) the preponderance of high negative velocities. Giovanelli (1977) argued that 4) the narrowness of the lines observed in most HVCs can not be reconciled with very large distances, as the velocity dispersions would be too low for the large masses implied. Currently the third argument can not be maintained, and for the VHVCs Oort's objections 1) and 2) are not valid: VHVCs are generally small, and Cohen & Mirabel (1979) and Wakker & Schwarz (1990) show that two VHVCs have a structure different from most other HVCs.

Three infall-type models are described by Wakker & Bregman (1990): The Pure Infall model, in which objects continuously flows in from intergalactic space; the Local Group model, in which clouds are assumed to have the velocity distribution of nearby (D<300 kpc) dwarf galaxies; and the Circular Motion model, in which clouds can remain in orbit around the Galaxy for 10 Gyr. In each of these models VHVCs are predicted to occur. No model gives a fully satisfactory fit to the sky and velocity distributions, however. It is possible to reproduce the Anticenter complexes only if one of the more massive clouds happens to be at the proper place at the present time. Also, the observed North-South asymmetry is never reproduced. Therefore, the limit that can be put on the amount of material falling in from the Local Group is $0.1\,M_\odot\,yr^{-1}$.

A simple model in which the Magellanic Stream is seen as the source of HVCs (Fig. 3b) is able to reproduce the North-South asymmetry very easily. This model is too simplistic to provide a proper fit, but indicates that a connection with the Stream is likely. This explanation was put forward also by Giovanelli (1981), because VHVCs are not exclusively seen toward the central regions of the Local Group, their velocity distribution does not match that of the galaxies, and the

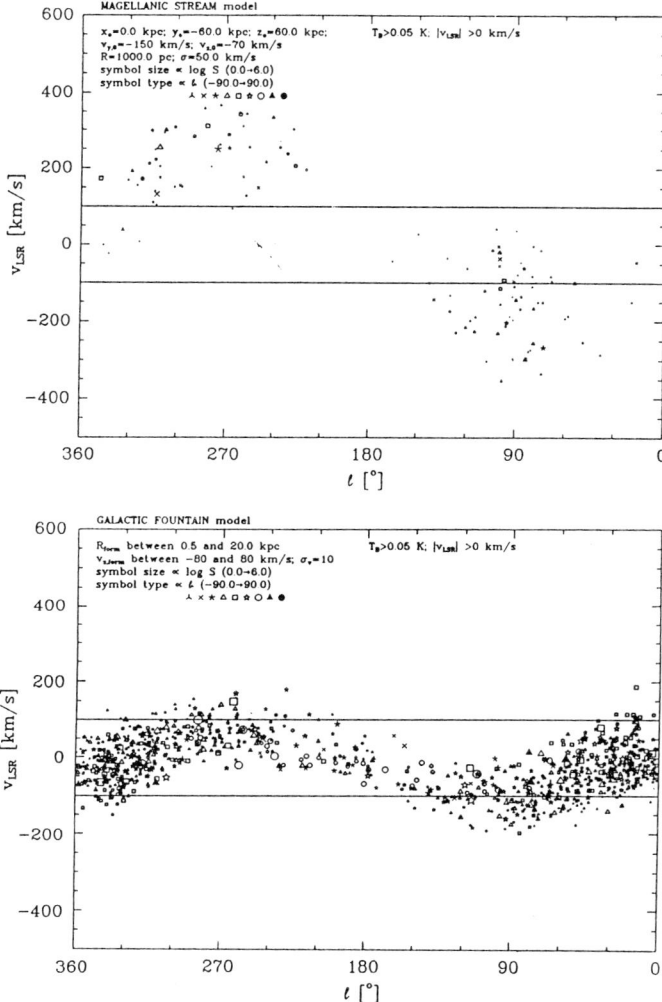

Fig. 3. Model predictions for the longitude-velocity distribution for the galactic fountain model and the one in which Magellanic Stream is the origin of the VHVCs.

internal motions imply instability, which is hard to accept if they are long-lived objects.

g) For the Outer Arm, Habing (1966) and Kepner (1970) made it likely that it is a spiral arm going up to very large heights above the disk. Davies (1972) and Verschuur (1973, 1975) suggested a similar explanation for the complexes A, C and M as relatively nearby high-z extensions of spiral arms. The deviating velocities are explained as due to instabilities (Davies 1972) or acceleration imparted by an intergalactic wind (Oort & Hulsbosch 1978). The positive-velocity HVCs were not properly included in this model. The "streams" of clouds that were defined are not confirmed by the later surveys.

h) Lynden-Bell (1976) proposed a connection of HVCs with distant globular clusters and the nearest dwarf galaxies in the Local Group, as some of these are seen projected on top of the large HVCs. However, the map that he used to find these coincidences is wrong and misleading. The idea of an association has stuck, however. Haud (1988) suggested that HVCs form a (broken) polar ring, mainly because the planes defined by the dwarf galaxies and the Magellanic Stream are at a small angle with each other and inclined by almost 90° to the galactic plane. As the northern counterpart of the Magellanic Stream he included complex C and many IVCs. A major problem with this model is that the plane defined by complex C is strongly inclined to that defined by the Magellanic Stream. More work is necessary to see if the new survey data can be reconciled with the polar ring model.

i) Oort & Hulsbosch (1978) considered the possibility that complex A was tidally separated from the Magellanic Clouds during the previous passage of these galaxies past our Galaxy. They rejected the hypothesis because of the very unlikely orbit it would have to follow.

The model calculations of Wakker & Bregman have shown conclusively that the fact that in the the veloicity asymmetry between the eastern and western galactic hemisphere is due to a reflection of the rotation of the LSR around the galactic center. Further, the existence of clouds with $|v_{lsr}| < -200 \,\mathrm{km\,s^{-1}}$ near the Anticenter is very hard to reproduce. The North-South asymmetry in the distribution of VHVCs is only found when the Magellanic Stream is the source of the gas. In all infall-type models VHVCs are predicted to occur in large numbers also in the northern hemisphere. To strengthen these conclusions and obtain better fits more work is clearly needed.

It seems very likely that the observed distributions of l, b, v_{lsr}, flux and size can only be reproduced by a combination of two or more models, and hence that there are several sorts of HVCs, with different origins.

CONCLUSIONS

Based on the data and models, the following set of explanations seems the most promising: The Magellanic Stream is a tidal tail drawn out of the Magellanic Clouds by the Galaxy during their last passage (see Murai & Fujimoto 1980). The VHVCs are shreds of the Stream that were disconnected long ago (as proposed by Giovanelli 1981). Other clouds are part of a galactic fountain (Bregman 1980). Complex C is either part of the fountain or a high-z spiral arm (Davies 1972, Verschuur 1973) (these two explanations need not be mutually exclusive). A number of clouds near the galactic anticenter and center probably consist of material streaming toward the Galaxy (Mirabel 1989). The source of this infalling material still remains unclear, as it may be gas coming from nearby intergalactic space, from the Local Group, or even from the Magellanic Stream.

Many problems remain unsolved. The major problem is still that no direct distance determinations are available. To solve this it will be necessary to correlate the whole-sky maps with possible probes, to measure good values of the column density in the direction of the probe using high-resolution data and to obtain high-resolution optical or ultra-violet spectra to search for absorption lines from heavy elements at the HVC velocity.

It would be useful to make in-depth studies of several complexes. Especially a possible connection of the Anticenter complexes with the low-velocity gas in the Taurus Molecular Cloud needs attention. Also, the survey data could be used to unravel the velocity field of complex C. In both cases it will be necessary to include data on intermediate-velocity clouds, as many of these overlap in position with the HVCs (and sometimes even show structural similarities).

To understand the small-scale structures better, hydrodynamical modelling is in order. This should aim at the prediction of observable parameters, like linewidths, temperatures and structure. Such models may then also be used to give constraints on the galactic fountain model, which can provide the initial conditions for these models. Probably it is necessary to include magnetic forces. Therefore, attempts at measuring values of or limits to magnetic field strengths within HVCs are useful.

The ballistic models described by Wakker & Bregman (1990) still need improvements and extensions. A better description of the l-b-v_{lsr} distribution of the VHVCs must be searched for. Further, the influence of drag forces on cloud motion should be studied, and we need an improved description of cloud evolution.

REFERENCES

Bajaja E., Cappa de Nicolau C.E., Cersosimo J.C., Loiseau N., Martin M.C., Morras R., Olano C.A., Pöppel W.G.L., 1985, *Astrophys. J. Suppl. Ser.* **58**, 143
Bregman J.N., 1980 *Astrophys. J.* **236**, 577
Cohen R.J., 1981, *Monthly Notices Roy. Astr. Soc.* **196**, 835
Cohen R.J., 1982, *Monthly Notices Roy. Astr. Soc.* **200**, 391
Cohen R.J., Mirabel I.F., 1979, *Monthly Notices Roy. Astr. Soc.* **186**, 217
Cohen R.J., Ruelas-Mayorga R.A., 1980, *Monthly Notices Roy. Astr. Soc.* **193**, 583
Colgan S.W.J., Salpeter E.E., Terzian Y., 1989, preprint
Cram T.R., Giovanelli R., 1976, *Astron. Astrophys.* **48**, 39
Danly L., 1989, *Astrophys. J.* **342**, 785
Davies R.D., 1972, Nature **237**, 88
Dickey J.M., Garwood R.W., 1989, *Astrophys. J.* **341**, 201
Dieter N.H., 1964, *Astron. J.* **69**, 288
Giovanelli R., 1977 *Astron. Astrophys.* **55**, 395
Giovanelli R., 1980, *Astron. J.* **85**, 1155
Giovanelli R., 1981, *Astron. J.* **86**, 1468
Giovanelli R., Verschuur G.J., Cram T.R., 1973, *Astron. Astrophys. Suppl. Ser.* **12**, 209
Giovanelli R., Haynes M.P., 1976, *Monthly Notices Roy. Astr. Soc.* **177**, 525
Habing H.J., 1966, *Bull. Astr. Inst. Neth.* **18**, 323
Haud U., 1988, *Astron. Astrophys.* **198**, 125

Hulsbosch A.N.M., 1975, *Astron. Astrophys.* **40**, 1
Hulsbosch A.N.M., 1978, *Astron. Astrophys.* **66**, L5
Hulsbosch A.N.M., Raimond E., 1966, *Bull. Astr. Inst. Neth.* **18**, 413
Hulsbosch A.N.M., Wakker B.P., 1988, *Astron. Astrophys. Suppl. Ser.* **75**, 191
Kepner M.E., 1970, *Astron. Astrophys.* **5**, 444
van Kuilenburg J., 1970, *Astron. Astrophys. Suppl. Ser.* **5**, 1
van Kuilenburg J., 1972, *Astron. Astrophys.* **16**, 276
Kutyrev A.S., 1985, *Astron. Tsirk.*, No. 1396, p3
Kutyrev A.S., Reynolds R.J., 1989, *Astrophys. J.* **344**, L9
Lilienthal D., Meyerdierks H., de Boer K.S., 1990, *Astronomy & Astrophysics*, in press
Lynden-Bell D., 1976, *Monthly Notices Roy. Astr. Soc.* **174**, 695
Mathewson D.S., Meng S.Y., Brundage W.D., Kraus J.D., 1966, *Astron. J.* **71**, 863
Mathewson D.S., Cleary M.N., Murray J.D., 1974, *Astrophys. J.* **190**, 291
Meng S.Y., Kraus J.D., 1970, *Astron. J.* **75**, 535
Mirabel I.F., 1989, in *Structure and Dynamics of the Interstellar Medium*, eds. G. Tenorio-Tagle, M. Moles, J. Melnick, IAU Colloquium **120**, Springer Verlag, Berlin
Mirabel I.F., Morras R., 1984, *Astrophys. J.* **279**, 86
Morras R., Bajaja E., 1983, *Astron. Astrophys. Suppl. Ser.* **51**, 131
Muller C.A., Oort J.H., Raimond E., 1963, *C.R. Acad. Sci Paris* **257**, 1661
Murai T., Fujimoto M., 1980, *Publ. Astron. Soc. Japan* **32**, 581
Oort J.H., 1966, *Bull. Astr. Inst. Neth.* **18**, 421
Oort J.H., 1978, in *Problems of Physics and Evolution of the Universe*, ed. L. Mirzoyan, Armenian Acad. Sci., Yerevan, p259
Oort J.H., Hulsbosch A.N.M., 1978, in *Astronomical papers dedicated to B. Strömgren*, eds. A. Reiz and T. Andersen, Copenhagen University Observatory, p409
Payne H.E., Salpeter E.E., Terzian Y., 1980, *Astrophys. J.* **240**, 499
Reynolds R.J., 1987 *Astrophys. J.* **323**, 553
Robertson G., Schwarz U.J., van Woerden H., Murray J.D., Morton D.A., Hulsbosch A.N.M., 1990, in preparation
Schwarz U.J., Sullivan W.T., Hulsbosch A.N.M., 1976, *Astron. Astrophys.* **52**, 133
Schwarz U.J., Oort J.H., 1981, *Astron. Astrophys.* **101**, 305
Schwarz U.J., van Woerden H., Wakker B.P., 1990, in preparation
Scoville N.Z., Sanders D.B., 1987, in *Interstellar Processes*, eds. D.J. Hollenbach and H.A. Thronson Jr., (Dordrecht, Reidel), p21
Shapiro P.R., Field G.B, 1976, *Astrophys. J.* **205**, 762
Smith G.P., 1963, *Bull. Astr. Inst. Neth.* **17**, 203
Songaila A., 1981, *Astrophys. J.* **243**, L9
Songaila A., Cowie L.L., Weaver H., 1988, *Astrophys. J.* **329**, 580
Verschuur G.L., 1973, *Astron. Astrophys.* **22**, 139
Verschuur G.L., 1975, *Ann. Rev. Astron. Astrophys.* **13**, 257
Wakker B.P., 1990a, thesis Groningen, Chapter 3
Wakker B.P., 1990b, thesis Groningen, Chapter 6
Wakker B.P., Boulanger F., 1986, *Astron. Astrophys.* **170**, 84
Wakker B.P., van Woerden H., 1990a, thesis Groningen, Chapter 4
Wakker B.P., Bregman J.N., 1990a, thesis Groningen, Chapter 5
Wakker B.P., Schwarz U.J., 1990a, thesis Groningen, Chapter 8
Wakker B.P., Vijfschaft B., Schwarz U.J., 1990, thesis Groningen, Chapter 9
Wannier P., Wrixon G.T., Wilson R.W., 1972, *Astron. Astrophys.* **18**, 224
West K.A., Pettini M., Penston M.V., Blades J.C., Morton D.C., 1985, *Monthly Notices Roy. Astr. Soc.* **215**, 481

van Woerden H., Schwarz U.J. & Wakker B.P., 1989, in *Structure and Dynamics of the Interstellar Medium*, eds. G. Tenorio-Tagle, M. Moles, J. Melnick, IAU Colloquium **120**, Springer Verlag, Berlin
Wright M.C.H., 1974, *Astron. Astrophys.* **31**, 317

MOLECULAR CLOUDS AT HIGH Z

LEO BLITZ
Astronomy Program
University of Maryland
College Park, Maryland, U.S.A. 20742

ABSTRACT. The evidence for the existence of molecular clouds at large distances from the Galactic plane is reviewed. The molecular clouds at high Galactic latitudes are shown to be largely confined to the Galactic plane. There is evidence for one giant molecular cloud as much as four scale heights from the Galactic plane, but given the sample size from which the cloud is drawn, it is reasonable to suppose that it is part of the tail of the thin disk population. There is weak evidence that one star-forming molecular cloud may have originated in the Galactic halo. On the basis of *kinematic* evidence however, it is shown that there are three molecular clouds identified at high galactic latitude that, if not at high z, are likely to have resulted from interaction with gas in the halo. Understanding how these clouds have formed is likely to be an important key to understanding how the halo interacts with the disk gas.

1. INTRODUCTION

The existence of molecular gas at large distances from the Galactic plane is of considerable interest because of the unique information the gas would provide about the physical conditions and processes that take place in the halo. As a first step, let us define "large distances" quantitatively. About 100 clouds have been identified at high galactic latitude (see Magnani, Blitz and Mundy 1985, for the initial listing of 57 objects) and several hundred of the appoximately 4000 giant molecular clouds (GMCs) in the Milky Way have been catalogued (*e.g.* Scoville *et al.* 1987). It is therefore expected statistically that a few catalogued clouds might be seen at distances as large as 3 σ_h from the Galactic plane, where σ_h is the rms scale height of the gas. Furthermore, one GMC at 4 σ_h is not highly improbable. We therefore require that a molecular cloud be at least 3 σ_h from the plane to qualify as high-z gas. Because the scale height of the atomic gas is about twice that of the molecular gas (*e.g.* Knapp 1987, Wouterloot *et al.* 1990), even gas at four or five σ_h from the plane may not be in contact with the halo gas. Nevertheless, it will be shown below that nearly all of the molecular clouds so far identified appear to be at distances less than 3 σ_h from the Galactic midplane, and only one is known to be as much as 4 σ_h.

High z molecular gas from the ensemble of high latitude clouds can also be identified kinematically. Because the distribution of radial velocities is roughly gaussian (Magnani, Blitz and Mundy 1985, hereafter MBM), clouds that have velocities in excess of 3 σ_v, where σ_v is the one dimensional velocity dispersion of the ensemble, are potentially at high z. On the other hand, the high velocity gas may be locally accelerated in the plane, so detection of such gas requires additional information to place it with certainty at high z. In any event, three clouds have been detected in CO with velocites from 4 to 8 σ_v, which suggests that they might have had some interaction with the halo gas. These clouds are discussed below.

2. SCALE HEIGHT OF THE MOLECULAR LAYER

The scale height of the molecular gas is determined primarily by three methods. 1) The angular scale height of CO is measured at the tangent points of inner Galaxy surveys. The Galactic rotation curve is then used to convert the measurements to a linear scale height. 2) The velocity dispersion of the CO in the solar vicinity and the midplane density of matter determine the local scale height of CO if the gas is in equilibrium with the gravitational potential of the disk. 3) A direct measurement of the scale height can be made from HII region/CO complexes with known distances. The first method can be extrapolated to the solar vicinity for comparison with the second. The third method is the only method that can be used in the outer Galaxy. The distances can, in principle, be either spectrophotometric or kinematic, but the latter introduces large uncertainties near the Sun and in regions of known gas streaming such as the Perseus Arm.

The table below gives a number of independently determined values of the local CO scale height extrapolated from inner Galaxy CO surveys.

TABLE 1. Molecular (CO) Scale Height at R_0

Reference	σ_h (pc)
Clemens et al. (1988)	68 (85)
Dame et al. (1987)	64
Knapp (1987)	48
Fich and Blitz (1984)	88 (101)
Wouterloot et al. (1990)	~85

The first three values are from inner Galaxy CO surveys in which the angular scale height of the CO is measured at the tangent point along the line of sight (the location of the largest radial velocity) and is then extrapolated to the solar distance. The value quoted for Clemens *et al.* is the mean of values near the solar circle; the parenthetical value is the measured value at R_0. The Fich and Blitz value is the scale height of the HII region/molecular cloud complexes within 500 pc of R_0. The quoted value is determined from those objects for which optical distances are available; the parenthetical value is the one quoted in the publication and includes kinematic distances. The Wouterloot *et al.* value is from their survey of molecular clouds associated with IRAS selected embedded stars, and is extrapolated to R_0 from measurements at larger Galactic radii.

Taking a straight average of the five determinations gives a gaussian scale height of 71 ± 16 pc. The local scale height of the molecular gas layer may then be taken to be about 70 pc with an uncertainty of about 25%. Molecular clouds at high z in the solar vicinity are therefore required to be at distances greater than about 200 pc from the plane.

In the inner Galaxy, the results of all of the CO surveys show that there is a monotonic increase in the CO scale height with increasing galactocentric distance. Similarly, in the outer Galaxy, Fich and Blitz (1984) and Wouterloot *et al.* (1990) also find that the scale height increases with distance. Fich and Blitz find that at 1.4 R_0, the scale height is more than twice the local value, and nearly three times the local value when all objects with distances greater than 1.5 R_0 are averaged together; the Wouterloot *et al.* observations give reasonable agreement with these values. It is therefore necessary to consider the scale height of the molecular gas at the location in the Galaxy where a particular cloud is located in order to determine whether it is in the plane or not. For example, a molecular cloud located 300 pc from the plane at R = 4 kpc may be reasonably considered to be at the disk-halo interface, whereas a similar object at R = 15 kpc may not.

2. ARE THERE MOLECULAR CLOUDS AT HIGH Z?

2.1 *GMCs in the Solar Vicinity*

A plot of the distances of GMCs from the midplane was given by Blitz (1978), collected from numerous published CO observations. The more recent all sky CO survey of Dame *et al.* (1987) has not appreciably added to the compilation. A number of GMCs are found at distances as much as 150 pc from the Galactic plane including the L1641 cloud in which the Orion Nebula is embedded. None though is farther than 2 σ_h from the plane. Thus, not only are the GMCs in the solar vicinity in the Galactic plane, but all are within one scale height of the atomic gas (Falgarone and Lequeux 1973, Knapp 1987).

2.2 High Latitude Molecular Clouds (HLCs)

The distances to the ensemble of HLCs was initialy determined in two ways (MBM). The first was to *assume* that they are Galactic plane objects, and assign them a scale height comparable to the other molecular clouds in the plane (70 pc - see above). By then assuming that they are all at one mean distance from the Sun, the distribution in b will give the mean distance to the ensemble. The second method assumes that the HLCs are in equilibrium with the z-component of the gravitational potential of the disk stars. If the measured radial velocity dispersion of the disk stars is equal to the velocity dispersion in z, then using the Oort limit to the density of matter in the local midplane of the Galaxy (Bahcall 1984) will give the scale height of the HLCs. The angular distribution then gives the mean distance from the Sun. The second method gives a value similar to the one assumed in the first, which provided initial confidence that the mean distance to the HLC ensemble had been reasonably well determined.

Subsequently, a number of direct distance determinations were made to individual HLCs. The first was a measurement of the distances by the classical method of Wolf diagrams (Magnani and deVries 1986). These found the distances to three clouds and upper limits to five more. The distances determined to the eight clouds were consistent with the mean distance of 100 pc. Next, in a series of three papers, Hobbs *et al.* (1986,1988) and Welty *et al.* (1989) determined the distances to 6 HLCs using the presence or absence of strong Na D and Ca H+K absorption toward stars along the line of sight to the HLCs. The stars all have known spectroscopic parallaxes and the distance determinations identified the nearest known molecular cloud (MBM12). Table 2 gives the distances to the clouds for which measurements have been obtained to date.

With the possible exception of MBM 41-44, none of the measured distances imply z-heights that are more than 3 σ_h from the plane. The distance to MBM 41-44 is very uncertain and is discussed separately below. Furthermore, the direct distance measurements imply that the scale height of the HLCs is the same as that determined for the overall population of molecular gas; the measured velocity dispersion of the clouds implies that they are in equilibrium with the gravitational potential of the stars in the disk. Therefore, the molecular clouds in the vicinity of the Sun are found to be Galactic plane objects.

2.3 Anomalous Velocity Clouds

Three clouds in the solar vicinity have been found to have velocities well outside the range 3 σ_v for the HLCs, where σ_v has been measured to be 5.6 ± 1.2 km s^{-1}(MBM). These are G90+38, also known as the Draco Cloud (Georigk *et al.* 1983) and MBM 41-44 (MBM), with a radial velocity of -23 km s^{-1}, G211+63 with a velocity of -39 km s^{-1}(Désert, Bazell and Blitz 1989), and G135+55 with a velocity of -45 km s^{-1}(Heiles, Reach and Koo 1988). These clouds have velocities that range from 4-8 σ_v, and are clearly different from the remainder of the HLC

TABLE 2. Distances to HLCs

MBM	Distance (pc)	z (pc)	(ref.)
7	125 ±50	-75	1
12	60 - 70	-35	2
16	100 ±50	-60	1
16	65 - 95	~ -50	3
18	≤ 175	≥ -100	1
20	≤ 125	≥ -75	1
26	175 ±50	95	1
32	≤ 275	≤ 180	1
40	≤ 140	≤ 100	4
41-44	(> 800)	(> 500)	5
53	110 - 155	~ -70	4
54	145 - 260	~ -120	4
55	≤ 175	≤ -110	1
55	30 to 265	-20 to -175	4
55A	≤ 275	≥ -180	1

refs:
1 Magnani and deVries (1986)
2 Hobbs, Blitz, and Magnani (1986)
3 Hobbs et al. (1988)
4 Welty et al. (1989)
5 Mebold et al. (1985)

population. If the velocities were representative of motions in equilibrium with the stellar gravitational potential, these clouds would be in the halo of the Galaxy. On the other hand, that all of the velocities are negative suggests that the velocities may represent some other kind of motion, perhaps related to the motions of the classical high-velocity clouds. Whatever their origin, these three clouds are the best candidates for the interaction of molecular gas in the disk-halo connection in the solar vicinity. They will be discussed in more detail below.

2.4 Galactic Plane GMCs

Nyman et al. (1987) have called attention to a large molecular cloud in Lupus that appears to be at a much larger distance from the Galactic plane than most of the molecular emission observed in the Dame et al. (1987) CO survey. At a distance of 2.9 kpc, the cloud has a distance of 220 pc from the plane. The cloud has a galactocentric distance of 6.7 kpc, assuming a value of 8.5 kpc for R_0. The CO scale height at that distance is ~ 60 pc (using the same surveys discussed in Section 2). The CO cloud in Lupus is therefore about 4 σ_h from the plane.

Nyman et al. (1987) suggest that because of its large distance from the plane, the Lupus cloud required an extraordinary event in order to form it, and they point to a deficiency in molecular gas at a nearby position in the plane as the site of the postulated explosive event. However, this cloud was identified because it stood out in the survey of inner Galaxy CO. It may therefore be the molecular cloud with the very largest z height relative to σ_h, and its position might simply be a statistical fluctuation. Recall that for an estimated 4000 GMCs in the Galaxy (Blitz 1978, Solomon and Sanders 1980), one expects $\sim$ 3 GMCs in the Galactic ensemble to be at distances that are as large as 4 σ_h from the plane. So, from its position alone, one does not necessarily require an unusual event to form the cloud. Note, however, that the cloud is at a distance of approximately 1.5 atomic scale heights from the plane (Knapp et al. 1987), putting it in the outer portion of the atomic layer but not in what would consider the halo gas. Whatever its origin, the Lupus cloud is unusual, but it nevertheless probably should still be considered a normal Galactic plane object.

2.5 Young Galactic Clusters

All of the youngest galactic star clusters are still associated with the gas out of which the stars formed (Bash, Green and Peters, 1977; Leisawitz, Bash and Thaddeus 1989). The latter authors have shown that there are no clusters with ages $< 5 \times 10^6$ y without associated CO. Furthermore, the clusters that are $< 5 \times 10^6$ y old have associated clouds with masses $> 10^4$ $M_\odot$, and the clusters that are $> 10^7$ y old have clouds with masses $< 10^3$ $M_\odot$ if they have any associated CO at all. It is therefore reasonable to conclude that clusters with ages of $\sim 10^7$ y, are destroying or have recently destroyed the clouds out of which the stars formed.

There are two star clusters, probably associated with one another that provide tantalizing evidence for molecular clouds at high z. Some of the parameters for the two clusters are tabulated below.

TABLE 3. Two Open Clusters at High Z

	NGC 457	NGC 281
l	126°56	123°13
b	-4°35	-6°24
d	2 - 3 kpc	2 - 2.5 kpc
z	-190 pc	-240 pc

NGC 281 contains an O6.5 main sequence star, thus an upper limit to its age is $\sim 3 \times 10^6$ y. The cluster also has an associated CO cloud (Leisewitz 1988). NGC 457, with nearly the same distance and position in the sky, contains an O9.5 star, but no associated CO. The estimated age of the cluster is 10^7 y from the combined presence of the late O star and absence of CO. What makes the second cluster so unusual, however, is that it has a measured proper motion of 65 km s^{-1} *toward the Galactic plane*. The proper motions are thought to be reliable because they have been tied to the FK 5 system (de Vecht, personal communication). If the proper motion measurement is correct, it implies that the cluster had an associated molecular cloud that was at one time much farther from the Galactic plane than the cluster is now. If we may take all of the parameters at face value, then 5×10^6 y ago, the cluster and molecular cloud were 500 pc from the plane! The existence of a molecular cloud associated with NGC 281 lends support to the notion that NGC 457 had a preexisting molecular cloud. However, there are no accurate proper motions for NGC 281.

The tantalizing prospect that a large star forming molecular cloud has formed so far from the Galactic plane rests on the accuracy of the proper motion measurements of the stars in NGC 457 and the correctness of the FK5 system. Confirmation of these proper motions would be good evidence that molecular clouds can exist in the Galactic halo, even though there is no direct evidence among the currently available CO data sets. Even if the proper motions turn out to be wrong, understanding the reason for the large z distance of the molecular cloud associated with NGC 281, at more than three scale heights below the plane are as much of a challenge as it is for the Lupus cloud.

4. THE ANOMALOUS VELOCITY HLCS

To summarize the previous section, we may conclude that 1) There is no evidence that any local GMCs are not in the thin disk of the Galaxy. 2) There is evidence that an occasional molecular cloud, such as the Lupus cloud, and the NGC 281 cloud are at 3-4 σ_h of the plane, but there is no strong evidence that these clouds are not at the tail of the thin disk population of molecular clouds. 3) The observational data for the NGC 457 cluster implies that it was formed from a molecular cloud at least 500 pc from the plane. The principal evidence comes from proper motions that need to be confirmed. 4) Direct distance measurements to the HLCs imply that they are part of the thin disk population. 5) There are three HLCs with highly anomalous velocities. Their high negative velocities suggest that they have either fallen from large z distances, or that they have been accelerated by collisions with infalling gas from the halo.

It appears that the best evidence for a disk-halo interaction among the molecular clouds comes from the anomalous velocity clouds. All three clouds have been the subject of previous studies, and I will summarize the observational data for each of these clouds.

G90+38; The Draco Cloud; MBM 41-44

Georigk *et al.* (1983) first called attention to this object because they noticed a high latitude HI feature corresponding to the position of a faint high latitude nebulosity. Furthermore, star counts indicated that there was more extinction than could be accounted by the dust associated with HI; CO was then sought and detected. Subsequently, Mebold *et al.* (1985), and MBM independently observed CO associated with the object. MBM called attention to the kinematic anomaly of the CO clouds, and Mebold *et al.* have attempted to establish a distance to the cloud with the specific aim of trying to ascertain whether the Draco cloud is a molecular cloud in the halo. Using UBV photometry and estimates of the extinction to the clouds, Georigk and Mebold (1986) argued that the distance to the clouds is > 800 pc implying that z is > 500 pc. If correct, it would place this molecular cloud unequivocally at the disk-halo interface.

The existing evidence is, however, weak. On the basis of two lines of sight for the CO, as well as the HI observations, Mebold *et al.* argue that the extinction (A_v) to the region is $\gtrsim 2$ mag. The star counts imply much lower extinctions and Mebold *et al.* concluded that the cloud is more distant than the stars used for the counts. However, the HI data of Georigk *et al.* (1983) indicate that the mean extinction due to HI is only 0.3 mag. Data given in MBM suggest that the *mean* extinction due to H_2 is no greater than 0.7 mag. The mean extinction over the region in which the star counts were made is probably no greater than 0.5 because of the limited extent of the CO emission. Since the distance of > 800 pc was derived by comparing the star count data to $A_v = 2$ mag, it is probably true that the true distance is less than that derived by Mebold *et al.*

Subsequently, Georigk and Mebold (1986) did photometry to 56 stars projected on the Draco atomic hydrogen cloud, and found that the reddeneing to the stars is inconsistent with an extinction of 2 mag, and concluded that the stars must be objects in the foreground of the Draco cloud. However, inspection of their color-color plot and the reddening line indicates that the locations of the stars may not be inconsistent with a mean extinction of 0.5 mag, and thus many of the stars, whose distances are determined from spectrophotometry and argued to be in the foreground, may indeed be background objects.

In any event, the question of the distance can be decided by interstellar absorption line studies of stars projected along the line of sight to the cloud, especially in the direction of the molecular emission. [author's note: A comment from the audience at the meeting mentioned that such observations have, in fact, been completed; the results are eagerly awaited]. Regardless of whether this cloud is in the halo or not, the morphology resembles that of the other two anomalous velocity clouds found to date. That is, the clouds have a cometary morphology suggesting that the highest density portion of the clouds is at the leading portion of the feature.

Odenwald and Rickard (1987), and Odenwald (1988) have catalogued high-latitude cometary objects gleaned from the IRAS data base in order to find other objects that resemble G90+38. They find that the data can be eplained by assuming that dense clouds are moving through the ambient gas with different Reynolds numbers. For G90+38, Odenwald argues that the data suggest that the dense portion of the cloud is moving subsonically through an ambient interstellar medium of low Reynolds number, which suggests that the cloud is indeed at the disk-halo interface and is moving through coronal gas. Other objects appear to be moving through ambient gas of higher viscosity producing turbulent wakes suggestive of

Mach cones. In the case of G90+38 the CO has been extensively mapped, the strongest lines are near the ends of elephant trunks (Mebold, et al. 1985; Rohlfs, et al. 1989) that point in the general direction of the Galactic plane. Odenwald's work suggests that much can be learned about the ambient gas and the dynamics of the disk-halo interface if one can obtain good estimates of the densities in the ambient gas of the general class of cometary structures seen at high Galactic latitudes.

G135+55

This object has the highest radial velocity, -45 km s^{-1}, of any molecular cloud detected at high galactic latitude and was discovered by Heiles, Reach and Koo (1988). The Heiles, Reach and Koo study sought to identify the properties of a sample of 26 isolated clouds at high galactic latitude from the IRAS survey. In their kinematic analysis they concluded that at least two of the clouds in their sample have undergone shocks, and that the shock has modified the grain size distribution. Specifically, they suggest that the very small grains responsible for 12 μm emission are destroyed by shocks with velocities of $\sim$ 10 km s^{-1} and formed in shocks with somewhat higher velocities. Fast shocks, $\gtrsim 40$ km s^{-1}, are argued to preferentially destroy large grains, thereby elevating the 60/100 μm flux ratio in clouds. Thus one would expect that the anomalous velocity clouds as a group would exhibit 60/100 μm flux ratios with values significantly greater than those found in the general interstellar medium. The G135+55 cloud was not otherwise explicitly analyzed in their study.

G211+63

This cloud has been detected by Désert, Bazell and Blitz (1990), but was independently observed by Heiles, Reach and Koo (personal communication). The cloud was found by Désert et al. in a CO survey of interstellar clouds that exhibit infrared excesses (Désert, Bazell and Boulanger 1988). This cloud is the only one of about 30 new detections of HLCs that has radial velocity more than 2 σ_v (Blitz, Bazell and Désert (1990).

Désert, Bazell and Blitz obtained CO (1-0) and CO (2-1) spectra at the position of infrared emission peak. The 2-1 line observation was made with high frequency resolution and shows a striking double peaked structure. However, no CO mapping of the source was made. Instead, the authors made carefully constructed maps at 12 μm, 60 μm, and 100 μm, and they compared the color ratios of the resulting maps. They conclude that like the cometary globules catalogued by Odenwald (1988), the interaction with the ambient gas is observed as plumes trailing from the back side of the two main dense clumps from which the most intense radiation is detected. The colors suggest that there has been grain processing as the result of shocks that have destroyed the small grains at the leading edge of the object. The brighter clump of dust shows an elevated 60/100 μm flux ratio as suggested by Heiles, Reach and Koo (1988) for fast shocks. The mass of the dense portion of the cloud is estimated to be 1.7 M$_\odot$ and the mass of the associated HI (Verschuur 1971) is estimated to be 8.7 M$_\odot$ at an assumed distance of 100 pc. The diameter of the cloud at this distance is 1.2 pc.

Based on their velocities, their morphologies, and their infrared properties, the anomalous velocity HLCs seem to be the best candidates to be molecular clouds at the interface between the Galactic disk and the halo. Even if they should turn out not to be at high z, their interaction with the ambient gas can provide useful information about the shock processing of dust and the production of molecules in shocks. On the other hand, if they do turn out to be high z objects, the most important question of all has not been addressed by any of the observations made to date: how did they get there? It is the answer to that question that is most likely to provide the important clues to what is happening at the disk-halo interface.

ACKNOWLEDGEMENTS

Funding for this work is partially provided by the USNSF grant AST-8918912, and the contribution of the State of Maryland to the Laboratory for Millimeter-wave Astronomy.

REFERENCES

Bahcall, J.N., 1984, *Ap. J.*, **276**, 169.
Bash, F.N., Green, E., and Peters, W.L., 1977, *Ap. J.*, **217**, 464.
Blitz, L., 1978, Ph.D. Dissertation, Columbia University.
Blitz, L. Bazell, D, and Désert, F.X., 1990, *Ap. J. (Letters)*, **352**, L13.
Clemens, D.P., 1985, *Ap. J.*, **295**, 402.
Clemens, D.P., Sanders, D.B., and Scoville, N.Z., 1987, *Ap. J.*, , **327**, 139.
Dame, T.M., Elmegreen, B.G., Cohen, R.S., and Thaddeus, P., 1986, *Ap. J.*, **305**, 892.
Dame, T.M.. *et al.*, 1987, *Ap. J.*, **322**, 706.
Désert, F.X., Bazell, D., and Boulanger, 1988, *Ap. J.*, **334**, 815.
Désert, F.X., Bazell, D., and Blitz, L., 1990, *Ap. J. (Letters)*, **355**, L51.
Falgarone, E., and Lequeux, J., 1973 *Astron. Ap.*, **25**, 253.
Fich, M. and Blitz, L., 1984, *Ap. J.*, **279**, 125.
Georigk, W., and Mebold, U., 1986, *Astr. Ap*, **162**, 279.
Georigk, W., Mebold, U., Reif, K., Kalberla, P.M.W., and Velden, L., 1983, *Astr. Ap*, **120**, 63.
Heiles, C., Reach, W.T., and Koo, B.-C., 1988, *Ap. J.*, **322**, 313.
Hobbs, L.M., Blitz, L. and Magnani, 1986, *Ap. J. (Letters)*, **306**, L109.
Hobbs, L.M., Blitz, L., Penprase, B.E., Magnani, L., and Welty, D.E., 1988, *Ap. J.*, **327**, 356.
Knapp. G.R., 1987, *Pub. A.S.P.*, **99**, 1134.
Leisawitz, D., 1988, NASA Reference Publication 1202, *Catalog of Open Clusters and Associated Interstellar Matter*.
Leisawitz, D., Bash, F.N., and Thaddeus, P., 1989, *Ap. J. Suppl.*, **70**, 731.

Magnani, L., Blitz, L., and Mundy, L., 1985, *Ap. J.*, **295**, 402.
Magnani, L. and de Vries, C.P., 1986, *Astr. Ap*, **168**, 271.
Mebold, U., Cernicharo, J., Velden, L., Reif, K., Crezelius, C., and Georigk, W., 1985, *Astr. Ap*, , **151**, 427.
Nyman, L.-Å., Thaddeus, P., Bronfman, R.S., and Cohen, R.S., 1987, *Ap. J.*, **314**, 374.
Odenwald. S.F., 1988, *Ap. J.*, **325**, 320.
Odenwald, S.F., and Rickard, L. J, 1987, *Ap. J.*, , **318**, 703.
Rohlfs, R., Herbstmeier, U., Mebold, U., and Winnberg, A., 1989, *Astr. Ap*, 211, 402.
Scoville, N.Z., Yun, M.S., Clemens, D.P., Sanders, D.B., and Waller, W.H., 1987, *Ap. J. Suppl.*, **63**, 821.
Solomon, P.M., and Sanders, D.B., 1980, in *Giant Molecular Clouds in the Galaxy*, P.M. Solomon, and M.G. Edmunds, eds., Pergamon: Oxford, p.41.
Verschuur, G.L., 1971, *Astron. J.*, **76**, 317.
Welty, D.E., Hobbs, L.M., Blitz, L., and Penprase, B.E., 1989, *Ap. J.*, **346**, 232.
Wouterloot, J.G.A., Brand, J., Burton, W.B., Kwee, K.K., 1990, *Astr. Ap*, **230**, 21.

OPTICAL AND ULTRAVIOLET ABSORPTION STUDIES OF COOL GAS
IN THE MILKY WAY HALO

L. Danly
Space Telescope Science Institute
3700 San Martin Drive
Baltimore, MD 21218

I. Introduction

Of the many and various means for observing the interstellar medium, and halo gas in particular, optical and ultraviolet absorption techniques provide both unique opportunities and unique limitations. As its title indicates, this article speaks specifically to the contributions from absorption methods to our knowledge of halo gas with temperatures below 10^5 K. A brief description of the trade-offs and benefits of the methods is therefore useful to set the stage for interpreting the observations of halo gas. Briefly stated, the trade-offs can be described as follows:

absorption vs emission studies. One obvious benefit from absorption studies is that they provide limits on the distance to gas. In addition, while both H^0 and H^+ can be studied well in emission, lines from most other elements are simply too weak to be studied extensively in emission. Absorption data, on the other hand, can be obtained on many different species with a wide range of ionization state. More quantitative information is therefore available on physical properties such as abundances, depletion and ionization of gas studied in absorption. A major drawback of absorption studies is the limitation imposed by curve-of-growth effects. Quantitative determinations can be very difficult under conditions where the instrumental resolution is too poor to permit separate components to be individually resolved. Finally, absorption studies require the presence of suitable background sources against which to measure the interstellar absorption lines. Finding suitable sources can lead to limited sky-coverage, and the selection of the gas that gets studied is determined solely by the luck of the positional alignment. Furthermore, the distribution of background sources can lead to selection effects: for example, the strong stellar UV continuum required for ultraviolet absorption studies is found only in very early type stars which have a distribution strongly confined to the galactic plane.

optical vs. ultraviolet studies. Optical absorption programs have the advantage that the data is gathered from ground based telescopes which, due mostly to their larger size, can provide higher resolution spectra from shorter exposures. In general, more observing time is available from ground based observatories, permitting broader surveys of halo sightlines. The species available for study with optical telescopes (primarily Na I, Ca II, and Ti II) are tracers of cool gas and provide information on the depletion of gas phase atoms, but they are weak lines from elements which have low abundances relative to hydrogen and they provide no information on the presence of warm or hot gas. The currently observable ultraviolet region of the spectrum, on the other hand, contains many strong (even resonance) lines from important elements such as C, N, O, and Fe and also have lines arising from ionization states as high as N V (which is found in gas as hot as 3×10^5 K). Unfortunately, ultraviolet observations suffer from the drawback that they must be obtained from space and are therefore limited by the characteristics of the available instrumentation and the stiff competition for its time. At the time of writing, this has essentially meant IUE (the International Ultraviolet Explorer).

cool vs. hot gas. Although data are available on resonance transitions of Al III, Si IV, C IV and N V in the halo, these will not be discussed here. Strictly speaking, these species could well arise in cool gas that is subjected to a strongly ionizing radiation flux. Indeed, whether the highly ionized species observed in the galactic halo arise from photo-ionization or from collisions with thermal electrons is still widely debated. However the nature of the highly ionized species will be discussed elsewhere in this volume.

The discovery of the highly ionized species in the halo focused a great deal of attention on the possible presence of a galactic "corona" with temperatures greater than 10^6 K or cooling through 10^5 K. Relatively little work has been done on what happens to the gas after it cools below 10^4 K and how it settles back to the plane. However, even the earliest observations of halo gas by Savage and de Boer (1979) showed that there is at least a factor of ten times more cool gas by mass in the halo than the highly ionized gas, as observed through lines such as C I, O I, Mg I, Mg II, Si II, Al II, Fe II, C II, and so on.

We will discuss the results from observations of the neutral and singly ionized species on the nature of cool gas in the halo, it structure and its kinematics. Ultimately, one hopes that the data will provide insight into the origin and history of halo gas: Is the cool gas former hot gas from a galactic fountain? Or was it removed from the disk to high z? If so, how? Or is cool halo gas simply a high-z tail to the disk distribution of the interstellar medium?

In the next section we present a brief overview of past and current optical and ultraviolet observational studies of halo gas. Section III highlights some interesting problems, and finally, Section IV discusses the most recent efforts to get more quantitative information from the absorption spectroscopy, particularly in light of new, higher quality instrumentation available. We will not discuss (1) the lines of sight toward the LMC/SMC/SN1987a or (2) Lyman-α absorption studies à la Lockman, Hobbs and Shull (1986); these subjects are quite broad and in some cases controversial, and they will be covered elsewhere in this volume.

II. Resources and Results from Optical and Ultraviolet Studies

This section does not aim to present the history of optical and ultraviolet absorption studies of halo gas. Many extensive reviews of the history and overall properties of halo gas can be found in the literature, including Savage (1989), de Boer (1989), and references therein. Instead, we merely list some of the most comprehensive sources, and briefly summarize some of the salient results. The list of sources here is *by no means* exhaustive: it is merely intended to provide a set of references which can be used as a resource for more in-depth work.

TABLE 1 RESOURCES FOR OPTICAL DATA	
Munch and Zirin (1961)	1st comprehensive halo survey
Albert (1983)	comp. halo survey (19 ☆s; 5 km/s res.)
Morton and Blades (1986)	halo and X-gal sightlines
Edgar and Savage (1989)	complilation from lit : refract. elem.
Albert et al. (1991)	comp. halo survey (66 ☆s; 5 km/s res.)

A. Optical data. Table 1 provides a partial list of references for optical surveys of absorption from halo gas. Next to each reference, comments indicate either the observational techniques, sample selection criteria, or other attributes which make that particular study unique or noteworthy.

To summarize the results of these and other studies, the optical data show that the number of cool clouds continues to increase as the distance above the galactic plane increases, indicating that such clouds can indeed be found in the halo. Interestingly, halo clouds possess velocities that are distinctly different from disk velocities, ranging from about -50 km/s to +20 km/s. Authors have noted the propensity for negative velocities to be found in halo components.

There is no significant increase in the number of components beyond the lower halo, i.e., there is not any substantial difference in the appearance of absorption spectra against extra-galactic sources compared to halo stars beyond 2-3 kpc above the disk. Halo clouds demonstrate the Routley-Spitzer effect; i.e., the ratio of Na I to Ca II decreases in high z -- and larger velocity -- clouds. In keeping with this, the refractory elements are seen to have higher scale heights than the non-refractory elements. Interestingly, refractory elements display a smoother distribution: the *scatter* in the observed correlation between distance (or z-distance) and column density is smaller for refractory elements.

Finally, mention should be made of efforts to detect molecules at high latitudes. Absorption from CH, CH+, and CN has been detected (DeVries and Van Dishoeck 1988; Welty et al. 1989; Penprase 1991) in stars lying in the direction of high b clouds from the survey of Magnani, Blitz and Mundy (1985). While none of the MBM clouds has been detected at distances beyond about z=250 pc, they may play an important role in the disk-halo interaction. The absorption data can provide important information on the presence of and the ionization conditions in high latitude molecular clouds.

TABLE 2 RESOURCES FOR ULTRAVIOLET DATA

Savage and de Boer (1979, 1981)	1st UV halo study
Van Steenberg & Shull (1988a,b)	disk and halo survey
Danly (1989)	halo kinematics survey (34 ☆s; b>40°)
Kinney et al. (1990)	X-gal sightlines
Danly et al. (1991)	halo survey (57 ☆s; data plotted)

B. Ultraviolet data. In a manner similar to Table 1, Table 2 lists references for ultraviolet absorption studies of halo gas. Some of the best known references on UV absorption data, such as Pettini and West (1983) or Savage and Massa (1987) are not listed because their focus was heavily weighted toward the high ionization species. Again, the samples in some of the surveys included both disk and extra-galactic sources, and again, the list is not exhaustive.

As mentioned earlier, an important result from the ultraviolet studies is that low ionization material is about a factor of ten times more abundant than the high ionization material. Observations at lower resolution toward extragalactic sources also clearly indicate that the low ionization species are more prevalent than the higher ionization species. Interestingly, the Fe II abundance may be enhanced in halo gas.

As was found with the weaker, optical components, halo absorption is commonly found with large negative velocities. However, the velocity extent of the ultraviolet absorption always exceeds the velocity extent of the optical absorption lines, usually by a significant amount: the very lowest column density halo gas (with column densities below the detection limit of the optical lines, but large enough to be detected with the most sensitive UV lines) can commonly be found with total velocity ranges of $\Delta v \sim 130 - 150$ km/s (see Figure 1). The full velocity range of the low ionization UV lines also always exceeds that of the higher ionization lines observed in the same direction, even when comparing lines of similar f-values.

One very important result from the ultraviolet studies is that the gas toward the North Galactic Pole (NGP) displays very different characteristics than the gas toward the South Galactic Pole (SGP). This asymmetry may have important implications for all models for halo gas production and maintenance. It is discussed in more detail in the next section. Finally, the data show considerable differences along different lines of sight suggesting structure on all scales. Comparison with 21-cm observations of clouds at high latitudes shows very good correlation and suggest that the interpretation of data along individual lines of sight may be best understood through a multi-frequency approach.

III. Some Interesting Problems

A. Infalling Gas Toward the NGP. The preference for negative over positive velocities toward the NGP is clearly observed in both the optical and ultraviolet data (see Figure 1). Toward the the NGP, absorption from infalling gas with velocities more extreme than v=-70 km/s is found toward nearly all stars with z>2 kpc, while toward the SGP, none of the stars show absorption beyond -70 km/s. In fact, toward the SGP, there is no evidence at all for any preference toward infalling gas (i.e. with negative velocities) over outflowing gas (i.e. with positive velocities). Based on the SGP data alone, one would be hard pressed to find evidence for the return of cool gas to the plane in a galactic fountain flow.

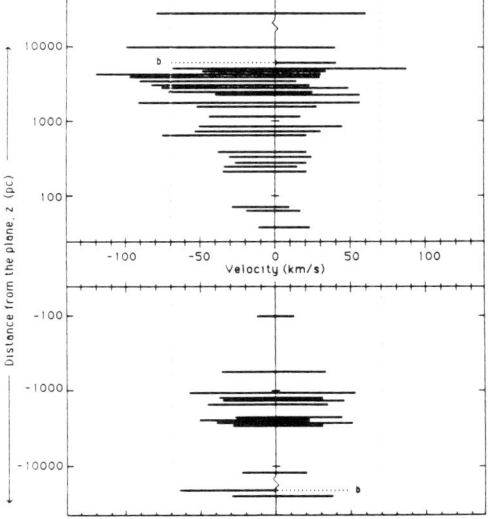

Figure 1. A plot of the absorption widths of the most sensitive low ionization lines as a function of position of the background object. The velocities are measured from composites of the Si II λ1260, C II λ1334, and the O I λ1302 lines. Absorption beyond z<-70 km/s is *only* seen toward the NGP, primarily in stars beyond z=2.4 kpc. The only stars with z<2.4 kpc that show absorption in this range lie behind the IV arch (see §IIIc). Lines of sight toward the SGP show no preference for infall of cool gas from the halo.

Most of the infalling gas in the north is concentrated in structures located primarily in the second quadrant of galactic longitude (i.e. l=90° to 180°). Some of the clouds possess very large velocities (v<-100 km/s) and make up the well known HVC Complexes C I, C III, and A (see Figure 2a; see also Wakker 1989 and his review in this volume). Also in this direction lie the IVC Complexes C II and M II which have velocities -90 km/s < v < -50 km/s (Figure 2b). Interestingly, the velocities of the infalling clouds appear to correlate roughly with distance in the sense that the most distant gas has the most extreme velocities. This is just the opposite of what would be expected if the gas were accelerating on ballistic orbits in the galactic potential. The properties of the infalling cloud complexes present unique problems in themselves and are discussed separately in the following two sections.

If any progress is to be made on the theory of galactic halo gas, one must first understand the nature of the infalling gas toward the NGP: its presence dominates our view of halo gas kinematics. The confirmation and refinement of a galactic fountain model strongly depends on whether this feature is a product of such a flow, or whether it can be understood in terms of an entirely separate, independent mechanism.

B. Where are the Elusive High Velocity Clouds? A complete and up-to-date review of all attempts to detect *any* HVCs in absorption can be found in Van Woerden, Schwarz and Wakker (1989) and will not be repeated here. In summary, while HVCs have been detected in absorption toward extragalactic sources, no positive, confirmed detection of an HVC has yet been made in absorption toward a galactic source which can provide an upper limit on its distance. In this instance, "high velocity" refers to $|v|>100$ km/s.

Songaila, Cowie and Weaver (1988; hereafter SCW) have claimed absorption detections of both Complex A and Complex C. They place an upper limit on the z-distances to the complexes at 1.1 kpc (Complex A) and 1.7 kpc (Complex C). Their claim has been disputed for several years. New evidence casting more doubt on their result has surfaced since the last review (Van Woerden, Schwarz and Wakker 1989) and will therefore be summarized here.

Lilienthal, Meyerdierks and de Boer (1990) show that the feature claimed by SCW as high velocity absorption toward Complex A is probably a misidentified stellar Ti II line blended with the stellar Ca II K line (see contribution by de Boer, this volume).

SCW claim absorption from Complex C can be seen in the spectrum of the RR Lyrae star BT Dra. However, the saturated stellar lines in RR Lyrae photospheres (including Ca II lines) have long been known to be highly complicated, demonstrating bumps, wiggles and splitting depending on the phase of the variable (Preston and Paczynski 1964). For example, the saturated Ca K spectrum shown in SCW is about 10 times broader than the Na D spectrum toward the same star published earlier (Songaila et al. 1985), and and the dips and troughs seen over the entire stellar line are the result of complicated optical depth effects as a function of the variable's phase, averaged over the length of the exposure. Since SCW give no information about exposures or phase information about the variables it is impossible to interpret their observations. No claims as to the distance to the HVCs based on these data can be substantiated unless a more careful analysis of the stellar spectrum as a function of phase and/or using other, less ambiguous species is carried out.

Evidence that in fact Complex C lies beyond 1.7 kpc comes from the non-detections observed by Danly (1989; data can be found in Danly et al. 1991) in the UV spectra of two stars which also lie toward Complex C at about that same distance. The UV lines are about two orders of magnitude more sensitive than the optical lines and should show strong absorption if the Complex indeed lies in front of the stars.

One way in which a non-detection might occur even if the Complex lies in front of the stars is if the gas distribution is patchy on scales smaller than the smallest radio telescope beam used to observe the 21-cm emission from the HVC. If such small scale structure exists, it is possible that the line-of-sight to the background source passes through a local "hole" in the gas distribution. Preliminary results from new observations using the Westerbork Radio Synthesis Telescope of the gas toward stars from Danly (1989) lacking HVC detections show that patchiness is not likely to be able to account for the non-detection (Danly, Wakker and Schwarz 1991).

Another possible explanation for the non-detections is that through some combination of abundance, depletion and ionization, the number of absorbing ions relative to N_H is too low to be detected in absorption. One way to test this alternative is by observing an extra-galactic source which lies near a galactic source showing a non-detection. An interesting example of this is seen in the case of the non-detection in the z=2.9 kpc halo star BD +56 1411 (Danly and Kuntz, 1991) and the detection toward QSO MK 106 (Schwarz, Wakker and Van Woerden 1991) which lies less than 5° from BD +56 1411 in the direction of Complex A. Significant variations in abundance over relatively small scales within a cloud would be required to reconcile these observations if the HVC resided in front of the star. While this cannot be ruled out, abundance variations over such scales would be remarkable, considering the mixing that is likely to occur in the clouds. Complex A is probably beyond z=2.9 kpc, although the question of patchiness still needs to be investigated along this line of sight.

A more general statement of all of the above arguments can be made on statistical grounds: the growing numbers of detections of HVCs in QSOs and the growing numbers of non-detections in galactic sources lying several kpc above the disk make it highly improbable that the non-detections toward halo stars are merely chance super-positions along lines of sight with unfavorable conditions.

The uncertainty in stellar distance determinations remains a serious limitation of absorption studies. Any firm, quantitative determinations will be limited by the quality of the stellar analysis. But at the present time, we

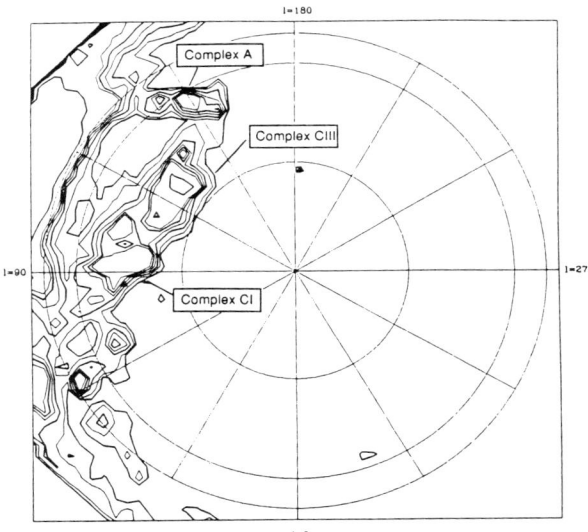

Figure 2a. Contour plot of 21-cm emission seen toward the NGP over the velocity ranges of -130 km/s to -160 km/s (HVCs). Contour levels are a, 2, 4, 6, 8, 10, 20, 40, and 60×10^{18} cm^{-2}. The HVC Complexes are clearly marked, as are the stars from Danly (1989) which provide lower limits from non-detections and BT Dra, the star claimed by SCW to show HVC absorption.

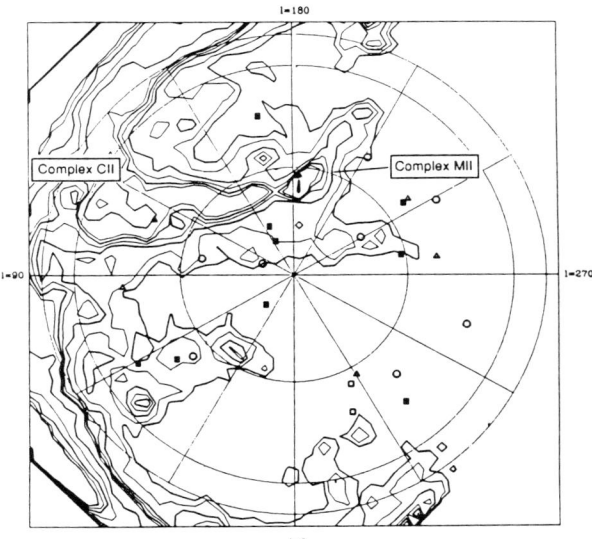

Figure 2b. Same as Figure 2a except it shows the velocity range of -70 km/s to -85 km/s. The symbols show the positions of stars from the survey by Danly (1989). The solid symbols show absorption over this range of velocities and the open symbols do not. The symbol shapes represent distance: circles have z<700 pc, triangles have 700 pc < z < 2400 pc, and squares have z>2400 pc.

conclude that no absorption detection of an HVC has yet unambiguously been made which provides a distance estimate for HVCs -- their distances are still unknown. We further conclude that, in fact, there is growing evidence to support the view that the NGP HVC complexes are beyond the lower halo: i.e. beyond about 2 kpc. Any advancement of our understanding of the origin of the infalling HVC complexes and their relation to overall halo gas models will rest on determining their distances; most further analyses of their physical properties rely on a knowledge of this important and undetermined parameter.

   C. *Infalling Gas with Intermediate Velocities.* The first evidence that the northern galactic sky was covered with infalling IVCs was from the 21-cm survey of Wesselius and Fejes (1973). Amazingly, in the 17 years since its publication, Wesselius and Fejes' paper has only been cited 25 times, and of those, only a handful advance any new results on the gas. The IVCs have been terribly overlooked.

   One fascinating property of the northern IVCs is the strong positional anti-correlation with the low-velocity disk gas: the IVCs are most concentrated in directions where emission from local disk gas is weakest. This anti-correlation is suggestive of gas removal or displacement, but the details have never been fully worked out nor have the implications been clearly understood.

   Individual IVCs or IVC complexes are not as well known or well studied as the HVCs. However, most of the NGP IVCs are contained in Complexes C II and M II which have velocities between roughly -50 km/s and -90 km/s -- their names were originally coined when they were studied as part of the HVC complexes. We refer to these Complexes and their associated emission as the "IVC arch" (see Figure 2b). The distance to the arch is somewhere between z=300 pc and z=1700 pc, although M II may be as close as 700 pc (Danly 1989 for complete discussion). Fortunately, its proximity suggests that, with further study, the distance to the arch will be even better constrained.

   The IVC arch has many intriguing characteristics. It contains most of the mass in the local halo, of order 10^5 M_o (Danly 1989). It contains more than 10^{51} ergs of kinetic energy; some unusual and highly energetic event must have occurred if it was kicked out from the disk. Indeed, its morphological structure suggests it may have been related to the disk. On the other hand, perhaps the IVCs are former HVCs which have been decelerated as they fall back to the disk. As such the study of their physical characteristics, such as abundances and ionization, may provide important insights into the HVC phenomenon. Again, the spatial coincidence of the HVCs and the IVCs suggests there may indeed be a link. In any case, the unique situation of the IVC arch between local disk gas and the HVCs, in both space and velocity, makes it particularly fascinating.

   An in-depth study of the physical properties of the IVC arch has only recently begun (Danly and Kuntz 1991), but early results suggest that the abundances are near solar. Further, the ionization structure is different in different directions along the complex. Interestingly, local densities along the arch are high enough to permit the excitation of the fine structure lines of C II. Further study, including investigations using other observational techniques, is needed for a more complete understanding of this feature.

Equally intriguing is the IV gas *not* in the IV arch. Absorption near -70 km/s can be seen in nearly all stars (and in all four quadrants of galactic longitude) beyond z=2.4 kpc, but lying well off the $N_H=10^{18}$ cm^{-2} contour. This is probably because the UV lines are sensitive to column densities of gas that are too low to be detected in H I. Absorption at -70 km/s toward stars with z<2.4 kpc can *only* be seen in stars behind the arch.

Equally remarkable is the 21-cm emission near -50 km/s that covers the *entire* sky at a level of $N_H=10^{18}$ cm^{-2}! Absorption measurements show that the distance to the -50 km/s gas is well determined to between 370 and 800 pc above the disk. The -50 km/s gas alone contains an additional 10^5 M_o and of order 10^{51} ergs of kinetic energy.

In addition to being intrinsically interesting, the IVCs between about 0.4 and 2 kpc above the disk can serve as "passive" probes of the environment in the lower halo and the disk-halo interaction. For example, at these distances, the IVCs lie well above the protective shielding of the disk gas and are subjected to both the escaped galactic and the inter-galactic radiation fields (see next section). The IVCs may play an important role in searching for evidence for the existence of chimneys or other galactic fountain-related phenomena. Additionally, the IVCs may be perfectly situated to interact with streaming galactic cosmic rays and might be observable through gamma-ray production.

Finally, the IVCs may be used in the study of the dynamics of the gas deceleration. Probably in the case where gas slows from HVC velocities to -70 km/s, and certainly in the case of the slowing from -70 km/s to -50 km/s to the near-0 km/s velocities of the disk, there is not enough material to provide sufficient viscous drag to slow the gas. Some other dynamical processes (such as magnetic fields) may be at work. The impact of the falling clouds as they arrive at the disk will be very different depending on whether they shock the disk with their full 10^{51} ergs of kinetic energy (Tenorio-Tagle 1981; Tenorio-Tagle et al. 1988) or whether they slowly settle on to the disk.

D. The ionization problem. Most photo-ionization models for galactic halo gas require an ionization rate of about $\Gamma \sim 10^{-11}$ sec^{-1} in order to produce the observed C IV and Si IV. An ionization rate on this order is supported by observations of the vertical column of H$^+$ integrated along the entire pathlength toward the NGP (Reynolds 1989) and by the C IV diffuse emission observed by Martin and Bowyer (1989) toward the NGP (G. Field, private communication). Looking outside the galaxy, the sharp edges of galaxies observed by Van Gorkom et al. (1991; see also Sancisi 1987) at limits of N_H of about 2×10^{19} cm^{-2} can also be accounted for by an ionization field on this order.

Recent observations of H-α fluxes from the HVCs lying > 2 kpc above the disk (see above) have shown that the ionization rate of hydrogen in the clouds is roughly two orders of magnitude *lower* -- about $\Gamma \sim 10^{-13}$ sec^{-1} (Songaila, Bryant and Cowie 1989, Kutyrev and Reynolds 1989). A lower ionization rate would also be required to be consistent with the observations by Danly (1989) of O I observed toward stars beyond 2.4 kpc. Since the ionization potential of Oxygen is very similar to Hydrogen (Φ = 13.6 eV) the two species track each other very well. The observation of O I in the halo in directions where 21-cm observations are not sensitive enough to detect the neutral gas suggest that neutral clouds with column densities between about 8×10^{16} cm^{-2} and 10^{18} cm^{-2} reside well above the shielding galactic disk. The presence of neutral clouds at these low column densities is inconsistent with the view that an extragalactic radiation field strong enough to account for the sharp edges observed on external galaxies reaches the lower halo (~2 kpc).

The controversy raised by these observations reminds us that very little is understood about the nature of the cool and neutral gas in the galactic halo: how can the observed amounts of the cool gas form and survive in an environment subjected to either (or perhaps even *both*) ionization from the extra-galactic radiation field and/or coronal gas from a fountain? Any complete theory of galactic halo gas production must account for all the observed phases.

IV. Future Work: The Quest for More Quantitative Results

The greatest limitations on absorption studies are imposed by curve-of-growth (c.o.g.) effects coupled with low instrumental resolution. The problems are not as severe for the Li-like ions (such as the high ionization species of C IV, Si IV, N V, O VI) which have doublet transitions of $2s^2S - 2p^2P^0$ or $3s^2S - 3p^2P^0$ for which fairly good quantitative estimates of column densities can be achieved using the doublet ratio method. But both line saturation and the uncertainties in f-values for species commonly found in neutral or weakly ionized gas make it difficult for quantitative measurements of interstellar column densities to be achieved with much accuracy.

Some authors have attempted to derive quantitative information from IUE data despite its comparatively low resolution (R=10^4). Van Steenberg and Shull (1988a,b) derive column densities using a single-component c.o.g. analysis, relying on the result by Jenkins (1986) that such an approximation is valid provided the absorbing ensemble of lines has a distribution of strengths that is not highly irregular (e.g. bi-modal). The derived column densities have large associated uncertainties. Additionally, a single-component c.o.g. analysis provides no information on component-to-component variations that are almost certain to be present along lines of sight which pass through regions possessing different physical conditions.

Danly and Blades (1989) have selected pairs and groups of stars for which very high quality, high resolution optical Ca II data are available from the survey of Albert et al. (1991). The selected lines of sight show clear multi-component structure. Using the optical data as a guide to model the cloud velocity structure, they are able to determine column densities for the low ionization species with an accuracy of a factor of two, in all but the very central, near-0 km/s portion of the profile. They find clear variations in ion ratios from one component to the next along a single line of sight.

Using a different technique, Savage, Massa and Sembach (1990) have carried out an in-depth analysis of the line of sight toward HD 163522 which lies 9 kpc away in the direction l=350° and b=-9°. In this direction, galactic rotation spreads out the absorption profile over a range of about 160 km/s. They find that under these conditions, saturation in the highly ionized species is minimized and direct estimations of column densities are obtainable by employing what is effectively a doublet ratio method carried out as a function of velocity. They find that applying the same method to many Fe II lines (where the f-values are less accurately determined) provides useful numbers in the line wings, though saturation in the core is still a problem.

Minimizing the limitations imposed by c.o.g. effects requires several lines of the same species, spread over a wide range in f-values. Observing in the UV part of the spectrum helps in this regard because of the large numbers of lines available. Even greater opportunities will be made available with the launch of Lyman-FUSE which will probe the far-UV portion of the spectrum at resolutions of 3×10^4 (although Lyman-FUSE will be best suited for the study of more highly ionized species). Finally, all quantitative measures will only be as accurate as the atomic data (i.e. f-values) available.

Overcoming the difficulties introduced by poor instrumental resolution will be made better by the next generation of space instruments: GHRS on HST will be able to perform spectroscopy on halo stars at a resolution of 10^5. However, GHRS is only able to record one order of the echelle spectrum at a time, requiring a separate observation for nearly every species desired. For example, a 10th magnitude B0 star (which with IUE takes about 2 hours to obtain an $R=1.2 \times 10^4$ spectrum with S/N~10:1) will require about 3.5 hours of exposure time to get 10 selected species with a S/N of about 30:1. The situation will improve dramatically, however with the second generation Space Telescope Imaging Spectrograph (STIS) which will be able to observe with R~140,000 and which will have a large imaging detector capable of capturing most of the important interstellar species with one exposure. The same 10th magnitude B0 star will require a 1-hour exposure to produce a spectrum complete over the range $\lambda\lambda 1050-1700$ with a S/N of about 100:1.

As observational techniques improve, progress in the study of halo gas will be made along two fronts: (1) the more quantitative analysis of interstellar data toward known, bright halo stars, and (2) an expansion of the roster of background sources to much fainter magnitudes. The number of bright, blue halo sources is limited, and many of the most suitable objects have been observed with IUE. Future progress can be made by re-analyzing those lines of sight at higher resolution to provide significantly more accurate column densities *on a component by component basis*. Importantly, advancements in the theory of halo gas are also needed in order to derive physical characteristics of the gas from the diagnostic line strengths and column density ratios.

Progress can also be made by significantly increasing the sampling of halo sightlines by identifying additional background sources. They can be either galactic or extra-galactic, although galactic sources have the added benefit of providing distance information for halo clouds. Surveys reveal a significant population of faint, blue halo stars whose characteristics are not well-understood and whose distances are therefore unknown. Improvements in their photospheric analyses could lead to a substantial increase in the number of halo stars identified which can serve as probes. The ability to observe to fainter magnitudes also means that blue horizontal-branch stars in globular clusters can be employed. Finally, extragalactic sources, such as QSOs, could provide a considerable increase in the sightlines available to sample halo gas. For example, STIS will be capable of obtaining a spectrum toward a B=16^m QSO with S/N~15:1 in its R=2×10^4 mode in about an hour. At a surface density on the sky of about 0.02 QSOs per square degree for B<16^m (Schmidt and Green, 1983), the use of QSOs could dramatically increase the available data on halo gas over the longer term.

Future developments in absorption studies of halo gas rely on very sensitive instruments outside the Earth's atmosphere. A serious trade-off is made between instrumental capabilities and long exposure times: it is unlikely that large, multi-purpose space telescopes would be used to survey the galactic halo the way IUE was employed in the past. Therefore, planning observations to make most efficient use of the spacecraft time is essential: the most effective programs will formulate predictions very carefully, identifying the most critical lines to be observed which will discriminate among competing theories.

ACKNOWLEDGEMENTS The author thanks Elise Albert, Keith Ashman and Colin Norman for careful readings of the manuscript. Richard Green also provided valuable information regarding the performance of STIS.

REFERENCES

Albert, C. E., 1983, Ap.J., **272**, 509
Albert, C.E., Blades, J.C., Morton, D.C., and Proulx, M., 1991, in preparation
Danly, L.,1989, Ap.J., **342**, 785
Danly, L., and Blades, J.C., 1989, in Proc. from the IAU Colloq. #120 *The Structure and Dynamics of the Interstellar Medium*, G. Tenorio-Tagle, M. Moles, and J. Melnick eds., p. 408

Danly, L. and Kuntz K.D., 1991, in preparation
Danly, L., Savage, B.D., Lockman, F.J., and Meade, M.R. Ap.J. Suppl., in press
Danly, L., Wakker, B., and Schwarz, U., 1991, in preparation
de Boer, K.S., 1989, IAU Colloq. #120 *op. cit.*, p. 432
de Vries, C.P. and Van Dishoeck, E.F., 1988, Astron. Ap., **203**, L23
Edgar, R.J. and Savage, B.D., 1989, Ap.J., **340**, 762
Jenkins, E.B., 1986, Ap.J., **304**, 739
Kutyrev, A.S. and Reynolds, R.J., 1989, Ap.J., **344**, L9
Kinney, A., Bohlin, R., Blades, J.C., and York, D.G., 1990, Ap.J., in press
Lilienthal, D., Meyerdierks, H., and de Boer, K.S., 1990, Astr. Ap, in press
Lockman, F.J., Hobbs, L.M., and Shull, M., 1986, Ap.J., **301**, 380
Magnani, L., Blitz, L., and Mundy, L, 1985, Ap.J., **339**, 244
Martin, C. and Bowyer S., 1989, Ap.J., **350**, 242
Morton, D.C. and Blades, J.C., 1986, M.N.R.A.S., **220**, 927
Munch, G., and Zirin, H. 1961, Ap. J., **133**, 11
Penprase, B.E., 1991, Ph.D. Thesis, University of Chicago
Pettini, M. and West, K.A., 1982, Ap.J., **260**, 561
Preston, G.W. and Paczynski, B., 1964, Ap.J., **140**, 181
Reynolds, R.J., 1989, in Proc. IAU Symp. #139, in press
Sancisi, R.,1987, in STScI Symposium Series #2 on *QSO Absorption Lines*
Savage, B.D., 1989, in Proc. ASP Centennial, Berkeley, in press
Savage, B.D. and deBoer, K.S., 1979, Ap.J., **230**, L77
Savage, B.D. and deBoer, K.S., 1981, Ap.J., **243**, 460
Savage, B.D. and Massa, D.M., 1987, Ap.J. **314**, 380
Savage, B.D. Massa, D.M., and Sembach, K., 1990, Ap.J., **355**, 114
Schmidt, M., and Green, R., 1983, Ap.J., **269**, 352
Schwarz U., Wakker B. and Van Woerden H.,1991, in preparation
Songaila, A., Bryant, W., and Cowie, L.L., 1989, Ap.J., **345**, L71
Songaila, A., Cowie, L. L., and Weaver H. 1988, Ap. J., **329**, 580
Songaila, A., York, D. G., Cowie, L. L., and Blades, J. C.,1985, Ap. J., **293**, L15
Tenorio-Tagle, G. 1981, Astr. Ap., **94**, 338
Tenorio-Tagle, G., Franco, J., Bodenheimer, P., and Rozyczka, M., 1988, Astr. Ap., **179**, 219
Van Gorkom J., Van Albada, T.S., Cornwell, T., and Sancisi, R., 1991, in prep.
Van Woerden H., Schwarz U., and Wakker B., 1989, IAU Colloq. #120 *op. cit.* p. 389
Wakker, B., 1989, Ph.D. Thesis, University of Groningen
Van Steenberg, M.E., and Shull, J.M., 1988a, Ap.J. Suppl., **67**, 225
Van Steenberg, M.E., and Shull, J.M., 1988b, Ap.J., **330**, 942
Welty, D.E., Hobbs, L.M., Blitz, L., and Penprase, B.E., 1989, Ap.J., **346**, 232
Wesselius, P.R., and Fejes, I. 1973, Astr. Ap., **24**, 15

IONIZED DISK/HALO GAS: INSIGHT FROM OPTICAL EMISSION LINES AND PULSAR DISPERSION MEASURES

R. J. REYNOLDS
University of Wisconsin-Madison, Department of Physics
1150 University Avenue, Madison, WI 53706, USA

ABSTRACT. Warm ($\approx 10^4$ K), diffuse H^+ is a significant component of the interstellar medium within the Galactic disk and lower halo. This gas accounts for about one quarter of the interstellar atomic hydrogen, consumes a large fraction of the interstellar power budget, and appears to be the dominant state of interstellar matter 1 kpc above the midplane. The origin of this ionized gas is not yet established; however, of the known sources of ionization only O stars and perhaps supernovae produce enough power to balance the "cooling" rate of the gas. If O stars are the source of the ionization, then the interstellar HI, including the extended "Lockman layer", must have a morphology that allows about 14% of the Lyman continuum photons emitted by the stars to travel hundreds of parsecs within the Galactic disk and up into the lower halo.

1. INTRODUCTION

Pulsar dispersion measures and faint, diffuse optical line emission from the interstellar medium have firmly established the existence of a warm ($\approx 10^4$ K), ionized medium that is distributed throughout the Galactic disk and lower halo (see reviews by Kulkarni and Heiles 1986, 1987; and Reynolds 1989b). The dispersion measures have revealed directly the column densities and space averaged volume densities of the free electrons (and H^+), while the emission lines have provided information about the emission measures, temperature, ionization state, clumpiness, and kinematics of the gas. (In order to avoid confusion between this diffuse, ionized gas and the discrete regions of ionized gas traditionally referred to as "HII regions", the diffuse gas will be denoted as H^+ rather than HII.) The following is a review of some of the observations that have provided insight into the nature of the H^+ at large distances (up to ~ 1 kpc) above the Galactic midplane and its possible relationship to stars and gas near the plane.

2. THE SCALE HEIGHT AND COLUMN DENSITY OF THE H^+

The existence of a widespread, ionized component of the interstellar medium was established approximately two decades ago by observations of pulsar dispersion measures (Hewish et al. 1968) and faint optical line emission from the Galaxy (Reynolds 1971; Reynolds, Scherb, and Roesler 1973). The results revealed gas with a temperature T ≈ 6000-8000 K occupying a large fraction (f ≳ 20%) of the interstellar volume. However, because the mean electron density of 0.03 cm^{-3} (e.g., Guelin 1974) amounted to only a few percent of the total midplane hydrogen density, and because probes of this gas tended to concentrate on its properties near the midplane, the total z-extent and column density of this component was not readily appreciated.

Recent discoveries of pulsars in globular clusters located far ($|z|$ ≳ 4 kpc) from the midplane have provided the opportunity to measure directly the total vertical extent and column density of the H$^+$. The vertical distribution of the gas is revealed in Figure 1, which is a plot of the component of the dispersion measure perpendicular to the Galactic disk against the distance $|z|$ from the midplane for pulsars with distances measured independently of their dispersion measures (e.g., Weisberg, Rankin, and Boriakoff 1987; Reynolds 1989a, and references therein; Frail 1989). The disk pulsars ($|z|$ ≲ 300 pc) with distances ranging from 130 pc to more than 10 kpc from the sun sample a large area of the Galaxy near the solar circle.

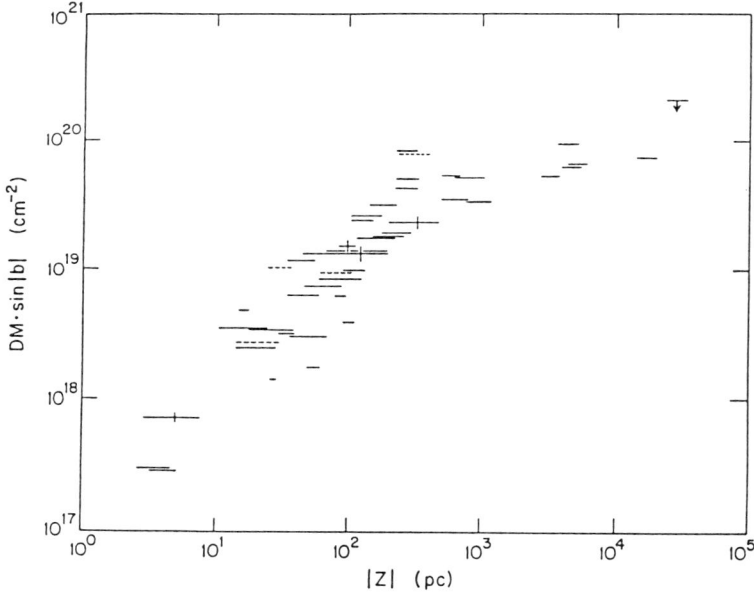

Fig. 1 - The component of the pulsar dispersion measure (i.e., the column density of H$^+$) perpendicular to the Galactic plane is plotted against distance $|z|$ from the midplane. Pulsars behind the Gum nebula are indicated by dashed lines; the pulsar in the LMC is indicated by an upper limit arrow.

Also included in Figure 1 are the globular cluster pulsars and the pulsar in the Large Magellanic Cloud (LMC). The data for these high $|z|$, non-disk pulsars are from Anderson et al. (1989 a,b,c,d, 1990); Biggs et al. (1990); D' Amico et al. (1990); Lyne et al. (1987, 1988); Manchester et al. (1989a,b); McCulloch et al. (1983); and Wolszczan et al. (1989a,b). The H^+ column density toward the LMC is plotted as an upper limit because of the possibility that some of it may be due to ionized gas within the LMC itself. On the other hand the fact that multiple pulsars in the same globular cluster have the same dispersion measure to within the measurement errors (e.g., Anderson et al. 1989a,b) implies that very little of the dispersion measure is associated with either the globular clusters or the pulsars themselves.

Figure 1 shows the extension of the H^+ layer up to $|z|$ heights of more than 1 kpc. Between $|z| \approx 0.8$ kpc and $|z| \approx 3-5$ kpc, for example, the mean column density increases from about 4.4×10^{19} cm^{-2} to 7.0×10^{19} cm^{-2}. The pulsars in the highest $|z|$ globular clusters and the LMC also reveal a "knee" in the distribution, implying a maximum extent $|z| < 4$ kpc and a total column density from the midplane approaching 10^{20} cm^{-2}. The mean value of DM $\cdot$ sin$|b|$ for the five highest $|z|$ pulsars (excluding the LMC) indicates an H^+ column density N ≈ 7.0 (+2.4,-1.5) $\times 10^{19}$ cm^{-2}. This value is about 40% lower than that derived previously (Reynolds 1989a) because the high $|z|$ data set has increased and there have been changes in the reported values of some of the dispersion measures (e.g., for 47 Tuc; Manchester et al. 1989a).

The data in Figure 1 can be fitted well by a two-component electron density distribution given by

$$\langle n_e \rangle = 0.015 \exp(-|z|/70) + 0.025 \exp(-|z|/900) \text{ cm}^{-3}, \quad (1)$$

where the first term represents, statistically, the contribution from discrete, classical HII regions near the midplane (see Lyne 1981; Manchester and Taylor 1981; Vivekanand and Narayan 1982; Harding and Harding 1982) and the second term represents the diffuse, extended component. This model is not strictly applicable to the data set in Figure 1 because the highest $|z|$ pulsars are all at high latitude and intersect no classical HII regions (i.e., have no contribution from the first term). However, this two-component fit clearly shows that diffuse ionized gas accounts for nearly all of the ionized gas in the Galactic disk, overwhelming the mass in the classical HII regions by 20 to 1.

Since the space averaged density of the extended, H^+ component at the midplane is $\langle n_e \rangle_o = 0.025 \pm 0.005$ cm^{-3} (Weisberg, Rankin, and Boriakoff 1980), its scale height H can be derived simply from the quotient $N/\langle n_e \rangle_o$, which has a value of 910 (+620, -320) pc. A similar analysis has recently been carried out by Lyne, Salucci, and Sciama (1990). A possible source of systematic error, which would tend to underestimate the value of H derived in this manner, is due to the fact that the high $|z|$ globular cluster pulsars used to determine N are all at high Galactic latitude and thus sample a relatively limited region of the disk near the sun, a cylinder of radius $\approx$ 1 kpc for a maximum z-height of $\pm$ 1 kpc, whereas the disk pulsars used to derive $\langle n_e \rangle_0$ are all at low latitudes and sample a much larger region of the disk, out to about 10 kpc from the sun. High latitude observations at 21 cm (Dickey and Lockman 1990; Lockman 1985) and at Hα (Reynolds 1984) indicate deficiencies of about a factor two in both the neutral hydrogen and the ionized hydrogen, respectively, near ($<$ 1 kpc) the sun compared to the more distant regions of the disk sampled by lower latitude observations. Therefore, values of N and H that are not biased by this local deficiency could be as much as twice the values derived above.

An independent measurement of the scale height has been made from observations of the diffuse interstellar Hα emission associated with the Perseus spiral arm located at a distance of 3 kpc from the sun (Reynolds 1985a). The emission component associated with this spiral arm (identified by its -40 km s^{-1} radial velocity with respect to the LSR) extends more than 20° from the plane, corresponding to a $|z|$ height greater than 1 kpc. The rate of decrease in the Hα intensity with distance above the arm indicates an H$^+$ scale height of 600-1200 pc (Reynolds 1985a plus additional data; in preparation). As the accuracies of these two methods of determining H improve, it may become possible to compare the scale height above a spiral arm with the disk-wide average.

3. IMPLICATIONS OF THE LARGE SCALE HEIGHT

3.1 The H$^+$/HI Mass Ratio

The large scale height of the H$^+$ implies that most of this ionized gas is located well outside the traditional disk of the Galaxy. For H = 0.9 kpc, approximately 70% of the H$^+$ is located at $|z| >$ 300 pc and half is at $|z| >$ 600 pc. Therefore, although the average volume density of H$^+$ at the Galactic midplane is small compared to the HI, the H$^+$ nevertheless accounts for a relatively large fraction of the interstellar matter. The significance of this ionized component is illustrated in Table 1, which compares H$^+$ with HI column densities toward the five globular cluster pulsars that are located more than 3 kpc above the midplane (and thus sample nearly all the H$^+$). For these five "random", high latitude lines of sight the amount of ionized hydrogen ranges from 23% to 63% of the neutral hydrogen. In Figure 2 the average density of the H$^+$ as a function of distance $|z|$ above the midplane is compared with the average density of the HI (Dickey and Lockman 1990). The best-fit data for the H$^+$ and the HI thus suggest that at $|z| >$ 700 pc the warm, ionized medium is the dominant state of the interstellar gas.

3.2 The Power Requirement and Source of the Ionization

The intensity of the Galactic Hα background at high latitudes provides a direct measure of the hydrogen recombination (and ionization) rate r_G in a cm^2 column perpendicular to the disk through

TABLE 1

Comparisons of H^+ and H I column densities at high galactic latitude*

ℓ	b	N_{H^+} $(10^{20} cm^{-2})$	N_{HI} $(10^{20} cm^{-2})$	N_{H^+}/N_{HI}
4°	+47°	0.91	4.0	0.23
59	+41	0.94	1.5	0.63
65	−27	2.1	6.2	0.34
306	−45	0.77	2.75	0.28
333	+80	0.74	2.0	0.37

*Toward globular cluster pulsars with $|z|$ distances > 3 kpc, which place them above > 90% of the H^+. Values for N_{H^+} are from Manchester et al. (1989a), Wolszczan et al. (1989a,b), and Anderson et al. (1989c,d). Values for N_{HI} are from McCammon et al. (1983) and Stark et al. (1989).

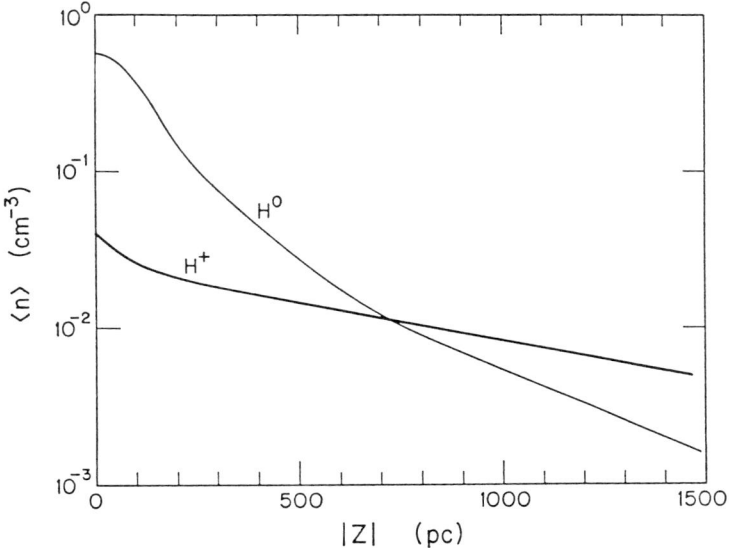

Fig. 2 − The best fit mean volume densities of H^+ from eq. 1 and H^0 from Dickey and Lockman (1990) are plotted against distance $|z|$ from the midplane.

the relation
$$r_G = \frac{8\pi}{\varepsilon} I_\alpha \cdot \sin|b|, \qquad (2)$$
where ε is the average number of Hα photons produced per recombination (≈ 0.46; case B), I_α is the Hα intensity in units of photons cm^{-2}s^{-1}sr^{-1}, and b is the Galactic latitude of the Hα observation. An all sky average of the Hα data implies that $\langle r_G \rangle \approx 4 \times 10^6$ s^{-1} cm^{-2} for the region within 2-3 kpc of the sun (Reynolds 1984, 1987a). At 13.6 eV per ionization this rate corresponds to a power consumption of 1×10^{-4} ergs s^{-1} cm^{-2} just to keep the gas ionized. Approximately the same power is deduced independently from an examination of the pulsar dispersion measure data (Reynolds 1990b).

Of the known sources of ionization and heating within the Galactic disk, only the O stars, which produce 3×10^7 ionizing photons s^{-1} cm^{-2} (Abbott 1982), and supernovae, which inject about 1×10^{-4} ergs s^{-1} cm^{-2} (e.g., Abbott 1982), meet or surpass the power requirements of the H$^+$. Hot evolved stars (Panagia and Terzian 1984), cosmic rays (van Dishoeck and Black 1986), and the diffuse X-ray background (McCammon et al. 1983) fail to produce the required ionization by factors ranging from 20 to more than 100. Furthermore, Hα observations of an intergalactic HI cloud (Reynolds et al. 1986) and of high velocity clouds (Kutyrev and Reynolds 1989; Reynolds 1987b; Songaila, Bryant, and Cowie 1989), which probe the ionizing radiation field within and outside the Galactic halo, imply that at most 10% of the ionization rate can be due to radiation coming from outside the Galaxy. If supernovae are the source, they must be extremely (~ 100%) efficient at producing warm, ionized hydrogen, and if O stars are the source, then 14% of their Lyman continuum photons must somehow escape the immediate vicinities of the stars.

In the McKee and Ostriker (1977) model of the interstellar medium it is proposed that the diffuse H$^+$ is located in transition regions between the cold HI clouds and the hot, very low density "coronal" gas, and that it is photoionized by a very dilute Lyman continuum flux originating from luminous stars and supernova remnants. However, this picture appears to be incompatible with the fact that most of the H$^+$ is located well above the thin layer of young stars and HI clouds that define the traditional disk of the Galaxy. The presence of the H$^+$ at high $|z|$ and along lines of sight that are far from ionizing stars (Reynolds 1990a), coupled with the existence of an opaque layer of HI that exends to high $|z|$ (Lockman 1984; Lockman, Hobbs, and Shull 1986), seem to require either the existence of some as yet unidentified source of ionization at high $|z|$ or a special morphology of the HI that allows the ionizing photons originating near the Galactic plane to travel a kiloparsec or more up into the lower halo (see Cox 1989; Bregman and Harrington 1986). For example, to account for the ionization at $|z| > 300$ pc with sources near the plane the flux of Lyman continuum photons flowing outward at $|z| = 300$ pc must be 2×10^6 photons cm^{-2} s^{-1}, or 7% of the total O star production rate. There is no known source of ionization with this strength located at such high $|z|$.

4. THE TEMPERATURE AND IONIZATION STATE OF THE GAS

The power required to sustain ionized gas against cooling and hydrogen recombination losses has a minimum near 10^4 K and increases enormously for temperatures that are significantly hotter or colder (Reynolds 1990b). Since the power requirements of the gas are large relative to the power that is available from the known sources, it can be deduced independently of the emission line observations that nearly all of the ionized gas, even at high $|z|$, must be at a temperature near 10^4 K. The emission line data confine the temperature to the range 6000-20,000 K (Reynolds 1989b).

High [SII]λ6716/Hα and low [OIII]λ5007/Hα intensity ratios relative to those observed in O and B star HII regions imply ionization/excitation conditions in the diffuse, ionized gas that differ significantly from conditions within classical HII regions (Reynolds 1985b,c 1988). Models of photoionized gas suggest that the observed line intensity ratios could be the result of a dilute radiation field (e.g., Mathis 1986; J. Bland Hawthorne 1990). However, measurements of the [SII]/Hα intensity ratios in very faint (emission measures down to 5 cm^{-6} pc), extended (diameters up to 260 pc) HII regions that have been identified around O and early B stars have shown that the superposition of such regions cannot be the source of the diffuse background (Reynolds 1988). The observed tendency of the [SII]/Hα ratio to increase gradually with increasing dilution of the ionizing radiation suggests that, if the diffuse ionization consists of dilute, photoionized HII regions, then the regions must have an extremely low emission measure (EM < 1 cm^{-6} pc) and must be ionized by a Lyman continuum flux F_{LC} < 6 × 10^5 photons cm^{-2}s^{-1} (the equivalent of an unattenuated flux from a typical O star at a distance d > 400 pc). The [SII],[NII], and [OIII] intensities also suggest a low state of excitation with few ions present that require an ionization energy greater than about 23 eV. On the other hand the absence of [NI] and [OI] emission implies that the hydrogen is nearly fully ionized, with an ionization ratio n(H$^+$)/n(H^0) > 2 within the ionized regions (Reynolds 1989c). However, this result was obtained from observations at the Galactic equator, where the line intensities are bright enough to set useful limits. A probe of the hydrogen ionization ratio at higher latitudes will have to await the construction of more sensitive instrumentation (see below).

5. CONCLUSIONS AND FUTURE DIRECTIONS

Warm, diffuse H$^+$ is a major component of the interstellar medium of our Galaxy, extending from the midplane into the lower Galactic halo, where it could be the dominant state of the interstellar gas. The existence of this component appears to have an important bearing upon the composition and topology of the interstellar medium and the principal processes of ionization and heating within the disk and halo. The diffuse H$^+$ also could contribute significantly to the total pressure at the midplane (Cox 1989) and have an important influence on the dynamics of hot (10^6 K) gas far above the plane (Heiles 1990). Furthermore, because the H$^+$ is a significant fraction of the interstellar matter, warm dust within it could be a non negligible source of IR continuum, and cosmic rays within it could produce a significant fraction of the diffuse γ-ray background, particularly at high Galactic latitudes (Bloemen 1989). The existence of the H^+ clearly needs to be taken into consideration in models of the interstellar medium and galactic halo and in the analysis of interstellar medium data.

While many of the basic properties of the H^+, such as its scale height, surface density, temperature, and power consumption, have been measured (though somewhat crudely), there is yet very little understanding about how this component fits together with the other components of the medium--the cold clouds, the warm HI, and the hot gas. The morphology of the H^+ is not known. Is this gas the ionized portion of the extended, warm HI (Lockman) layer, existing only where ionizing radiation from the disk leaks out between the clouds, like rays of sunlight through partial cloud cover (Cox 1989)? Or is this gas located on the inner surfaces of hot supernova created bubbles, chimneys, and worms? Also not understood is the support of the gas at high $|z|$ and, of course, the mechanism of ionization. Can O stars really be the source of the ionization 700 pc above the midplane, or are _in situ_ sources required such as radiative cooling of hot gas (Martin and Bowyer 1990), a population of faint, hot stars at high $|z|$, or perhaps something exotic (e.g., Sciama 1990; Melott et al. 1989)?

It is possible that many of these questions will be answered in the near future. An extended H^+ component has recently been identified in the Sb galaxies NGC 891 (Rand, Kulkarni, and Hester 1990; Dettmar et al. 1989) and M31 (Walterbos and Braun 1990), and in some irregular galaxies (Hunter and Gallagher 1990). This has provided the opportunity to study the nature of this component from entirely new perspectives. Also, the continued discovery of pulsars in additional globular clusters around our Galaxy promises to improve column density and scale height determinations and probe the uniformity of the H^+ layer. Finally, the development of a more efficient spectrometer could soon make it possible to carry out a velocity resolved, all-sky survey of the optical line emission with an angular resolution comparable to that of the the 21 cm surveys (Reynolds et al. 1990). This new instrumentation would provide for the first time a clear, detailed picture of the distribution and kinematics of the H^+ within the interstellar medium and make possible the detection and study of additional, extremely faint diagnostic emission lines that probe the temperature and ionization conditions within the gas.

This work has been supported by the NSF grant AST 88-13467.

REFERENCES

Abbott, D. C. 1982, Ap. J., 263,723.
Anderson, S., Gorham, P., Kulkarni, S., Prince, T., and Wolszczan, A., 1989a, IAU Circular No. 4762.
Anderson, S., Gorham, P., Kulkarni, S., Prince, T., and Wolszczan, A.,

1989b IAU Circular No. 4772.
Anderson, S., Kulkarni, S., Prince, T., and Wolszczan, A., 1989c IAU Circular No. 4819.
Anderson, S., Kulkarni, S., Prince, T., and Wolszczan, A., 1989d IAU Circular No. 4853.
Anderson, S. Kulkarni, S., Prince, T., and Wolszczan, A. 1990, IAU Circular No. 5013.
Biggs, J. D., Lyne, A. G., Manchester, R. N., and Ashworth, M. 1990, IAU Circular No. 4988.
Bland Hawthorn, J. 1990, private communication.
Bloemen, H. 1989, Ann. Rev. Ast. Ap., 27, 469.
Bregman, J. N., and Harrington, J. P. 1986, Ap. J., 309, 833.
Cox, D. P. 1989, in IAU Colloquim No. 120, Structure and Dynamics of the Interstellar Meduim, ed. G. Tenorio-Tagle, M. Moles, and J. Melnick (New York: Springer) in press.
D'Amico, N., Lyne, A. G., Bailes, M., Johnston, S., Manchester, R. N., Staveley-Smith, L. 1990, IAU Circular No. 5013.
Dettmar, R. J., Keppel, J., Roberts, M. S., and Gallagher, J. S. 1989, preprint.
Dickey, J. M., and Lockman, F. J. 1990, Ann. Rev. Astr. Ap., Vol. 28, in press.
Frail, D. A. 1989, Ph. D. Thesis, Univ. of Toronto.
Guelin, M. 1974, in Galactic Radio Astronomy, IAU Symposium No. 60, ed. F. J. Kerr and S. C. Simonson (Dordrect:Reidel), p. 51.
Harding, D. S., and Harding, A. K. 1982, Ap. J., 257, 603.
Heiles, C. 1990 Ap. J., 354, 483.
Hewish, A., Bell, S. J., Pilkington, J. D., Scott, P. F., and Collins, R. A. 1968, Nature, 217, 709.
Hunter, D. A. and Gallagher, J. S., III 1990, preprint.
Kulkarni, S. R., and Heiles, C. 1986 preprint of Chapter 3 in Galactic and Extragalactic Radio Astronomy, 2nd Ed., eds. K. I. Kellerman and G. L. Verschuur (New York: Springer-Verlag), p. 95.
Kulkarni, S. R., and Heiles, C. 1987, in Interstellar Processes eds. D. J. Hollenbach and H. A. Thronson, Jr. (Dordrect:Reidel), p. 87.
Kutyrev, A. S., and Reynolds, R. J. 1989, Ap. J. (Letters), 334, L9.
Lockman, F. J. 1984. Ap. J., 283, 90.
Lockman, F. J. 1985, in Gaseous Halos of Galaxies, proceeding of NRAO Workshop No. 12, ed. J.N. Bregman, and F.J. Lockman (NRAO: Green Bank), p. 63.
Lockman, F. J., Hobbs, L. M., and Shull, J. M. 1986, Ap. J., 301, 380.
Lyne, A. G. 1981 in Pulsars, IAU Symposium No. 95, eds. W. Sieber and R. Wielebinski (Dordrecht:Reidel), p. 423.
Lyne, A. G., Biggs, J. D., Brinklow, A., Ashworth, M., and McKenna, J. 1988, Nature, 322, 45.
Lyne, A. G., Brinklow, A., Middelditch, J., Kulkarni, S. R., Backer, D. C., and Clifton, T. R. 1987, Nature, 328, 399.
Lyne, A. G., Salucci, P., and Sciama, D. W. 1990, in preparation.
Manchester, R. N., Lyne, A. G., Johnston, S., D'Amico, N., Lim, J., and Kniffen, D. A. 1989a, IAU Circular No. 4892.
Manchester, R. N., Lyne, A. G., Johnston, S., D'Amico, N., Lim, J., Kniften, D.A., Fruchter, A. S., and Goss, W. M. 1989b, IAU Circular No. 4905.

Manchester, R. N., and Taylor, J. H. 1981, A. J., 86, 1953.
Martin, C., and Bowyer, S. 1990, Ap. J., 350, 242.
Mathis, J. S. 1986, Ap. J., 301, 423.
Melott, A. L., McKay, D. W., and Ralston, J. P. 1988, Ap. J. (Letters), 324, L43.
McCammon, D., Burrows, D. N., Sanders, W. T., and Kraushaar, W. L. 1983, Ap. J., 269, 107.
McCulloch, P. M., Hamilton, P. A., Ables, J. G., and Hunt, A. J. 1983, Nature, 303, 307.
McKee, C. F., and Ostriker, J. P. 1977, Ap. J., 218, 148.
Panagia, N. and Terzian, Y. 1984, Ap. J., 287, 315.
Rand, R. J., Kulkarni, S. R., and Hester, J. J. 1990, Ap. J. (Letters), 352, L1.
Reynolds, R. J., 1971 Ph.D. thesis, University of Wisconsin - Madison.
Reynolds, R. J. 1984, Ap. J., 282, 191.
Reynolds, R. J. 1985a, in Gaseous Halos of Galaxies, Proceedings of a Workshop held at the NRAO, Green Bank, W.V., eds., J.N. Bregman and F.J. Lockman (NRAO:Green Bank), p.53.
Reynolds, R. J. 1985b, Ap. J., 294, 256.
Reynolds, R. J. 1985c, Ap. J. (Letters), 298, L27.
Reynolds, R. J. 1987a, Ap. J., 323, 118.
Reynolds, R. J. 1987b, Ap. J., 323, 553.
Reynolds, R. J. 1988, Ap. J., 333, 341.
Reynolds, R. J. 1989a, Ap. J. (Letters), 339, L29.
Reynolds, R. J. 1989b, in Galactic and Extragalactic Background Radiation, IAU Symposium No. 139, ed. S. Bowyer and C. Leinert (Dordrecht: Kluwer).
Reynolds, R. J. 1989c, Ap. J., 345, 811.
Reynolds, R. J. 1990a, Ap. J., 348, 153.
Reynolds, R. J. 1990b, Ap. J. (Letters), 349, L17.
Reynolds, R. J., Magee, K., Roesler, F. L., Scherb, F., and Harlander J. 1986, Ap. J. (Letters), 309, L9.
Reynolds, R. J., Roesler, F. L., Scherb, F., and Harlander, J., 1990, in Instrumentation in Astronomy VII, SPIE Proceedings No. 1235, 1990, in press.
Reynolds, R. J., Scherb, F., and Roesler, F.L. 1973, Ap.J., 185, 869.
Sciama, D. W. 1990, M.N.R.A.S., 244, 1p.
Songaila, A., Bryant, W., and Cowie, L. L. 1989, Ap. J. (Letters), 345, L71.
Stark, A. A., Gammie, C., Bally, J., Linke, R. A., Heiles, C., and Wilson R. A. 1989, in preparation.
van Dishoeck, E. F., and Black J. N. 1986, Ap. J. Suppl., 62, 109.
Vivekanand, M., and Narayan, R. 1982, J. Ap. Astr., 3, 399.
Walterbos, R. and Braun, R. 1990, preprint.
Weisberg, J. M., Rankin, J. M., and Boriakoff, V. 1980, Astr. Ap., 88, 84.
Weisberg, J. M., Rankin, J. M., and Boriakoff, V. 1987, Astr. Ap., 186, 307.
Wolszczan, A., Anderson, S., Kulkarni, S., and Prince, T. 1989a, IAU Circular, No. 4880.
Wolszczan, A., Kulkarni, S. R., Middleditch, J., Backers, D. C., Fruchter, A. S., and Dewey, R. J. 1989b, Nature; 337, 531.

DARK MATTER DECAY AND THE HEATING AND IONISATION OF HI REGIONS

D.W. SCIAMA
International School for Advanced Studies
International Centre for Theoretical Physics
Strada Costiera 11, 34014 Trieste, Italy

I have proposed (Sciama 1989) that the heating and ionisation observed in HI regions from pulsar dispersion measure data and Hα emission is mainly produced by u-v photons emitted by decaying dark matter particles. If a particle P_1 of rest mass m_1 decays into a photon and a particle P_2 of rest mass m_2 one has:

$$P_1 \longrightarrow \gamma + P_2.$$

The energy E_γ of the photon in the rest-frame of P_1 is given by

$$E_\gamma = \frac{m_1^2 - m_2^2}{2m_1}. \qquad (c=1)$$

If $m_2 \ll m_1$, we have the simple relation

$$E_\gamma \sim \frac{1}{2}m_1.$$

If the particles are neutrinos we know that their cosmological density will have the critical value if

$$\sum_i m_i \sim 100h^2 \text{ eV},$$

where h is the Hubble constant in units of 100 km sec^{-1} Mpc^{-1} ($\frac{1}{2} < h < 1$) and the sum is over the three types of neutrino (τ, μ, e). Presumably P_1 is ν_τ. We notice immediately that if the neutrino density is critical and if m_{ν_τ} dominates, then $E_\gamma > 13.6$ eV unless h is almost exactly 1/2.

My proposal is speculative but was made because there is growing evidence that it is difficult to account for the observed ionisation using conventional astronomical sources. This is true near the sun (Reynolds 1990) and at heights ~ 1 kpc above the galactic plane (Reynolds 1989). It is also true of the local interstellar medium (LISM) (Cox and Reynolds 1987), of NGC 891 at heights several kpc above its plane

(Rand, Kulkarni, and Hester 1990), of the intergalactic medium at red shifts ~3-4 (Shapiro and Giroux 1987) and of Lyman α clouds at similar red shifts (Bajtlik, Duncan, Ostriker 1988). All these anomalies would be resolved at one stroke if the neutrino decay lifetime τ were given by

$$\tau \sim 1.5 \times 10^{23} \text{ sec.}$$

In a series of papers (Sciama 1989a-e, Salucci and Sciama 1990, Sciama and Salucci 1990) I have explored some of the consequences of this hypothesis. These consequences turn out to be rather remarkable. One can show that

(i) the rotation curve of the Galaxy can be derived from $n_e(r)$ as deduced from pulsar data.
(ii) $13.6 < E_\gamma < 14.5$ eV independently from (a) the observed upper limit on the extragalactic flux of ionising photons, (b) m_{ν_τ} derived from the Tremaine-Gunn phase space argument applied to our Galaxy, (c) the constraint that the decay photons must be unable to ionise nitrogen in the LISM.
(iii) the density of the universe is close to the critical value (and $m_{\nu_\tau} = 27.7 \pm 0.5$ eV).
(iv) the Hubble constant H must be 54.5 ± 1 km sec^{-1} Mpc^{-1},
(v) $n_e(r,z)$ for NGC 891 can be derived from its rotation curve via its dark matter and agrees with observation.
(vi) the distance of NGC 891 and H can be derived independently of our previous determination of H, with consistent results.

The hypothesis can be tested directly by searching for a line in the u-v background at $E_\gamma \sim 14$ eV. The predicted flux of this line is $\sim 2 \times 10^3$ cm^{-2} sec^{-1}. I am planning such a search in collaboration with S. Bowyer and R. Stalio.

CONSTRAINTS ON GALACTIC INFALL FROM STUDIES OF CHEMICAL EVOLUTION

M.TOSI
Osservatorio Astronomico
Via Zamboni 33
I-40100 Bologna, Italy

ABSTRACT. It is shown that, in order to reproduce the observational properties of our Galaxy, models of galactic chemical evolution require a conspicuous amount of gas falling on the disk with very slow time decay. The constraints on the characteristics of this gas derived from the models are presented.

1. INTRODUCTION

The first interpretation of the observational phenomenon of High Velocity Clouds in terms of external gas falling on the galactic disk was given by Oort (1965). A few years later, Larson (1972) computed the first galactic evolution model taking infall explicitely into account, with the aim of preventing a too rapid gas consumption due to the disk star formation. The infalling gas in this case was supposed to originate from the collapsing halo. Since then, infall has become one of the most important but controversial issues in the field of galaxy evolution. It is generally believed that the galactic disk has formed via gas accretion from the halo: what will be discussed here is the existence of gas infall when the disk has already settled in a roughly stable configuration. Several authors consider this infall necessary for a realistic interpretation of the observed galactic properties, whereas others question even its existence or, at least, its relevance in the evolution of our Galaxy.

There are, in fact, some basic open questions about infall which have not received yet a definite answer because of the large uncertainties in the observational data. If any gas actually falls into the disk, what is its amount ? Is the gas of extragalactic origin or is it the residual of the halo collapse ? What are its space and time distributions and its metallicity ? We will see in the following that by comparing the observational properties of our Galaxy with the corresponding predictions by models of galactic chemical evolution we can derive useful indications on these issues and reduce the range of possible values of the infall parameters.

2. MODEL REQUIREMENTS

The comparison between model predictions and empirical data provided the first information about the infall properties in the early eighties. Tinsley (1980) and

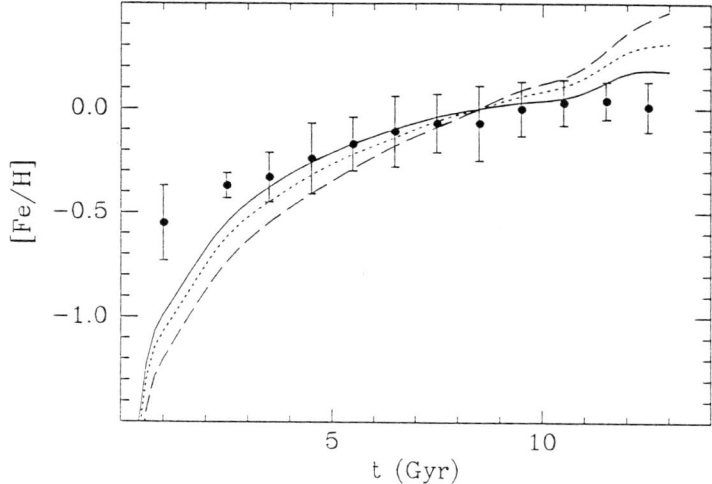

Fig.1. AMR in the solar neighbourhood. The dots with error bars represent Twarog's (1980) data, the dashed line a model with constant SFR and no infall, the dotted line a model with constant SFR and constant infall rate of 4 10^{-3} $M_\odot kpc^{-2} yr^{-1}$, the solid line a model with slowly decreasing SFR (e-folding time: 15 Gyr) and the same infall as the dotted line.

Twarog (1980) showed that to better reproduce the major observational constraints of the solar neighbourhood (age-metallicity relation - AMR - and scarcity of metal poor stars - G-dwarf problem) an infall rate roughly half of the star formation rate - SFR - had to be assumed, and that both rates were approximately constant during the entire Galaxy lifetime. The same conclusions on both infall and SFR were reached by Tosi (1982) to reproduce the abundance gradients and the gas and total mass distributions observed in the disk.

Figure 1 shows the AMR derived by Twarog from a sample of ∼1000 F stars in the solar neighbourhood. The error bars correspond to his estimated uncertainties on the data. The dashed curve represents a model with constant SFR and no infall after the disk formation: it is apparent that it predicts too low abundances in the early epochs and too large abundances at the present time. A model with the same SFR and constant infall (dotted curve) is more consistent with the data because the lower initial gas content of the region is more readily polluted by the metals ejected by evolving stars, thus allowing a faster rise of the AMR curve in the early epochs and because the infall of metal poor gas dilutes the chemical abundance of the interstellar medium in the recent phases. A model with the same infall and slowly decreasing SFR is in better agreement with the data because it provides more star (and metal) formation in the early epochs and less in the late epochs than the previous two models. Had we assumed for the initial disk metallicity a finite value instead of Z=0, we would have obtained a good fit also for the left part of the diagram (see e.g. Matteucci and François 1989).

Most chemical evolution models do not consider at all the dynamics of the system. Hence, the effect of infall simply corresponds to an increase of the gas reservoir and, since the infalling gas is usually assumed to be metal poor, its major result is to dilute the ambient metallicity. However, an important caveat for models with infall was pointed out by Mayor and Vigroux (1981), who showed that the reduction of angular momentum suffered by the infalling gas when reaching the disk causes an inevitable radial (inward or outward) motion of this gas. Radial flows must then be taken into account if infall is assumed. Models taking radial flows into account have later been computed by Lacey and Fall (1985), Tosi (1988a) and Pitts and Tayler (1990). Despite the different approaches and approximations, they all found that inward radial flows steepen the metallicity gradients but that infall is still required to reproduce all the galactic properties even when radial flows are included in the models.

During the last ten years, there have been some attempts to demonstrate that infall is not necessary to reproduce the galactic properties, if particular choices of the other parameters (SFR, initial mass function and stellar parameters, see Tosi 1988a) are adopted. For instance, Gusten and Mezger (1982) proposed as an alternative solution a bimodal SFR with only stars more massive than 3 $M_\odot$ forming on spiral arms. Rana and Wilkinson (1986, 1988), on the other hand, suggested that the SFR be proportional only to the molecular hydrogen density, rather than to the total gas density or mass. These proposed SFRs share the property of being larger in the inner than in the outer galactic regions, thus producing easily the radial abundance gradients. However, up to now the only models in satisfactory agreement with *all* the observational constraints of our Galaxy are those assuming a conspicuous, long lasting infall (Lacey and Fall 1985, Tosi 1988a, Matteucci and François 1989, Pagel 1989). For instance, Figure 2 shows some of the results derived for the Galaxy from recent calculations by Tosi and Diaz (1990). The top left panel represents the current radial distribution of the SFR as derived from Lacey and Fall's collection of observational data available in the literature. The top right panel represents the radial distribution of the molecular hydrogen density given by Talbot (1980), and the bottom panel the oxygen abundance gradient derived by Shaver et al. (1983) from optical and radio observations of HII regions. The solid line in all panels corresponds to a model with very slowly decreasing SFR (e-folding time: 15 Gyr) and gas falling on the disk at a constant rate and with uniform density (hereinafter: reference model). All the other curves refer to models *à la* Rana and Wilkinson, i.e. with SFR proportional to the molecular hydrogen density through a metallicity-molecular hydrogen relation (see Tosi and Diaz 1990, for details). All the models reproduce fairly well the observed SFR and oxygen distributions, but the infall model is significantly better than the others as for reproducing the molecular hydrogen data. It seems inevitable to reject models which assume a SFR proportional to molecular hydrogen but are unable to reproduce its observed distribution !

By focusing our attention on the oxygen abundance distribution, we can get some more detailed information on the infall requirements. Figure 3 shows again Shaver's et al. data and the reference model of the previous graphs (bottom solid

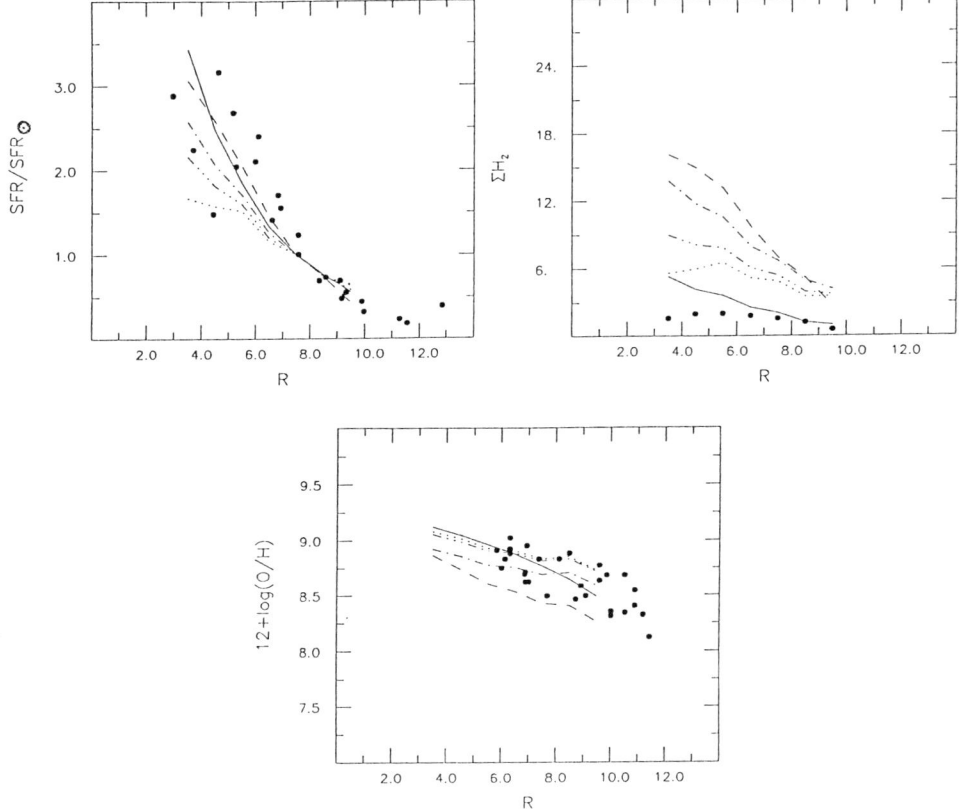

Fig.2. Comparison between observed and predicted radial distribution of the SFR (top left), the molecular hydrogen density (top right) and oxygen abundance (bottom). The solid line represents the reference model, the others refer to models à la Rana and Wilkinson.

line). The dotted line corresponds to a similar model with more rapidly decreasing SF and infall rates and gives also a good fit to the data. If we keep unchanged all the parameters adopted for the reference model but assume no infall, the resulting oxygen abundances are given by the dash-two-dotted line: clearly too flat and too metal rich with respect to the observations, due to the lack of diluting mechanisms for the metals synthesized by the stars. If we keep unchanged all the parameters, reinclude infall, but assume that its density instead of being uniform increases inwards (for instance as a function of the total mass of the region) we obtain the dash-dotted line of Fig.3. The flatness of this curve is due to the fact that the larger metal production of the inner galactic regions is now overcompensated by their larger infall rate, and is inconsistent with the observed gradient. Finally, the dashed

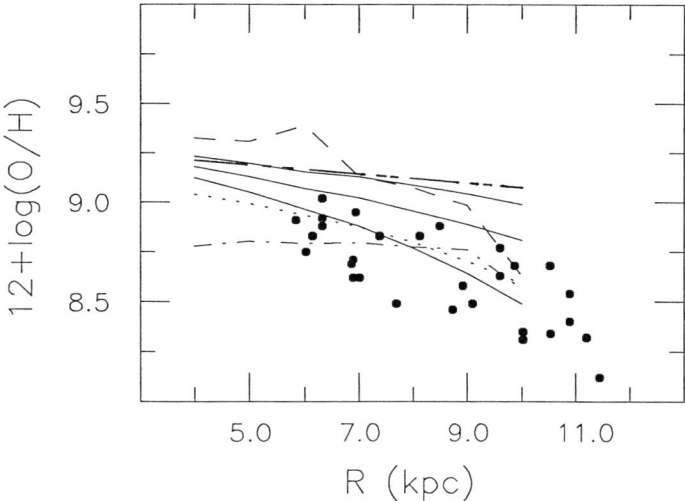

Fig.3. Oxygen abundance distribution in the galactic disk as derived from HII region observations (dots) and model predictions. The solid lines represent the reference model (see text) with increasing infall metallicity from bottom to top. The dash-dotted line corresponds to the same model as the bottom solid line, but with infall density increasing inwards; the dash-two-dotted line refers to the same model, but without infall. The dotted line corresponds to a model similar to that of the solid line but with different e-folding times of the SFR and the infall rate, and the dashed line to a model à la Gusten and Mezger.

line represents a model à la Gusten and Mezger. Since oxygen is mostly produced by massive stars, their larger fraction on spiral arms produces the bump shown in the Figure in correspondence with the Sagittarius arm. Again the morphology of the resulting oxygen distribution and its large abundances are inconsistent with the observed data.

The same set of data provides information also on the infall metallicity. Till now, the infalling gas has been assumed to be primordial (i.e. to contain no metals), but this assumption is perhaps unrealistic not only in the case of halo gas, which should have typical Population II abundances, but even in the case of extragalactic gas, which may have been polluted through galactic winds. If we consider in Figure 3 the reference model represented by the bottom solid line, leave all the parameters unchanged but consider the infall metallicity Z_f to be as large as 0.1 $Z_\odot$ and the various element abundances to be in solar proportions, the resulting line is in practice undistinguishable from the reference line. For increasing Z_f, the metal dilution by the infall gas is less effective and the resulting oxygen abundances become larger and larger and their gradient increasingly flat. We can see from Fig.3 that the middle solid line (corresponding to $Z_f=0.5$ $Z_\odot$) is already at the upper limit of the observed oxygen distribution and the top solid line ($Z_f=Z_\odot$) is definitely inconsistent with the data. Equivalent results are obtained also for the other models (Tosi 1988b).

To generalize as much as possible the model constraints on galactic infall, we have computed a large number of models with very different assumptions about the major parameters and compared their results with the largest number of observed galactic properties (Tosi 1988a) available in the literature. The χ^2 resulting from the comparison of each independent set of observational constraints with the corresponding model predictions has been computed. For each class of models (i.e. models where all the other parameters were fixed except the infall rate density F and e-folding time θ), the combination of infall parameters leading to the minimum χ^2 in the comparison with each of the observational constraints was found. In the F-logθ plane of Figure 4, the projection of the regions defined by $\chi^2 < \chi^2_{min}+4$ on each axis yields the 95% confidence limit on the infall rate density and e-folding time parameters (Avni 1976). An example of the method is given in Fig.4 for models with exponentially decreasing SFR (e-folding time: 15 Gyr) and exponentially decreasing infall rate with uniform density. For sake of simplicity, Fig.4 shows only the surfaces relative to the stellar metallicity distribution with galactocentric distance (dotted curves), the oxygen abundance gradient derived from HII regions (dashed curves) and the AMR derived by Twarog (1980) for the solar neighbourhood (solid curves). The χ^2, however, has been computed also for the gas and total mass distributions, the radial profile of the current SFR and the gradients in the abundances of other elements. Only the combination of parameters falling in the intersecting portion of all the surfaces provides results simultaneously consistent with the three sets of data. The shaded areas correspond to forbidden combinations of the infall parameters leading to unphysical conditions (e.g. negative initial mass of the disk or too rapid gas consumption). The left panel refers to models without radial flows and the acceptable values of infall parameters in this case are F $\sim$ 4 10^{-3} M$_\odot$kpc^{-2}yr^{-1} and $\theta \geq 100$ Gyr (i.e. almost constant rate). The right panel refers to models where all the infallen gas is forced to move inward. No intersection exist in this case, thus showing that this kind of extreme models do not apply to our Galaxy and that radial flows should be more moderate. Note that in any case the surfaces corresponding to a good agreement between data and predictions lie at large θ: in practice, the time decay of the infall rate must have been quite slow.

The results of applying the χ^2 method to a significantly large sample of models with very different assumptions on the various parameters and the arguments presented above can be schematically summarized with the following model constraints on galactic infall:

a) the current infall rate ranges between 0.3 and 1.8 M$_\odot$yr^{-1} for the whole disk;

b) its e-folding time should be longer than that of the SFR. Since the SFR has decreased quite slowly with time (e-folding time $\geq$ 5 Gyr), this latter result implies that the infall rate must have been almost constant too;

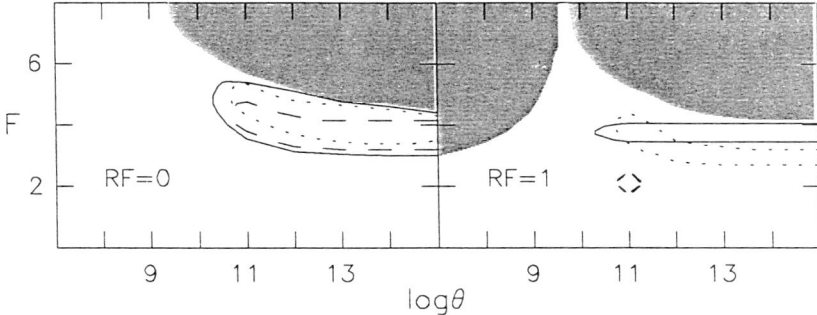

Fig.4. Contours of 95% confidence level for models with exponentially decreasing SFR, exponentially decreasing infall and uniform infall density F. RF indicates the amount of radial flows normalized to the infalling gas. The contours correspond to the AMR (solid curves), radial distribution of the stellar metallicity (dotted curves) and of the oxygen abundance of HII regions (dashed curves). F is in units of 10^{-3} $M_\odot \text{kpc}^{-2}\text{yr}^{-1}$ and θ in Gyr.

c) the infall rate density ($M_\odot \text{kpc}^{-2}\text{yr}^{-1}$) must have been uniform or increasing outwards, certainly not inwards;

d) the required diluting effect is consistent with infall metallicities up to 0.3 $Z_\odot$.

3. DISCUSSION

It is important to emphasize that items $a) \to d)$ of the previous section, do not provide any clue to understand the origin of the infalling gas. Both an intergalactic gas and a residual of a slow halo collapse can fulfill all the above requirements. Very recent calculations (Matteucci and François 1989, Pagel 1989, Ferrini et al. 1990) have modeled together the disk and halo phases of the Galaxy evolution and, despite the different approaches, derive infall properties very similar to those described above. They all suggest that the initial rapid collapse of the halo gas on the disk is followed by a slow gas capture by the disk. However, the suggested source of this captured gas varies from one author to another: some consider it as pure halo gas, some as a mixture of halo and thick disk gas, others as a mixture of halo and extragalactic gas.

The situation in other galaxies is even more uncertain. The paucity of accurate and extensive observational data reduces the reliability of the chemical evolution models. For instance, there are only eight nearby spirals for which the gas and total mass distributions are well defined and the oxygen abundances accurately derived from HII region observations. For these galaxies Tosi and Diaz (1985) have computed models of the type of the reference model described above. What we generally found is that an infall of the same kind as that appropriate for our

Galaxy appears to be needed to dilute the metallicity of more massive galaxies, whilst in smaller objects it does not seem to be necessary. Perhaps for these lower mass spirals galactic winds would improve the agreement between predictions and empirical data, as already suggested for irregular galaxies (Matteucci and Tosi 1985).

The consistency of our derived infall properties with the current observational scenario of High and Very High Velocity Clouds (HVCs and VHVCs) is not completely settled. There seems to be a general agreement on an overall infall rate of 0.1 $M_\odot yr^{-1}$ from VHVCs. However, the controversy over the interpretation of HVCs (e.g. Mirabel 1981, Mirabel and Morras 1984, Danly 1989, Wakker 1990) and the corresponding current infall rate (0.1 or 1 $M_\odot yr^{-1}$?) does not allow yet to understand if our theoretical estimates are in agreement with the observational evidence. Certainly, up to now no model of galactic chemical evolution with current infall rate as small as 0.1 $M_\odot yr^{-1}$ has been successful in reproducing all the Galaxy properties.

As for the metallicity, instead, the observational estimates (e.g. De Boer and Savage 1983, 1984) are consistent with our upper limit of $Z_f \leq 0.3\ Z_\odot$. Since both the halo and the intergalactic gas have apparently an average metallicity lower than this limit, our model results cannot discriminate between these two possible origins of infall. Viceversa, galactic fountains can be excluded as its major sources (but they can be present if some other source of metal poor gas is found). Their metal content, in fact, will mostly be that expelled by Supernovae, much higher and enhanced in some particular elements (e.g. oxygen for SNe II and iron for SNe I) than the average interstellar medium. The diluting effect of infall in this case would be completely lost.

It is of fundamental importance that the big effort currently made by several observers to reach more accurate information find soon a satisfactory achievement. The major points that require an observational answer are the following: a) what are the clouds (HVCs or only VHVCs ?) that can safely be taken as evidence of gas infall, b) what current infall rate do they imply, c) what is their metallicity and the relative proportions (solar or not ?) of the various elements, d) are the features of the gas recently discovered around other spirals consistent with those of an infall ?

REFERENCES

Avni, Y. 1976, *Astrophys.J.* **210**, 642
Danly, L. 1989, *Astrophys.J.* **342**, 785
De Boer, K.S. and Savage, B.D. 1983, *Astrophys.J.* **265**, 210
De Boer, K.S. and Savage, B.D. 1984, *Astr.Astrophys.Letters* **136**, L7
Ferrini, F., Matteucci, F., Pardi, C., Penco, U. 1990, in preparation
Gusten, R. and Mezger, P.G. 1982, *Vistas Astron.* **26**, 159
Lacey, C.G. and Fall, S.M. 1985, *Astrophys.J.* **290**, 154
Larson, R.B. 1972, *Nature* **236**, 21
Matteucci, F. and François, P. 1989, *Mon.Not.Roy.Astr.Soc.* **239**, 885
Matteucci, F. and Tosi, M. 1985, *Mon.Not.Roy.Astr.Soc.* **217**, 391
Mayor, M. and Vigroux, L. 1981, *Astr.Astrophys.* **98**, 1

Mirabel, I.F. 1981, *Rev.Mex.Astron.Astrophys.* **6**, 245
Mirabel, I.F. and Morras, R. 1984, *Astrophys.J.* **279**, 86
Oort, J.M. 1965, *Trans.IAU* **12A**, 789
Pagel, B.E.J. 1989, *Rev.Mex.Astron.Astrofis.* in press
Pitts, E. and Tayler, R.J. 1990, *Mon.Not.Roy.Astr.Soc.* **240**, 373
Rana, N.C. and Wilkinson, D.A. 1986, *Mon.Not.Roy.Astr.Soc.* **218**, 497
Rana, N.C. and Wilkinson, D.A. 1988, *Mon.Not.Roy.Astr.Soc.* **231**, 509
Shaver, P.A., McGee, R.X., Newton, L.M., Danks, A.C. and Pottasch, S.R. 1983, *Mon.Not.Roy.Astr.Soc.* **204**, 53
Talbot, R.J.Jr. 1980, *Astrophys.J.* **235**, 821
Tinsley, B.M. 1980, *Fund.Cosmic Phys.* **5**, 287
Tosi, M. 1982, *Astrophys.J.* **254**, 699
Tosi, M. 1988a, *Astron.Astrophys.* **197**, 33
Tosi, M. 1988b, *Astron.Astrophys.* **197**, 47
Tosi, M. and Diaz, A.I. 1985, *Mon.Not.Roy.Astr.Soc.* **217**, 571
Tosi, M. and Diaz, A.I. 1990, *Mon.Not.Roy.Astr.Soc.* in press
Twarog, B.A. 1980, *Astrophys.J.* **242**, 242
Wakker, B.P. 1990, this conference

INFALL OF HVC'S AND THE ORIGIN OF HI SUPERSHELLS

I.F. MIRABEL
Service D'Astrophysique, CEA-CEN. Saclay,
91191 Gif sur Yvette, FRANCE

ABSTRACT. Surveys for atomic hydrogen at very high velocities ($|V| \geq 140$ km s^{-1}) in the galactic center and anticenter regions of the sky reveal a net inflow of gas toward the Milky Way. In the anticenter, the collisions of infalling clouds with galactic material trigger the most energetic structural disturbance in the Galaxy, the "anticenter supershell".

1. INTRODUCTION

Oort (1970) envisioned for the first time that the penetration into the Galaxy of extragalactic neutral gas with Very High Velocities (VHVC's; $|V| \geq 140$ km s^{-1}) could trigger the formation of High Velocity Clouds at lower velocities (HVC's; 80 km s$^{-1} \leq |V| \leq 140$ km s^{-1}). Some years later, Tenorio-Tagle (1980) proposed that the collective action of infalling HVC's subsequently causes the large-scale disturbances at lower velocities observed by Heiles (1979) as HI supershells. To avoid the complications implied by models of differential galactic rotation, I consider only VHVC's in the center and anticenter regions of the sky, namely, those clouds with truly anomalous motions that have not interacted with galactic material and therefore, do not participate in Galactic rotation. A review on infall of neutral gas toward the Milky Way was published in the Proceedings of the IAU Colloquium No 120 (Mirabel, 1989); the observational evidences of this phenomenom were presented in a series of publications (Mirabel and Morras 1984, 1990, and references therein).

2. EVIDENCE FOR INFALL OF VERY HIGH VELOCITY CLOUDS

The main difficulties regarding the question of the infall of neutral gas toward the Galaxy reside in two unknowns: the distance to the HI gas clouds, and the effects of differential galactic rotation on the observed radial velocities. The latter difficulty can be overcome by selecting data only from regions of the sky where the line-of-sight component of the solar motion about the galactic center is small.

Surveys of high velocity HI in the galactic center and anticenter provide compelling evidence for the galactic accretion of HI at Very High Velocities. More than 99% of the VHV gas in the anticenter region has negative velocities (Hulbosch and Wakker, 1988 and references therein). In addition, Mirabel and Morras (1984) found that in the inner Galaxy, 84% of the VHV gas has negative velocities. The striking preponderance of clouds with the same extreme inward motions in the center and anticenter is the strongest evidence for a high velocity inflow of gas toward the Milky Way.

In the most simple infall scenario the infalling clouds are likely to move preferentially toward the central regions of the Galaxy. In this context, most of the VHV gas seen in the direction of the inner Galaxy must be at distances greater than 10 kpc from the Sun, beyond the galactic center. Hence, for a given physical size, the clouds detected in the inner Galaxy should appear smaller than their counterparts in the anticenter. Maps by Mirabel and Morras (1984) show that in fact the clouds in the inner Galaxy are about 5 times smaller than clouds in the outer Galaxy.

3. CLOUD-MILKY WAY COLLISIONS AND THE ORIGIN OF HI SUPERSHELLS

A natural consequence of the high velocity infall of neutral gas must be the collision with Milky Way gas at high z. Oort (1970) had proposed that most high velocity streams of HI with moderate velocities ($|V| \leq 140$ km s^{-1}) are galactic clouds which have been set in motion by gas of extragalactic origin. However, no direct observational evidence for the physical interaction between extragalactic VHVC's and galactic material had been found. Physical associations between gas at VHV and Milky Way gas may be scarce because once the supershell is formed, the impinging cloud should have been shocked and incorporated into the accelerated structures.

The observations by Mirabel and Morras (1990) support the idea by Tenorio-Tagle (1981) that HI supershells result from the interaction between infalling clouds at very high velocities and galactic material. Their observations show the physical association between a stream of impinging high-velocity clouds at -200 km s^{-1} and the most remarkable HI supershell in the anticenter of the Galaxy (Heiles, 1979).

Taking the Galaxy as a whole, the energy injected by VHVC's will be about 2×10^{47} ergs yr^{-1}, only 1% of the energy injected by supernovae of 10^{51} ergs exploding at a rate of 1/50 yr^{-1}. However, outside the solar circle, the energy due to the infall of gas may be relatively important, since the supernova rate decreases with galactocentric distance. Comparatively, few molecular gas clouds and massive stars populate the outer regions of the Galaxy, and in those regions, supernovae explosions may not dominate completely the energy balance of the interstellar medium.

REFERENCES

Heiles, C. (1979) *Astrophys. J.* **55**, 585
Hulsbosch, A.N.M., Wakker, B.P. (1988) *Astron. Ap. Supp* **75**, 191
Kulkarni, S.R., Mathieu, R. (1986) *Ap. Space Sci.* **118**, 531
Mirabel, I.F. (1989) in *Structure and Dynamics of the Interstellar Medium*, eds. G. Tenorio-Tagle, M. Moles, J. Melnick, p. 396
Mirabel, I.F., Morras, R. (1984) *Astrophys. J.*, **279**, 86
Mirabel, I.F., Morras, R. (1990) *Astrophys. J.* **356**, 130
Oort, J.H. (1970) *Astr. Ap.* **7**, 381
Tenorio- Tagle, G. (1980) *Astron. Ap.* **88**, 61

WAVE STRUCTURE WITHIN HI FILAMENTS AT HIGH GALACTIC LATITUDE AND THE NATURE OF "CLOUDS" IN INTERSTELLAR SPACE

GERRIT L. VERSCHUUR
4802 Brookstone Terrace, Bowie, MD 20720

ABSTRACT. Large amplitude waves have been found in the morphology and velocity patterns of several long filaments of HI at high latitude. HI in the filaments is controlled by magnetic fields and the velocity patterns and morphology bear the hallmarks of Alfvén waves. Enhanced emission features (EEFs), traditionally referred to as"clouds," are seen wherever a segment of flux tube is viewed end-on. This suggests that HI emission structure teaches us about field geometry and not about cloud physics. Similar effects have been recognized in other regions mapped with high-resolution as well as in completely mapped high-velocity "clouds."

1. THE DATA

Figure 1a shows a sketch of the filaments recognized in the data given by Verschuur (1974) for the region under consideration. Figure 1b shows the velocity of gas associated with the major Filament A as a function of position along its axis. A first-order slope, part of a larger scale effect, was removed and the velocity residuals show a clear wave-like pattern, as is seen in Figure 1c. Comparison of these data with the morphology of the filament suggests a three dimensional wave structure, with amplitude 5 km/s and $1/4$ λ. This is probably an Alfvén wave, consistent with a field of 5 μG and a density of 5 cm^{-3}.

Figure 2 is a detailed view of a sample of three channel maps compared with a schematic that shows the location of the filaments in a region of overlap. Enhanced emission features are indicated with labels giving RA, dec., and velocity. The curve labelled "simulation" is a projection of a helix viewed from a suitable angle to account for the filament that curves into and out of EEF H0936+00.0+11. Enhancement in 21-cm emission appears to be the result of viewing a segment of the helical flux tube end-on. The inset shows the area in which Guhathakurta and Tyson (1989) found optical emission associated with an enhancement of 100μm IR radiation.

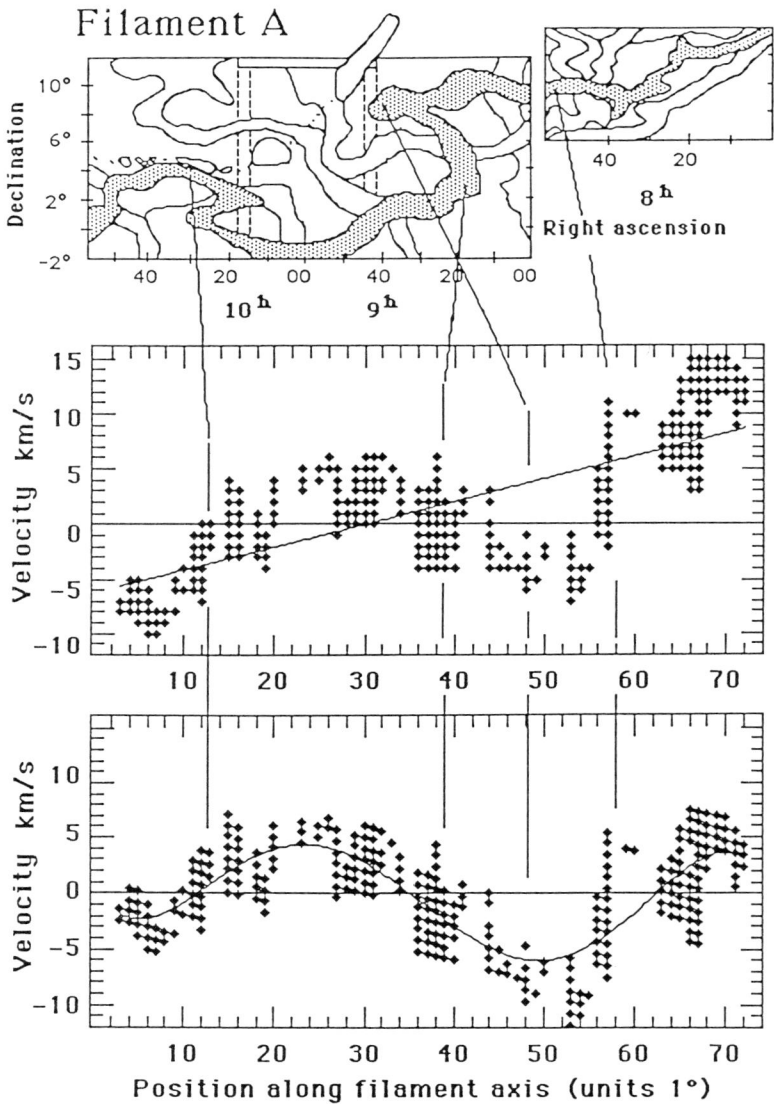

Figure 1. (a) Schematic of the filaments identified. The velocity of emission associated with Filament A (shaded) is shown in (b) After removal of a first-order slope, the velocity residuals (c) show a wave pattern that is closely related to the morphology of the filament.

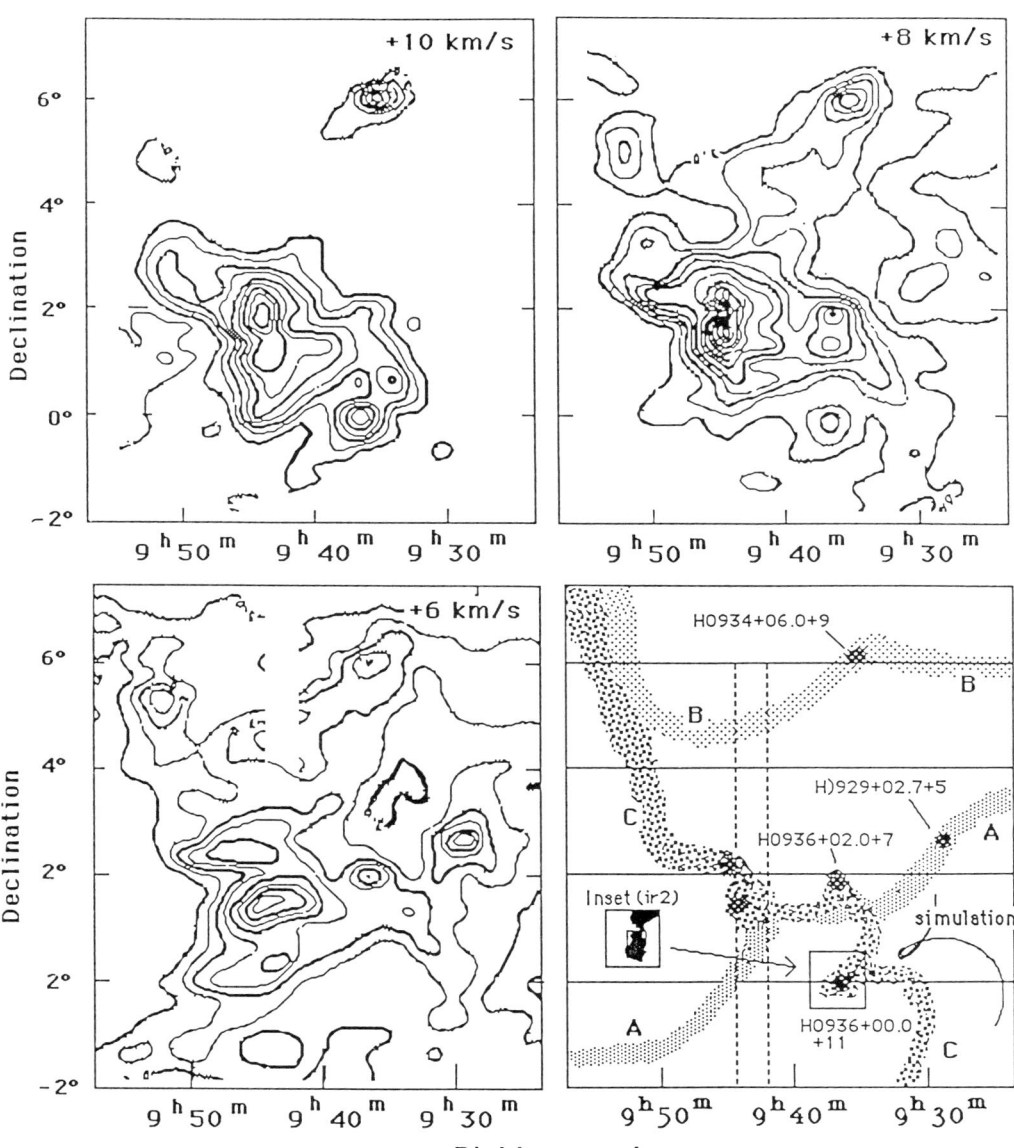

Figure 2. Three channel maps of a portion of area mapped (Figure 1a). Use of 29 channel maps allowed the filament pattern summarized in bottom-right diagram to be determined. The brightest EEFs occur where the filament axes show the most pronounced change in direction. The simulation was produced by rotating a helix and viewing it from an angle that gave good agreement with the pattern in the HI emission around H0936+00.0+11.

2. CONCLUSIONS

The following statements summarize extensive analysis to be published elsewhere. Interstellar HI is found in filaments that twist in three dimensions. Gas motion is controlled by magnetic fields whose magnetic pressure (for B=5 to 10 μG) controls thermal pressure of the HI (with density 5 cm^{-3} at 200 K or less). Wave structure of wavelength 30° and amplitude 8° in the morphology of these filament is associated with a velocity wave, amplitude 5 km/s. This is probably an Alfvén wave that is three dimensional. This pattern shows fine structure on several scale-lengths. Every enhanced emission feature ("cloud") in the region is part of one of seven filaments identified and is produced where the line-of-sight is along a local segment of twisted flux tube. EEFs thus teach us about field geometry and not about the physics of interstellar HI "clouds." Observations normal to a filament axis give column densities of 3 x 10^{19} cm^{-2}. The enhancement of emission in an EEF with respect to this value is an indicator of the depth of the segment of flux tube with respect to its width. These ratios are found to be 5 to 10. Assumptions usually made regarding the depth of HI "clouds" being the same as their width may be seriously in error. Also, masses for nearly 400 HI "clouds" found in the literature are probably incorrect. In the region studied, every enhancement in 100μm emission is associated either with a specific HI EEF and/or with the overlap of HI filaments in the line-of-sight. This suggests that cirrus structures at high latitudes may also be heavily modulated by geometric effects. Preliminary examination of other high-resolution HI data suggests that these phenomena are common. Structure at intermediate and high latitudes may be part of a single phenomenon involving a mass of twisted filaments, possibly remnants of an old supernova inside which the sun is located. Without exception, high-resolution HI data for high-velocity clouds show that they, too, are filamentary and within them EEFs occur where the line-of-sight is along a filament axis. Within such high-velocity filaments the structure is complex and preliminary interpretation of the data sugggests interesting constraints on the relative values of field strength and density in the HVCs. This, in turn, has a bearing on the peculiar phenomenon of the morphological similarity between CO clouds and HVCs reported by Verschuur (1990). Finally, it is noted that when a segment of a twisted HI filament is seen along the axis of the twist it will appear as a ring on the sky. Based on the data in Figure 1, such rings may have diameters as large as 10°. This phenomenon may be responsible for many of the apparent shell-like structures found in low resolution 21-cm surveys.

REFERENCES

Guhathakurta, P. and Tyson, J. A. (1989) *Ap. J.* **346**, 773
Verschuur, G. L. (1974) *Ap. J. Suppl.* **27**, 65
Verschuur, G. L. (1990) *Ap. J.* September. In press.

THE STELLAR DISK–HALO CONNECTION

ROSEMARY F. G. WYSE
The Johns Hopkins University, Baltimore, MD 21218, USA

GERARD GILMORE
Institute of Astronomy, Cambridge CB3 0HA, England

ABSTRACT. The thick disk is the stellar disk-halo connection. At least near the solar circle, this component is on average as old as the system of disk globular clusters, or ~ 12 Gyr. This implies that it most probably formed early in the process of Galaxy formation, so that its properties – chemical abundances, stellar kinematics and spatial distribution – contain clues to the physics of these stages of Galaxy evolution. Its present-day importance for the interstellar disk-halo connection lies in the evolution of its constituent stars – gas loss through winds on the red giant and asymptotic giant branches, through planetary nebulae prior to white dwarf formation, and through supernovae. This gas loss results in mass injection, momentum injection and energy injection into the interstellar medium from a stellar population with a scale height of ~ 1 kpc.

1. THE STELLAR THICK DISK

The canonical disk galaxy of the 1970s consisted of two stellar components, *viz.* a metal-poor, low-angular momentum, high velocity dispersion, extended spherical halo on the one hand, and a metal-rich, low dispersion, high-angular momentum thin disk on the other hand. The disk-halo connection was tenuous. Recent observations have revealed the existence of the thick stellar disk, which provides the stellar disk-halo interface, and whose importance lies in the clues it contains to the physics of Galaxy formation. The properties of the thick disk, as presently understood, are discussed below, in the context of the disk-halo transition.

2. KINEMATICS AND METALLICITY

The best available description of the properties of the stellar disk-halo connection is obtained by observations of stars which currently occupy the region a few kiloparsecs above the plane of the thin disk. This description can be derived from studies of nearby but high-velocity stars, or by study of distant stars currently *in situ* at the disk-halo interface.

2.1 Kinematically Selected Samples

Samples of nearby stars with high space motions provide an observationally convenient probe of the structure of the Galaxy far from the Galactic plane. The large proper-motion selected stellar samples of Sandage & Fouts (1987) and of Carney, Latham and collaborators have proved particularly valuable for studying the kinematics and chemical abundances expected a few kiloparsecs from the Sun. Figure 1, which shows the data of Laird, Carney & Latham (1988), illustrates the difference between the thick disk and the halo in terms of kinematics and metallicity. The mean value of the rotation velocity decreases from $\sim 160\,\mathrm{km\,s^{-1}}$ at [Fe/H] $= -0.8$ (these values are evidently still poorly determined) to near zero at [Fe/H]$= -1.5$, and shows no significant correlation at lower abundances. The broad distribution of properties of the different components of the galaxy is evident, illustrating the difficulties in distinguishing between continuous and discrete multi-component descriptions of the Galaxy.

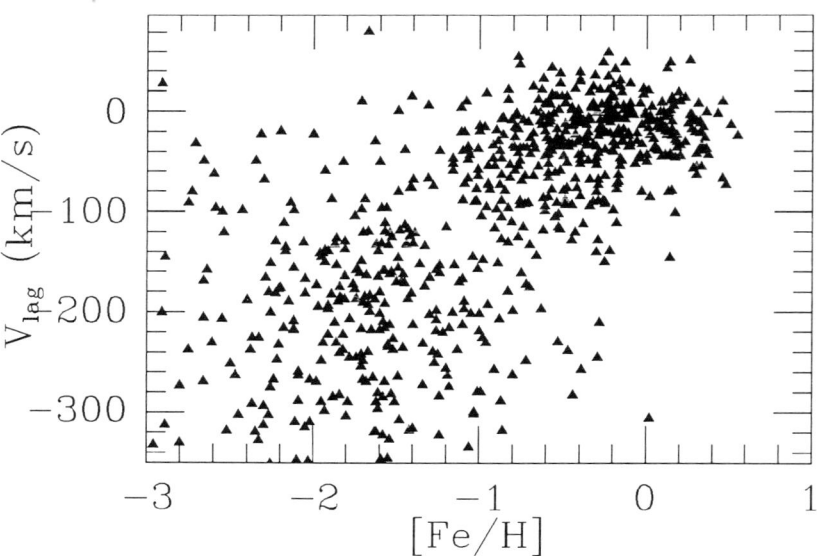

Figure 1. The relation between rotation velocity relative to the Sun, V_{lag}, and metallicity, for the sample of proper motion stars studied by Laird, Carney & Latham (1988).

A question of some interest is whether or not the appearance of Figure 1 is consistent with a continuous trend – indicative perhaps of significant star formation *during* the period when the protogalaxy was collapsing and spinning-up, or represents a superposition of relatively discrete sub-systems – indicative perhaps of the later merger of subsystems which retained a recognisable identity during the early stages of Galactic formation. Some suggestive rather than conclusive evidence in favour of the picture of discrete substructure in phase space comes from the existence of several apparently intermediate groups of tracers which are identifiably discrete using astrophysical criteria. These include the metal-rich RR Lyrae stars ($\Delta S \lesssim 3$, $V_{rot} \sim 110 \, \text{km s}^{-1}$, Strugnell *et al.* 1986); c-type RR Lyrae stars ($V_{rot} \sim 100 \, \text{km s}^{-1}$, Strugnell *et al.* 1986); long period variables with 150d $\lesssim$ Period $\lesssim$ 200d, ($V_{rot} \sim 115 \, \text{km s}^{-1}$, Osvalds & Risley 1961); the metal rich (G-type) globular clusters ([Fe/H]$\gtrsim -1$, $V_{rot} \sim 100 - 200 \, \text{km s}^{-1}$, Armandroff 1989); and the Arcturus moving group ($V_{rot} \sim 110 \, \text{km s}^{-1}$, Eggen 1987). The field type-II Cepheids (Harris 1981) are another closely related tracer sample, but with less well-known kinematical properties at present.

2.2 Photometrically Selected Samples

Although solar-neighbourhood, proper-motion selected samples provide valuable clues to the earliest phases of Galaxy formation, the kinematic biases inherent in these samples require careful modelling. It is therefore desirable to have available an *in situ* sample, truly representative of the dominant stellar population. A variety of such studies exist or are in progress, and are summarised by Gilmore, Wyse & Kuijken (1989). The available data provide the following description of the thick stellar disk:

The thick disk has a scale-height of ~ 1 kpc, at least at the solar radius, and of order 2% of local stars belong to this component (the other $\sim 98\%$ being members of the thin disk, with the $r^{1/4}$ spheroid accounting for merely a fraction of a percent). The first evidence for the global applicability of these thick disk parameters (Gilmore & Reid 1983) is found in the analysis by Fenkart (1989 and refs therein) of the extensive star-count data set available from the Basel Halo Program, in many fields at both high and low Galactic latitudes.

The total luminosity of the thick disk can at present best be constrained either by analogy with external galaxies (van der Kruit 1987; Wyse & Gilmore 1988) or by theoretical considerations, such as the chemical evolution of the Galaxy (Gilmore & Wyse 1986). These imply that the thick disk dominates over the $r^{1/4}$ spheroid out to many kpc away from the plane, and probably has a total luminosity several times that of the metal-poor $r^{1/4}$ spheroid, or a few $\times 10^9 L_\odot$. The detailed values of the descriptive parameters remain poorly determined however, primarily because the offset in the mean values characterising the thick disk distribution function over age, metallicity, and kinematics from those characterising the oldest thin disk stars is much less than the dispersions in these quantities.

The vertical velocity dispersion corresponding to a scale-height of ~ 1 kpc, ~ 45 km s^{-1}, has now been observed in many samples (Hartkopf & Yoss 1982; Sandage &

Fouts 1987; Carney, Latham & Laird 1989; Ratnatunga & Freeman 1985). Figure 2 shows the run of vertical velocity dispersion with height above the plane, from the photometrically-selected samples of Gilmore & Kuijken (Kuijken & Gilmore 1989) and of Gilmore & Wyse (in preparation). Typical thick disk stars appear to be on high-angular-momentum orbits (Sandage & Fouts 1987; Carney, Latham & Laird 1989; Ratnatunga & Freeman 1985), lagging behind the Sun by only ~ 40 km s^{-1}. The thick disk is apparently kinematically discrete from the subdwarf system to an adequate approximation. This means simply that the rate of dissipation in the vertical direction was relatively high, compared to the star formation rate, as the proto-disk collapsed. The thick disk is probably also kinematically discrete from the Galactic old thin disk, though the data remain inadequate for robust conclusions (Norris 1987; Gilmore, Wyse & Kuijken 1989). Determination of the kinematic relationship between the thin and the thick disks is of interest primarily as a test of the dynamical history of high angular momentum material early in the evolution of the Galaxy, and thus is a topic of considerable current activity.

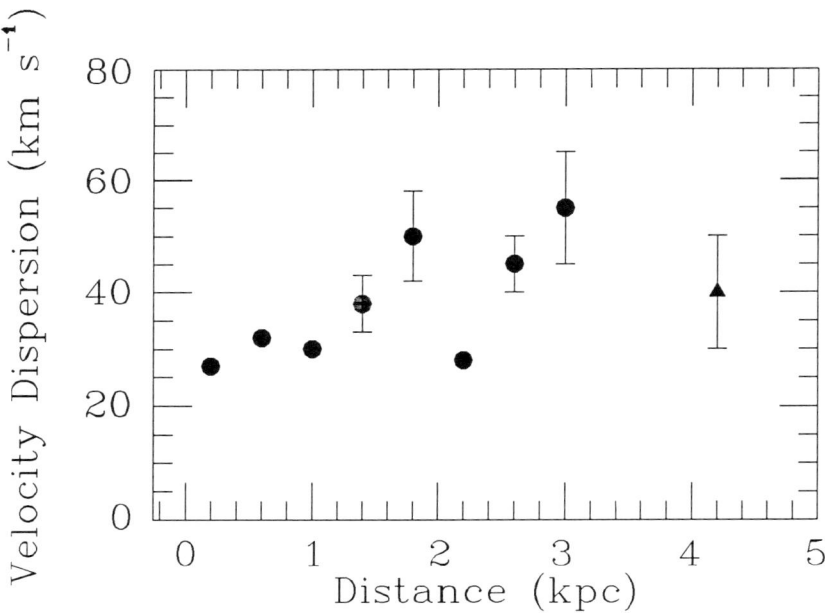

Figure 2. Vertical velocity dispersion for samples of stars *in situ* at height z above the plane of the Galaxy. Circles: some data from the ongoing Wyse/Gilmore survey; triangle Hartkopf & Yoss (1982).

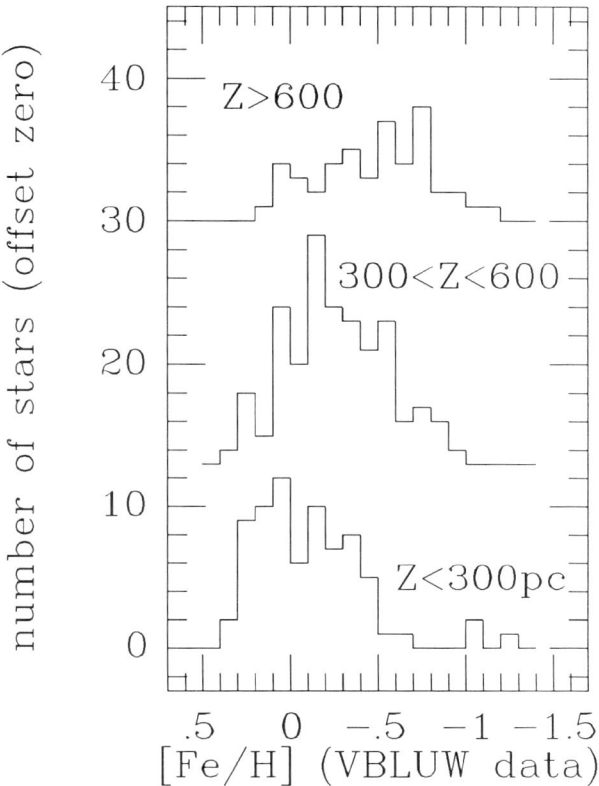

Figure 3. Metallicity distributions for F/G stars observed at different heights above the plane.

The mean metallicity of thick disk stars, which dominate 1 − 2 kpc above the Galactic plane, is about one-quarter of the solar metallicity (Gilmore & Wyse 1985) *i.e.* [Fe/H] ∼ −0.7 dex. Their abundance distribution is important in understanding the chemical evolution of the solar neighbourhood, which requires knowledge of the abundance distribution for long-lived stars in a *representative* volume of the Galaxy. The large scale-height and low local normalisation of the thick disk meant that these stars were not found in the earlier small surveys of stars in the solar neighbourhood, even though they contribute significantly to the abundance distribution in a column through the Galactic disk, and may provide an elegant solution to the G-dwarf problem (Gilmore & Wyse 1986).♠ Figure 3 shows the

───────────────

♠ The 'G-dwarf problem' is the observed paucity of metal-poor G-dwarfs in the solar-neighbourhood, relative to the predictions of the most naive model of chemical evolution (van den Bergh 1958; Pagel & Patchett 1975) Since late G-dwarfs live for a data of Trefzger *et al.* (1990) for their sample of G-type stars selected from the Basel photographic RGU survey in three fields *viz.* the SGP, SA 94 and SA 107. Thus their sample is selected on a purely photometric basis, with metallicities and gravities estimated *via* Walraven VBLUW photometry. Their data are displayed in three bins of distance above the Galactic plane (the sample is complete only to $z = 600$pc). As is clear from the Figure, the data are better characterized not by a steady decrease in mean metallicity with increasing height above the plane, but rather by increasing dominance of a peak with mean metallicity [Fe/H]~ -0.7 dex, the thick disk. Similar results are apparent from the studies by Gilmore & Wyse (1985); Friel (1987); Sandage & Fouts (1987); Laird, Carney & Latham (1988).

3. FORMATION OF THE THICK DISK

Possible formation mechanisms for the thick disk include:

♡ A slow, pressure-supported collapse phase following formation of the extreme Population II system, similar to the sequence of events in Larson's (1976) hydrodynamical models of disk galaxy formation. The thick disk is then simply part of the thin disk, since the physics of formation is the same, but the oldest part. It should be remembered that Larson's models represent state-of-the-art for the 1970s rather than the 1990s. More numerically sophisticated analyses are required – and are being developed – to treat such parameters as viscosity, which Larson found to be a crucial parameter determining the final morphology of the galaxy, before one could confidently compare such models with observation.

♡ Violent dynamical heating of the thin disk by *(a)* satellite accretion (*c.f.* Hernquist & Quinn 1990), or by *(b)* violent relaxation of the non-spherical Galactic potential (Jones & Wyse 1983). The thick disk is then some part of the once-thin disk. For satellite accretion models the range of ages of the stars in the thick disk is determined by the epoch(s?) of satellite accretion, while the structural parameters of the thick disk (both radial profile and vertical profile) are determined by the details of the initial orbits of the satellites. Simulations of satellite accretion that contain all the known important physical effects have yet to be completed (Quinn 1990). In case *(b)* the present thick disk obviously consists of the oldest once-thin-disk stars, and the structural paremeters may be expected to be more uniform.

♡ An extended period of enhanced kinematic diffusion of stars formed in the thin disk to higher-energy orbits. The details of the heating process will determine the

---

Hubble time all late G-dwarfs ever born should still be around today. They thus can form a convenient tracer of the time-integrated chemical abundance distribution, provided only that one determines abundances for an unbiased subset of them within a suitably large volume. This volume must include the G-dwarfs in the thick disk.

properties of stars in the thick disk. As yet there is no identified heating mechanism that can produce the required vertical structure.

♡ A rapid increase in the dissipation and star-formation rates due to enhanced cooling once the metallicity is above ~ -1 dex (*c.f.* Wyse & Gilmore 1988).

The importance of the thick disk in terms of the disk-halo transition clearly differs among these models; discrimination amongst these several types of model is possible from appropriate age, metallicity, and kinematic data.

4. THE AGE OF THE THICK DISK

In practise only stars near the main-sequence turnoff have surface gravities which change sufficiently rapidly and monotonically that reliable comparison with evolutionary tracks is possible, although some useful information on a combination of age and chemical abundance can be derived from the colour of field giant stars (*e.g.* Sandage 1987). For *single* stars near the turnoff the comparison of $uvby\beta$ photometry with theoretical isochrones is by far the most reliable and precise age-dating technique available. If independent abundance estimates are available, then any photometric measure of the temperature of the hottest turn-off stars will measure the age of the *youngest* star in a tracer population. It is this method which is utilised to determine ages for globular clusters, where it also seems that all the member stars are coeval. A similar technique can be applied to field stars (*c.f. e.g.* Gilmore & Wyse 1987), and is illustrated in Figure 4. This figure shows the colour-metallicity data for the high proper-motion stars studied by Laird, Carney, & Latham (1988), as well as the turn-off points of all those globular clusters with recent CCD photometry, and a representative isochrone for old metal-rich stars (VandenBerg & Bell 1985).

The important conclusion from Figure 4 is that essentially all of the stars with [Fe/H] $\lesssim -0.8$ are, insofar as is measurable, the same age as the globular cluster system. Stars more metal rich than ~ -0.8 dex have a bluer turnoff, implying that *at least some* of these stars are younger. The *distribution* of ages is however unmeasurable from a turnoff colour. Stromgren photometry has been obtained by Nissen *et al.* (1990) for a photometrically-selected sample of evolved (turnoff) F/G stars, from which metallicities and ages can be derived. Their results support the conclusions from Figure 4, and further show that there is little spread in ages of stars more metal-poor than ≈ -0.5 dex, the scatter being 2–3 Gyr. Thus, stars of metallicity typical of the thick disk are as old as the (younger) disk globular clusters *i.e.* ~ 12 Gyr.

The distinct age and metallicity of the thick disk are severe constraints on the formation mechanism of the thick disk. In particular, they argue against kinematic diffusion of thin disk stars, and against recent satellite accretion causing heating of the thin disk, without considerable *post hoc* fine-tuning.

The absence of a younger turnoff shows that no substantial continuing star formation has taken place in the spheroid – though note the existence of some apparently young high latitude metal–rich A stars whose place of formation remains

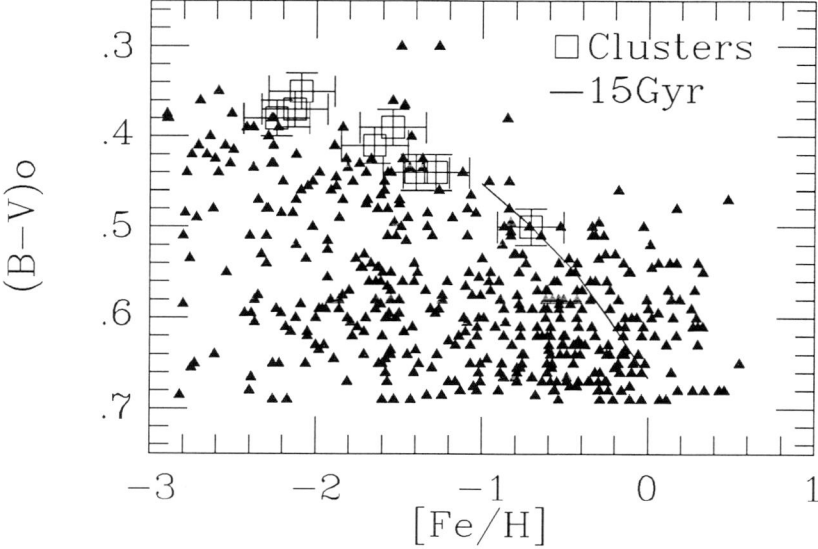

Figure 4. The B−V vs [Fe/H] relation for all stars observed by Laird, Carney, & Latham (1988; points), and for those globular clusters with recent turnoff colours from CCD photometry (Stetson & Harris 1988; boxes). The photometric data are corrected for interstellar reddening. The solid line is a 15Gyr isochrone calculated with oxygen-enhanced element ratios, and scaled in B−V to match the turnoff colour of 47Tuc. The blue edge of the stars with [Fe/H] $\lesssim -0.7$ is adequately defined by the isochrone and by the globular cluster data, showing that effectively all stars more metal poor than ~ -0.7 dex are as old as the globular clusters. At higher abundances the trend for the data to move to the blue of the isochrone shows that at least some stars are younger than the globular clusters.

an extremely important mystery (Lance 1988) and some distant B stars which cannot have travelled from the thin disk in their main sequence lifetimes (Keenan et al. 1986: Conlon et al. 1988).

5. CHEMICAL ELEMENT RATIOS

In attempting to deduce the rate of star formation and dynamical evolution in a proto-galaxy, it is desirable to have available a clock which is able to resolve dynamical evolutionary timescales. Such a clock is provided by stellar evolution of high-mass stars, while the fossil record of the clock is observable in the chemical abundance enrichment patterns in long-lived low-mass stars. Fortunately, there exists a subset of common elements (most importantly oxygen) whose creation sites are restricted to very massive stars (Type II supernovae), and another subset (most importantly iron) which is also created during the evolution of lower mass stars (Type I supernovae). Since the evolutionary timescales for high- and low-mass stars span the timescale range of interest in galaxy formation, the differential enrichment of elements such as oxygen and iron provides an ideal clock to calibrate the rate of star formation in the proto-Galaxy.

Element ratios have been now been measured for a sufficient number of stars to define the systematic trends in the data (Wheeler, Sneden & Truran 1989; Nissen 1990). A significant change of slope occurs in the relationship between the element ratios [O/Fe] and [Fe/H] close to the iron abundance where there also occurs a change in the stellar kinematics, that is at [Fe/H] ~ -1. The 'alpha' elements, so-called since their synthesis involves the addition of a helium nucleus to an extant nucleus, are believed to have similar stellar nucleosynthesis sites to those of oxygen, and so may be expected to show the same elemental ratio patterns. However, as emphasised by Lambert (1989), and most recently confirmed by Nissen (1990), magnesium shows a further break, in that the trend of [Mg/Fe] with [Fe/H] changes slope at [Fe/H] ~ -0.5. It should be remembered that the two values of [Fe/H] where the element ratios show breaks essentially bracket the thick disk – which is kinematically distinct certainly from the $r^{1/4}$ spheroid, and most probably from the thin disk. Indeed, Nissen's results mean that the thick disk may be defined equivalently in terms of [Mg/Fe], or in terms of of [Fe/H], or of kinematics.

The coincidence of the value of [Fe/H] at which the Galaxy changed from a pressure-supported system to an angular momentum-supported system, with the value of [Fe/H] at which the interstellar medium became diluted by the products of long-lived stars, provides a diagnostic of the relative star-formation and dissipation rates in the proto-Galaxy. A possible explanation is that at metallicities [Fe/H] $\gtrsim -1.5$ the efficiency with which a gas cloud cools from $\sim 10^6$ K (a typical galactic virial temperature) increases markedly, due to a transition of the dominant cooling mechanism from free-free radiation, independent of metallicity, to line radiation, proportional to the number density of metals. Thus a rapid increase in the dissipation rate and collapse to a disk-like angular-momentum supported structure is not implausible at a metallicity of ~ -1 dex. It is not crucial for these arguments that the breaks in kinematics and element ratios occur at *exactly* the same metallicity.

As mentioned above, the breaks are generally explained by appeal to the onset of an additional source of one element, while keeping the sources of the other element constant. For example, the break in [O/Fe] at [Fe/H] ~ -1 can be understood by postulating that the stars more metal-rich than this formed subsequently to the explosion of a significant number of low mass and long-lived supernova progenitors (Type I). This timescale is rather difficult to estimate precisely, due to uncertainties in the mechanism of Type I supernovae and the fraction of all stars formed which are in binaries of the type that may be expected to be precursors (*c.f.* Iben 1986); the lowest mass, and hence most numerous, progenitors of CO white dwarfs have main-sequence masses and lifetimes of $\sim 5\mathcal{M}_\odot$ and $\sim 2.5 \times 10^8$ yr respectively. Thus a reasonable estimate for the characteristic time after which one expects dominance of iron from Type I supernovae is $\lesssim 10^9$ yr (but bearing in mind that some Type I systems will take a Hubble time to evolve). This general argument appears to be the strongest direct evidence for a rapid formation timescale for the Extreme Population II stars in the Galaxy.

The second break, seen in [Mg/Fe] but not in [O/Fe], means that we do not understand the nucleosynthetic sites of these elements, since this is not predicted by present models (Wheeler, Sneden & Truran 1989).

The inhomogeneity and temporal variability of the star-formation process in the Galaxy is constrained by the important fact that there is apparently *no intrinsic scatter* in the value of the ratio [Mg/Fe] at a given [Fe/H]; all the scatter found is consistent with observational uncertainties. This lack of scatter contrasts with the recently-established but large (factor of several) intrinsic scatter in the [Fe/H]-age relationship for thin disk stars ([Fe/H] $\gtrsim -0.6$; Nissen *et al.* in preparation). Consistency between these observations suggests that the different elements are synthesised in the same enrichment 'event', but that enriched material can be transported a significant distance.

6. THE PRESENT EVOLUTION OF THE THICK DISK

Assuming that the age of the thick disk stars reflects the formation epoch of the thick disk leads to the conclusion that the thick disk is a fossil of the interstellar disk-halo connection during the early evolution of the Galaxy. The importance of the thick disk at the present epoch for the interstellar medium lies in the later stages of stellar evolution of its constituent stars. The mass loss from these stars provides a source of unenriched material from quiescent stellar winds and a source of enriched material from more violent events such as planetary nebula formation and Type I supernovae. The momentum and energy injection into the interstellar medium can also be significant, contributing to the existence of a tenuous, warm gaseous component above the plane, remembering that the sources are distributed vertically with a scale-height of $\gtrsim 1$ kpc (*c.f.* Heiles 1987, who first discussed the importance of Type I supernovae to the energy balance of the interstellar medium above the plane, but without taking account of the thick disk).

7. CONCLUSIONS

Metallicity may be used with reasonable confidence to distinguish the stellar populations in the Galaxy : [Fe/H]$= -1$ is a definable metal-rich end of the 'halo', while [Fe/H] $= -0.5$ is the metal-poor end of the thin disk. The stars with metallicity between these two limits belong to the stellar disk-halo connection, the thick disk. This population was first identified by star counts, and thus was originally characterized in terms of its vertical structure, but can be equivalently defined by kinematics – σ_W or V_{rot} – or by chemical abundances – [Fe/H] or [Mg/Fe] – and possibly also by age.

The thick disk stars are old, and hence provide a fossil record of the interstellar medium in the early galaxy, mapping its enrichment and mixing history. They are now relevant as a source of new interstellar gas, and as a source of energy and of momentum, at high vertical distance.

ACKNOWLEDGMENTS

RFGW thanks all at the Astronomy Department, University of California, Berkeley, for hospitality during a visit while this paper was prepared, and acknowledges support from the NSF (grant AST 88-07799).

REFERENCES

Armandroff, T. (1989) *Astron. J.*, **97**, 375.
Carney, B., Latham, D.W. & Laird, J.B. (1989) *Astron. J.*, **97**, 423.
Conlon, E.S., Brown, P.J.F., Dufton, P.L., & Keenan, F.P. (1988) *Astron. Astrophys.*, **200**, 168.
Eggen, O.J. (1987) in *The Galaxy*, eds G. Gilmore & B. Carswell, Dordrecht, Reidel, p211.
Fenkart, R.P. (1989) *Astron. Astrophys. Suppl.*, **81**, 187.
Friel, E.D. (1987) *Astron. J.*, **93**, 1388.
Gilmore, G. & Reid, I.N. (1983) *Mon. Not. Roy. astr. Soc.*, **202**, 1025.
Gilmore, G. & Wyse, R.F.G. (1985) *Astron. J.*, **90**, 2015.
Gilmore, G. & Wyse, R.F.G. (1986) *Nature*, **322**, 806.
Gilmore, G. & Wyse, R.F.G. (1987) in *The Galaxy*, eds G. Gilmore & B. Carswell, Reidel, Dordrecht, p247.
Gilmore, G., Wyse, R.F.G., & Kuijken, K. (1989) *Ann. Rev. Astron. Astrophys.*, **27**, 555.
Harris, W. (1981) *Astron. J.*, **86**, 719.
Hartkopf, W.I. & Yoss, K. M. (1982) *Astron. J.*, **87**, 1679.
Heiles, C. (1987) *Astrophys. J.*, **315**, 555.
Hernquist, L. & Quinn, P.J. (1990) *Astrophys. J.*, in press.
Iben, I. (1986) in *Cosmogonical Processes*, eds W.D. Arnett, C.J. Hansen, J.W. Truran & S. Tsuruta, Utrecht, VNU Science Press, p155.
Jones, B.J.T. & Wyse, R.F.G. (1983) *Astron. Astrophys.*, **120**, 165.
Keenan, F.P., Lennon, D.J., Brown, P.J.F. & Dufton, P.L. (1986) *Astrophys. J.*, **307**, 694.
Kuijken, K. & Gilmore, G. (1989) *Mon. Not. Roy. astr. Soc.*, **239**, 605.
Laird, J.B., Carney. B.W. & Latham, D.W. (1988) *Astron. J.*, **95**, 1843.
Lambert, D.L. (1989) in *Cosmic Abundances of Matter*, ed C.J. Waddington, Amer. Inst. Phys., New York, p168.
Lance, C.M. (1988) *Astrophys. J.*, **334**, 927.
Larson, R.B. (1976) *Mon. Not. Roy. astr. Soc.*, **176**, 31.
Nissen, P.E. (1990) in proceedings of *New Windows to the Universe*, ed C. Molaro, Cambridge University Press, Cambridge, in press.
Nissen, P.E. *et al.* (1990) in preparation.
Norris, J. (1987) in *The Galaxy*, eds G. Gilmore, B. Carswell, Reidel, Dordrecht, p297.
Osvalds, V. & Risley, A.M. (1961) *Publ. McCormick Obs.*, **11**, part 21.
Pagel, B.E.J. & Patchett, B.E. (1975) *Mon. Not. Roy. astr. Soc.*, **172**, 13.

Quinn, P.J. (1990) talk given at Aspen Summer Workshop.
Ratnatunga, K.U. & Freeman, K.C. (1985) *Astrophys. J.*, **291**, 260.
Sandage, A. (1987) in *The Galaxy*, eds G. Gilmore & B. Carswell, Reidel, Dordrecht, p321.
Sandage, A. & Fouts, G. (1987) *Astron. J.*, **92**, 74.
Stetson, P.B. & Harris, W.E. (1988) *Astron. J.*, **96**, 909.
Strugnell, P., Reid, I.N. & Murray, C.A. (1986) *Mon. Not. Roy. astr. Soc.*, **220**, 413.
Trefzger, C. *et al.* (1990) in preparation.
VandenBerg, D. & Bell, R.A. (1985) *Astrophys. J. Suppl.*, **58**, 711.
van den Bergh, S. (1958) *Astron. J.*, **63**, 492.
van der Kruit, P.C. (1987) in *The Galaxy*, eds G. Gilmore, B. Carswell, Reidel, Dordrecht, p27.
Wheeler, J.C., Sneden, C. & Truran, J.W. (1989) *Ann. Rev. Astron. Astrophys.*, **27**, 279.
Wyse, R.F.G. & Gilmore, G. (1988) *Astron. J.*, **95**, 1404.

STAR FORMATION AT LARGE GALACTIC z?

WILLIAM TOBIN
Department of Physics and Mt John University Observatory,
University of Canterbury, Christchurch 1, New Zealand

ABSTRACT. It seems very probable that young, luminous B stars occur in the galactic halo. An origin in the disc followed by ejection may account for many of these stars; certain proposed formation-ejection mechanisms involve the infall of halo material. The kinematics and supposed locations of some halo B stars seem incompatible with an origin in the disc, suggesting that these stars may have formed in the halo.

1. INTRODUCTION

In this review I present evidence which I believe fairly strongly suggests that young, massive stars occur in the galactic halo, and that these stars are normal in every respect save their location and kinematics. If this 'normal' interpretation is correct, these hot, luminous stars can be used to probe the halo interstellar medium, and indeed have been used for this purpose (e.g. Pettini & West, 1982; Savage & Massa, 1985; Lockman, Hobbs & Shull, 1986; Danly, 1989). Obviously astronomers who use these stars as halo probes should understand the arguments concerning their nature and location, so it is appropriate to discuss them at this Symposium.

Many of these halo stars can plausibly somehow have acquired sufficient velocity for them to be transported from disc birthplaces to their present locations within their evolutionary lifetimes. Some of the formation and/or ejection mechanisms involve the interaction of disc and halo material, and again are appropriate to discuss here.

However the kinematics and apparent locations of some of the apparently-normal B stars are such that, if the stars really are of normal age, then they must surely have formed in the halo of the Galaxy, and it is from these arguments that my review takes its title.

This paper updates my earlier review on the existence of apparently-young stars apparently-far from the galactic plane (Tobin, 1987). Also relevant are reviews by Lambert (1987), Trimble (1988), and Conlon (1989). Studies of blue halo stars all depend heavily on, and develop from, the pioneering paper by Greenstein & Sargent (1974) 'The nature of the faint blue stars in the halo'.

2. THE PHENOMENA: APPARENTLY-YOUNG STARS WITH HIGH VELOCITIES AND/OR APPARENTLY LOCATED IN THE HALO OF THE GALAXY

Normal, Population I O stars have lifetimes of only a few million years. No apparently-normal O stars are known far outside the galactic disc, but disc O stars do have a bimodal velocity distribution with half of them in a 'high'-velocity peak of 30 km s^{-1} dispersion, and the remainder in a low-velocity peak of 10 km s^{-1} dispersion (Stone, 1979).

Normal, Population I B star lifetimes range from 10^7 to 10^8 years. Neglecting the complication of Gould's Belt, they have a scale height perpendicular to the galactic plane $\beta_z \simeq 60$ pc.

B-type stars have long been known apparently located at very much greater values of $|z|$ (up to many kiloparsecs in some cases). Spectroscopically these B stars are almost always completely normal, having within the errors of analysis the gravities, temperatures, helium and metal abundances, and projected rotations typical of main-sequence and slightly evolved stars (e.g. Tobin & Kilkenny, 1981; Tobin & Kaufmann, 1984; Keenan et al., 1987; Conlon et al., 1988, 1989). The distances are calculated using standard but secondary methods, such as spectral-type–absolute-magnitude relations. The B stars in question are *not* the hot, high-gravity subdwarfs, helium-poor horizontal branch stars, or other spectroscopically peculiar stars that are also found – and expected – at high galactic latitudes.

Normal A stars have somewhat longer lifetimes, from 10^8 to 10^9 years. During these lifetimes, the disc-heating mechanism can increase the scale height to $\simeq 120$ pc. However in a study of the South Galactic Pole, Rodgers, Harding & Sadler (1981, hereafter 'RHS') found a population of apparently main-sequence 12th-15th magnitude A stars at distances 1-4 kpc from the galactic plane. Unlike the case of the B stars, these stars' abundances are often below solar ([Ca/H] $\simeq$ 0.0 to -0.5).

A young B star in the galactic halo will evolve into a later-type supergiant. The UU (or 89) Herculis class of stars are variable halo A-F supergiants. At one time, they were suspected of being massive objects (e.g. Sasselov, 1983), but they are systematically metal-deficient (e.g. Bond & Luck, 1987), and are probably mass-losing stars of $M < 1 M_\odot$ in a brief, pre-planetary nebula phase. Many similar objects are now being discovered amongst cold *IRAS* sources (e.g. Likkel et al., 1987; Trams et al., 1989). I shall not discuss the UU Herculis stars any further.

3. ELEVEN POSSIBLE EXPLANATIONS

3.1 Subluminous Hypotheses

The principal parameters furnished by atmospheric analyses are the stellar effective temperature, T_{eff}, and surface gravity, $\log g$. Normal stars have gravities lower than the $\log g \approx 4.2 - 4.4$ (c.g.s.) of the Zero-Age Main Sequence (ZAMS). This can also be the case for highly-evolved stars with masses $M = 0.5 - 1.0 M_\odot$. A normal-helium, low-mass star could conceivably mimic a Population I object, and be confused with it.

(I) POPULATION II EXPLANATION. Low-luminosity B stars with roughly normal helium do exist, and are seen in globular clusters ~ 1 magnitude above the horizontal branch. Barnard 29 in M 13 and von Zeipel 1128 in M 3 are famous examples. However the abundance analyses cited earlier show quite clearly that many halo B stars have Population I metallicities and cannot be Population II objects.

(II) OLD DISC POPULATION EXPLANATION. If the halo B stars were evolved members of an Old Disc Population (ODP), then their metallicities might be closer to solar and their spectra indistinguishable from normal. Astronomers at UNAM have proposed that the high-velocity O stars are ODP objects (Carrasco et al., 1980), but the supporting evidence in a recent paper is not strong (Carrasco, Costero & Stalio, 1987). One high-velocity O star is a double-lined spectroscopic binary for which the values of $m \sin^3 i$ show unequivocally that the components are massive (16.2 and 16.9 $M_\odot$ for HD 198846 $\equiv$ Y Cyg, $V_{LSR} = -45$ km s^{-1}; Batten, Fletcher & Mann, 1978).

The case of the high-latitude star PB 166 is pertinent here, and disquieting for supporters of normal luminosity hypotheses. The metal lines in a low-resolution IUE spectrum suggested that this 12th-magnitude early B star was a young object, implying $z > 6.2$ kpc (Tobin, 1986). An analysis of an intermediate-dispersion optical spectrum by de Boer, Heber & Richtler (1988) found a normal helium abundance, but deduced $\log g = 4.8 \pm 0.2$ from the H I and He I absorption profiles. A single, normal star should have a surface gravity rigorously lower than that of the ZAMS, whereas no such constraint would apply to a low-mass, ODP object. However Conlon et al. (1989) found $\log g = 4.6 \pm 0.3$ for PB 166 (albeit from Hγ and Hδ profiles that were affected by various instrumental problems) and derived essentially normal abundances for He, C, N, O, Mg, Al, Si and S. They concluded that a normal, main-sequence nature was not excluded for PB 166, and the star's rotation, $v \sin i = 50 - 60$ km s^{-1}, is certainly high for the horizontal branch. If PB 166 is a subluminous object with $M \sim 0.5 M_\odot$ and $M_V \sim 2.4$, then $z \sim 600$ pc.

3.2 Normal Luminosity Hypotheses

Even if not yet completely conclusive, the evidence for the normal luminosity of many halo B stars seems overwhelming. The halo B helium and metal abundances are normal, as is the distribution of rotations. There is a spatial correlation with spiral arms if $|z|$ is not too extreme (Kilkenny, Hill & Schmidt-Kaler, 1975). Further, minor peculiarities usual in disc B stars have been found in the halo B stars, including Be stars (Kilkenny, 1989), and two β Cephei stars—HD 129929 (Waelkens & Rufener, 1983; $z = 540$ pc) and PHL 346 (Waelkens & Rufener, 1988; Kilkenny & van Wyk, 1990; $z = -5.3$ kpc). (Incidentally, the variability of PHL 346 presumably accounts for the discrepant IUE SWP flux reported for this star by Heber & Langhans, 1986.)

For many of the halo B stars, the radial velocities are such that the stars could plausibly have been born in the galactic disc and have reached their present locations within the lifetimes available to them. B star radial velocities can be much greater than for the high-velocity O stars, and can exceed 200 km s^{-1} (e.g. Tobin & Kaufmann, 1984; Keenan *et al.* 1987; de Boer, Heber & Richtler, 1988).

3.2.1. Ejection from the Galactic Plane soon after Birth

(III) EJECTION AS THE RESULT OF A SUPERNOVA EXPLOSION IN A BINARY STAR. This idea was first developed by Blaauw (1961), following a suggestion by Zwicky. It supposes that a supernova explodes in a binary star and that the secondary is freed to move off with its previous orbital velocity. For a large velocity, the binary needs to have been tight and massive, but then the binary evolution is complicated because the pre-supernova primary will spill matter onto the secondary as the radius of the primary increases during its evolution away from the main sequence. Stone (1982) has considered all this, and found that only the most massive stars would be accelerated, $M > 11 M_\odot$. Stone also found that most binaries should not be disrupted by the supernova explosion, but the incidence of binaries amongst the high-velocity OB stars is much less than in the field (Gies, 1987). Further, two of the known high-velocity O stars are double-lined spectroscopic binaries, so cannot contain a compact object (HD 3950 and HD 198846, Gies & Bolton, 1986). Two high-velocity Wolf-Rayet stars do however present evidence for low-mass companions (209 BAC: Moffat, Lamontagne & Seggewiss, 1982; HD 143414: Isserstedt, Moffat & Niemela, 1983). That supernova explosions *can* produce high velocities is indicated by the large velocities of many pulsars [Arnaud & Rothenflug (1981) estimate $< |v_z| >= 130$ km s^{-1} at ejection].

(IV) STAR FORMATION AND EJECTION RESULTING FROM A COLLISION BETWEEN THE GALAXY AND A MAGELLANIC CLOUD-TYPE OBJECT $\sim 6.5 \times 10^8$ YEARS AGO. Pier (1983) analysed photometry and spectroscopy of AB stars in the southern galactic halo. He found no A stars with near-solar metallicities beyond 2-3 kpc from the plane, in conflict with the results of RHS. Working afresh at the South Galactic Pole, Lance (1988a) upheld the RHS findings, attributing Pier's result to a selection effect. (The most luminous and thus most distant main sequence A stars can be confused with horizontal branch stars because the horizontal branch crosses the main sequence near early A.)

For the distant A stars with near-solar calcium ([Ca/H]> -0.5), Lance (1988b, 1989) finds a scale height $\beta_z \simeq 1000 - 1600$ pc (similar to that of the thick disc). However none of these stars with $|z| > 1000$ pc is older than 650×10^6 years, whereas A stars with $|z| < 500$ pc range from zero to 2×10^9 years old. Lance suggests that 'at around 6.5×10^8 years ago, a major source of relatively low abundance hydrogen was accreted by the Galactic disk,' so forming these stars. This mechanism cannot account for the present-day halo and high-velocity B stars which (i) must have formed much more recently, and (ii) show no evidence of reduced metallicity.

(V) STAR FORMATION AND EJECTION RESULTING FROM COLLISION BETWEEN DISC MATERIAL AND INFALLING INTERMEDIATE- AND HIGH-VELOCITY HI CLOUDS. This is similar to (IV), except that the mechanism can act repeatedly. A collision of this sort appears to be happening in the Draco Nebula, and will be discussed later under hypothesis (X).

(VI) MASSIVE STARS FORMED FROM ACCELERATED INTERSTELLAR MATERIAL. Matter might be accelerated (a) due to increase of pressure by ionisation and heating by a massive O star on one side of a cloud (Oort & Spitzer, 1955), or (b) by being swept up by a supernova explosion (Herbst & Assousa, 1977). However when sufficient mass has been swept up for star formation, the velocity may be relatively small (<60 km s^{-1}).

3.2.2. Ejection from the Galactic Plane at a Random Time

(VII) EJECTION DURING THE DYNAMICAL EVOLUTION OF A STELLAR CLUSTER. In my opinion, this is at present the most promising of the ejection mechanisms, and can account for both the high-velocity O and B stars. Numerical simulations by Poveda, Ruiz & Allen (1967) suggested that massive stars could be ejected with velocities up to ~ 200 km s^{-1} during the initial dynamical relaxation of a small, dense cluster of massive stars. Recent N-body simulations by Leonard & Duncan (1988, 1990) have shown that binary-binary collisions within young *low-density* clusters can at any time result in the ejection of stars. The properties of the ejected stars match observations, at least qualitatively. Less massive stars are ejected faster; the maximum velocity is $\simeq 200$ km s^{-1}; and for $M > 10 M_\odot$, binaries can be ejected, but their frequency is reduced, perhaps to 10%. Finally, the very uncertain estimated number of actual high-velocity O and B stars is similar to the number to be expected according to the Leonard & Duncan simulations, given the surface density of young clusters in the galactic disc. The simulations are restricted to clusters of massive stars. An obvious priority is to extend the calculations to predict to what extent lower mass stars with high velocities might be expected.

As will be discussed in Section 3.2.3, some of the apparently normal B stars in the halo *cannot* plausibly have originated in the plane, and the Leonard & Duncan mechanism cannot be the sole explanation of the high-velocity O and B stars.

(VIII) GRAVITATIONAL EFFECTS OF A HALO OF VERY MASSIVE OBJECTS. Ipser & Semenzato (1985) and Lacey & Ostriker (1985) have investigated the dynamical effects on the galactic disc of a galactic halo of very massive ($\sim 10^5 - 10^6 M_\odot$), fast-moving objects, such as might be a remnant of Population III. Weak gravitational encounters warm the disc, while rarer strong encounters can eject disc stars with high velocity. If the disc is warmed in this way at the observed rate (Wielen, 1977), any individual disc star has a probability of $\sim 10^{-11}$ per year of being ejected at high velocity, irrespective of its mass. The fraction of ejected stars would then be (ejection rate) $\times$ (lifetime), and the mechanism could eject the Gilmore & Reid (1983) thick disc (20% of G stars). The mechanism could not produce the 50% of high-velocity O stars, but it could produce occasional 250 km s^{-1} B stars. However a fundamental objection to this mechanism is that if it acted, it would act on all stars at all times. Although it would produce the correct number of RHS A stars (0.5% of A stars), it could not produce the observed upper limit on ages. The mechanism would also produce a thick disc of F stars, but no such disc is seen.

3.2.3 Star Formation in situ *in the Galactic Halo*

The suspicion that this is occuring arises because the kinematics of certain individual halo B stars seem incompatible with an origin in the disc.

Many authors have compared the evolutionary ages of halo B stars with the kinematic times needed to get them to their apparent present locations, assuming an origin in the disc (e.g. Greenstein & Sargent, 1974; House & Kilkenny, 1978; Conlon *et al.*, 1989). A rigorous comparison is not possible, because halo B star proper motions are too small to measure from the ground. Only the radial velocities are available observationally. For any particular star it is always possible to assume that the necessary velocity for a disc origin is hidden in the transverse velocity, but this becomes more and more difficult to sustain for stars near the galactic poles, or when the required transverse velocity greatly exceeds the ≈ 250 km s^{-1} maximum of observed radial velocity. Let me cite two stars for which birth in the plane is very difficult to entertain.

PHL 346, mentioned earlier as a β Cephei star ($V = 11.5$), is found at galactic latitude $b = -58°$; has normal abundances; has an apparent age of $\sim 11 \times 10^6$ yr; and is apparently located at $z = -8.7 \pm 1.5$ kpc (Keenan *et al.*, 1986). A velocity perpendicular to the plane $v_z > 800$ km s^{-1} would be necessary to take PHL 346 to its present location within 11×10^6 yr. (There is no need to compute galactic gravitational deceleration at this v_z.) The v_z component of the radial velocity is a mere 56 ± 10 km s^{-1}, so the required transverse velocity would be greater than 1400 km s^{-1}! (This would actually lead to an observable proper motion $\mu > 0.03$ arcsec yr^{-1}, which would be worth searching for.)

It is even worse for another early B star, PG 0832+676 (Brown *et al.*, 1989). Spectra from the Hale 5 m and Isaac Netwon 2.5 m telescopes indicated normal helium and metal abundances for this $V = 14.5$ star at $b = +35°$. The inferred distance is $z \sim 18$ kpc (galactocentric distance $R = 37$ kpc!), and the inferred age is $\sim 13 \times 10^6$ yr. The radial velocity is negative: -73 km s^{-1}. A transverse velocity greater than 1800 km s^{-1} ($\mu > 0.012$ arcsec yr^{-1}) would be needed for an origin in the plane (at $R \sim 45$ kpc).

These enormous transverse velocities, unprecedented in any halo B star radial velocity, have led to the suggestion that stars can form in the halo of our Galaxy.

However PHL 346, and more especially PG 0832+676, are faint. Thus their nature might well be different from that of the brighter halo B stars: but even if they were ODP objects, they would still be located rather far from the galactic plane ($z \sim -1.7$ and $z \sim 3.5$ kpc, respectively, if $M = 0.5 M_\odot$). A fact possibly favouring a smaller distance for PG 0832+676 is that it lies in the direction of High-Velocity Cloud (HVC) A, but van Woerden, Schwarz & Wakker (1990) have failed to detect any interstellar absorption at the velocity of HVC A.

(IX) FORMATION OF STARS WITHIN INTERMEDIATE- AND HIGH-VELOCITY HI CLOUDS IN THE HALO. Dyson & Hartquist (1983) suggested that shock-induced star formation might occur in the halo as a result of supersonic collisions between individual cloudlets within larger intermediate- and high-velocity H I clouds, assuming that HVCs are themselves halo objects. Shock-induced star formation is believed to favour the production of more massive stars, and a collision between cloudlets of characteristic mass $M_c \approx 1.2 \times 10^4 M_\odot$ (and density $n_H \sim 0.1$ cm^{-3}) might produce $N_{OB} \approx 7 - 70$ O-B0 stars, and $N_{BA} \approx 70 - 700$ B1-A5 stars ($M > 2 M_\odot$). If HVCs are the infalling material of a galactic fountain, they might well spawn stars with Population I metal abundances. Assuming ~ 10 cloudlets per HVC, Dyson & Hartquist estimate a cloudlet-cloudlet collision would occur every $\sim 10^8$ yr within any particular HVC. With a total of ~ 100 HVCs, and OB lifetimes $\leq 10^7$ yr, there should thus be $\sim 10^3$ OB stars (and $\sim 10^4$ B-A5 stars) in the galactic halo. Since the cloudlet-cloudlet velocities are typically only ~ 30 km s^{-1}, these stars should be clumped in fairly close association with their parent HVC, both in position and velocity. This does not seem to be the case (e.g. Brown et al., 1989), though a systematic search would be desirable. The absence of any apparently-normal O star in the halo must be a further objection to the Dyson & Hartquist mechanism, since there should have been ~ 10 cloudlet-cloudlet collisions within the last $\sim 10^7$ yr from which O stars should still be visible.

It is pertinent here to note that automated measurements of Schmidt plates have clearly confirmed the existence between the two Magellanic Clouds of a 'bridge' of blue, 15-20th magnitude stars (Irwin, Demers & Kunkel, 1990). The radial velocity of several stars reinforces the association with the Magellanic Clouds. There is a general correlation of the stellar positions with the H I envelope around the Clouds, though the north-south spatial extent of the bridge stars is still unexplored. Since the H I bridge may be $\sim 2 \times 10^8$ yr old (Fujimoto & Murai, 1983) and the Irwin, Demers & Kunkel bridge stars could range in age up to $\sim 1 \times 10^8$ yr (colour index > 0), it would appear that B stars can form in interstellar material similar to that in the bridge. (It would be interesting, however, to know the spatial distribution of the *youngest* bridge B stars.) The H I column densities in the bridge are $\sim 4 \times 10^{20}$ cm^{-2} (Irwin, Kunkel & Demers, 1985). Assuming a distance of 50 kpc, and that the north-south angular extent of the H I is indicative of its line-of-sight depth, this corresponds to an H I density $n_H \sim 0.02$ cm^{-3}, not greatly different from that in HVC cloudlets (if at halo distances).

Even though there are objections to the Dyson & Hartquist mechanism, the evidence of the Magellanic bridge suggests that HVCs must nevertheless be considered serious candidates for sites of halo star birth.

(X) STAR FORMATION IN MOLECULAR CLOUDS OUTSIDE THE GALACTIC PLANE. Mebold et al. (1985) have reported finding molecular emission and absorption associated with an extended, faint optical and H I 21 cm nebula (the 'Draco nebula', $l \approx 91°, b \approx +38°$). From UBV photometry of foreground stars, and star counts, Goerigk & Mebold (1986) limit the location of the Draco nebula to $500 < z < 1500$ pc. Under a number of assumptions (including $z > 500$ pc), Mebold et al. find that two of the molecular concentrations (nicknamed 'Fang' and 'Wart') could plausibly be gravitationally bound. Somewhat boldly, these authors remark that their result *'presents the first clear indication that sites of star formation are located more than 500 pc above the galactic plane'* (their italics). Mebold, Heithausen & Reif (1987) suggest that star formation in bound molecular clumps at high z could be the origin of the halo B stars. A score or more *IRAS* Point Sources are associated with the Draco complex, and may be protostellar Bok globules (Johnson, 1986)—or perhaps artifacts of the Point Source Catalogue detection algorithm.

Other authors (Odenwald & Rickard, 1987) have found it difficult to accept $z > 500$ pc for the Draco cloud. However all authors agree that in Draco halo material is ploughing into matter situated nearer the disc. Odenwald & Rickard believe that the main Draco nebula and nearby, morphologically-similar 'comet-like' structures seen in the *IRAS* 100 μm maps are falling subsonically into disc gas at $z \sim 120$ pc. From a detailed study of the Fang, Rohlfs *et al.* (1989) believe that the Fang molecular clump ($v \sim -25$ km s^{-1}) has resulted from the collision of a HVC ($v \sim -180$ km s^{-1}) with the $v \sim -18$ km s^{-1} Draco material. Odenwald (1988) has conducted optical and infra-red studies on 14 other comet-like structures seen at 100 μm. The morphologies of five of Odenwald's objects 'are found to be consistent with objects moving supersonically through the interstellar medium. These clouds are also active in forming B-type stars in their nuclei...'

Irrespective of its distance, it seems clear that Draco provides evidence of halo material interacting with near-disc material in a way that might produce stars. But for the problem of the halo stars, it is disappointing that the velocity of the Fang molecular clump is so low, and a quick *SIMBAD* check showed that the velocities of the B stars in Odenwald's comet-like structures are also low.

Other evidence which suggests that stars may form at high z are the Orion-Nebula–like H II region seen ~ 800 pc from the disc of the edge-on spiral galaxy NGC 4244 (Walterbos, this Symposium), and Blitz's report (this Symposium) that Leisawitz & de Geus have found that the space velocity of the open cluster NGC 457 is directed towards the galactic plane.

Greater velocity after interaction, or molecular clouds well into the halo, will be necessary if high-$|z|$ molecular clouds are the origin of stars such as PHL 346 and PG 0832+676.

3.3 Extended Lifetime Hypothesis

(XI) THE HALO STARS HAVE EXTENDED EVOLUTIONARY TIMESCALES. Blue stragglers are stars seen above the main-sequence turnoff in star clusters of all ages. Consequently, their ages are thought to be greater than those implied by their absolute magnitude and colour. Shields & Twarog (1988) suggested that the RHS A stars might be A-type blue stragglers: their larger scale height would result from a greater number of scatterings in the disc. Lance (1988b, 1989) notes, *inter alia*, that the age cut-off she found is incompatible with this hypothesis.

A blue straggler explanation also seems unlikely for the halo B stars. Early-B blue stragglers are predominantly Be stars, and late-B blue stragglers mostly present Ap/Bp abundance peculiarities (Mermilliod, 1982).

4. CONCLUDING REMARKS

The problem of young halo and/or high-velocity stars has been evident since at least the 1950s (e.g. Blaauw & Morgan, 1954; Münch, 1956). Resolution of the problem still seems very distant. It seems likely that several processes may be acting simultaneously. In my 1987 review I stressed the need for (and difficulty of!) a systematic search for halo B stars over a wide area, over a wide range of colour, and to deepish magnitudes ($V \sim 14$). Surveys such as the Edinburgh-Cape Town one are now under way (e.g. Kilkenny & Pauls, 1990), and the TYCHO package aboard Hipparcos should discover many new halo B stars.

An immediate priority must be to establish unequivocally the nature of the halo B stars, and thus the validity of arguments based on Population I metallicities. PB 166 must be reobserved and reanalysed, both absolutely, and differentially with respect to some clearly disc stars, so that all doubt about its gravity is removed. Should the high gravity be upheld, the normal luminosity hypothesis for the halo B stars will be severely undermined, because a major argument for this hypothesis is the Population I abundances.

The cameras aboard the Hubble Space Telescope have sufficient sensivity—provided the images are good—to detect any blue stars of Population I luminosity that may exist in the halos of nearby edge-on spiral galaxies.

Viton et al. (1991) appear to have discovered a $2 \times 10 M_\odot$ double-lined spectroscopic binary located at $z = -10$ kpc. Further observations of this star (already underway) may produce unequivocally large lower mass limits, though the required observations may be lengthy, because an orbital period of as long as 30 yr could still result in observable velocity variations.

Magnitudes fainter than $V \sim 10$ and the later B spectral subtypes have barely been explored, and interesting results may well arise from individual objects haphazardly found. (Irwin, Demers & Kunkel report a B7 V star, $V = 14.25, v_r = +268$ km s$^{-1}, z \sim -6$ kpc.) The strong selection effect that most analysed halo B stars are Henry Draper objects makes any analysis of group properties as yet unconvincing. In addition, it is only at fainter magnitudes that any further stars clearly indicative of halo star formation will be found. Deep, multi-colour searches for blue stars in the Magellanic Stream would be appropriate. The Draco nebula merits further study for evidence of actual star formation, and a better distance limit.

For interstellar astronomers seeking new sight lines, it should be noted that the fraction of clearly abnormal B stars increases at fainter magnitudes (e.g. the absence of metal lines in most stars reported by Kilkenny & Lydon, 1986; or the peculiar helium lines found by Möhler et al., 1990), and the distance to each probe star should be evaluated with care. Savage (private communication) notes that interstellar Ti II appears smoothly distrubuted in both the Galaxy's disc and halo, and can provide an additional indication of distance. It would be interesting to see if a consistent picture of halo gas would arise if it were assumed that the B star probes used to investigate it were subluminous with $M \sim 0.5 M_\odot$.

REFERENCES

Arnaud, M., Rothenflug, R. (1981) *Astron. Astrophys.* **103**, 263
Batten, A.H., Fletcher, J.M., Mann, P.J. (1978) *Publ. Dom. Ap. Obs. Victoria* **15**, 121
Blaauw, A. (1961) *Bull. Astron. Inst. Netherlands* **15**, 265
Blaauw, A., Morgan, W.W. (1954) *Ap. J.* **119**, 625
Bond, H.E., Luck, R.E. (1987) in *The Second Conference on Faint Blue Stars*, IAU Colloquium No. 95, A.G. Davis Philip, D.S. Hayes, J.W. Liebert (Eds), L. Davis Press, Schenectady, p.149
Brown, P.J.F., Dufton, P.L., Keenan, F.P., Boksenberg, A., King, D.L., Pettini, M. (1989) *Ap. J.* **339**, 397
Carrasco, L., Bisiacchi, G.F., Cruz-González, C., Firmani, C., Costero, R. (1980) *Astron. Astrophys.* **92**, 253
Carrasco, L., Costero, R., Stalio, R. (1987) *Rev. Mexicana Astron. Astrof.* **14**, 301
Conlon, E.S. (1989) *Irish Astron. J.* **19**, 59
Conlon, E.S., Brown, P.J.F., Dufton, P.L., Keenan, F.P. (1988) *Astron. Astrophys.* **200**, 168
Conlon, E.S., Brown, P.J.F., Dufton, P.L., Keenan, F.P. (1989) *Astron. Astrophys.* **224**, 65
Danly, L. (1989) *Ap. J.* **342**, 785
de Boer, K.S., Heber, U., Richtler, T. (1988) *Astron. Astrophys.* **202**, 113
Dyson, J.E., Hartquist, T.W. (1983) *Monthly Notices Roy. Astron. Soc.* **203**, 1233
Fujimoto, M., Murai, T. (1983) in *Structure and Evolution of the Magellanic Clouds*, IAU Symposium No. 108, S. van den Bergh, K.S. de Boer (Eds), Reidel, Dordrecht, p.113
Gies, D.R. (1987) *Ap. J. Suppl. Ser.* **64**, 545
Gies, D.R., Bolton, C.T. (1986) *Ap. J. Suppl. Ser.* **61**, 419
Goerigk, W., Mebold, U. (1986) *Astron. Astrophys.* **162**, 279
Gilmore, G., Reid, N. (1983) *Monthly Notices Roy. Astron. Soc.* **202**, 1025
Greenstein, J.L, Sargent, A.I. (1974) *Ap. J. Suppl. Ser.* **28**, 157
Heber, U., Langhans, G. (1986) in *New insights in astrophysics: 8 years of UV astronomy with IUE*, E.J. Rolfe (Ed.), ESA SP-263 p.279
Herbst, W., Assousa, G.E. (1977) *Ap. J.* **217**, 413
House, F., Kilkenny, D. (1978) *Astron. Astrophys.* **67**, 421
Ipser, J.R., Semenzato, R. (1985) *Astron. Astrophys.* **149**, 408
Irwin, M.J., Demers, S., Kunkel, W.E. (1990) *Astron. J.* **99**, 191
Irwin, M.J., Kunkel, W.E., Demers, S. (1985) *Nature* **318**, 160
Isserstedt, J., Moffat, A.F.J., Niemela, V.S. (1983) *Astron. Astrophys.* **126**, 183
Johnson, H.M. (1986) *Ap. J.* **309**, 321
Keenan, F.P., Lennon, D.J., Brown, P.J.F., Dufton, P.L. (1986) *Ap. J.* **307**, 694
Keenan, F.P., Brown, P.J.F., Conlon, E.S., Dufton, P.L., Lennon, D.J. (1987) *Astron. Astrophys.* **178**, 194
Kilkenny, D. (1989) *Monthly Notices Roy. Astron. Soc.* **237**, 479
Kilkenny, D., Hill, P.W., Schmidt-Kaler, Th. (1975) *Monthly Notices Roy. Astron. Soc.* **171**, 353
Kilkenny, D., Lydon, J. (1986) *Monthly Notices Roy. Astron. Soc.* **218**, 279

Kilkenny, D., Pauls, L. (1990) *Monthly Notices Roy. Astron. Soc.* **244**, 133
Kilkenny, D., van Wyk, F. (1990) *Monthly Notices Roy. Astron. Soc.* in press
Lacey, C.G., Ostriker, J.P. (1985) *Ap. J.* **299**, 633
Lambert. D.L. (1987) in *Stellar evolution and dynamics in the outer halo of the Galaxy* M. Azzopardi, F. Matteucii (Eds), ESO Conf. Workshop Proc. No. 27, p47
Lance, C.M. (1988a) *Ap. J. Suppl. Ser.* **68**, 463
Lance, C.M. (1988b) *Ap. J.* **334**, 927
Lance, C.M. (1989) *Nature* **337**, 513
Leonard, P.J.T., Duncan, M.J. (1988) *Astron. J.* **96**, 222
Leonard, P.J.T., Duncan, M.J. (1990) *Astron. J.* **99**, 608
Likklel, L., Omont, A., Morris, M., Forveille, T. (1987) *Astron. Astrophys.* **173**, L11
Lockman, F.J., Hobbs, L.M., Shull, J.M. (1986) *Ap. J.* **301**, 380
Mebold, U., Cernicharo, J., Velden, L., Reif, K., Crezelius, C., Goerigk, W. (1985) *Astron. Astrophys.* **151**, 427
Mebold, U., Heithausen, A., Reif, K. (1987) *Astron. Astrophys.* **180**, 213
Mermilliod, J-C. (1982) *Astron. Astropl.ys.* **109**, 37
Möhler, S., Richtler, T., de Boer, K.S., Heber, U. (1990) *Astron. Gessel. Abstract Ser.* **4**, 44
Moffat, A.F.J., Lamontagne, R., Seggewiss, W. (1982) *Astron. Astrophys.* **114**, 135
Münch, G. (1956) *Publ. Astron. Soc. Pacific* **68**, 351
Odenwald, S.F. (1988) *Ap. J.* **325**, 320
Odenwald, S.F., Rickard, L.J. (1987) *Ap. J.* **318**, 702
Oort, J.H., Spitzer, L., Jr. (1955) *Ap. J.* **121**, 6
Pettini, M., West, K.A. (1982) *Ap. J.* **260**, 561
Pier, J.R. (1983) *Ap. J. Suppl. Ser.* **53**, 791
Poveda, A., Ruiz, J., Allen, C. (1967) *Bol. Obs. Tonantzintla y Tacubaya* **4**, 86
Rodgers, A.W., Harding, P., Sadler, E. (1981) *Ap. J.* **244**, 912 (RHS)
Rohlfs, R., Herbstmeier, Mebold, U., Winnberg, A. (1989) *Astron. Astrophys.* **211**, 402
Sasselov, D.D. (1983) *Inf. Bull. Var. Stars* No. 2314
Savage, B.D., Massa, D. (1985) *Ap. J. Lett.* **295**, L9
Shields, J.C., Twarog, B.A. (1988) *Ap. J.* **324**, 859
Stone, R.C. (1979) *Ap. J.* **232**, 520
Stone. R.C. (1982) *Astron. J.* **87**, 90
Tobin, W. (1986) *Astron. Astrophys.* **155**, 326
Tobin, W. (1987) in *The Second Conference on Faint Blue Stars*, IAU Colloquium No. 95, A.G. Davis Philip, D.S. Hayes, J.W. Liebert (Eds), L. Davis Press, Schenectady, p.149
Tobin, W., Kaufmann, J-P. (1984) *Monthly Notices Roy. Astron. Soc.* **207**, 369
Tobin, W., Kilkenny, D. (1981) *Monthly Notices Roy. Astron. Soc.* **194**, 937
Trams, N.R., Waters, L.B.F.M., Waelkens, C., Lamers, H.J.G.L.M., van der Veen, W.E.C.J. (1989) *Astron. Astrophys.* **218**, L1
Trimble, V. (1988) *Nature* **336**, 111
van Woerden, H., Schwarz, U.J., Wakker, B.P. (1990) in *Structure and Dynamics of the Interstellar Medium*, IAU Colloquium No. 120, G. Tenorio-Tagle (Ed.) Springer, p.379
Viton, M., Deleuil, M., Tobin, W., Prévot, L., Bouchet, P. (1991) *Astron. Astrophys.* (submitted)
Waelkens, C., Rufener, F. (1983) *Astron. Astrophys.* **119**, 279
Waelkens, C., Rufener, F. (1988) *Astron. Astrophys.* **201**, L5
Wielen, R. (1977) *Astron. Astrophys.* **60**, 263

MOLECULAR CLOUDS AND STAR FORMATION AT LARGE R[†]

J. BRAND
Osservatorio Astrofisico di Arcetri
Largo Enrico Fermi 5, Firenze, Italia

J.G.A. WOUTERLOOT
I. Physikalisches Institut
Zülpicher Strasse 77, Köln, B.R.D.

1. INTRODUCTION

In the outer Galaxy (defined here as those parts of our system with galactocentric radii $R>R_0$) the HI gas density (Wouterloot et al., 1990), the cosmic ray flux (Bloemen et al, 1984) and the metallicity (Shaver et al., 1983) are lower than in the inner parts. Also, the effect of a spiral density wave is much reduced in the outer parts of the Galaxy due to corotation. This changing environment might be expected to have its influence on the formation of molecular clouds and on star formation within them. In fact, some differences with respect to the inner Galaxy have been found: the ratio of HI to H_2 surface density is increasing from about 5 near the Sun to about 100 at $R\approx 20$kpc (Wouterloot et al., 1990). Because of the "flaring" of the gaseous disk, the scale height of both the atomic and the molecular gas increases by about a factor of 3 between R_0 and $2R_0$ (Wouterloot et al., 1990), so the mean volume density of both constituents decreases even more rapidly than their surface densities. The size of HII regions decreases significantly with increasing galactocentric distance (Fich and Blitz, 1984), probably due to the fact that outer Galaxy clouds are less massive (see section 3.3), and therefore form fewer O-type stars than their inner Galaxy counter parts. There are indications that the cloud kinetic temperature is lower by a few degrees (Mead and Kutner, 1988), although it is not clear to what extent this is caused by beam dilution.
A study of the properties of molecular clouds at large R could be used to put constraints on their formation mechanisms, such as reviewed e.g. by Elmegreen (1990).

[†] Partly based on observations collected at the European Southern Observatory, Chile

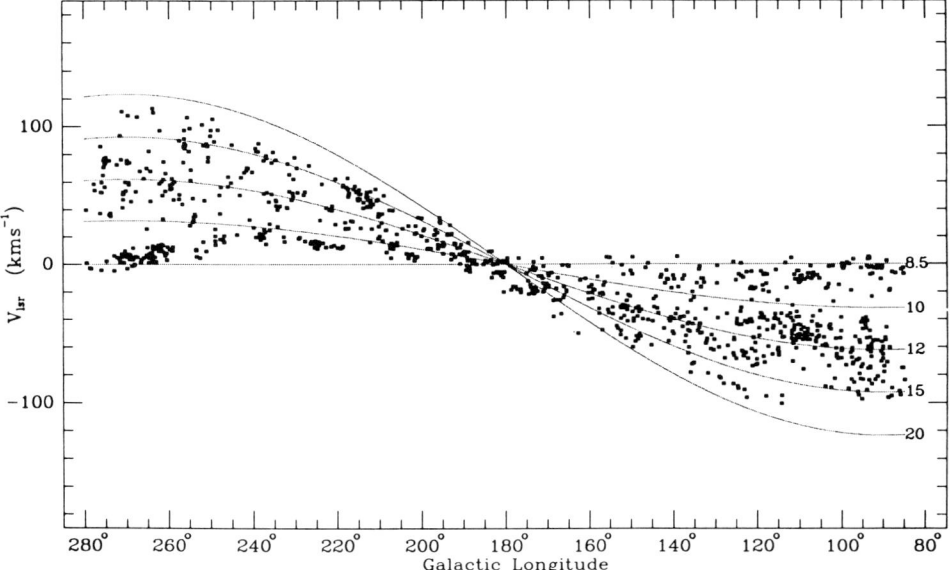

Figure 1. Longitude-velocity diagram of the CO emission components which were judged to be physically associated with the IRAS point sources. The lines indicate constant galactocentric radii (based on the rotation curve given in the text).

2. STAR FORMATION AT LARGE R

From optical studies of the outer Galaxy (Fich and Blitz, 1984; Brand,1986) it has become clear that the Galaxy contains HII region/molecular cloud complexes at distances R from the galactic centre of up to 20kpc. Because of interstellar extinction, these surveys found only a few very distant regions, most of them seen in certain directions of low extinction. In order to get a more representative view of star-forming molecular clouds in the outer Galaxy, we have looked for CO emission towards a large number (1302) of IRAS sources in the second and third galactic quadrant. These were selected from the Point Source Catalogue, using colour criteria that discriminate in favour of those sources that are frequently associated with H_2O masers and NH_3 cores (Wouterloot and Walmsley, 1986; Wouterloot et al., 1988b). As all star formation takes place in molecular clouds, these IRAS sources were used as flags for their location. In the direction of 1077 (83%) of the IRAS sources CO emission was detected, with the ESO 15-m SEST and the IRAM 30-m telescope (Wouterloot and Brand, 1989). The longitude-velocity diagram for this sample is shown in figure 1. Comparison with data from regular-grid surveys of the outer Galaxy (e.g. May et al., 1988) shows that these have missed almost all the emission with velocities in excess of 50-70 kms^{-1}, due to a combination

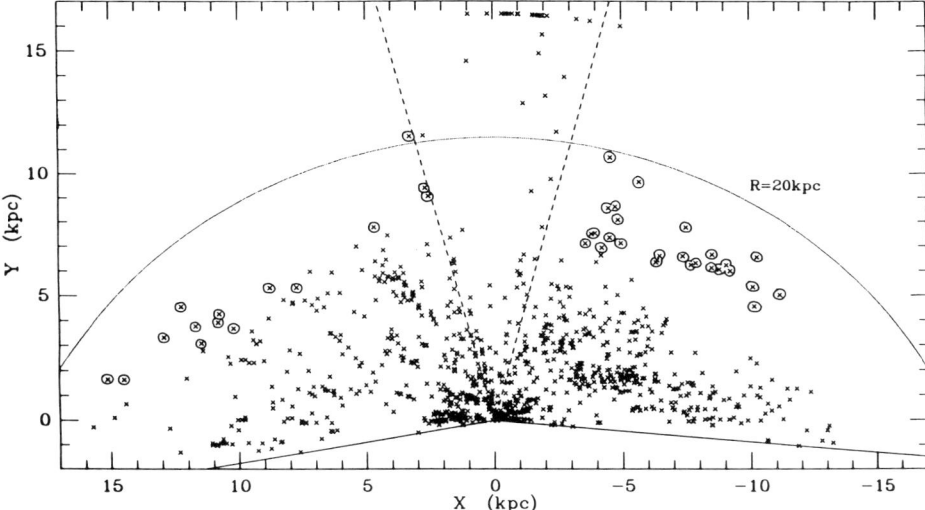

Figure 2. Distribution projected onto the galactic plane of all CO emission components shown in figure 1. The Sun is at (0,0), the galactic centre at (0,−8.5). The full-drawn lines show the longitude limit of the point source sample. The dashed lines mark the region within ±15° of the galactic anticentre where kinematic distances are particularly uncertain. The circle segment represents the galactocentric distance R=20kpc. The clouds that are discussed in the present paper (section 3.1) are identified by an open circle.

of coarser sampling, lower sensitivity, and insufficient coverage in galactic latitude. Kinematic distances to the clouds in our survey were derived using a rotation curve $\Theta=220(R/8.5)^{0.0382}$ (Brand, 1986; Wouterloot et al., 1990). The distribution of these clouds, projected onto the galactic plane, is shown in figure 2.

Molecular clouds with embedded IRAS sources are found up to about 20kpc from the galactic centre (i.e. $R\approx2.5R_0$), and in contrast to the optical surveys, distant objects are distributed more or less evenly over galactic longitude. Because of the selection criteria used, this implies that star formation is taking place in the outer Galaxy up to R=20kpc, the radius defining the edge of the molecular disk of the Galaxy. Towards alsmost 600 of the IRAS sources in figure 2 we have searched for H_2O emission, finding an overall detection rate of 17%; for clouds with $R=R_0$-$1.5R_0$ 16.5% is detected; for those with $R> 1.5R_0$, 18.5%. This is another indication that star formation is active in this part of the Galaxy. Most of the IRAS sources in figure 2 have luminosities between 10^3 and $5\ 10^4$ $L_\odot$, corresponding to (ZAMS) stars of spectral types B3 to O8.5 (Panagia, 1973). From VLA data (Wouterloot et al., 1988a; Mead et al., 1990) and optical studies (Georgelin, 1975; Moffat et al., 1979; Brand, 1986) there is no indication that the average spectral type of the stars that ionize distant HII regions differs from that in comparable (i.e. of the same size) regions closer to the Sun. Therefore the most massive stars formed in the outer Galaxy are not as massive as those near the Sun, explaining the smaller size of the largest HII regions in the outer parts. The cause of this may be in the properties of the molecular clouds.

3. MOLECULAR CLOUDS AT LARGE R

3.1 Observations

We have used the results of Wouterloot and Brand (1989) to select those clouds at the largest distances from the galactic centre (that are located at the edge of the molecular disk of the Galaxy) to obtain their properties (e.g. sizes and masses).
A sample of 56 clouds with 16kpc<R<20kpc has been studied using the 3-m KOSMA telescope at Gornergrat for the northern hemisphere clouds and the 15-m SEST for the southern hemisphere clouds. The distance from the Sun of these clouds is about 10kpc, which translates to linear beamsizes of the two telescopes at 115GHz of about 12pc (KOSMA) and 2.5pc (SEST). This means that if these IRAS sources would be embedded in giant molecular clouds such as found in the inner Galaxy, these clouds would be resolved with the 3-m telescope.

At Gornergrat we have observed 32 clouds in ^{12}CO(J=1-0) and 5 clouds in ^{13}CO(J=1-0) at the IRAS position. The sensitivity was 0.05-0.10K for ^{12}CO(J=1-0) and 0.01-0.03K for ^{13}CO(J=1-0), respectively. Observations were made with a resolution of 0.4kms^{-1} and in frequency-switching mode. Subsequently we have mapped 13 of these clouds in ^{12}CO(J=1-0) on a 4' grid, tracing the clouds until the signal disappeared.

With the SEST we have mapped 14 clouds in ^{12}CO(J=1-0) on a 40" or 80" grid. We used a resolution of 0.11kms^{-1} and frequency-switching, and obtained spectra with an rms of about 0.5K. Some positions have been observed in ^{13}CO(J=1-0) (rms≈0.1K). In addition to these maps, ^{12}CO and ^{13}CO were observed at the IRAS position in 11 clouds.

At the IRAS position the ratio $T_A^*(^{12}CO)/T_A^*(^{13}CO)$ is 7.1±1.0 for the Gornergrat clouds, and 8.9±7.6 for the SEST clouds (the latter number is greatly influenced by two sources with ratios of 21 and 33, respectively). For the inner galaxy, Solomon and Sanders (1980) found values between 3 and 20 with an average value of 5.5 and possibly an increase for larger R. Before any conclusions can be drawn from this, we need to obtain ^{13}CO data at positions in the clouds, away from the IRAS sources.

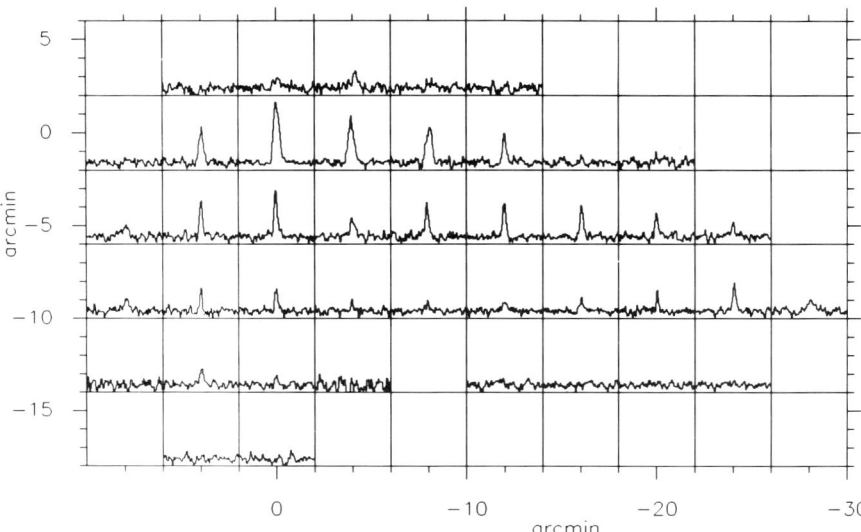

Figure 3. KOSMA-spectra of the cloud associated with IRAS01037+6504 (L=2.8 $10^3 L_\odot$) and IRAS01045+6505 (L=1.1 $10^5 L_\odot$). The sources are at (−5.4,−0.6) and (0,0), respectively. Offsets are in galactic coordinates. For each spectrum the x-range is −110 to −65kms^{-1} and the y-range is −0.2 to 1.8K.

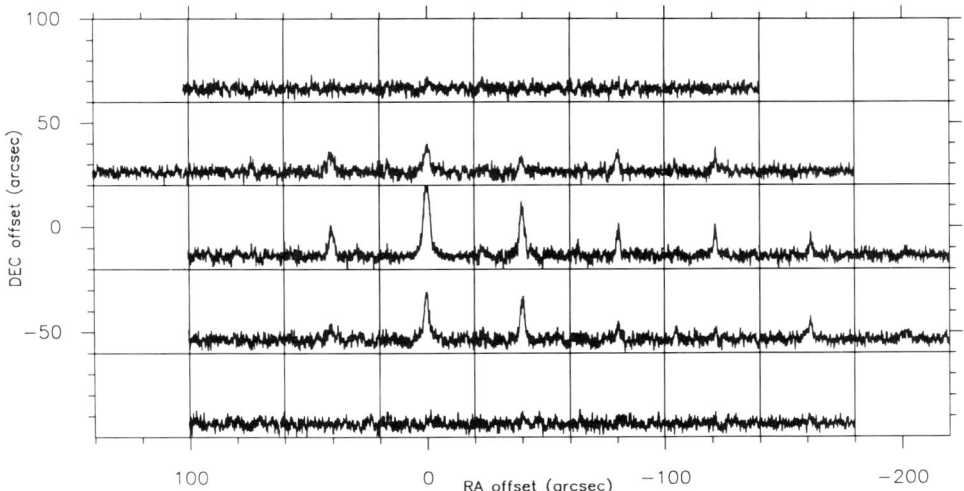

Figure 4. SEST-spectra of the cloud associated with IRAS06145+1455 (L=1.9 $10^4 L_\odot$), which is at (0,0). For each spectrum the x-range is 10 to 55kms^{-1} and the y-range −1.5 to 8K.

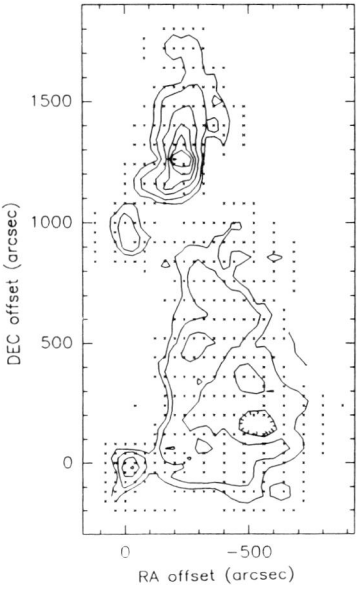

Figure 5. Contour map for the cloud complex associated with IRAS07257−2033 (L=8.7 $10^3 L_\odot$) of $\int T_A^* dv$ between 77 and 84kms^{-1} (SEST data). The IRAS source is at (0,0). Offsets are in arcseconds. Contour levels are 2.5, 5,10,15,20,25Kkms^{-1}.

3.2 Sizes

With the 3-m telescope we detected 30 of the 32 clouds. The size of the clouds that were not, or only marginally, detected is expected to be less than 10pc. The T_A^* at the IRAS position is 0.2 to 2.5K and typically 1K, which is about one order of magnitude lower than when observed with the 30-m IRAM telescope. The reason is that the IRAS source usually is associated with a relatively warm, small molecular cloud (or clump) as is shown by some maps made by Wouterloot *et al.* (1989) and by our SEST maps. Also the beam filling factor for the 3-m telescope of the extended lower temperature emission of the cloud surrounding this clump will generally be less than one.

The 13 molecular clouds mapped at Gornergrat were detected at between 2 and 22 positions. The largest clouds are elongated, reaching lengths of 80-100pc. All clouds have a maximum in T_A^* (or $\int T_A^* dv$) at, or close to, the IRAS position. Examples of the spectra for two clouds are given in figures 3 (Gornergrat) and 4 (SEST), respectively.

Contour maps of two cloud complexes mapped with the SEST are shown in figures 5 and 6. In figure 5, the source IRAS07257−2033 at (0,0) may be displaced from the integrated CO peak by 1pc (less than half a beam). The complex in figure 5 consists of several clouds that are separated by a few parsec, similar to the situation in Orion or Mon OB1. The northern cloud contains an IRAS source that

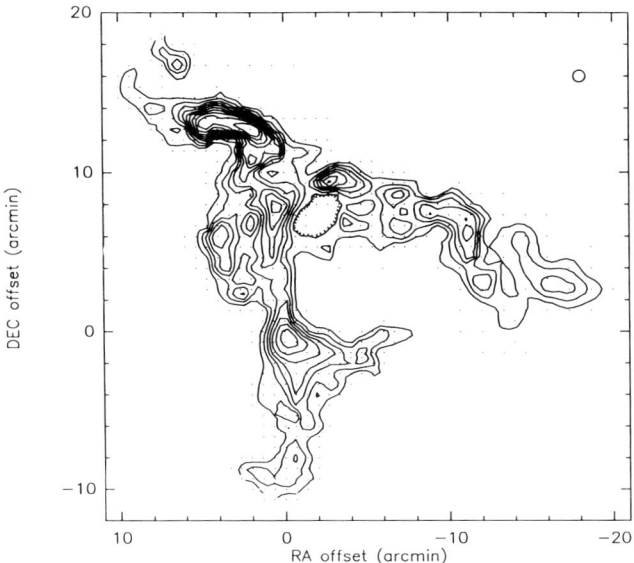

Figure 6. Distribution of $T_A^*(^{12}CO)$ for the cloud associated with IRAS06158+1506 (L=3.9 $10^3 L_\odot$; at (0,0)) and IRAS06158+1517 (L=6.4 $10^4 L_\odot$; at (0.75,11.72)); SEST data. Contour levels are 1 to 10K in steps of 1K. The circle in the upper right-hand-side corner indicates the HPBW.

is equally strong at 60μm as the one at (0,0), but it was not included in our survey because its flux density has an upper limit at 100μm.

Figure 6 shows the distribution of $T_A^*(^{12}CO)$ for the cloud associated with IRAS06158+1506 and IRAS06158+1517. With a maximum length of 80pc, this is one of the largest clouds in our sample. However, as this object is at a galactic longitude of 195°, its kinematic distance is uncertain, and it may well be closer by than the derived 9.6kpc, and consequently its size may be smaller.

Observing two small areas of sky, Terebey et al. (1986) found several clouds at an average R of 12.9kpc (when using the same rotation curve as here), and derived that the size spectrum there is the same as that near the Sun, and in the inner Galaxy.

A rough estimate of the diameter of a cloud is determined from the square root of the surface area where emission is detected, rather than from the area above some value of T_R^*, as is usually done. For the clouds observed with the 3-m telescope, it ranges from 14 to 55pc with a median value of about 25pc. The diameter of the clouds mapped with the SEST ranges between 3 and 50pc. These values are smaller than what is found for inner Galaxy clouds, but similar to the outer Galaxy clouds at around R=13kpc studied by Mead and Kutner (1988). In figure 7 we show the distribution of cloud radii.

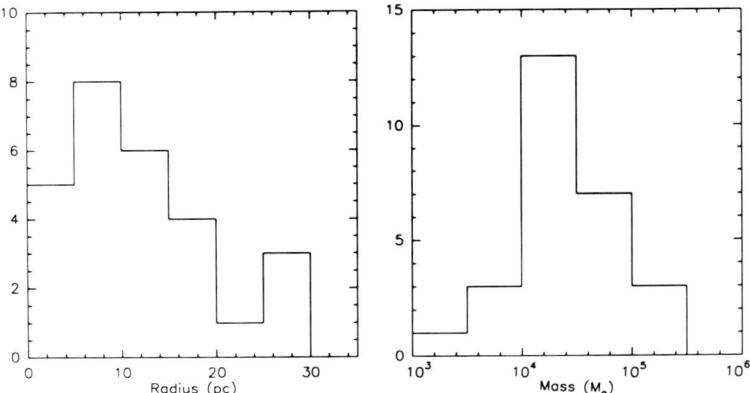

Figure 7. (left panel) Distribution of radii of the 27 clouds discussed here.
Figure 8. (right panel) Distribution of masses of the 27 clouds.

3.3 Masses

We derive virial masses assuming spherical clouds, with density falling off with cloud radius as r^{-1}: $M_{vir}=363r(\Delta v)^2$. Here Δv is the line width, for which we take the value at the IRAS position. Masses then range from $1.3\ 10^4 M_\odot$ to $1.5\ 10^5 M_\odot$ for the clouds observed at Gornergrat, and from $2.8\ 10^3 M_\odot$ to $1.6\ 10^5 M_\odot$ for those observed with the SEST.

Figure 8 shows the distribution over cloud masses; it is essentially identical to that for the sample of Mead and Kutner (1988). Almost all clouds in the latter sample are at R=13kpc (and in the first galactic quadrant), while our clouds are at R>16kpc. We find almost no clouds with a mass higher than $10^5 M_\odot$, which are often found in the inner Galaxy. It is not yet clear whether the lack of very high mass clouds (M>$10^6 M_\odot$) is real, or due to the relatively small number of objects investigated, or perhaps a consequence of the definition of cloud sizes and the calculation of masses.

The sum of the masses of the clouds is $7.2\ 10^5 M_\odot$ (Gornergrat) and $5.2\ 10^5 M_\odot$ (SEST) for about equal intervals (about 60°) in galactocentric azimuth investigated. The results of Wouterloot *et al.* (1990) give a total molecular mass within these sectors of about $1.2\ 10^7 M_\odot$, which would mean that the mapped clouds contain only 10% of this mass. Because we mapped only clouds associated with 27 of the 50 IRAS sources in the interval 16-20kpc, we estimate that 10-20% of the extrapolated H_2 mass is associated with IRAS sources. It is possible that because clouds are smaller (i.e. less massive) at large R, Wouterloot *et al.* (1990) have overestimated the molecular surface density in the outer Galaxy when extrapolating the value near the Sun, scaling it with the surface density of IRAS sources; but only unbiased surveys can show whether in the outer Galaxy there are many clouds which have no embedded IRAS sources with luminosities above 10^3-$10^4 L_\odot$. This may be the case if there exists an external triggering mechanism of star formation that is unrelated to cloud formation. The surveys of Mead *et al.* and others however do not suggest that there is a large population of such clouds. It seems then, that in spite of the differences between the inner- and outer parts of the Galaxy, star formation occurs at large R, but cloud formation is much more inefficient than at smaller R.

Elmegreen and Elmegreen (1987) conclude from a study of HI superclouds in the inner Galaxy, that the total hydrogen mass (being about $10^7 M_\odot$) of these objects does not vary with R, but that their molecular fraction is decreasing with R, probably because of a systematic variation with R of the pressure, metallicity, and radiation field within the galactic disk. HI measurements of the most massive molecular clouds in our sample will show whether these objects are enveloped by $10^7 M_\odot$ HI clouds.

ACKNOWLEDGEMENTS

The KOSMA radiotelescope at Gornergrat-Süd observatory is operated by the University of Cologne and supported by the Deutsche Forschungsgemeinschaft through grant SFB-301, as well as by special funding from the Land Nordrhein-Westfalen. The observatory is administered by the Hochalpine Forschungsstationen Jungfraujoch und Gornergrat, Bern, Switzerland.

REFERENCES

Bloemen, J.B.G.M., et al. (1984) *Astron. Astroph.* **135**, 12
Brand, J. (1986) *Ph.D. Thesis, University of Leiden*
Elmegreen, B.G. (1990) in *The Evolution of the Interstellar Medium*, (Blitz, ed.), *in press*
Elmegreen, B.G., Elmegreen, D.M. (1987) *Astroph. J.* **320**, 182
Fich, M., Blitz, L. (1984) *Astroph. J.* **279**, 125
Georgelin, Y.M.: 1975, *Ph.D. Thesis, Université de Provence*
May, J., Murphy, D.C., Thaddeus, P. (1988) *Astron. Astroph. Suppl.* **73**, 51
Mead, K.N., Kutner, M.L. (1988) *Astroph. J.* **330**, 399
Mead, K.M., Kutner, M.L., Evans II, N.J. (1990) *Astroph. J.* **354**, 492
Moffat, A.F.J., FitzGerald, M.P., Jackson, P.D. (1979) *Astron. Astroph. Suppl.* **38**, 179
Panagia, N. (1973) *Astron. J.* **78**, 929
Shaver, P.A., McGee, R.X., Newton, L.M., Danks, A.C. Pottasch, S.R. (1983) *Mon. Not. R.A.S.* **204**, 53
Solomon, P., Sanders, D. (1980) in *Giant Molecular Clouds in the Galaxy*, (Solomon, Edmunds eds.), p41
Terebey, S., Fich, M., Blitz, L., Henkel, C. (1986) *Astroph. J.* **308**, 357
Wouterloot, J.G.A., Brand, J. (1989) *Astron. Astroph. Suppl.* **80**, 149

Wouterloot, J.G.A., Brand, J., Burton, W.B., Kwee, K.K. (1990) *Astron. Astroph.* **230**, 21
Wouterloot, J.G.A., Brand, J., Henkel, C. (1988a) *Astron. Astroph.* **191**, 323
Wouterloot, J.G.A., Henkel, C., Walmsley, C.M. (1989) *Astron. Astroph.* **215**, 131
Wouterloot, J.G.A., Walmsley, C.M. (1986) *Astron. Astroph.* **168**, 237
Wouterloot, J.G.A., Walmsley, C.M., Henkel, C. (1988b) *Astron. Astroph.* **203**, 367

UV ABSORPTION AND EMISSION LINES FROM HIGHLY IONIZED GAS IN THE GALACTIC HALO

BLAIR D. SAVAGE
Department of Astronomy
University of Wisconsin
Madison, Wisconsin, USA

ABSTRACT. Highly ionized gas in the galactic halo has been detected through UV absorption and emission lines. In absorption the species studied include Si IV, C IV and N V. The UV emission studies have recorded C IV and O III]. Absorption measurements toward galactic stars reveal that the $|z|$ distribution of the gas is roughly exponential with a scale height of approximately 3 kpc and has column densities perpendicular to the galactic plane of N ~ 2×10^{13}, 1×10^{14} and 3×10^{13} atoms cm^{-2}, for Si IV, C IV and NV, respectively. Similar absorption line profiles for these species suggests a common process for their origin. The presence of N V absorption implies the existence of some gas with a temperature near T ~ 2×10^5 K. The highly ionized absorbing gas toward distant stars in direction b < -50° has simple and relatively narrow line profiles (FWHM ~ 45 to 70 km^{-1}) and small average LSR velocities while the gas in the direction b > 50° reveals a complex pattern of motions with substantial inflow and outflow velocities. Galactic rotation has an appreciable effect on the absorption line profiles to very distant stars located in the low halo. C IV emission has been seen at greater than a 3σ level of significance in 4 of 8 directions. The emission brightens toward the galactic poles and has a polar intensity I(C IV) ~ 5000 photons cm^{-2}s^{-1}ster^{-1}. If the emitting and absorbing gas coincide in space the measurements imply n_e ~ 0.01 cm^{-3} and P/k ~ 2000 cm^{-3} K for gas with T ~ 10^5 K. This phase of the gas fills only a small volume of the space (f ~ 0.03) and accounts for only a small fraction of the total column density of gas perpendicular to the galactic plane [~3×10^{18} atoms cm^{-2} vs 3.5×10^{20} atoms cm^{-2} for H I and 1×10^{20} atoms cm^{-2} for H$^+$]. However, the gas provides a large EUV/UV emission line flux (~1×10^{-5}erg cm^{-2} s^{-1}) which corresponds to a H I ionizing flux of ~2×10^5 ionizations cm^{-2} s^{-1}. Gas with T near 2×10^5 K cools very rapidly. Its origin may be associated with the cooling gas of a galactic fountain flow or with thermal condensations in cosmic ray driven fountains. In the nonequilibrium cooling of a Galactic fountain, a flow rate of 4 M$_\odot$/ year to each side of the Galaxy is required to produce the amount of N V absorption found in the halo while a flow rate 5x larger is required to produce the observed level of C IV emission.

1. INTRODUCTION

Highly ionized atoms in the general interstellar gas of the galactic disk were first detected through interstellar absorption line observations of O VI with the Copernicus satellite (Rogerson et al. 1973). Survey measurements by Jenkins (1978) of interstellar O VI absorption toward 72 stars demonstrated the general presence of O VI in the interstellar medium of the galactic disk. This research and parallel observational studies of the soft X-ray background (Williamson et al. 1974; McCammon et al. 1983; Marshall and Clark 1984) provided direct evidence for the existence of hot low density gas in the interstellar medium of the galactic disk.

The extension of the absorption line studies to the distant gas of the galactic halo required the launch of the International Ultraviolet Explorer (IUE) satellite in 1978. The first measures of highly ionized gas in the galactic halo were obtained with the IUE when it was used to record high resolution spectra of bright stars in the Large Magellanic Cloud (Savage and de Boer 1979). Those early spectra revealed the presence of absorption by Si IV and C IV in the galactic halo and have been followed by a number of surveys with IUE of highly ionized gas in the galactic disk and halo (Savage and de Boer 1981; Pettini and West 1982; Savage and Massa 1987).

The study of UV emission from highly ionized gas in the halo has progressed more slowly because of the intrinsic faintness of the emission. However, very important recent measurements were obtained by Martin and Bowyer (1990) using a low resolution nebular spectrograph flown on the space Shuttle in 1986.

In this review we discuss both absorption and emission line measurements with emphasis on the most recent work. Other recent reviews of halo gas and the hot ISM include those of Savage (1987, 1990), Jenkins (1987), and Spitzer (1990). The organization of this paper is as follows: The properties of the highly ionized atoms are briefly overviewed in §2. The results from interstellar absorption line spectroscopy are discussed in §3. Emission line studies are the subject of §4 and several current theoretical ideas concerning the origin of highly ionized gas in the halo are discussed in §5.

2. PROPERTIES OF THE HIGHLY IONIZED ATOMS

Table 1 lists information about the lithium-like resonance lines of the abundant highly ionized atoms which have transitions at wavelengths longward of the photoionization edge of H I at 912 Å. The table lists species, wavelengths, f values, ionization energy required to produce (IP_{x-1}) and to destroy (IP_x) the ions in eV. $T_{max}(K)$ is the temperature at which a particular ion reaches maximum abundance assuming conditions of coronal ionization according to calculations of Shull and Van Steenberg (1982). For Si IV the effects of charge exchange are included according to the study of Baliunas and Butler (1980).

The species listed in Table 1 are ordered according to increasing energy required for their production. The production of Si IV, C IV, S VI, N V and O VI require approximately 34, 48, 73, 78 and 114 eV , respectively. With increasing energy it becomes less likely that photoionization in warm (T ~ 10^4 K) gas is the source of ionization. The energy required to convert He^+

TABLE 1.
Resonance lines for various highly ionized interstellar species

species	λ(Å)	f	IP_{x-1}	IP_x	T_{max}(K)
Si IV	1402.77	0.262	33.5	45.1	0.6×10^5
Si IV	1393.76	0.528			
C IV	1550.76	0.097	47.9	64.5	1.0×10^5
C IV	1548.19	0.194			
S VI	944.52	0.210	72.5	88.0	2.0×10^5
S VI	933.38	0.426			
N V	1242.80	0.0757	77.5	97.9	1.8×10^5
N V	1238.81	0.152			
O VI	1037.63	0.0648	113.9	138.1	3.0×10^5
O VI	1031.95	0.130			

to He^{+2} (54 eV) is of particular importance in determining which species might be created by photoionization in warm gas versus electron collisional ionization in hot gas. Most hot stars containing helium have strong discontinunities at 54 eV, and are unlikely to be strong sources of radiation more energetic than 54 eV. Therefore, among the species listed in Table 1, S VI, N V and O VI stand out as the best for diagnostic information on the hot phase of the interstellar medium. These ions, if created under conditions of equilibrium collisional ionization, will probe gas in the temperature range from about 1×10^5 to 4×10^5 K . Unfortunately , the resonance lines of both S VI and O VI occur at wavelengths that are shortward of the 1150 Å limit of IUE and the spectrographs aboard the Hubble Space Telescope. Therefore, data on these important ions in halo gas will need to wait for future satellites such as Lyman/FUSE. In the wavelength region for which halo gas has been probed in absorption by IUE ($\lambda > 1150$Å), the most important hot gas diagnostic ion is N V.

Hot collisionally ionized gas will produce a UV emission line spectrum that will include most of the resonance lines listed in Table 1. In addition the gas will produce forbidden emission lines and lines due to transitions between excited electronic states. As an illustration of what might be expected, Table 2 provides a list of the line flux for the strongest lines produced by a cooling conductive interface with solar abundances during two times in its evolution. The photon fluxes are from Borkowsky, Balbus, and Fristrom(1990) and were calculated by following the full nonequilibrium time evolution for a medium evolving from 7.5×10^5 K with an initial number density of $n(H^+) = 2.3 \times 10^{-3}$ cm^{-3}. The strongest UV lines produced by the interface are C III λ977, C III] λ1909 , C IV $\lambda\lambda$1550, 1551 and O VI $\lambda\lambda$1032, 1038.

TABLE 2.
UV emission line fluxes produced during the nonequilibrium
cooling of gas in a conductive interface[a]

species	λ(Å)	(photons cm^{-2} s^{-1}) t = 2.8x10^5 yr	(photons cm^{-2} s^{-1}) t = 2.2x10^6 yr
He II	1640	101	13
C II	1335	143	255
C II]	2324-2328	237	478
C III	977	995	1107
C III]	1909	414	680
C IV	1548,1551	1110	525
N II]	2140,2142	39	71
N III	990	31	64
N IV]	1486	35	53
N V	1239,1243	174	106
O III]	1661,1667	46	123
O IV]	1402-1413	95	226
O V]	1218	183	269
O VI	1032,1038	606	684
Si III	1207	139	139
Si III]	1892	118	182
Si IV	1394,1403	104	100

a. From the nonequilibrium calculations of Borkowsky, Balbus, and Fristrom (1990).

3. UV ABSORPTION LINE STUDIES

Most of the UV absorption line observations of halo gas have been obtained with the short wavelength spectrograph aboard the IUE satellite which is capable of recording absorption produced by Si IV, C IV and N V in addition to the absorption arising in atoms found in moderately ionized and neutral gas (e.g. H I, O I, C II, Fe II, Si II, Mg II, S II, Si III, Al III, etc.). The earliest IUE measurements generally consisted of spectra based on single exposures and typically had rather low signal to noise ratios. In contrast for some of the most recent spectra, which are based on averaging four or more images, the data quality is quite good. The higher signal to noise is crucial for the detection of absorption by N V which is the best hot gas diagnostic in the IUE wavelength range. The principal results relating to IUE observations of highly ionized gas in the halo are found in: Savage and de Boer (1979, 1981), Pettini and West (1982), Fitzpatrick and Savage (1983), Savage and Massa (1987), and Savage , Massa and Sembach (1990). A summary of results follows:

Absorption by Si IV, C IV , N V , and O VI is found in the general interstellar medium of the galactic disk away from pronounced photoionized H II regions. The gas is patchy and exhibits a spread in the average line of sight density, n(ion) [cm^{-3}], of about ±3x. The average midplane density of this gas is listed in Table 3. Measures of the distribution of the gas away from the galactic plane have been inferred from studies of the shape of curves of N(ion)|sinb| versus |z| for measurements toward large numbers of stars in the disk and halo. The data are consistent with an exponential distribution of density with scale height, h ~ 3 kpc for Si IV and C IV and ~2 kpc for N V (Savage and Massa 1987). A similar estimate for O VI does not exist because the sample of Jenkins (1978) only included 3 stars with |z| > 1 kpc. These scale heights are substantially larger than the 0.5 kpc scale height for the extended component of H I (see Lockman 1984 and this volume) or for the 1 kpc scale height for the electrons (see Reynolds 1989 and this volume).

The total observed column densities for the high ions through the halo on one side of the galaxy, N_∞|sinb|, are listed in Table 3. Note that for N V, N_∞|sinb| ~ 3×10^{13} atoms cm^{-2}. If the N V were produced by equilibrium collisional ionization balanced by radiative recombination in gas having solar abundances near 2×10^5 K, the amount of N V implies the existence of ~3×10^{18} atoms cm^{-2} of hot gas. This number should be compared to the column densities of H I and electrons perpendicular to the galactic plane [N_∞(HI)|sinb| ~ 3.5×10^{20} atoms cm^{-2} (Lockman, Hobbs and Shull 1986) and N_∞(e)|sinb| ~ 1×10^{20} atoms cm^{-2} (Reynolds, this volume)]. Although the 2×10^5 K gas represents about only 1% of the total mass column density, it provides important information about those galactic processes that cause gas to be found a such large distances from the galactic plane.

TABLE 3.
Absorption line column densities and midplane densities for highly ionized halo gas.

Ion	midplane density[a] n_0(cm^{-3})	Observations of Halo Stars[a] <N_∞|sinb|>	Photoionized Halo[b] N_∞|sinb|	Cooling Fountain[c] N_∞|sinb|
Si IV	2×10^{-9}	~2×10^{13}	1.3×10^{14}	$(3.3-6.4)\times10^{12}$
C IV	7×10^{-9}	~1×10^{14}	1.2×10^{14}	$(4.3-7.9)\times10^{13}$
N V	3×10^{-9}	~3×10^{13}	7.3×10^{11}	$(2.8-3.6)\times10^{13}$
O VI	2×10^{-8}	>3×10^{13}	1.4×10^{11}	$(5.8-6.0)\times10^{14}$

a. Observations are from Savage and Massa (1987) for Si IV, C IV and N V and from Jenkins (1978) for O VI.
b. Photoionized halo model of Hartquist, Pettini and Tallant (1984). Assumes, n_0(H$^+$) = 0.003 atoms cm^{-3}, a gas scale height, h = 3 kpc and an estimate of the density of ionizing Lyman continuum photons from the QSO background radiation impinging on the outer region of the halo of n(γ) = 1×10^{-6} cm^{-3}.
c. Nonequilibrium cooling fountain calculation of Edgar and Chevalier (1986). Assumes a mass flow rate of 4 M$_\odot$/year to each side of the galactic plane. This flow produces a C IV emission intensity of I(C IV) = 890 photons cm^{-2} s^{-1}ster^{-1}. The observed intensity is I(C IV) = 5000 photons cm^{-2} s^{-1}ster^{-1}. Increasing the flow rate by a factor of 5.6 to produce agreement with the emission line data would result in a 5x overproduction of N V.

The evidence for a jump or substantial increase in n(Si IV) or n (C IV) near $|z| \sim 1$ kpc as proposed by Pettini and West (1982) and hinted at in the data of Savage and Massa (1987) is not strong. Such a jump is predicted by models involving the origin of Si IV and C IV through photoionization by extragalactic radiation. However, the existence of such a jump is inconsistent with the galactic rotational analysis of profiles to very distant stars at low latitudes ($|b| < 15°$) with $|z| \sim 1$ to 3 kpc (Savage, Massa and Sembach 1990). Jumps are difficut to see in plots of N(ion)$|\sin b|$ versus $|z|$ because of the degree of variability or patchyness of the absorption. Local galactic structure may also influence our view. The column densities of Pettini and West (1982) mostly refer to stars near the sun while the more recent measures include many stars at very large distances from the sun (up to 11 kpc) and thus involve substantial averaging over the galaxy. More work is needed to better define the true $|z|$ distribution of the highly ionized atoms.

In those cases where absorption line profiles for the highly ionized species have been well measured by obtaining multiple IUE spectra, the profiles for Si IV, C IV and NV are quite similar in shape and very different from profiles for the low ions or for species directly produced by photoionization such as Al III (see Savage et al 1989; Savage, Massa and Sembach 1990). This result suggests that Si IV, C IV and N V may be created in the same regions of interstellar space by a similar process.

The kinematical information provided by the high ionization lines is quite interesting. However, the interpretation of the results have been hampered by the general patchyness of the absorption, by the relatively low signal to noise ratios of individual IUE spectra, and by the complexity of the kinematics. In a sample of stars situated in the general direction of the south galactic pole, ($b < -50°$), Danly (1987) found the high ionization line profile structure for stars at $|z| \sim 1$ to 2 kpc to be relatively simple. For example, the lines of Si IV and C IV typically have widths (FWHM) of about 45 to 70 km s^{-1} and average LSR velocities $<v> \sim -10$ to $+20$ km s^{-1}. The view in this direction suggests a relatively quiescent region of space. The exact opposite behavior is found when looking at distant stars toward the north galactic pole, ($b > +50°$), where the profiles are complex, involving positive and negative velocities and multiple components with a preference for negative velocities. In this direction the profile widths (FWHM) vary from about 45 to 100 km s^{-1}. The direction of the north galactic pole suggests a disturbed region with substantial downflow. Clearly, local galactic structure is greatly influencing our view of the z motions of the gas.

The effect of differential galactic rotation on the appearance of interstellar absorption lines of the highly ionized gas in the halo was first studied by Savage and de Boer (1979, 1981) and by Savage and Massa (1987). The data from the 1987 study of 40 stars suggested that substantial deviations from corotation occur in highly ionized halo gas and that rotation may cease completely at about $|z| \sim 3$ kpc. However, much additional work will be needed to confirm that result. In a new observing program Savage, Massa and Sembach (1990) are obtaining high quality IUE line profiles for selected halo stars for which galactic rotation effects are expected to be large. In a detailed analysis of the profiles toward HD 163522, a B1 Ib halo star at a line of sight distance of 9 kpc and a z distance of -1.5 kpc in the direction l = 350° and b = -9°, it was found that the profiles of interstellar species known to have large scale heights (e.g. Si IV, C IV and N V) are significantly more affected by galactic rotation than the profiles of species having smaller scale heights (e.g. H I, Fe II, Mg II, S II, etc). Simple model calculations were performed to understand this result. In the case of gas with a small scale height, the sight line simply runs out of gas before the effects of galactic rotation become appreciable. Such studies demonstrate that once the galactic rotation curve is known for matter away from the galactic plane, it will be possible to infer from the observed line profiles the actual run of density with distance away from the galactic plane for a large number of interstellar species.

4. UV EMISSION LINE STUDIES

The study of UV emission lines from highly ionized atoms in halo gas was significantly advanced with the flight of the Berkeley EUV/FUV Nebular Grating Spectrophotometer on the space shuttle in January 1986 (Martin and Bowyer 1990). The 0.1 x 4.0 degree field of view of the spectrophotometer recorded spectra of the UV background in eight directions with 13 Å resolution from 1350 to 1900 Å. Emission from C IV λ1550 Å at a greater than 3σ level was found for 4 of 8 directions while O III] λ 1663,1667 emission (3σ) was found in 2 of 8 directions. In addition, the summed high latitude spectrum reveals the 3σ detection of the O IV/Si IV λ1401 and N III] λ1750 emission lines. The data are limited but suggest several trends. The C IV emission appears to be galactic pole brightened and anti-correlated with the column density of neutral hydrogen. The intensity of C IV emission toward the galactic polar directions is, I(C IV) ~ 5000 photons cm^{-2} s^{-1} ster^{-1}. Martin and Bowyer (1990) demonstrated that the emission is not likely due to detector fixed pattern noise, earth atmospheric airglow, zodiacal light, cool star chromospheres, or hot stars. They proposed that the emission is probably from collisionally ionized interstellar gas with T near 10^5 K and suggested that the gas is likely distant halo gas. A possible complication is that some of the emission may be associated with the negative velocity complex that extends over much of the north galactic pole (Wesselius and Fejes 1973; Danly 1989). In this case the emission process might be radiative shocks produced when the clouds strike the H I disk. To produce the observed features a shock velocity of about 100 km s^{-1} is required, while the complex has radial velocities of about 50 to 70 km s^{-1}. However, projection effects may play a role and the actual velocities may be substantially larger than the observed radial velocities. An inspection of Bell Lab 21 cm data reveals that strong H I emission between -40 and -70 km s^{-1} is seen for all the north galactic polar pointings of the Berkeley spectrophotometer. The data for the one southern pointing (near l = 216º and b = -39º) is in a direction for which the 21 cm emission is restricted to the velocity range from about -40 to +40 km s^{-1}. For this direction C IV emission is possibly detected with an intensity of 2700 (+1000,-1500) photons cm^{-2} s^{-1} ster^{-1}. It is important to obtain additional emission measurements for other directions.

If the observed emission is due to collisional excitation of C IV and OIII] in gas under conditions of equilibrium collisional ionization near 10^5 K, the required emission measure of 10^5 K gas is ~0.01 cm^{-6}pc, assuming cosmic abundances. The observed emission has important implications for the properties of the highly ionized gas found at large distances from the galactic plane. In collisionally excited gas the C IV emission line intensity, I(C IV) ∝ ∫ γ(T) n(C IV) n$_e$ dx, where γ(T) is the electron collision excitation rate coefficient. Absorption data gives the C IV column density, $N(C\ IV) = \int n(C\ IV)\ dx$, integrated out to the distance of the background star. If the emitting and absorbing gas are assumed to coincide in space, it is possible to estimate the electron density in the region containing C IV. Assuming $T \sim 10^5$ K and using $I(C\ IV) = 5000$ photons cm^{-2} s^{-1}ster^{-1} and $N_\infty(C\ IV) |\sin b| = 1\times10^{14}$ atoms cm^{-2}, the result is $n_e \sim 0.01$ cm^{-3}. This implies a thermal pressure $P/k = 2nT \sim 2000$ cm^{-3} K. Both the electron density and the pressure increase if the assumed gas temperature is increased. The filling factor, f, of the highly ionized gas can also be obtained if the path length occupied by the emitting and absorbing region is known. Taking the path length to be the exponential scale height estimated from the absorption line data (3 kpc) and assuming equilibrium ionization we obtain $f = 0.01$ for $T \sim 10^5$ K. The filling factor increases to 0.06 for the nonequilibrium ionization calculations performed by Martin and Bowyer(1990). In either case the gas phase sampled in the CIV absorption and emission lines seems to occupy only a small fraction of the sight line. The cooling time for gas with the conditions estimated above is short (about 3×10^5 yrs.). The assumption of collisional ionization equilibrium is only valid if the cooling time is much longer than the time required to establish ionization equilibrium which is about 5×10^6 yrs. Clearly nonequilibrium ionization effects must be considered (see §5).

The C IV emission results would seem to have important implications for the energy budget of the ISM if they are characteristic of the galaxy as a whole. Scaling from the C IV data, the implied emission line flux integrated over the entire spectrum is approximately 1×10^{-5} erg cm^{-2} s^{-1} for $T \sim 1\times10^5$ K. If this flux is typical of the entire galaxy, it implies a luminosity of $L \sim 4 \times 10^{40}$ erg s^{-1} or $\sim 13\%$ of the estimated injection power of supernovae (Martin and Bowyer 1990). In addition, the H I ionizing flux from this radiation is estimated to be 2×10^5 ionizations cm^{-2} s^{-1} which is about 20 times larger than the ionization rate estimated by Fransson and Chevalier (1985) to be associated with the QSO EUV background and about 20x smaller than the rate required to produce the diffuse Hα background (Reynolds 1984).

5. ORIGIN OF THE HIGHLY IONIZED ATOMS

In galactic fountain models, gas is found in the halo because of dynamic phenomena which result in the ejection of gas from the disk. In magnetic and cosmic ray supported halo models, the pressure of cosmic rays interacting with the galactic magnetic field is employed to support the gas found at large distances from the galactic plane. In the following, we examine how these two classes of theories are able to explain some of the existing observations of highly ionized gas in the halo.

5.1 Galactic Fountain Models

The term "galactic fountain" describes a process in which hot gas rises above the galactic plane before cooling and condensing to form clouds which then fall to the plane (Shapiro and Field 1976). The height to which hot gas will rise and the expected velocities of the condensations in the cooling gas depend on the temperature of the gas at the base of the fountain, the rate of cooling of the upflowing gas, and whether or not there are heating processes occurring at large z. Estimates of the expected velocities show that fountains driven by hot gas (1 to 2×10^6 K) can roughly reproduce the pattern of motions observed in the high velocity cloud phenomena seen in the neutral hydrogen 21 cm line (see Bregman 1980) while fountains driven by cooler gas (2 to 3×10^5 K) have velocities which are more compatible with the existing optical and ultraviolet data for gas in the low halo (Houck and Bregman 1990).

The filling factor of the hot gas in the galactic disk is currently very uncertain. If the filling factor is large, the flow of hot bouyant gas into the halo may occur quite freely. Gas with a temperature of 10^6 K has a thermal scale height of 6 kpc in the galactic gravitational field and it will attempt to assume a distribution in $|z|$ compatible with that scale height. If the filling factor of hot gas in the galactic disk is small, individual bubbles of hot gas may experience difficulty in pushing the cooler matter away. With the realization that the cooler gas of the Milky Way also has an extended component with a scale height of ~0.5 kpc for the neutral phase (Lockman, this volume) and ~1 kpc for the ionized phase (Reynolds this volume), it has become clear that the flow of hot gas into the halo may not occur as freely as previously imagined (Cox, this volume and references therein). Recent models for the flow of gas into the halo therefore have generally considered the phenomena occurring in regions of multiple supernovae which create superbubbles of hot gas that may have a chance of breaking through the cooler matter of the galactic disk (e.g. see MacLow and McCray 1988 ; Norman and Ikeuchi 1989). In this new type of model referred to by Norman and Ikeuchi (1989) as "chimney model," it is proposed that the connection between gas in the disk and halo is through "chimneys" which are the consequence of superbubbles bursting out of the galactic disk, forming collimated structures through which hot gas flows into the halo. Cox (this volume) has argued, however, that even this chimney model is likely to be suppressed by the high cosmic ray and magnetic pressures which are generally believed to be present in the halo. Models by Tomisaka (this volume) have tended to confirm this expectation but Cesarsky (also this volume) points out that the containment may be subject to instabilities.

The required circulation rate of gas from the disk into the halo and back can be estimated from the measurements of N V absorption and C IV emission. Cooling gas of a galactic fountain can explain the IUE observations of N V absorption (see Table 3) , provided the fountain flow rate is about 6×10^{-9} $M_\odot$ yr^{-1} pc^{-2} (Edgar and Chevalier 1986). This corresponds to a galactic flow rate of 4 $M_\odot$yr^{-1} to each side of the galactic plane. A flow rate 5 times larger is required to explain the C IV emission observations of Martin and Bowyer (1990). A difference this large appears to point toward a definite problem in the way the measurements have been interrelated or in the basic assumptions of the cooling fountain calculations. Perhaps 100 km s^{-1} shocks are enhancing the C IV emission over the north galactic polar region as discussed in §4. Another possiblility is that the inclusion of more details in the cooling gas models will modify the various line ratios. The work of Benjamin and Shapiro presented at this meeting suggests that more realistic nonequilibrium cooling models including the effects of photoionization provide better representations of the measurements.

5.2 Magnetic and Cosmic Ray Supported Galactic Halo Models

Another explanation for the support of gas at large distances away from the galactic plane is found in those models which involve pressure support from the galactic magnetic field and cosmic rays. Such models, have been proposed by Hartquist, Pettini and Tallant (1984), Chevalier and Fransson (1984), Hartquist and Morfill (1986) and Bloemen (1987). In one model the B field is parallel to the galactic plane and the support is via magnetic pressure which is affected by the cosmic ray pressure (see Bloemen 1987). In the other models the B field is perpendicular to the galactic plane and the pressure support is from the streaming motions of cosmic rays along the B field (see Hartquist and Morfill 1986).

In some of the models, it is proposed that the ionization of the gas, and in particular the production of the highly ionized species, is by photoionization from radiation produced by hot galactic stars (Bregman and Harringtion 1986) or from the extragalactic EUV background (York 1982; Hartquist, Pettini and Tallant 1984; Fransson and Chevalier 1985). Photoionized halo models have been successful at providing a possible explanation for the observed amounts of Si IV and C IV in gas at large distances from the galactic plane (see Table 3). However, the photoionization models have not been successful in explaining the observed amount of N V which appears to require the existence of collisionally ionized gas near 200,000 K.

5.3 Composite Models

In a recent extension of the photoionization models by Ito and Ikeuchi (1988), it was found necessary to include three gas phases: 1) A neutral gas phase with T $<10^4$ K. 2) A photoionized gas phase containing C IV and Si IV with T near 10^4 K. 3) A hot collisionally ionized gas phase with T $\sim 2 \times 10^5$ K to explain the existence of N V. The support of gas in this model was proposed to be a galactic fountain driven by superbubble phenomena. However, another possibility is that the gas is supported by diffusing cosmic rays and that the high temperatures required to produce N V arise from cosmic ray heating of the gas (Hartquist and Morfill 1986). If the processes producing Si IV, C IV and N V were as different as those proposed in such models it is difficult to understand why these three ions have such similar absorption line profile shapes.

ACKNOWLEDGEMENTS

Helpful comments about draft versions of this manuscript were provided by Donald Cox and Kenneth Sembach. The author acknowledges support for research on galactic halo gas from NASA grant NAG-186.

REFERENCES

Baliunas, S.L., and Butler, S.E. (1980) *Ap.J. (Letters)*, **235**,L45.
Bloemen, J.B.G.M. (1987) *Ap.J.*, **322**,694.
Borkowsky,K.J., Balbus, S.A., and Fristrom, C.C. (1990) *Ap.J.*, **355**, 501.
Bregman,J.N. (1980) *Ap.J.*, **236**,577.
Bregman, J.N., and Harrington, P.J. (1986) *Ap.J.*, **309**,833.
Chevalier, R. A., and Fransson, C. (1984) *Ap.J. (Letters)*, **279**,L43.
Danly, L. (1987) Ph.D. Thesis, University of Wisconsin-Madison.
_____. (1989) *Ap.J.*, **342**, 785.
Edgar, R.J., and Chevalier, R.A. (1986) *Ap.J. (Letters)*, **310**, L27.
Fitzpatrick, E.L., and Savage, B.D. (1983) *Ap.J.*, **267**,93.
Fransson, C., and Chevalier, R.A. (1985) *Ap.J.*, **296**, 35.
Hartquist, T.W., and Morfill, G.E. (1986) *Ap.J.*, **311**, 518.
Hartquist, T.W., Pettini, M., and Tallant, A. (1984) *Ap.J.*, **276**, 519.
Houck, J.C., and Bregman, J.N. (1990) *Ap.J.*, **352**, 506.
Ito,M., and Ikeuchi, S. (1988) *Publ. Astron. Soc. Japan*, **40**, 403.
Jenkins, E. B. (1978) *Ap.J.* , **220**, 107.
_____. (1987) in *Exploring the Universe with the IUE Satellite*,
    (Dordrecht:D.Reidel.Pub.Co.),p.531.
Lockman, F. J. (1984) *Ap.J.*, **283**, 90.
Lockman, F.J., Hobbs, L.M., and Shull,M. (1986) *Ap.J.* , **301**, 380.
Mac Low,M.-M., and McCray , R.C. (1988) *Ap.J.*, **324**,776.
Martin, C., and Bowyer, S. (1990) *Ap.J.*, **350**, 242.
Marshall,F.J., and Clark,G.W.(1984) *Ap.J.* , **287**, 633.
McCammon, D., Burrows, D.N., Sanders, W.T., and Kraushaar, W.L. (1983)
    Ap.J., **269**, 107.
Norman, C.A., and Ikeuchi,S. (1989) *Ap.J.* , **345**, 372.
Pettini, M., and West, K.A. (1982) *Ap.J.*, **260** , 561.
Reynolds, R.J. (1989) *Ap.J.(Letters)*, **339**, L29.
_____.(1984) *Ap.J.*, **282**, 191.
Rogerson, J. B.,York, D.G., Drake, J.F., Jenkins, E.B., Morton, D.C., and
    Spitzer, L. (1973) *Ap.J. (Letters)*, **279**,L43.
Savage, B.D. (1987) in *Interstellar Processes*, eds. D..Hollenbach and
    H.A.Thronson,Jr., (Dordrecht:D.Reidel Pub.Co.), p.123 .
Savage, B.D. (1990) in *Evolution of the Interstellar Medium* , ed, L. Blitz,
    (PASP Conference Proceedings), in press.
Savage, B.D., and deBoer, K.S. (1979) *Ap.J. (Letters)* , **230**, L77.
_____. (1981) *Ap.J.* , **243**, 460.
Savage, B.D.,Jenkins, E.B., Joseph, C. L., and de Boer, K.S. (1989) *Ap.J.* , **345**,
    393.
Savage, B.D., and Massa, D. (1987) *Ap.J.* , **314**, 380.
Savage, B.D., Massa, D., and Sembach, K. (1990) *Ap.J.* , **355** ,114.
Shapiro, P.R., and Field, G.B. (1976) *Ap.J.* , **205**, 762.
Shull, M.J, and van Steenberg, M. (1982) *Ap.J. (Supplment)*, **48**, 95.
Spitzer, L. (1990) *Ann. Rev. Astr. Ap.*, *(in press)*.
Wesselius, P.R., and Fejes,I. (1973) *Astr.Ap.*, **24**,15.
Williamson, F.O., Sanders, W.T., Kraushaar, W.L., McCammon, D.,
    Borken,R., and Bunner, A.N. (1974) *Ap.J.(Letters)* , **193**, L133.
York, D.G. (1982) *Ann. Rev. Astr. Ap.*, **20**, 221.

HOT GAS IN THE DISK, HALO, AND DISK-HALO INTERACTION

D. P. COX
Department of Physics
University of Wisconsin-Madison
1150 University Ave.
Madison, Wisconsin 53706
U.S.A.

ABSTRACT. The interstellar medium and its hot gas component are very different from common conception.

1. INTRODUCTION

For a wide variety of reasons it has seemed likely that hot gas might be common in the disk of the Galaxy (e.g. Cox and Smith, 1974), possibly in the form of a pervasive "coronal" phase (e.g. McKee and Ostriker, 1977). Similarly, hot gas might be expected in the galactic halo (e.g. Spitzer, 1956), possibly deriving from infall (e.g. Sciama, 1972), from a fountain rising out of the disk (e.g. Shapiro and Field, 1976), or from the blowout of OB association bubbles, or "chimneys" (see review by Tenorio-Tagle and Bodenheimer, 1988; Norman and Ikeuchi, 1989).
   This paper presents a critical review of some of the evidence for such hot gas, and the corresponding disk-halo connection mechanisms.

2. THE SOFT X-RAY BACKGROUND

Attempts to understand the soft X-ray background led to early rejection of many source mechanisms, eventually leaving "thermal" emission of a "collisional" plasma at a temperature $T \approx 10^6$ K. (As yet we have no confirming spectra.) Comparison between simple models and the observed surface brightness suggest that a thermal pressure $p/k \approx 10^4$ cm^{-3} K and path ≈ 100 pc are sufficient (see McCammon et al., 1983; McCammon and Sanders 1990 for reviews).
   The apparent observation that the Solar System is immersed in a large region of temperature 10^6 K was sufficient to suggest that gas of this temperature could be common is the interstellar medium (ISM) and led to consideration of supernova generation of the hot gas (Cox and Smith, 1974). Since this early work, however, much has been learned about details. These details have rather consistently shown that most of the X-rays observed below 1/4 keV arise in a single Local

Bubble, nearer than $N \sim 10^{19}$ cm^{-2} of absorbing material. This bubble corresponds to a local hole in the HI distribution and, to everyone's surprise and no one's satisfaction, much of the observed anticorrelation between the X-ray surface brightness and N_{HI} appears to derive not from absorption, but from the scale of the hole being comparable to that of the scale height of the narrow HI components (e.g. Snowden, et al. 1990).

Two points appear worth stressing. The origin of the Local Bubble is not well understood; several options have been discussed (e.g. Cox and Reynolds, 1987), but none are particularly satisfying. More relevant to this discussion, the soft X-ray background does not directly imply that 10^6 K gas is common in the ISM, only that it is present right around us. More particularly, there seem to be a few directions in which absorption is low for quite some distance, without the X-ray surface brightness being large in those directions.

The limit placed on X-ray emission from the galactic halo (or beyond) is rather severe: The total emission measure at 10^6 K is no greater than that in the Local Bubble, which is to say that over the 10 kpc scale of the galactic potential, no more emission is present than in the first $\sim$ 100 pc. The net surface brightness has been estimated at $\lesssim$ 1% of the available supernova power.

The X-ray background is much less stringent in its restrictions on material at lower temperature (e.g. 3×10^5 K) or at higher temperature (e.g. 3×10^6 K). In fact, the slightly higher energy M band ($\sim$ 0.5 to 1 keV) is not yet understood. Away from bright features associated with specific bubbles, about half of the M band appears to derive from a combination of extragalactic and stellar sources, while the rest is sufficiently isotropic that it requires extra components from both inside and outside the Galaxy (c.f. McCammon and Sanders, 1990). The emission coefficient in this band would be highest at $T \simeq 3 \times 10^6$ K, but even at that temperature is sufficiently small that very high volume occupation of the plane would be possible at typical pressure estimates. A corresponding peculiarity is that even though dynamically significant amounts of very hot gas are allowed and possibly even suggested by the M band data, both in the disk and in the halo, the total radiated power of this material is again only of order 1% of the available supernova power. If there is widespread material at this temperature, the supernova energy deposited within it almost certainly is thermally conducted to colder regions where it is radiated, probably in the EUV. (Cox, 1986, argued that the alternative of a powerful galactic wind would shorten the cosmic ray escape time too severely.) The necessary conductivity is available, even if saturated, but would have to occur only on large scales; conduction to an embedded cloud population could lower the temperature as much as a factor of $\sim$ 10, bringing on a McKee and Ostriker scenario.

In short, the soft X-ray background indicates that our own environment is 10^6 K gas, but that such gas may be unusual. Much of the medium energy X-ray background is unexplained and could be interpreted as indicating large amounts of 3×10^6 K gas in and around the Galaxy, though not radiating the available power. Models of the ISM with large filling factor of such high temperature gas are so far in very short supply, but would undoubtedly become very common if a spectrum of the low intensity M band were to indicate a diffuse thermal origin. (The only example I know of so far is a picture explored by Sanders, et al., 1983.)

2. X-RAY OBSERVATIONS OF OTHER GALAXIES

The halo of our galaxy is radiating less than about 1% of the available supernova power at temperatures of $1-3 \times 10^6$ K. Similar limits have been placed on the emission from two edge-on spirals (NGC 3628 and 4244 by Bregman and Glassgold 1982) and the face-on M101 (McCammon and Sanders, 1984). In the latter galaxy, at least, there is no shortage of evidence for supernovae, OB associations, and large holes in the ISM. Evidently such activity at a moderate rate does not lead to a halo or fountain radiating the supernova power at high temperature.

As in the Galaxy, this probably does not rule out the presence of a dynamically significant amount of very hot gas, but requires that energy dissipation occur at lower temperatures.

3. HIGH IONIZATION STAGES IN THE DISK

High ionization stages have been found in UV absorption studies of the ISM. The mean densities have been estimated to be:

$\bar{n}$(O VI) ~ 2.8×10^{-8} cm^{-3}
$\bar{n}$(N V) ~ 3×10^{-9} cm^{-3}
$\bar{n}$(C IV) ~ 7×10^{-9} cm^{-3}
$\bar{n}$(Si IV) ~ 2×10^{-9} cm^{-3}

(Jenkins, 1978 for O VI, Savage and Massa, 1987). These ions are thought to indicate the presence of gas with $T \sim 10^5$ to 3×10^5K rather commonly in the lower disk, although the actual volume filling factor of the required gas is low (~ 1%).

Historically, these ions have been used extensively as evidence for the presence of substantial amounts of hot gas in interstellar space, the ions arising in boundary layers with cooler gas (e.g. McKee and Ostriker, 1977). The fact is that the ions also place constraints on the amount of hot gas present. For example, if the hot gas temperature were typically ~ 3×10^5 K occupying most of interstellar space, the implied O VI density contributed by the hot gas itself is one to two orders of magnitude higher than the observations. (This problem disappears at $T \gtrsim 10^6$ K where the oxygen is more highly ionized.) Thus it has long been common practice to suppose that turbulence in the hot gas hides the large intrinsic component of O VI, making the line too broad for the Copernicus satellite to have seen it and leaving only the weak boundary regions visible.

Recently, however, Slavin and Cox (1990) and Cox and Slavin (1990) have proposed an alternate source for these ions. By showing that supernova disruption of the interstellar warm gas was previously overestimated by roughly a factor of 30, they conclude that it is reasonable to investigate the long term evolution of SNR generated bubbles in an environment with typical intercloud density $n \sim 0.2$ cm^{-3}. Including a significant nonthermal pressure term, corresponding to B $\approx$ 5µG, they find remnants disturb the medium very little, creating bubbles of fairly hot gas ($\sim 4 \times 10^5$ K) lasting roughly 5×10^6 years. The collection of these bubbles is found to provide the observed mean densities of O VI, N V, and C IV (and maybe Si IV as well when photoionization is included), while occupying only about 10% of interstellar space.

With this model, the high ions in the disk no longer constitute strong evidence for there being a pervasive distribution of hot gas occupying most of interstellar space. A modest occupation by isolated bubbles is entirely sufficient.

(The reasons for the earlier overestimate of the porosity are that: a high SN rate was assumed, SNR bubble evolution was approximated to be adiabatic between shell formation and pressure equilibration with the surroundings -- a faulty assumption dating back to Cox, 1972 -- and the high interstellar pressure was represented only by its nearly inconsequential thermal component.)

4. HIGH ION STAGES IN THE HALO

Savage and Massa (1987) have estimated the scale height of the high ions to be roughly 3 kpc, a number which depends somewhat on whether the distribution at high z is directly related to the ions found in the plane. There has been some indication, in fact, that the high ion density has a rather sudden increase a few hundred parsecs off the plane. The evidence for the latter is controversial (see Savage, this volume) but it remains distinctly possible that much of the ion content in the plane has a separate origin from that found at high z.

Apart from the M band possibilities, the high ions found both in absorption and emission (again, Savage, this volume for review and references) at high z are the only direct evidence we have that there is high temperature gas in the halo. And once again, the total amount of that observed hot gas is small (Savage quotes a 3% filling factor in the 3 kpc path, for total of 100 pc along an average sight line out of the Galaxy.) Only if there are much larger amounts of much higher temperature gas is this material dynamically interesting.

In contrast to the X-ray results, however, the limited information we presently have on C IV emission (from Martin and Bowyer, 1990) implies that it is possible that this small quantity of fairly hot gas could be radiating at a rate comparable to the SNR input power in the disk. (An alternative not yet fully explored is that a significant amount of the C IV emission derives from the boundary of the Local Bubble or a few SNR bubbles along the line of sight.)

5. IMPLICATIONS OF THE THICK DISK

Evidence reviewed elsewhere (e.g. Boulares and Cox, 1990, Cox and Slavin 1990, Lockman, this volume, Reynolds, this volume) indicates that it is inappropriate to think of the disk of the Galaxy having a scale height of roughly 100 pc, beyond which lies the halo. Instead, measurements of neutral and ionized gas, cosmic rays, and magnetic field all show that the Galaxy has a much thicker disk, reaching to roughly 2 kpc above the plane. Within this thick disk, the dense cloud population is like a condensate, low in the gravitational potential.

By evaluating the weight of the material in the disk, one learns that the midplane pressure has been seriously underestimated. The value implied (Boulares and Cox find $p/k \gtrsim 25,000$ cm^{-3} K) is roughly consistent with recent upward trends in measurements of the interstellar magnetic field, the cosmic ray pressure, and the dynamical pressure of the warm HI.

This change in perspective has severe consequences for the evolution of large bubbles blown by OB associations. They continue to encounter appreciable matter densities a few hundred parsecs off midplane, and substantial pressures for a much greater distance beyond that. The smaller of these bubbles should be even smaller than previously calculated, while the larger ones will avoid breakout and could grow to larger size holes in the plane. In addition, the shells will be less compressed, owing to the large magnetic pressure, and will rebound sooner to erase the HI evidence for the bubble's having been there. A pioneering numerical model showing some of these effects is presented by Tomisaka in this volume.

Having an ISM disk whose scale height (for the warm components) is greater than the (probable) scale height for diffuse supernova explosions (i.e. those not clustered in OB associations) also alters one's thinking on whether supernova generation of diffuse hot gas can sustain low density channels through the disk to the halo. Hence, both common types of fountain models (general disk source, superbubble blowout source) are made much less likely by the presence of this thick disk.

6. DISCUSSION

It is possible that within the lower disk, hot gas is found primarily in individual bubbles generated by supernovae and in the larger but confined collective bubbles generated by OB associations. Estimates suggest these two types occupy $\approx 10\%$ and $\sim 20\%$ respectively of the midplane volume in the Solar neighborhood. That gas alone is sufficient to explain the high ion content of the midplane, and is consistent with the constraints implied by the soft x-ray background.

In this picture it is unlikely that either of the popular types of galactic fountain actually occurs, consistent with the weak halo emission in X-rays. For that reason an alternative source is needed for the high ion stages found in the outer parts and boundary of the thick disk. Two such alternatives that appeal to me rely on local high z energy dissipation in a "chromospheric" layer on the outer boundary of disk (Sciama, 1972; Hartquist 1983). The energy could be carried to that regime by thermal conduction from a much hotter halo, by large amplitude Alfven waves in the warm disk, or possibly even by thermal conduction along hot channels following flux tubes out of the lower disk.

In any case, nothing is so certain as we might like to have believed. The thick disk and its high pressure complicate all fountain/chimney models enormously. There is no airtight evidence for a quasicontinuous distribution of hot gas in the disk. My best guess at present for the high z high stage ions is that they arise from shocks in a region where nonlinear Alfven waves produce highly supersonic motions. Yet, at the same time that I am suggesting that most hot gas evidence can be explained in terms of discrete events rather than global behavior, I have a wary eye on the M band map and the nearly unexplored possibilites of the transmegakelvin regime.

ACKNOWLEDGEMENTS

I would like to thank Hans Bloemen and the rest of the S.O.C. for organizing this wonderful meeting and inviting me; also the other participants of the meeting for the quality of their presentations, discussions, and attitudes. Ron Reynolds provided a helpful reading of the manuscript. This work was supported in part by the National Aeronautic and Space Administration under grant number NAG5-629.

REFERENCES

Boulares, A. and Cox, D. P. (1990) Ap. J. in press (Dec. 20)
Bregman, J. N., and Glassgold, A. E. (1982) Ap. J. 263, 564
Cox, D. P. (1972) Ap. J. 178, 159
Cox, D. P. (1986) in Proc. of the Workshop on Halos of Galaxies, eds. J. N. Bregman and F. J. Lockman, NRAO Workshop 12
Cox, D. P. and Reynolds, R. J. (1987) Ann. Rev. Astron. and Astrophys. 25, 303
Cox, D. P. and Slavin J. D. (1990) Ap. J. submitted
Cox, D. P. and Smith, B. W. (1974) Ap. J. (Letters) 189, L105
Hartquist, T. W. (1983) M.N.R.A.S. 204, 997
Jenkins, E. B. (1978) Ap. J. 220 107
Martin, C. and Bowyer. (1990) Ap. J. 350, 242
McCammon, D., Burrows, D. N. Sanders, W. T. and Kraushaar, W. L. (1983), Ap. J., 269, 107
McCammon, D. and Sanders, W. T. (1984) Ap. J. 287, 167
McCammon, D. and Sanders, W. T. (1990) Ann. Rev. Astron. and Astrophys. in press
McKee, C. F. and Ostriker, J. P. (1977) Ap. J. 218, 148
Norman, C. A. and Ikeuchi, S. (1989) Ap. J. 345, 372
Sanders, W. T., Burrows, D. N., McCammon, D., and Kraushaar, W. L. (1983) in Supernova Remnants and their X-ray Emission, eds. J. Danziger and P. Gorenstein (Dordrecht: Reidel), p. 361
Savage, B. D. and Massa, D. (1987) Ap. J. 314, 380
Sciama, D. W. (1972) Nature 240, 456
Shapiro, P. R. and Field, G. B. (1976) Ap. J. 205, 762
Slavin, J. D. and Cox, D. P. (1990) Ap. J. submitted
Snowden, S. L., Cox, D. P., McCammon, D., and Sanders, W. T. (1990) Ap. J. 354, 211
Spitzer, L., Jr. (1956) Ap. J. 124, 40
Tenorio-Tagle, G. and Bodenheimer, P. (1988) Ann. Rev. Astron. and Astrophys 26, 145

THE HIGH-LATITUDE SKY AT IR, OPTICAL, AND UV WAVELENGTHS

F. X. DÉSERT
DEMIRM, Observatoire de Meudon
92195 Meudon Cedex, France

ABSTRACT. We review the current data on the not-so-dark sky covering infrared, visible and ultraviolet wavelengths. Here, we are mainly concerned with the emission from the interstellar gas and dust above and below the galactic plane. Zodiacal light is not discussed in detail and emission from unresolved stars is briefly mentioned. Recent improvements in these studies have been made with the use of new satellite UV data, the use of high-performance CCD in the visible spectrum and extensive analyses from the *Infrared Astronomical Satellite* (IRAS). We show that cirrus clouds which subtend a large solid angle at high galactic latitudes are made of neutral gas and dust, are within a few hundred parsecs of the Sun, and are almost optically thin up to UV wavelengths. The brightness of these clouds, expressed as $\nu I_\nu = \lambda I_\lambda$, is estimated to be within 10^{-8} and $10^{-7}\,\mathrm{W\,m^{-2}\,sr^{-1}}$ at almost all wavelengths from $\lambda = 0.1$ to $300\,\mu\mathrm{m}$ and peaks at $150\,\mu\mathrm{m}$, for a typical column density of $3 \times 10^{20}\,\mathrm{H\,cm^{-2}}$. They may yield the fundamental limitation to all extragalactic and halo studies.

1. INTRODUCTION

The high-galactic latitude sky has long been the domain for extragalactic studies. However, due to improvements in the sensitivity of the astronomical instruments, it has become possible to investigate the nature of the thin layer of galactic matter that separates us from intergalactic space (see this entire conference). Due to the proximity of the high latitude matter, one can observe phenomena with a better spatial resolution and less confusion problems than those usually occuring in studies of the galactic plane. Therefore, by generalisation, one can hope to gain a better physical picture of the whole galactic matter. In the following, we will review the present knowledge of the high latitude matter that has been gained so far and we will be mainly concerned with high-latitude interstellar (IS) matter: gas and dust, as revealed by its diffuse emission from IR to UV wavelengths. Sky specific brightnesses will be expressed as $\nu I_\nu = \lambda I_\lambda$ (*i.e.* per unit log of frequency or wavelength) in units of $1\,\mathrm{nW\,m^{-2}\,sr^{-1}} = 10^{-9}\,\mathrm{W\,m^{-2}\,sr^{-1}}$. In case of an unresolved line, we will indicate some adopted filter band $\Delta\nu$ and an equivalent brightness for the integrated line: νI_ν such that $(\nu I_\nu)\Delta(\log\nu) = \int d\nu I_\nu$.

2. THE EMISSION OF HIGH-LATITUDE GAS

The 21 cm emission from the galactic HI gas has been thoroughly discussed in this conference (see the articles in these proceedings by Burton, Lockman, Mirabel, Verschuur, and Wakker) as well as the 3 mm emission from the CO-molecular gas (Blitz), the UV emission lines from highly ionized gas (Savage), and the $0.6563\,\mu\mathrm{m} = H_\alpha$ emission from the H^+ gas (Reynolds). Here, for completeness, we would like to point out other emission processes from the gas ocurring at UV and IR wavelengths, mainly the H_2 fluorescence and the gas cooling via far infrared lines; 2-photon processes are unimportant: Deharveng, Joubert and Barge 1982.

2.1 H_2 emission

H_2 molecules cannot be directly photodissociated by the UV photons of the interstellar radiation field (ISRF). From the absorption of a UV photon (of wavelength $\lambda \geq 0.11\,\mu\mathrm{m}$) the hydrogen molecule is excited to the Werner or Lyman energy levels (Duley and Williams 1980). The molecule returns to the ground electronic level on a vibrationnally excited level $v \geq 0$ by the emission of a UV photon. It is only when $v \geq 15$ that the molecule is dissociated. On average 9 photons will be emitted before the H_2 dissociation occurs. The deexcitation UV photons have specific wavelengths within the Werner (around $0.10\,\mu\mathrm{m}$) or Lyman (around $0.16\,\mu\mathrm{m}$) bands. The emission of these photons from molecular hydrogen in the diffuse IS medium (Jura 1974) was predicted by Duley and Williams (1980). Suspected by Jakobsen (1982), the Lyman band photons have just been clearly detected at high latitude for the first time by Martin, Hurwitz and Bowyer (1990, hereafter MHB). The presence of molecular hydrogen is proved in direction where CO has previously been detected. Moreover, "halos" of H_2 fluorescence are shown to exist even where CO is not detectable. Other regions do not show any detectable H_2 emission. MHB suggest that the molecular halos are the result of photodissociation of dense molecular clumps. From the observed intensity of the Lyman band I_0 one can infer the speed at which H_2 is dissociated:

$$v_d = 8 \times 10^{-3}\,\mathrm{pc/Myr}\,(n_{H2}/50\,\mathrm{cm}^{-3})\,(I_0/3 \times 10^4\,\mathrm{phot.\,cm}^{-2}\,\mathrm{s}^{-1}\,\mathrm{sr}^{-1})$$

which implies a relatively slow process: a $10 M_\odot$ cloud of H_2 density $50\,\mathrm{cm}^{-3}$ will be destroyed in typically 100 Myr. The same Werner and Lyman photons have recently been observed in one reflection nebula by Witt et al. (1989) who concluded that it was UV pumping rather than shocks that can produce this fluorescence. Finally, let us remark that in reflection nebulae, an *infrared* $\simeq 2\,\mu\mathrm{m}$ rotation-vibration spectrum of lines from molecular hydrogen has been observed (Gatley et al. 1987, Sellgren 1986) that correspond to the end of the cascade to the ground state (Sternberg 1989, Sternberg and Dalgarno 1989). Hence one could try to observe this infrared counterpart of molecular fluorescence in the IS medium. However, from the clouds observed by MHB one can deduce an equivalent brightness of about $0.6\,\text{nW}\,\text{m}^{-2}\,\text{sr}^{-1}$ in lines covering 1 to $3\,\mu\text{m}$ which is quite weak.

2.2 Gas cooling emission

The neutral IS medium cools mainly via two fine-structure lines of oxygen and ionized carbon atoms at $63\,\mu\text{m}$ and $158\,\mu\text{m}$ with about the same brightness. It is far from certain that the oxygen line significantly contributes to the $60\,\mu\text{m}$ IRAS emission from the diffuse high-latitude medium (cirrus clouds) (see the propositions by Harwitt et al. 1986 and Stark 1990 and counter-arguments by Terebey and Fich 1986 and Verstraete 1990). This is a hot topic that should receive a clear answer within the next few years with the Kuiper Airborne Observatory, the *Infrared Space Observatory* (ISO) and maybe earlier with the now in orbit *Cosmic Background Explorer* (COBE). The FIRAS instrument on COBE covers a range of wavelengths from $100\,\mu\text{m}$ to $1\,\text{cm}$ with a spectral resolution of few percent and a beam of 7 deg. and has a sensitivity in νI_ν of about $1\,\text{nW}\,\text{m}^{-2}\,\text{sr}^{-1}$. Table 1 and Figure 1 give some estimated values for the gas emission.

3. THE DUST-LIGHT INTERACTION AT HIGH LATITUDES

The IS dust acts as a veil for the light emitted by stars. Studies of the way dust interacts with light allow us to deduce the properties of the incident stellar radiation field but also of the intervening matter. We can assume that the high-latitude dust is almost optically thin up to UV wavelengths since a typical cloud column density of $N_H = 3 \times 10^{20}\,\text{H}\,\text{cm}^{-2}$, which we will use as an example in the following, corresponds to a visible extinction $A_V \simeq 0.16\,\text{mag}$ and an extinction of about $0.70\,\text{mag}$ at $0.1\,\mu\text{m}$ (Savage and Mathis 1979). While in the seventies it was thought that dust could be observed only in the visible (scattering) and far infrared (FIR thermal emission) we would like to show that new observations lead us to a picture where the light reprocessed by dust occurs at almost all wavelengths from submillimeter to UV wavelengths. Table 1 and Figure 1 illustrate this point of view and are extensively discussed in the following for what concern: 1) the UV and visible scattering, 2) the optical luminescence, and 3) the IR emission.

3.1 UV and visible dust scattering in a high-latitude cloud

We consider first an example of UV observations taken by the *D2B* satellite (Joubert et al. 1983) and reanalysed by Pérault, Lequeux, Hanus and Joubert (PLHJ 1990). PLHJ have attempted to remove some striping that was present in the broad band data at $0.169\,\mu\text{m}$ by using the North Ecliptic Pole as a reference and correcting for gain variations with time. The effective beam is about $2° \times 2°$. One can compute the average at constant latitude ($|b| \geq 30°$) of the UV sky brightness as a function of $1/sin(b)$. This can then be compared with the same average (done

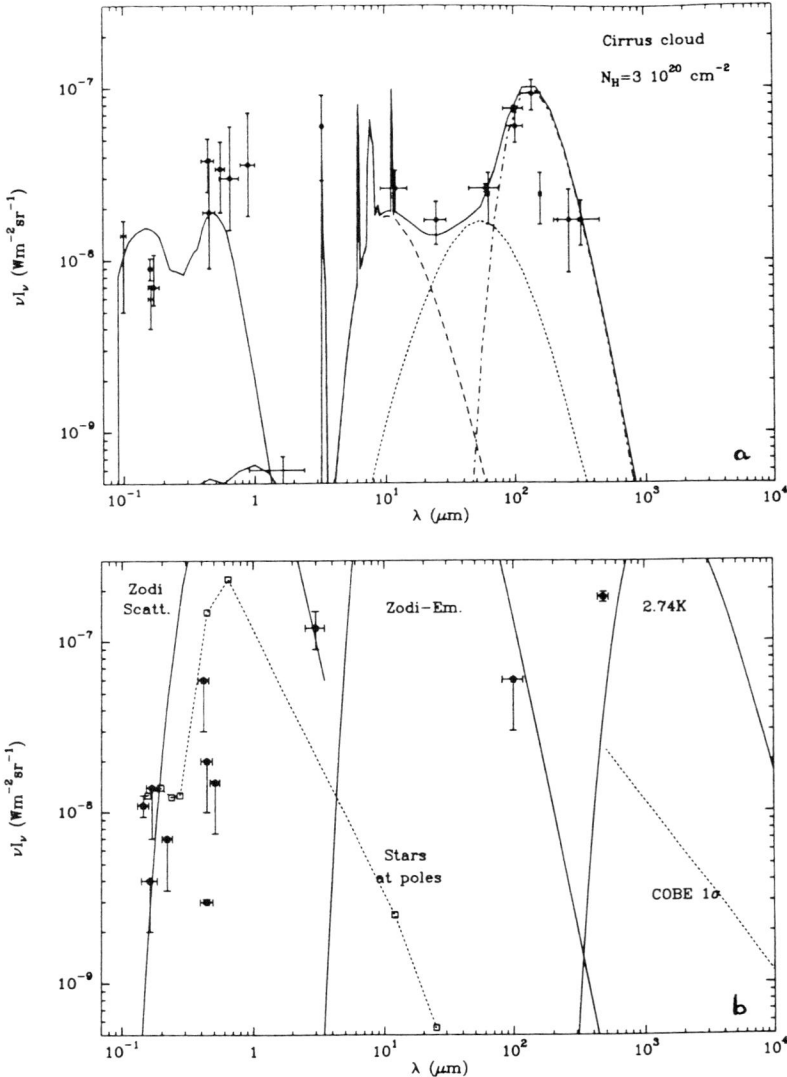

Figure 1. (a) Brightness of a neutral IS cloud (see Table 1). Continuous line is from a three component dust model (PAHs: long-dashed, very small grains: short-dashed, big grains: alternate-dashed) by Désert, Boulanger and Puget (1990). The visible and UV part is normalised with GT observation at B_J band. Observations of gas emission are noted with error bars without a central dot whereas dust emission observations have a central dot. (b) Backgrounds and foregrounds at the galactic poles in the same units and scales (see Section 4)

on exactly the same areas in the sky) of the 100 µm *IRAS* map made by Boulanger and Pérault (hereafter BP 1988). They both roughly follow a cosecant law that is typical of material distributed in a plane-parallel geometry. Whereas it is known (see BP and Section 3.3) that the FIR emission is associated to dust in the HI neutral medium, it is not enough that the UV brightness follows a cosecant law to be certain that the UV comes from dust scattering of the ISRF. Integrated light from faint UV stars could produce this effect as well. Therefore it is instructive to subtract the average cosecant law from both UV and IR data in order to see if "clumps" of emission correlate or not. The result is shown in Figure 2 (here for the North galactic pole). The map of the excesses are shown in Figure 3. PLHJ conclude that UV scattering by dust of the ISRF is at the origin of the UV brightness of the sky because Figure 2 shows a positive correlation and the deduced ratio of UV to IR brightness is the same as the ratio of the cosecant laws. Note that some UV excesses in Figure 2 and 3 are probably due to scattering in the telescope of the UV light from single stars and may not be real.

Other studies of the UV sky brightness by Jakobsen et al. (1984 and 1987), Morgan et al. (1978), Hurwitz et al. (1989) and Fix et al. (1989) have also revealed a correlation with HI gas. However, Murthy et al. (1989 and 1990) do not find such a correlation and the slope of the previously mentioned correlations do not agree with each other: *e.g.* PLHJ find a brightness at 0.169 µm of about $\nu I_\nu \simeq 7\,\mathrm{nW\,m^{-2}\,sr^{-1}}$ for $N_H = 3 \times 10^{20}\,\mathrm{H\,cm^{-2}}$ whereas Jakobsen et al. (1987) find at 0.214 µm a brightness of $36\,\mathrm{nW\,m^{-2}\,sr^{-1}}$. The standard interpretation for a correlation between the UV brightness of the sky and its FIR (or HI emission which is closely related, see BP) is that the IS dust produces both. The UV part is scattering of the ISRF and FIR is thermal dust emission (Jura 1979). For low optical depth, the UV brightness is proportional to the ISRF and the scattering optical depth, for a given dust asymmetry factor and a given latitude, whereas the FIR emission is proportional to the ISRF and the absorption optical depth. The correlation indicates a rather constant albedo which is the ratio of the scattering optical depth to the extinction (scattering plus absorption) optical depth.

Dust scattering at optical wavelegenths and at high latitudes has been discovered earlier and is more firmly established. de Vaucouleurs (1960) already mentioned some faint optical filamentary nebulosity in the vicinity of (but not associated with) the Large Magellanic Cloud. Sandage (1976) has shown the pervasiveness of these faint reflection nebulosities at high latitudes, later to be dubbed "cirrus" clouds with IRAS (Low et al. 1984). de Vries and Le Poole (1985) have clearly demonstrated from photographic plate analyses the correlations between the visible diffuse brightness of the sky, the visible extinction and the FIR emission.

3.2 Optical luminescence

Recently, Guhathakurta and Tyson (GT 1989) used a mosaic of CCD frames at B_J, R and I bands to observe these high-latitude nebulosities. Beside the general association of optical emission with FIR emission, GT find that there is a red excess of emission compared with the dust scattering expectations. This excess is

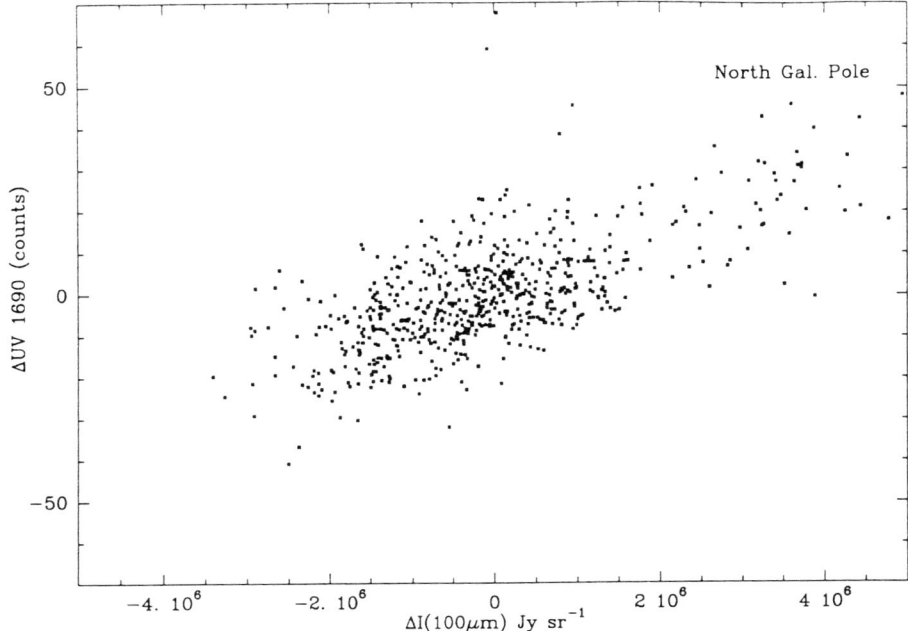

Figure 2. Correlation between UV and IR excesses over the cosecant law for the North Galactic polar cap ($b \geq 30°$) where each pixel (25 sq.deg. large) is shown as a dot. One UV count is $0.4\,\mathrm{nW\,m^{-2}\,sr^{-1}}$ in a bandwidth of $0.033\,\mu\mathrm{m}$.

identified with the red excess from 0.550 to 0.850 μm that is observed in reflection nebulae (e.g. Witt and Schild 1985) and even high latitude dark nebulae (Mattila 1979) and has a complex, varying, and as yet unidentified spectral structure (Schmidt, Cohen and Margon 1980, Witt and Boroson 1990). Candidates for this broad red excess all imply some form of carbon dust grains: diamond-like (Duley and Williams 1988), hydrogenated amorphous carbon grains (Witt and Schild 1988) or Polycyclic Aromatic Hydrocarbon (PAHs) molecules (d'Hendecourt et al. 1986). The involved emission mechanism is usually thought to be luminescence i.e. a cascade in low electronic levels of UV-excited very small grains.

3.3 Infrared emission

A review of the high-latitude IR sky has already been given by Boulanger (1989). Here, we merely want to stress the main points. The diffuse IR sky, as IRAS showed us wonderfully, is dominated by IS dust emission, once the zodiacal light is removed. BP have used a similar technique as that used in Section 3.1 to show that IR emission is closely associated with the HI neutral medium. The infrared sky has also revealed that a non-negligible fraction ($\sim 30\%$) of IS dust emission occurs at shorter wavelengths ($\lambda \leq 80\,\mu\mathrm{m}$) than grains can emit if they are at an equilibrium temperature. The current interpretation is that very small grains and/or large

TABLE 1.
Scattering and emission from a 3×10^{20} H cm^{-2} cirrus cloud

λ μm	νI_ν nW m^{-2} sr^{-1}	HWHM μm	σ_+ nW m^{-2} sr^{-1}	σ_- nW m^{-2} sr^{-1}	Comments, ref.
			Gas emissivity		
0.100	14.	0.005	3.	9.	Werner H$_2$ (1)
0.160	6.	0.008	1.2	2.	Lyman H$_2$ (2)
1.65	0.6	0.75	0.12	0.12	H$_2$ rot.-vib. (3)
63.	24.	1.6	8.	8.	Oxygen line (4)
158.	24.	3.95	8.	8.	C$^+$ line (4)
			UV-visible dust emissivity		
0.158	9.	0.0015	1.3	1.3	UVX (5)
0.169	7.	0.0165	3.8	1.5	D2B (6)
0.441	38.	0.049	13.	13.	Pioneer 10 (7)
0.450	19.	0.049	20.	10.	CCD (8)
0.550	34.	0.045	15.	15.	Photo. plate (9)
0.650	30.	0.110	30.	15.	CCD (8)
0.900	36.	0.120	36.	18.	CCD (8)
			IR dust emissivity		
3.3	60.	0.025	31.	31.	Balloon (10)
12.	26.	2.85	7.1	7.1	IRAS (11)
25.	17.	4.9	4.8	4.8	IRAS (11)
60.	26.	15.5	1.4	1.4	IRAS (11)
100.	76.	17.8	2.9	2.9	IRAS (11)
102.	60.	15.3	12.	12.	Rocket (12)
137.	93.	24.	19.	19.	Rocket (12)
262.	17.	47.	8.6	8.6	Rocket (12)
325.	17.	125.	5.	5.	Balloon (13)

Notes to Table 1: The νI_ν brightness unit is nW m^{-2} sr^{-1} = 10^{-9} W m^{-2} sr^{-1}. The brightness uncertainties σ_+ and σ_- have been estimated from the quoted references and include true variability. HWHM is the spectral coverage half-width at half maximum that is assumed for the brightness calculation (see Section 1 for definition).

References: (1) when detected, deduced from Lyman band (not yet observed); (2) when detected (MHB); (3) estimated using Sternberg 1989 (not yet observed); (4) estimated using a 5% spectral resolution and a ratio of the efficiency of heating the gas by small grains to IR emission of 6% (Verstraete 1990) (not yet observed); (5) Martin et al. 1989; (6) PLHJ see text; (7) Toller 1981; (8) GT 1989; (9) Sandage 1976; (10) Giard et al. 1989; (11) cosecant law from BP; (12) Lange et al. 1989; (13) Fabbri et al. 1986.

molecules have temperature fluctuations due to their low heat capacity and to the individual photon heating events (see Puget and Léger 1989 and references therein). About 10 to 20% in mass of IS dust is necessary to explain the short wavelength emission (Désert, Boulanger and Puget 1990). In addition, evidence for the existence of large molecules *e.g.* PAHs is given from observations of the so-called unidentified emission features (3.3, 6.2, 7.7, 8.6 and 11.3 μm) in reflection nebulae (Sellgren 1984) and the diffuse medium (Giard et al. 1989).

Questions that are still open concerning the IS dust IR emission are related to the dust content of the other phases of the IS medium. Is the warm ionised medium dust-deficient or spatially correlated with the HI medium as suggested by BP (see also Abraham 1990)? Is the large 60/100 μm ratio (0.3 to be compared with 0.2 on average outside the galactic disk) observed in the polar caps indicating some dust processing in the nearby IS medium? What is the distance to the cirrus clouds (see *e.g.* Magnani and de Vries 1986, Franco 1989, typical distances are in the 50 to 200 pc range)? What is the velocity structure of cirrus clouds (*e.g.* Deul and Burton 1990) and are the IR colors (hence dust composition and heating) varying (*e.g.* BP, Terebey and Fich 1986, Herter, Shupe and Chernoff 1990)?

4. FOREGROUNDS AND BACKGROUNDS

The picture that was drawn in the preceding sections would be misleading if we did not mention the presence of strong foregrounds and backgrounds which have to be understood in order to be subtracted. Figure 1b, which was purposely drawn on the same scale as Figure 1a, shows the various fore-backgrounds which are in general *one to two* orders of magnitude above the cirrus cloud brightness. The zodiacal scattering at the galactic poles is estimated by assuming constant reflectivity of interplanetary dust (continuous curve on the left). The unresolved stars (dashed curve and open squares) also produce a background at the same UV to IR wavelengths that has been observed by Gondhalekar et al. (1980) in the UV, Toller et al. (1987) in the visible and estimated by BP in the IR. The Zodiacal emission (second continuous curve) is the strongest foreground in the IR. It follows approximately a blackbody curve modified by an emissivity $\propto \lambda^{-1}$ (Hauser et al. 1984). The cosmic microwave background (continuous curve on the right) follows a perfect blackbody law with $T = 2.735\,\text{K}$ as demonstrated by the COBE satellite (Mather et al. 1990).

Looking at Figure 1b, one can see that there are three main extragalactic spectral windows ($\lambda \leq 0.15\,\mu$m, $\lambda \sim 3$, and $\sim 300\,\mu$m) where the various fore-backgrounds are relatively weak. It should be no surprise that there have been several claims (but as yet unconfirmed) of an extragalactic background in all these windows (points on Figure 1b with symmetric error bars) *e.g.* by Fix et al. (UV, 1989), Tyson (visible, 1988), Matsumoto et al. (near-IR, 1988a), Rowan-Robinson (far IR, 1986 treated as an upper limit by BP) and Matsumoto et al. (submillimetre, 1988b). Other observations have yielded upper limits to an extragalactic background (error bars pointing downwards) from UV to IR: Martin, Hurwitz and Bowyer (1989),

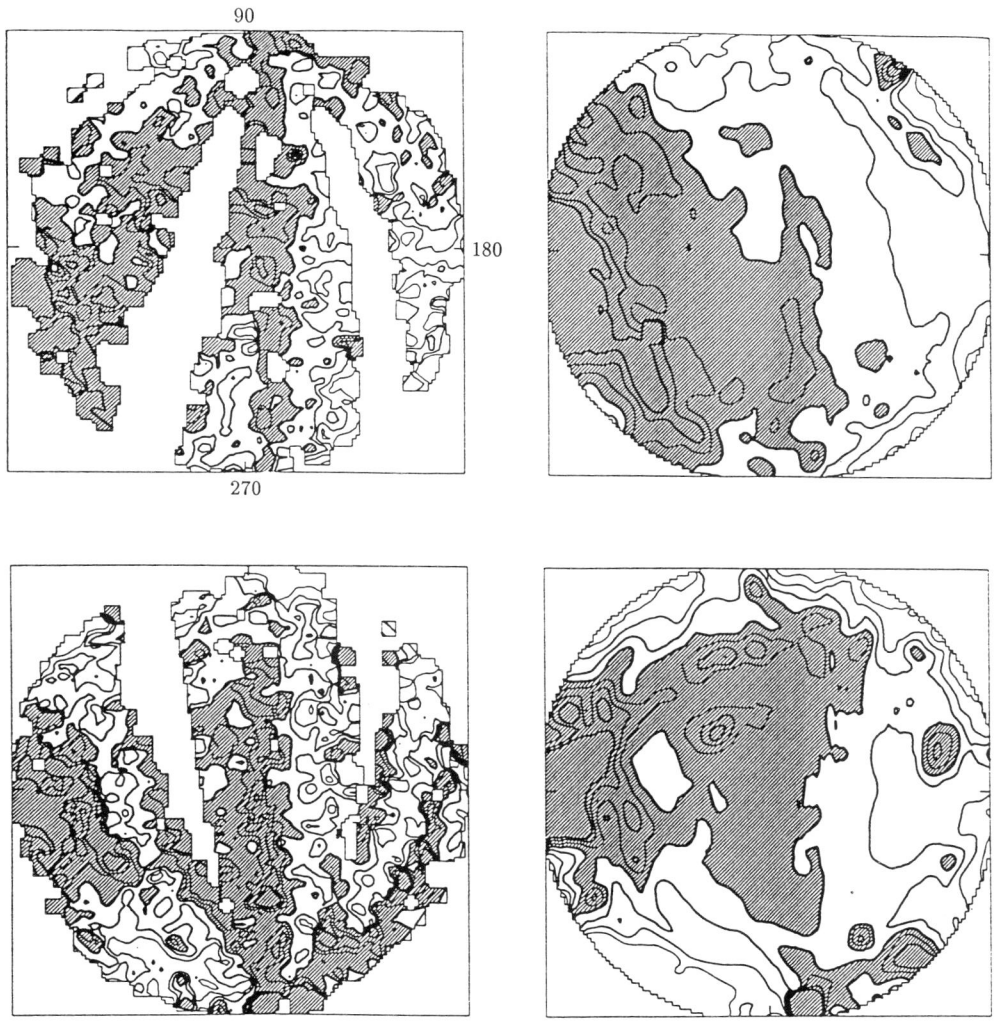

Figure 3. Contour and greyscale plot of the polar caps: top is North, bottom is South. Galactic poles are at the center of the maps, and radii are proportional to galactic latitudes from 90 to 30° or −90 to −30. Polar angle is galactic longitude. On the left is the UV brightness and on the right is the far IR (100 μm) brightness. Both are cosecant law subtracted. Therefore the small-scale features are enhanced. The thickest contour (thickness 4) is the zero level. Contour thickness goes as the sequence 1 2 3 4 1 2 3. The increment for the contours is $3\,\text{nW}\,\text{m}^{-2}\,\text{sr}^{-1}$ in a 0.033 μm band and 1 MJy/sr resp. Grey parts correspond to positive excesses over the cosecant law, and white parts to negative excesses.

Joubert et al. (1983), Spinrad and Stone (1978), Toller (1983), Dube, Wickes, and Wilkinson (1977), Mather et al. (1990 right dashed curve).

Spatially, one wants to find the extragalactic windows where local emission is the least bright. Let us mention the areas given by BP and by Jahoda, Lockman, and McCammon (1990). The latter authors find that any line of sight (with a beam of 21′) contains at least $\sim 0.5 \times 10^{20}$ H cm^{-2}. Therefore a scaling of Figure 1a downwards by a factor ~ 6 gives the minimum cirrus brightness anywhere in the sky and gives the fundamental limit for the study of diffuse extragalactic backgrounds. These windows may prove useful for the study of the galactic halo emissions as well.

ACKNOWLEDGMENTS

I am very much indebted to J. Lequeux for his help in the preparation of this review. Many thanks to M. Perault for discussions and data reduction.

REFERENCES

Abraham, P. (1990) preprint and poster paper at this conference
Boulanger, F., and Pérault, M. (1988) *Astrophys. J.*, **330**, 964 (BP)
Boulanger, F. (1989) in *Proc. of IAU Symp. 139, The Galactic and Extragalactic Background Radiation*, eds. S. Bowyer and C. Leinert, Kluwer, Dordrecht.
Deharveng, J. M., Joubert, M., and Barge, P. (1982) *Astron. Astrophys.*, **109**, 179
Désert, F.-X., Boulanger, F., and Puget, J. L. (1990) *Astron. Astrophys.*, in press
Deul, E. R., and Burton, W. B. (1990) *Astron. Astrophys.*, **230**, 153
de Vaucouleurs, G. (1960) *The Observatory*, **80**, 106
de Vries, C. P., and Le Poole, R. S. (1985) *Astron. Astrophys.*, **145**, L7
d'Hendecourt, L. B., et al. (1986) *Astron. Astrophys.*, **170**, 91
Dube, R. R., et al. (1977) *Astrophys. J. (Letters)*, **215**, L51
Duley, W. W., and Williams, D. A. (1980) *Astrophys. J. (Letters)*, **242**, L179
Duley, W. W., and Williams, D. A. (1988) *M.N.R.A.S.*, **230**, 1P
Fabbri, R., et al. (1986) in *Proceed. of Marcel Grossman Meeting*, Rome June 1985
Fix, J. D., Craven, J. D., and Frank, L. A. (1989) *Astrophys. J.*, **345**, 203
Franco, G. A. P. (1989), *Astron. Astrophys.*, **223**, 313
Gatley, I. et al. (1987) *Astrophys. J. (Letters)*, **318**, L73
Giard, M.,et al. (1989) *Astron. Astrophys.*, **215**, 92
Gondhalekar, P. M., Phillips, A. P., and Wilson, R. (1980) *Astron. Astrophys.*, **85**, 272
Guhathakurta, P., and Tyson, J. A. (1989) *Astrophys. J.*, **346**, 773 (GT)
Harwitt, M., Houck, J. R., and Stacey, G. J. (1986) *Nature*, **319**, 646
Hauser, M. G. et al. (1984) *Astrophys. J. (Letters)*, **278**, L15
Herter, T., Shupe, D. L., and Chernoff, D. F. (1990)J*Astrophys. J.*, in press
Hurwitz, et al. (1989) in *Proc. of IAU Symp. 139, The Galactic and Extragalactic Background Radiation*, eds. S. Bowyer and C. Leinert, Kluwer, Dordrecht.
Jahoda, K., Lockman, F. J., and McCammon, D. (1990) *Astrophys. J.*, **354**, 184
Jakobsen, P. (1982) *Astron. Astrophys.*, **106**, 375
Jakobsen, P.,et al. (1984), *Astron. Astrophys.*, **139**, 481
Jakobsen, P., de Vries, J. S., and Paresce, F. (1987) *Astron. Astrophys.*, **183**, 335

Joubert, M., et al. (1983) *Astron. Astrophys.*, **128**, 114
Jura, M. (1974) *Astrophys. J.*, **191**, 375
Jura, M. (1979) *Astrophys. J.*, **227**, 798
Lange, A. E. et al. (1989) in *IAU Symp. 135 on Interstellar Dust*, ed. Allamandola, L., and Tielens, A. G. G. M., Kluwer, Dordrecht, p.499
Low, F. J. et al. (1984) *Astrophys. J. (Letters)*, **278**, L19
Magnani, L., and de Vries, C. P. (1986) *Astron. Astrophys.*, **168**, 271
Martin, C., Hurwitz, M., and Bowyer, S. (1990) *Astrophys. J.*, **354**, 220 (MHB)
Martin, C., et al. (1989) in *Proc. of IAU Symp. 139*
Mather, J. S. et al. (1990) *Astrophys. J. (Letters)*, **354**, L37
Matsumoto, T., Akiba, M., and Murakami, H. (1988a) *Astrophys. J.*, **332**, 575
Matsumoto, T., et al. (1988b) *Astrophys. J.*, **329**, 567
Mattila, K. (1979) *Astron. Astrophys.*, **78**, 253
Morgan, D. H., Nandy, K., and Thompson, G. I. (1978) *M.N.R.A.S.*, **185**, 371
Murthy, J., et al. (1989) *Astrophys. J.*, **336**, 954
Murthy, J., et al. (1990) *Astron. Astrophys.*, **231**, 187
Pérault, M., Lequeux, J., Hanus, M., and Joubert, M. (1990) in preparation (PLHJ)
Puget, J. L., and Léger, A. (1989) *Ann. Rev. Astron. Astrophys.*, **27**, 161
Rowan-Robinson, M. (1986) *M.N.R.A.S.*, **149**, 365
Sandage, A. (1976) *Astron. J.*, **81**, 954
Savage, B. D., and Mathis, J. S. (1979) *Ann. Rev. Astron. Astrophys.*, **17**, 73
Schmidt, G. D., Cohen, M., and Margon, B. (1980) *Astrophys. J. (Letters)*, **239**, L133
Sellgren, K. (1984) *Astrophys. J.*, **277**, 623
Sellgren, K. (1986) *Astrophys. J.*, **305**, 399
Spinrad, H., and Stone, R. P. S. (1978) *Astrophys. J.*, **226**, 609
Stark, R. (1990) *Astron. Astrophys.*, **230**, L25
Sternberg, A. (1989) *Astrophys. J.*, **347**, 863
Sternberg, A., and Dalgarno, A. (1989) *Astrophys. J.*, **338**, 397
Terebey, S., and Fich, M. (1986) *Astrophys. J. (Letters)*, **309**, L73
Toller, G. N. (1981), PhD Thesis, State University of New York, Stony Brook
Toller, G. N. (1983) *Astrophys. J. (Letters)*, **266**, L79
Toller, G., Tanabe, H., and Weinberg, J. L. (1987) *Astron. Astrophys.*, **188**, 24
Tyson, J. A. (1988) *Astron. J.*, **96**, 1
Verstraete, L. (1990) PhD Thesis, and Verstraete et al. (1990) *Astron. Astrophys.*
Witt, A. N., and Schild, R. E. (1985) *Astrophys. J.*, **294**, 225
Witt, A. N., and Schild, R. E. (1988) *Astrophys. J.*, **325**, 837
Witt, A. N., et al. (1989) *Astrophys. J. (Letters)*, **336**, L21
Witt, A. N., and Boroson, T. A. (1990) *Astrophys. J.*, **355**, 182

METALS AND MOLECULES IN HALO CLOUDS

Klaas S. de Boer, Uwe Herbstmeier, Ulrich Mebold
Astronomical Institutes of the University of Bonn
Auf dem Hügel 71
D-5300 Bonn, F.R.Germany

The abundance of the elements in clouds of halo gas, as determined from observations, is an important parameter for the test of the validity of models explaining the existence of, e.g., the high latitutude high-velocity clouds (HVCs) of the Milky Way. Individual HVCs have been detected in absorption only on very few lines of sight so that the distance of the HVCs, another important parameter for the models, stays ill determined as well. We will follow here the more or less established convention by calling HVCs those with $/v/ > 100$ km s^{-1} and IVCs (intermediate-velocity clouds) those with $50 </v/ <100$ km s^{-1}. We will define halo as the space with $/z/ > 1$ kpc, although for $/b/ > 45°$ also $/z/ > 0.5$ kpc is used.

Only in very few cases have interstellar absorption lines of individual HVCs or IVCs been seen. In fact, to our knowledge, the only HVCs detected as well defined absorption structures are in the direction of the LMC (Savage and de Boer, 1981), in the direction of the galaxy Fairall 9 (Songaila, 1981), and toward HD 135485 (Albert et al., 1989). IVCs have been found in the direction of M3, M13, HD 93521, HD 97991, the LMC, and HD 215733 (see Table 1). Note that the HVC at 120 km s^{-1} in the direction of the LMC is definitely not associated with the LMC (de Boer, Morras, Bajaja, 1990). Note also, that the interstellar nature of the spectral structure seen in the spectra of stars towards Chain A and Complex C (Songaila et al., 1988) most likely is stellar (Lilienthal, Meyerdierks, de Boer, 1990). In many other directions interstellar lines were looked for but either no absorption was seen at velocities sufficiently different from the LSR or the absorption profiles were rather smooth without revealing individual clouds (see e.g. Pettini and West, 1982; Pettini et al., 1982; Albert, 1983; Danly, 1989). For a listing of papers with non-detections see de Boer (1989). Further IVCs will be given by Danly (these proceedings).

The absorption lines show that the detected HVCs and IVCs contain metals indeed (for Ca detected in HVCs against extragalactic probes see van Woerden et al., 1989). In some cases only limits to the abundance can be given due to saturation of the absorption lines. For vZ1128 and

B29 it is found that Ab(C)> -1.5 dex of the Solar value (de Boer and Savage, 1984) whereas the metals detected in the HVC and the IVC in the general direction of the LMC are close to Solar. In fact, combining the UV data from Savage and de Boer (1981) with HI from McGee et al. (1983) one finds that in the 60 and 130 km s^{-1} clouds Fe has an abundance of -4.8 and -4.6, respectively, within a factor of 2 of Solar. The Ti investigated by Albert (1983) is -1.0 dex of Solar, the Ca (from various studies) is between -2 and -3 of Solar abundance.

Intermediate velocity gas is present in a fair portion of the (well studied) northern sky at high galactic latitudes. A detailed summary of the earlier data has been given by Wesselius and Fejes (1973). They list two Ca-H coincidences, toward HD 93521 and HD 97991, where the Ca abundance ranges from -2 to -3 dex. Basing themselves on earlier ideas, they give two possible interpretations for the rather widespread intermediate velocity gas. The gas may be the material swept-up by the approaching shell of supernova gas while, alternatively, it may be gas collected in a collision with a big complex coming from l= 120°, b=+40°. This large IV gas structure then reemerged (of course) in all later observational surveys, such as those of Giovanelli (1981), Albert (1983), Hulsbosch and Wakker (1988), and Danly (1989).

Another piece of evidence for the nature of the metallicity of the halo gas is the scale height of the galactic z-distribution of the various elements. Such scale heights are, admittedly, a function of the

Table 1 Individual HVCs and IVCs detected in absorption at b>25°

Name	l	b	z kpc	v(abs) km s^{-1}	Element	v(HI) km s^{-1}	Ref •)
vZ1128 M3	42	+79	10	-60	CII,CIV		dBS84
B29 M13	59	+41	4.1	-80	CII,MgII	-80	dBS83
HD 203664	61	-28	-0.7	-55	"all"		DB89
				+70	"all",CIV		DB89
HD 215733	85	-36	1.5	-50	CaII,TiII	-50	A83
Mk 106	161	+43	x)	-156	CaII	-140	SWW89
HD 93521	183	+62	1.8	-55	CaII,TiII	-50	WF73, A83
LMC *)	270	-33	28	130	"all"	130	SdB81, MNM83
				60	"all"	60	B+88, S+89
Fairall 9	295	-58	x)	193	CaII	195	S81 MB86
HD 135485	347	+35	1.4	-128	CaII		A+89

*) LMC objects: R136, HD 36402, SN 1987A and many others; note that the N(HI) given by B+88 for the SN 1987A line of sight clouds is uncertain by at least a factor 2, a factor propagating into the abundances given by B+88.
x) Mk 106 and Fairall 9 are extragalactic.
•) references are explained in the final reference list.
"all" means all ions common in neutral gas.

spatial distribution of the gas, convolved with the ionization balance and influenced by the possible depletion of metals due to dust. In fact, some scale height values may suffer from selection effects in that data to the most distant stars not always have been or could be included in the determination. Due to the use by Edgar and Savage (1989) of FeII data from only the Copernicus satellite, very few large z sight lines could be included in their analysis. On the other hand, Ca and Ti data are available for very distant stars so that even few entries with enhanced column densities will result in enlarged scale heights. Determinations of the scale heights include those by Edgar and Savage (1989) for CaII ($\approx$1 kpc), TiII ($>$2 kpc), and FeII (0.5 kpc), by Savage and Massa (1987) for SiIV ($\approx$3 kpc), CIV ($\approx$3 kpc), and NV ($\approx$2 kpc), and by Reynolds (1989) for electrons ($\approx$2 kpc). The scale height of HI is approximately 0.5 kpc. Nevertheless, heavy metals such as C and Si, as well as Ti are present out to well above z= 2 kpc.

When halo clouds fall to the disk of the Milky Way, they will interact with the disk gas. Shock fronts will run into the gas complexes resulting in a change in the physical conditions of the gas, in particular compression. This may lead to a phase of enhanced formation of molecules which then show up with velocities of disk approach. A review of such scenarios is given by Mebold et al. (1989a).

In addition to the well known high latitude CO and cirrus at velocities near the LSR (Magnani et al., 1985), three molecule-containing clouds have up to now been detected at intermediate velocity (Table 2). The Draco cloud is well studied and shows all the characteristics of a cloud compressed and pierced by a HVC (Mebold et al., 1985, 1987). In particular, Herbstmeier (1990) could show from a principal component analysis that in the Draco interaction zone the so-called X-factor is well over an order of magnitude smaller than thought previously (de Vries et al., 1987; Heithausen and Mebold, 1989). In this zone, almost all C seems to have been processed into CO and H_2CO, while there is very little HI at the same velocities. A second high-latitude intermediate-velocity CO cloud is reported by Heiles et al. (1988); it is in a gas complex known already by Wesselius and Fejes (1973), who in general put intermediate velocity gas at z-distances of a few 100 pc. A third cloud of this kind has been discovered by Désert et al. (1990).

Summarizing, we have collected all information available on the abundances of metals in and on the existence of molecules induced by halo clouds.

Table 2. Molecules detected in the halo in IVCs

Object	l	b	v	molecules	ref.	d	ref
Draco	90	+39	-25	CO, H_2CO, NH_3	Mebold et al. 1989b	>500	M+89b
	135	+54	-45	CO	Heiles et al. 1988	>100	WF73
DBB 306	211	+63	-39	CO	Désert et al. 1990	<400	WF73

References

Albert, C.E.: 1983, Astrophys.J. 272, 509   (A83)
Albert, C.E., Blades, J.C., Morton, D.C., Proulx, M., Lockman, F.J.: 1989, in IAU Coll 120 "Structure and Dynamics of the Interstellar Medium", Eds. G.Tenorio-Tagle, M.Moles, J.Melnick; Springer; Lecture Notes in Physics; p.442   (A+89)
Blades, J.C., Wheatly, J.M., Panagia, N., Grewing, M., Pettini, M., Wamsteker, W.: 1988, Astrophys.J. 332, L75   (B+88)
Danly, L.: 1989, Astrophys.J. 342, 785
Danly, L., Blades, J.C.: 1989, in IAU Coll 120; op. cit., p.408   (DB89)
de Boer, K.S.: 1989, in IAU Coll 120; op. cit., p.432
de Boer, K.S., Savage, B.D.: 1983, Astrophys.J. 265, 210   (dBS83)
de Boer, K.S., Savage, B.D.: 1984, Astron. Astrophys. 136, L 7   (dBS84)
de Boer, K.S., Morras, R., Bajaja, E.: 1990, Astron. Astrophys. in press
Desert, F.-X., Bazell, D., Blitz, L.: 1990, Astrophys. J. 355, L 51
de Vries, H.W., Heithausen, A., Thaddeus, P.: 1987, Astroph. J. 319, 723
Edgar, R.J., Savage, B.D.: 1989, Astrophys.J. 340, 762
Giovanelli, R.: 1980, Astron.J. 85, 155
Heiles, C., Reach, W.T., Koo, B.-C.: 1988, Astrophys.J. 332, 313
Heithausen, A., Mebold, U.: 1989, Astron. Astrophys. 162, 279
Herbstmeier, U.: 1990, thesis, Univ. Bonn
Hulsbosch, A.N.M., Wakker, B.P.: 1988, Astron. Ap. Suppl.Ser. 75, 191
Lilienthal, D., Meyerdierks, H., de Boer, K.S.: 1990, Astr. Ap. in press
Magnani, L., Blitz, L., Mundi, L.: 1985, Astrophys. J. 295, 402
McGee, R.X., Newton, L.M., Morton, D.C.: 1983, Mon. Not. R. astr. Soc. 205, 1191   (MNM83)
Mebold, U., Cernicharo, J., Velden, L., Reif, K., Crezelius, C., Goerigk, W.: 1985, Astron. Astrophys. 151, 427
Mebold, U., de Boer, K.S., Wennmacher, L.: 1989a, in XI Regional European Astronomy Meeting, Ed. S. Mitton; Cambr. Univ.P.; in press
Mebold, U., Herbstmeier, U., Kalberla, P.W.M., Souvatzis, I.: 1989b, in IAU Coll. 120; op. cit. p.424   (M+89b)
Mebold, U., Heithausen, A., Reif, K.: 1987, Astron. Astrophys. 180, 213
Morton, D.C., Blades, J.C.: 1986, Mon.Not.R.astr.Soc. 220, 927   (MB86)
Pettini, M., West, K.M.: 1982, Astrophys.J. 260, 561
Pettini, M., et 12 alii: 1982, Mon.Not.R.astron.Soc. 199, 409
Reynolds, R.J.: 1989, Astrophys. J. 339, L 29
Savage, B.D., de Boer, K.S.: 1981, Astrophys.J. 243, 460   (SdB81)
Savage, B.D., Massa, D.: 1987, Astrophys.J. 314, 380
Savage, B.D., Jenkins, E.B., Joseph, C.L., de Boer, K.S.: 1989, Astrophys.J. 345, 393   (S+89)
Schwarz, U.J., Wakker, B.P., van Worden, H.: 1989, "in prep."
Songaila, A.: 1981, Astrophys.J. 243, L 19   (S81)
Songaila, A., Cowie, L.L., Weaver, H.: 1988, Astrophys.J. 329, 580
van Woerden, H., Schwarz, U.J., Wakker, B.P.: 1989, in IAU Coll 120; op. cit., p.389
Wesselius, P.R. Fejes, I.: 1973, Astron. Astrophys. 24, 15   (WF73)

GALACTIC WORMS

B.-C. KOO, C. HEILES, AND W. T. REACH
Astronomy Department, University of California,
Berkeley, CA 94720

ABSTRACT. We have found and cataloged over 100 vertical structures in H I, infrared, and radio continuum emission. These correspond to the H I worms detected by Heiles (1984). The infrared and the radio continuum properties of worms suggest that some worms have associated ionized gas. The area filling factor of superbubbles in the inner Galaxy is estimated to be greater than ~ 0.1.

1. INTRODUCTION

Heiles (1984), in his pioneering study of H I shells and supershells, found interesting features in the inner Galaxy, 'H I worms'. By applying a median filter, which enhances small scale structure, to the H I channel maps near the Galactic plane ($|b| < 10°$), he found that the filtered structure appears to be random in the outer Galaxy, while in the inner Galaxy it tends to run perpendicular to the Galactic plane and individual features tend to persist over a large velocity interval. He interpreted these vertical structures, or H I worms, as the walls surrounding superbubbles, which have broken through the thin gaseous disk inside the solar circle.

In this paper we report a systematically determined catalog of 118 similar structures, not only in H I but also in the infrared and the radio continuum, and present some preliminary results.

2. OBSERVATIONS AND DATA ANALYSIS

The H I 21-cm maps between $10 \leq l \leq 350$ are made from the Galactic plane survey of Weaver and Williams (1973) and of Kerr *et al.* (1986). The missing data near the Galactic center were obtained using the Hat Creek 85-foot. The infrared 60 and 100 μm maps are made from the *IRAS Zodiacal Observation History File* (1988). The radio continuum maps are made from the 408 MHz survey of Haslam *et al.* (1982). All the maps have $0.°5 \times 0.°5$ pixel size, and a median filter of $3.°5 \times 3.°5$ was applied to see small scale structure near the Galactic plane ($|b| < 10°$).

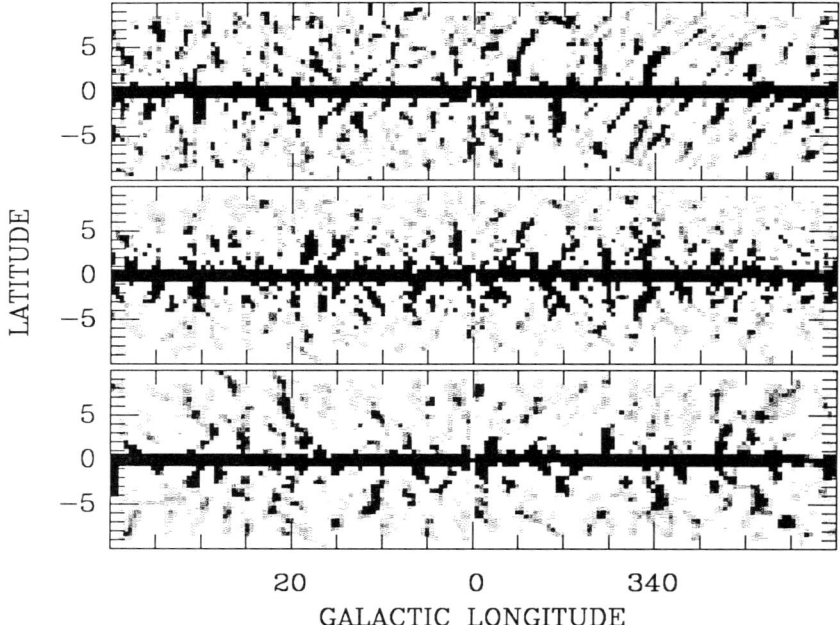

Figure 1. The median-filtered H I, 100 μm, and 408 MHz map (from top to bottom) in the inner Galaxy.

Figure 1 shows the median-filtered H I map integrated from −200 to +200 km s^{-1} in the inner Galaxy together with the corresponding 100 μm and 408 MHz maps as an example. The correlation between the H I and the 100 μm map is remarkable, which proves that most of these structures are real. By cross-correlating the 60 and 100 μm maps with the H I map, but excluding the very central region of the plane ($|b| < 1°$), we found a total of 118 isolated structures that are larger than 5 pixels and appear in both infrared and H I maps. The identification and the cross correlation of structures were made by 'unbiased' search technique based on a computer algorithm which will be described in a subsequent publication. We will call all 118 structures 'worms' even though some of them are small and at high latitudes, so that they are not worms in the sense that they do not 'crawl out of the Galactic plane'.

Not all the worms appear in 408 MHz emission. Among 118 worms, 35 have counterparts at 408 MHz.

3. INFRARED AND RADIO CONTINUUM PROPERTIES OF WORMS

Figure 2a shows the median value of the 60 to 100 μm intensity ratio, I_{60}/I_{100}, versus the median value of the 100 μm emissivity, $I_{100}/N_{\rm H\,I}$, of each worm (open circles). The ratio I_{60}/I_{100} is almost independent of $I_{100}/N_{\rm H\,I}$, although there may be a trend such that a worm with a larger $I_{100}/N_{\rm H\,I}$ has a larger I_{60}/I_{100}. This result is in good agreement with theoretical calculations of Draine and Anderson (1985). The average I_{60}/I_{100} of all worms is large 0.28 ± 0.06, and implies a grain size distribution with a relatively large number of smaller grains, which may be due to the population of shock-processed grains.

The large variation of $I_{100}/N_{\rm H\,I}$ is due to at least three factors: (1) the general increase of the diffuse interstellar radiation field toward the Galactic interior, (2) the worms with associated H II regions, e.g., GW6.5−3.7 (Galactic Worm centered at $l \approx 6.5$ and $b \approx -3.7$) with S25, GW14.9−1.6 with S45, GW17.8+3.0 with S54, and G31.6−5.9 with W43 in Figure 1, and (3) the worms with associated molecular gas, e.g., GW1.9+6.0 and GW19.5−6.4. The decomposition of the infrared emissivity into each contribution needs to be done for individual worm.

Figure 2b shows the average 408 MHz brightness temperature versus the average 60 μm intensity of each worm. There is a good correlation between the two. We are currently unsure of whether the 408 MHz continuum is thermal or nonthermal. One known example of nonthermal emission associated with H I is the North Polar Spur (Heiles *et al.* 1980), and a known example of thermal emission is GW17.8+3.0 (Müller, Reif, and Reich 1987). For the worms with associated H II regions, the 408 MHz emission is likely to be thermal. In Figure 2b, the straight line is an expected relationship for a diffuse H II region by assuming that all the Lyman α photons are converted to infrared photons. The apparent deficit of 60 μm emission possibly arises because some Ly α photons leak through the vertical direction. On the other hand, it may suggest that the 408 MHz emission is non-thermal.

4. SUMMARY AND DISCUSSION

We have found that there is a very good correlation between H I worms and infrared worms, and that many prominent worms have their counterparts at 408 MHz. We also found that some worms are very likely to have associated ionized gas, which needs to be confirmed by some direct observations, e.g., radio recombination lines.

One primary goal of this work is to answer the following question: "What is the filling factor of superbubbles in our Galaxy ?" We can make a very crude estimate based on our results. A superbubble that has broken through the *thin* gaseous disk occupies an area with radius comparable to or larger than the scale height $\sim$ 190 pc. In the first and the fourth quadrant, there are about 30 worms that actually rise *from* the Galactic plane. If we assume that most worms within 4 kpc are detected, then the filling factor of the inner Galaxy is greater than $\sim$0.1.

A superbubble that has broken through the thin gaseous disk does *not* necessarily

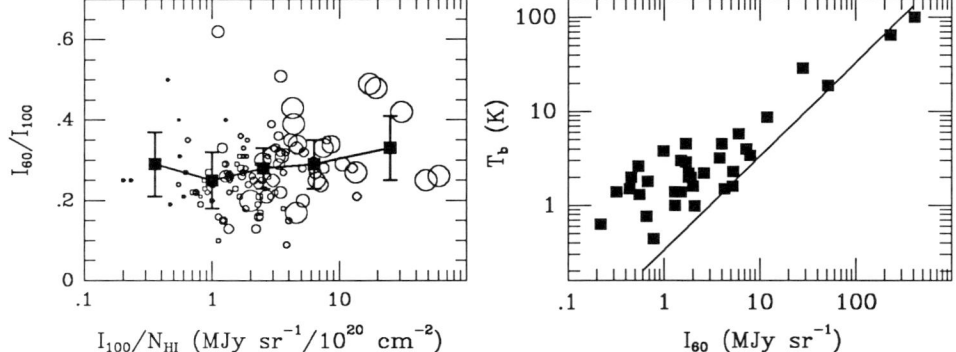

Figure 2. (a) The median value of I_{60}/I_{100} vs. the median value of I_{100}/N_{HI} of each worm (open circles). The area of the circle is proportional to the average 100 μm intensity. The average values per unit logarithmic interval are shown as squares with 1σ error bars. (b) The average 408 MHz brightness temperature vs. the average 60 μm intensity of each worm. The straight line is an expected relationship for a diffuse H II region.

inject hot gas into the halo, because there is a thick H I layer, and also a even thicker layer of ionized gas (Lockman 1984; Reynolds 1989). However, the cold fragments, which result from the Rayleigh-Taylor instability when the supershell accelerates, are likely to be injected into the halo. According to Mac Low *et al.* (1989), about 5% of the shell mass is in cold fragments, then, if we take 50 Myr as the characteristic lifetime of a superbubble, the filling factor of $\sim$0.1 implies a mass injection rate of $\sim$0.2 $M_\odot$/yr.

REFERENCES

Draine, B.T., and Anderson, N. (1985) *Ap. J.* **292**, 494
Haslam, C.G.T., Salter, C.J., Stofell, H., and Wilson, W.E. (1982) *Astr. Ap. Suppl.* **47**, 1
Heiles, C. (1984) *Ap. J. Suppl* **55**, 585
Heiles, C., Chu., Y.-H., Reynolds, R.J., Yegingil, I., and Troland, T. H. (1980) *Ap. J.* **242**, 533.
IRAS Zodiacal Observation History File (version 3) (1988), as described in the *IRAS Explanatory Supplement* (1988).
Kerr, F.J., Bowers, P.F., Jackson, P.D., and Kerr, M. (1986) *Astr. Ap. Suppl.* **66**, 373
Lockman, F.J. (1984) *Ap. J.* **283**, 90
Mac Low, M., McCray, R., and Norman, M.L. (1989) *Ap. J.* **337**, 141
Müller, P., Reif, K., and Reich, W. (1987) *Astr. Ap.* **183**, 327
Reynolds, R.J. (1989) *Ap. J. Letters* **339**, L29
Weaver, H., and Williams, D.R.W. (1973) *Astr. Ap. Suppl.* **8**, 1

MAGNETIC FIELDS IN THE DISK-HALO INTERFACE

Yoshiaki SOFUE

Institute of Astronomy, The University of Tokyo
Mitaka, Tokyo 181, Japan

ABSTRACT: A review is given of large-scale magnetic fields in disks and halos of spiral galaxies. A particular attention is given to vertical field structures, and we discuss their origin and implication on their interaction with halo gas. We point out that the disk-halo magnetic interface plays an important role in circulation of interstellar gas in galaxies, in particular a large-scale circulartion from the galactic center to outer disk regions.

1. INTRODUCTION

Large-scale magnetic fields in spiral galaxies have been measured by observing their synchrotron polarized radio emission. Magnetic field energy density in spiral galaxies is usually comparable to thermal and kinetic energies of interstellar gases, which indicates that magnetic field plays a siginificant role in the behaviour of interstellar matter. Existence of the large-scale fields suggests that the field may affect a galaxy-scale circulation of gas. In some galaxies magnetic fields in halos are as strong as those in disks, suggesting that their halo gases are driven by a magnetic mechanism. In some galaxy nuclei, magnetic energy density exceeds that of gaseous kinetic energy density, which suggests that the gas dynamics is controled by magnetic fields.

In this paper we review magnetic fields in disks and halos of spiral galaxies. We pay particular attention to vertical magnetic fields in spiral galaxies and their central regions. We discuss a possible evolution of the primordially trapped galactic field, particularly the vertical component. We also discuss implication of vertical fields in an evolution of primeval galaxies. We show that the frozen-in and amplified vertical field in turn plays a role to accelerate the accretion of gas toward the center through the magnetic angular momentum transfer. We suggest that the vertical field near the galactic centers of primordial galaxies may be related to formation of a dense nuclear disk with starburst.

2. DISK FIELDS IN SPIRAL GALAXIES

Field Strength: Strength of magnetic field can be approximately determined from synchrotron radio intensity by assuming an equipartition between magnetic and cosmic-ray energy densities. Using thus obtained field strengths B for galaxies, an empirical relation beteen magnetic energy density and molecular gas density ρ in disk galaxies has been derived as follows:

$$B^2/8\pi \propto \rho,$$

This relation can be rewritten to give field strength as a function of gas density as

$$B/B_0 \simeq \sqrt{\rho/\rho_0}$$

with $B_0 \simeq 3 - 5$ μG and $\rho_0 \simeq 1 m_H$ cm^{-3}.

Field Orientation and the Origin: Orientation of disk magnetic fields as measured from linear polarization observations and Faraday rotation analyses is either BSS (bisymmetric spiral) or a ring (see, e.g. Sofue et al 1986). In most cases, higher-order and/or random fields of comparable strengths are superposed.

The BSS fields may have their origin in a primordial magnetic field when galaxy formed: A larger-scale intergalactic/intrucluster field was trapped and wound up by differential rotation. The BSS field can be then maintained in a steady state by a dynamo mechanism (e.g., Sawa and Fujimoto 1986). If the primordial field had an anisotropy, even a ring field can be created by a reconnection in the inner disk (Sofue et al 1986).

It is difficult to creat a large-scale ring field only from small-scale local fields (such as due to supernovae) by dynamo amplification, because directions of dynamo-amplified ring fields should reverse frequently from a radius to another. Therefore we need a seed field which had originally large-scale ring configuration. Combination of the primordial-origin ring field, which was created from an anisotropic component of the primordial field, and a dynamo (amplification and maintaining) may explain the observed large-scale ring configuration.

3. HALO FIELDS IN SPIRAL GALAXIES

Our Galaxy: Magnetic fields in our Galactic halo can be derived from Faraday rotation analysis of external radio sources and pulsars. If we plot $|RM|$ (rotation measure) against $|\cot b|$ for radio galaxies and quasars with b being galactic latitude, the upper envelope of the plot can be fitted by a relation,

$$|RM|_{\text{RG,Quasar}} \simeq 30 |\cot b|.$$

On the other hand if we plot the same for pulsars, we obtain

$$|RM|_{\text{Pulsar}} \simeq 10 |\cot b|.$$

The difference between the coefficients for the two plots may be due to Faraday rotation in a space above a disk in which pulsars are distributed, namely it may be due to a halo beyond $0.5 - 1$ kpc from the galactic plane:

$$|RM|_{\text{Halo}} \sim 20 |\cot b|.$$

If we take an electron density and a thickness of the halo to be approximately 10^{-3} cm^{-3} and a few kpc, respectively, the field strength in the halo is estimated to be a few μG.

External galaxies: Observational data for halo fields in external galaxies are still crude. A few edge-on galaxies like NGC 4631 show extended nonthermal radio halo (e.g. Hummel et al 1988). NGC 4631 has a well ordered vertical halo field. The field strength from the radio emissivity is estimated to be a few μG. There are many galaxies for which no evidence for radio halo is seen.

4. VERTICAL FIELDS IN SPIRAL GALAXIES

Vertical Radio Spurs – Magnetic Disk-Halo Interface: The radio continuum maps of the Milky Way show vertical structures emerging from the galactic plane (Sofue 1988). Their appearance parallel to each other suggests their coherent origin, likely driven by magnetic lines of force emerging normal to the disk plane. The vertical spurs are concentrated toward the inner region of the solar circle, suggesting that they are not local objects. Moreover, spurs are more often found above spiral arms than in interarm regions. This fact suggests that the spurs are manifestation of ejection of matter, including cosmic rays, from intense star forming regions in the spiral arms (Sofue and Tosa 1974; Sofue 1976). Their vertical extents are a few hundred pc to a kpc, while their widths are of the order of 100 pc.

Vertical Dust Lanes in External Galaxies: External edge-on galaxies often show vertical structures, which are most clearly seen as dust lanes. Long, coherent and thin filaments running normal to the disk of some dust-rich spirals like NGC 253 suggest the exitence of a large-scale vertical field running across the disk plane (Sofue 1987; Sofue et al 1990). The vertical structures are found not only in the central regions but also at radii of a few to 10 kpc, and run for more than a kiloparsec toward the halo, in some cases more than 3 kpc.

Vertical Magnetic Fields in the Galactic Center: In the inner region of our Galaxy direct evidence for a vertical field has been found by high-resolution radio observations of the synchrotron radio emission as well as by polarization observations. A large number of straight filaments extending for a hundred pc scale run almost perpendicular to the disk plane near the Arc and are well understood as the trace of a magnetic field running vertical to the disk (Yusef-Zadeh et al 1984). The radio Arc shows strong linear polarization, and Farday-roatation corected data directly indicates a poloidal magnetic field (Tsuboi et al 1986; Sofue et al 1987).

Vertical Magnetic Fields in Nuclei of External Galaxies: Polarization observations of the nuclear radio source in M31 shows the magnetic efield orientation perpendicular to the major axis (Berkhuijsen et al 1987). Since the galaxy is highly inclined, this may be attributed to a poloidal magnetic field in the nucleus. Besides M31, however, no obvious magnetic structures are known for nuclei in external galaxies mainly because of a lack of observations.

5. EVOLUTION OF MAGNETIC FIELDS IN GALAXIES

Primordial Origin of Galactic Magnetic Fields: The BSS field configuration can be understood, if it is a fossil of an intergalactic and/or intracluster field wound up by the primordial galaxy disk. The disk field is then maintained in a steady state by the induction-dynamo mechanism (Fujimoto and Sawa 1987; Sawa and Fujimoto 1986). Even a ring field can be produced from the primordial one, if we allow for an initial asymmetry with respect to the center.

A vertical magnetic field trapped into a premeval galaxy is accreted toward the central region of the galaxy, producing there a strong vertical field (Sofue and Fujimoto 1987). It has been shown that a vertical field of strength of the order of a mG can be produced as the result of accretion of mass toward the center according to the viscous-inflow and on-going star formation model of primeval galaxies (Yoshii and Sommer-Larsen 1989; Yoshii and Saio 1990).

Vertical Primordial Magnetic Field: It is natural that a large-scale field component parallel to the rotation axis existed in addition to the disk field, when a galaxy formed. This field component is also trapped to the prime,val gas sphere. Since the disk radius is large enough and the diffusion time is longer than the galaxy evolution time, the vertical field is almost frozen into the disk gas. The vertical field then follows an evolution as described below (Sofue and Fujimoto 1987). Strength of a vertical field frozen into a protogalactic gas sphere varies with the radius as $B = B_0[\Sigma(r)/\Sigma_0]$, where B_0 is the primordial intergalactic magnetic field strength of the order of $B_0 \sim 10^{-9}$ G, Σ is the surface mass density of the galaxy, and Σ_0 its initial value before protogalactic contraction as givey by $\Sigma_0 \sim M_G/(\pi R_0^2)$. Here M_G is the total mass of the galaxy, and is given by $M_G \sim (4\pi/3)R_0^3\rho_0$ with ρ_0 being the intergalactic gas density and R_0 the initial radius of the pre-contraction gas sphere. Then we obtain $\Sigma_0 \sim (16/9\pi)^{1/3}\rho_0^{2/3}M_G^{1/3}$.

The surface density at radius r in the formed galaxy can be written as $\Sigma \sim V(r)^2/(\pi Gr)$, where $V(r)$ is the rotation velocity at r and G is the gravitational constant. We thus obtain a field strength B (in G) at the stage when the initial contraction of the gas sphere finished:

$$B \sim 4.1 \times 10^{-5}(\frac{B_0}{10^{-9}\text{G}})(\frac{\rho_0}{10^{-5}\text{cm}^{-3}})^{-2/3}(\frac{M_G}{10^{11}M_\odot})^{-1/3}(\frac{V}{200\text{km s}^{-1}})^2(\frac{r}{1\text{ kpc}})^{-1}.$$

We here confirm that the magnetic field is passive against the gravitational contraction in this stage of primeval galaxy formation: In the stage of primeval galaxy contraction, when nearly all material was gas, the gas density is given approximately by $M(r) \sim (4\pi/3)\rho r^3$, and $M(r) \sim rV^2/G$. Thus a possible maximum field (in G), below which the field is passive against gas motion, can be evaluated as

$$B_{\max} \sim \sqrt{3/G}V^2/r \sim 8.5 \times 10^{-4}(\frac{V}{200\text{km s}^{-1}})^2(\frac{r}{1\text{ kpc}})^{-1}.$$

This value is much larger than the value estimated from the frozen-in formula, and we may naturally assume that the magnetic field in the protogalactic sphere (disk) is passive against the gas in gravitational contraction. We also note that the time scale for diffusion of the fertical field from the protogalaxy is more than 10^{12} years (see section 6), and the assumption of frozen-in field is a good one.

Exponential Disk in Primeval Galaxies and Stronger Field near the Center: If we take into account the formation of an exponential disk according to the viscosity-driven angular momentum transfer and on-going star formation model (Yoshii and Sommer-Larsen 1989), the central part of a galaxy gets stronger magnetic field. The magnetic flux conservation results in a radial distribution of the field strength obeying the exponential law. In the central few hundred pc region the gas density attains an excess by an order of magnitude over the value given by a simple exponential disk, and the field strength becomes correspondingly stronger, or of the order of 0.1 – 1 mG. The initial star formation then finishes when the gas is fed into stars and the density decreases to a certain threshold value, after which the magnetic field is no more frozen into the stellar disk.

Strong Vertical Fields near the Galactic Nuclei: The vertical magnetic field is then frozen into the gas left behind the initial star formation. At this stage the gas may have a constant threshold density below which the initial star formation did not take place, and shares about ten percent of the total mass. The "interstellar" gas then follows its own evolution goveraned by the density wave shock, cloud-cloud collisions, star formation, and/or magnetic accretion as described in the next section. In particular, in the central region, a bar-induced shock and the magnetic accretion will enhance the accretion. Since the diffusion time of the vertical field is shorter than the dynamical time scale, this results in a formation of a stronger vertical field in the center. If the present intergalactic or intracluster magnetic field is of the order of $10^{-9 \sim -10}$ G, the vertical field strength in the central 1 kpc of normal spiral galaxies is expected to be of the order of a mG.

6. MAGNETIC ACCRETION AND GALACTIC MAGNETIC FOUNTAIN

Magnetic Accretion: The interstellar gas in turn suffers from a magnetic torque by the vertical field which is twisted by the galactic rotation. The time scale with which the rotating gas element loses angular momentum is given by $\tau = Vr/(B^2/4\pi\rho_{gas})$, where ρ_{gas} is the interstelar gas density. This time scale is then calculated to be $\tau \sim 10^{11}, 10^8$, and 10^5 years, respectively at $r = 5$, 1 and 0.1 kpc. This shows that the magnetic torque is not negligible in the inner region of thFe galaxy. The angular momentum of the disk gas is transfered and lost by the magnetic torque. Then the accretion of gas is accelerated, and as well the accretion of the vertical field is accelerated. This magnetic accretion proceeds until the magnetic energy density, $E_m = B^2/8\pi$, becomes comparable to the gravitational/rotational energy density of gas, $E_g \sim \rho_{gas} V^2/2$.

After separation of interstellar gas from stars, the magnetic field is not necessarily passive against the gas motion. In the inner 1 kpc, for example, where the galactic-shock and/or bar-shock cause further accretion coupled with the magnetic accretion, the accretion proceeds until the magnetic energy becomes comparable to the gravitational energy, or the accretion stops when $E_m \sim E_g$. For $\rho_{gas} = 100$ $m_H \text{cm}_{-3}$ and $V \sim 200$ km s^{-1} the magnetic accretion proceeds until the field strength becomes as strong as ~ 1 mG.

The rapid accretion of gas toward the nucleus makes a dense nuclear gas disk, and will be related to starburst activity near the center. We emphasize that the magnetic accretion is effective to create a dense gas disk near the nucleus even without a bar potential and disturbance from companion galaxies, which is often suggested for some interacting galaxies with starburst activity. We note that some non-interacting galaxies like NGC 253 show starburst activity.

Galactic Magnetic Fountain - Large-Scale Circulation: The twisted vertical magnetic field near the nucleus then accelerates a screwing outflow of gas, and results in a vertical jet from the nucleus (Shibata and Uchida 1988). This mechanism will explain many of the observed vertical radio features near the nuclei of spiral galaxies (Hummel et al 1983). The outflow transfers gases from the nuclear region into the halo. The gas in the disk is highly poluted by heavy elements as a result of atarburst activity, etc, so that this large-scale circulation may cause an overall polution of gas in a galaxy: by this large-scale outflow, which we call a "galactic magnetic fountain", heavy elements involved in the nuclear disk gas are spread over the entire galaxy via the halo magnetic fields.

References

Berkhuijsen, E.M., Beck, R., and Gräve, R. 1987, in *Interstellar Magnetic Fields*, ed. R. Beck and R. Gräve (Springer Verlag, Berlin), p.38.
Fujimoto, M., and Sawa, T. 1987, *Publ. Astron. Soc. Japan*, **39**, 375.
Hummel,E., Lesch,H., Wielebinski,R., Schilickerser,R. 1988, *Astron. Astrophys. Letters*, **197**, L29.
Hummel,E., Kotanyi,G.G., van Gorkom,J.H. 1983, *Astrophys. J. Letters*, **267**, L5.
Sawa, T., and Fujimoto, M. 1986, *Publ. Astron. Soc. Japan*, **38**, 133.
Shibata, K., and Uchida, Y. 1987, *Publ. Astron. Soc. Japan*, **39**, 559.
Sofue, Y. 1976, *Astron. Astrophys.*, **48**, 1.
Sofue, Y. 1987, *Publ. Astron. Soc. Japan*, **39**, 547.
Sofue, Y. 1988, *Publ. Astron. Soc. Japan*, **40**, 567.
Sofue, Y., and Fujimoto, M. 1987, *Publ. Astron. Soc. Japan*, **39**, 843.
Sofue, Y., Fujimoto, M., and Wielebinski, R. 1986, *Ann. Rev. Astron. Astrophys.*, **24**, 459.
Sofue, Y., Reich, W., Inoue, M., and Seiradakis, J.H. 1987, *Publ. Astron. Soc. Japan*, **39**, 359.
Sofue, Y.and Tosa, M. 1974, *Astron. Astrophys.*, **36**, 237.
Sofue, Y., Wakamatsu, K., Mallin, D.F., 1990, in this issue.
Tsuboi, M., Inoue, M., Handa, T., Tabara, H., Kato, T., Sofue, Y., and Kaifu, N. 1986, *Astron. J.*, **92**, 818.
Yoshii, Y., and Sommer-Larsen, J. 1989, *Monthly Notices Roy. Astron. Soc.*, **236**, 779.
Yusef-Zadeh, F., Morris, M., and Chance, D. 1984, *Nature*, **310**, 557

THE COSMIC-RAY HALO:
INSIGHT FROM GAMMA RAYS AND COSMIC-RAY OBSERVATIONS

V.A. DOGIEL
P.N. Lebedev Physical Institute
Leninsky pr. 53, SU-117924 Moscow, USSR

ABSTRACT. The physical processes that determine the size of the galactic cosmic-ray halo are discussed and observational information on the cosmic-ray halo is summarized. Based on theoretical investigations as well as observations, we conclude that the halo surrounding the galactic disk is extensive and that its extent in the direction perpendicular to the galactic plane is more than 10 kpc.

1. INTRODUCTION

A halo is generally defined as a quasi-spherical or even substantially flattened region surrounding the galactic disk of a spiral galaxy (see Figure 1). One can speak of many different halos, namely: the stellar halo, gaseous halo, cosmic ray halo, magnetic halo, radio halo, gamma-ray halo, and even the invisible (dark matter) halo (see for instance Ginzburg, 1988). All these halos can be absolutely different in their parameters, which are determined by different physical processes. On the other hand, these processes are not completely independent and very often influence each other. So, we have a complex situation with connections and interactions of halos of different types. In the following we shall mainly discuss the origin of the cosmic-ray halo and its different manifestations in observational data.

Historically, the first evidence for the existence of the galactic cosmic-ray halo came from radio observations. Measurements of the diffuse radio emission showed that its distribution on the sky is anisotropic, but the intensity of the radio emission does not significantly vary for directions perpendicular to the galactic plane. From this one may conclude that a considerable part of the radio flux is generated inside our Galaxy, but the radio emitting region is not limited to the galactic disk. Schklovskii (1952) proposed that the radio emission is generated in a quasi-spherical halo-type volume. However, he connected this radio halo to the stellar halo, suggesting that the radio emission is generated by hypothetic "radio stars" filling the halo region. Soon it became clear that the "radiostar" hypothesis is

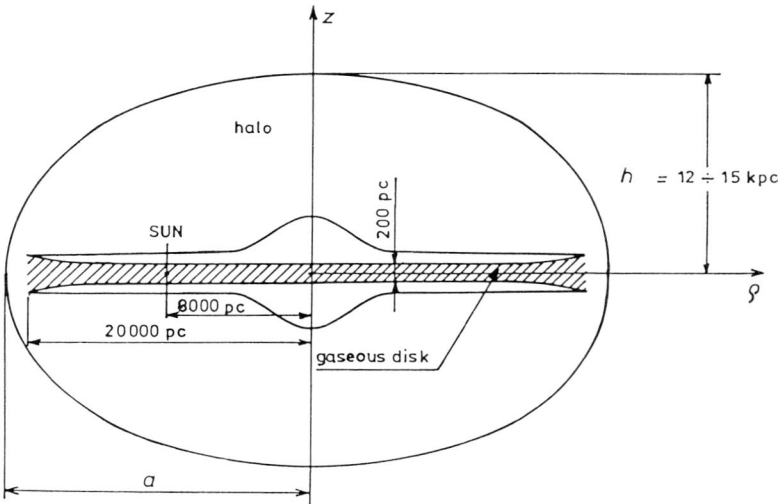

Figure 1. Schematic view of the galactic stellar and gaseous disks and the halo. The Sun is assumed to be at a distance of $\rho_\odot = 8$ kpc from the galactic centre.

invalid and that the observed radio emission is of synchrotron nature, i.e., is generated by relativistic cosmic-ray electrons propagating in the magnetic fields of the halo (Ginzburg, 1953). This was the first time the cosmic-ray halo was mentioned.

2. PHYSICAL GROUNDS FOR THE EXISTENCE OF A COSMIC RAY HALO

The existence of a cosmic-ray halo is expected on physical grounds. One can imagine the interstellar medium of the galactic disk as a mixture of thermal gas, magnetic fields and cosmic rays, with similar energy densities for each of these components. In such a scenario, confinement of cosmic rays inside the disk seems highly improbable, even if all cosmic-ray sources are located inside the disk. Indeed, both theoretical and experimental plasma studies testify to the fact that even in regular magnetic fields (of a special configuration), a plasma develops various instabilities and flows out of the traps. So there seems to be no doubt that cosmic rays flow out of the gas disk and fill up a more extended region. In this scenario, the cosmic-ray halo is an extended region surrounding the gas and stellar disks, filled with cosmic rays and magnetic fields.

Very often the existence of the cosmic-ray halo is connected to the development of so-called Parker instabilities (see, for example, Parker, 1966; Cesarsky, 1980; Kuznetsov and Ptuskin, 1983), which occur in a disk filled with a thermal gas, magnetic fields and cosmic rays. The equilibrium condition in the direction perpendicular to this disk is

$$\frac{\partial}{\partial z}\left(P_g + P_{cr} + \frac{H_o^2}{8\pi} + \frac{H_t^2}{24\pi}\right) = -\rho(z)g(z), \qquad (1)$$

where P_g is the gas pressure, P_{cr} the cosmic-ray pressure, ρ the gas density distribution, g the gravitational acceleration determined by the stellar disk, and H_o and H_t the regular and turbulent components of the interstellar magnetic fields.

The equilibrium state described by eq. (1) can be unstable due to the excitation of Rayleigh-Taylor instabilities, which leads to a situation in which an initially small curvature of a magnetic-field line is increased. As a result, large loops of magnetic fields, filled with cosmic rays, are generated. They may extend far above the galactic plane, with characteristic scales as large as several kiloparsecs.

The characteristics of cosmic-ray propagation are determined by the scattering of cosmic rays on small-scale magnetic-field fluctuations. The interaction of cosmic rays with such fluctuations can be described by a typical scattering frequency $\nu(r, E)$, where r is a space coordinate and E is the particle energy. It is realistic to suppose that the value of ν decreases sufficiently far away from the galactic disk. The kinetic equation for the cosmic-ray distribution function f is of the form (for more details see Dogiel et al., 1990)

$$\frac{\partial f}{\partial t} + v\frac{\partial f}{\partial r} = I(f) + Q(r, E), \tag{2}$$

where v is the particle velocity, $Q(r, E)$ is the distribution of cosmic-ray sources and $I(f)$ is the integral of particle collisions with magnetic inhomogeneities

$$I(f) = \nu(r, E)\frac{\partial}{\partial p_i}\left\{(\delta_{ik}p^2 - p_i p_k)\frac{\partial f}{\partial p_k}\right\}, \tag{3}$$

where p is the particle momentum.

The problem is characterized by the dimensionless parameter

$$\delta = \frac{\nu(r, E)\ell(r)}{v} \qquad \left(\ell = f/\frac{\partial f}{\partial r}\right). \tag{4}$$

The condition $\delta = 1$ determines the boundary $r = \bar{r}$, at which the character of particle propagation changes drastically.

In the inner region, $r \ll \bar{r}$ ($\delta \gg 1$), the distribution function f is quasi-isotropic and particle propagation is described by the diffusion equation with a diffusion coefficient $D \simeq c^2/\nu(r, E)$.

In the outer region, $r \gg \bar{r}$ ($\delta \ll 1$), the function f is highly anisotropic, because particle scattering is not effective there.

The average velocity of the flux of relativistic particles ($v \simeq c$), ejected in the central part of the system, increases at the boundary $r = \bar{r}$ from the value $u = D/\ell \ll c$ (for $r \ll \bar{r}$) to the value $u = c$ (for $r \gg \bar{r}$). So cosmic rays spend a relatively long time in the volume $r \ll \bar{r}$ (where particle scattering is effective), before escaping from the boundary at $r = \bar{r}$ to the metagalactic space.

We shall consider the volume bounded by the surface $r = \bar{r}$ as the cosmic-ray halo. It follows from eq. (4) that the scale height of the halo may a function of the particle energy, $\bar{r} = \bar{r}(E)$.

This type of halo, where particles are diffusively propagating along magnetic field lines, is usually called a "static halo". Another type of halo, "dynamic (convective) halo", in which cosmic rays are transported by a galactic wind with magnetohydrodynamic velocity $u \ll c$, is also probable for our Galaxy (Bulanov et al., 1972, Lerche and Schlickeiser, 1982). This wind is caused by the outward flux of MHD-waves and, according to Breitschwerdt et al. (1987), does probably occur at the periphery of the galactic halo (at a distance of several kiloparsecs from the galactic plane).

The cosmic-ray transport equation, that gives the cosmic-ray density for both halo types, is of the form (see Berezinsky et al., 1990)

$$\nabla(D(\nabla f) - uf) + \frac{\partial}{\partial E}(b(E)f) = Q(r,E), \tag{5}$$

where $b(E)$ is the rate of continuous energy losses.

It can be seen from Eq. (5) that the halo parameters may be different for different components of cosmic rays due to the continuous loss term. We remind the reader that we consider the cosmic-ray halo as a region surrounding the disk, in which the cosmic-ray density is significantly higher than in the intergalactic medium. Speaking of the cosmic-ray halo, it is most reasonable to refer to stable components of cosmic rays. These particles have the largest halo size. For heavy elements (starting from iron), for radioactive elements with a relatively small lifetime (Be^{10} and others), and for electrons which are under the influence of continuous energy losses, the halo scale height will be smaller and, generally speaking, energy dependent.

3. THE HALO OF COSMIC-RAY ELECTRONS AND THE RADIO HALO

As mentioned in the Introduction, the halo of cosmic-ray electrons (or the radio halo, connected with it) was discovered as early as 1952-53. However, the existence of this halo was still doubtful for a long time. There were several reasons for this. First of all it is very difficult to investigate the radio halo of our Galaxy, because we are inside and it is very hard to distinguish between local, metagalactic and halo components of radio emission. Secondly, the attempts to observe radio halos of other spiral galaxies were unsuccessful (see for instance van Woerden , 1967).

Nevertheless, good estimates of the radio flux generated by the cosmic-ray halo could be obtained if a reasonable cosmic-ray electron distribution in the Galaxy could be calculated in the framework of a model with a minimal number of free parameters. This is the case for the diffusion model discussed above (eq. 5). The intensity of the radio emission in a direction ℓ is given by

$$J_\ell(\nu) = \int_\ell d\ell \int_E P(E,\nu) f_e(E,r) dE, \tag{6}$$

where $P(E,\nu)$ is the intensity of synchrotron radiation emitted by a single electron with energy E at a frequency ν and $f_e(E,r)$ is the electron distribution calculated from eq. (5).

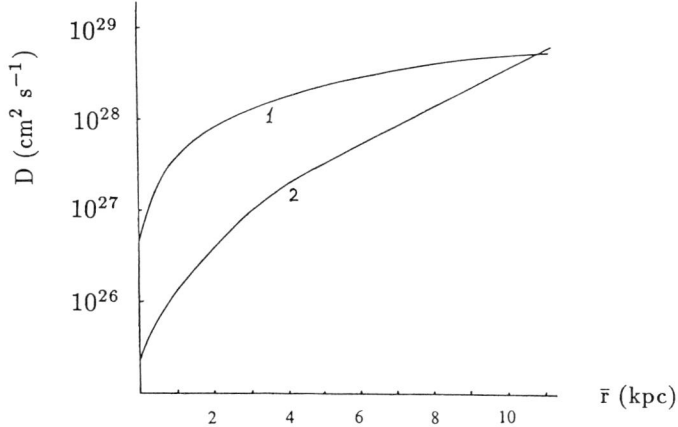

Figure 2. Dependencies of the diffusion coefficient D on the halo extent $\bar{r}$:
(1) obtained from cosmic-ray electron data and diffuse radio emission;
(2) determined from the chemical composition of cosmic rays.

It follows from eq. (5) that electrons with large energies fill only part of the cosmic-ray halo, because of the extensive energy losses. Their path length $\lambda(E)$, determined from the equation

$$\lambda^2(E) = \int_{E_o}^{E} \frac{D(E')}{b(E')} dE' \qquad (7)$$

(E_o is the initial energy of an electron), is smaller than the extent of the cosmic-ray halo, $\bar{r}$ ($\lambda(E) < \bar{r}$). As a result, the size of the cosmic-ray electron halo is energy dependent. Correspondingly, the scale height of the radio halo depends on the radio frequency.

The analysis of the observed diffuse radio emission, based on eqs. (5) and (6), indicated that the size of the cosmic-ray halo should be rather large, $\bar{r} > 5$ kpc (i.e., much larger than the thickness of the galactic disk), for reasonable parameters of the model (Bulanov et al., 1975). In fact, the result of this analysis is a relationship between the diffusion coefficient D and the halo extent $\bar{r}$. This dependence is shown by curve 1 in Figure 2. Thus it was shown that the Galaxy has an extensive cosmic-ray halo for reasonable values of the magnetic field strength.

An important success in the field of radio- and cosmic-ray–halo studies was the discovery in 1977 of pronounced radio halos in the edge-on galaxies NGC 4631 and NGC 891 (Ekers and Sancisi, 1977; Beck et al., 1979). For later observations of radio halos we refer to the review by E. Hummel presented at this Symposium. It is important to notice that the radio observations of edge-on galaxies provide most likely only a lower limit to the size of the cosmic-ray electron halo. Firstly, the sensitivity of the radio telescope is a limiting factor. Secondly, the size of the observed radio halo is not only determined by the electron distribution, but also by the magnetic-field distribution. As the energy density of the magnetic fields decreases towards the periphery of the Galaxy, the cosmic-ray electron halo may be larger than the radio halo.

4. CHEMICAL COMPOSITION OF COSMIC RAYS AND THE HALO

A small part of the nuclei observed in the flux of cosmic rays is not ejected by sources, but generated in the interstellar medium through collisions of cosmic rays with the interstellar gas. These are so-called secondary nuclei (primary nuclei are particles ejected by the sources). By measuring the flux of stable secondary nuclei one can estimate the matter thickness x passed by primary cosmic rays in the Galaxy, given by

$$x = \bar{n}vt, \qquad (8)$$

where $\bar{n}$ is the average gas density in the disk, v is the velocity of the primary nucleus and t is the time the particles spend in the galactic gaseous disk.

The analysis of the value x alone does not allow to estimate the scale height of the cosmic-ray halo. The reason is that the observed chemical composition of cosmic rays (and x) can be reproduced in very different models of cosmic-ray propagation (see e.g. Berezinsky et al., 1990). In other words, the value of x is not (or is weakly) model dependent for stable nuclei. As an example, we consider the equation for x derived in the framework of the diffusion model (Ginzburg and Ptuskin, 1976)

$$x \simeq \frac{\bar{n}vh_g\bar{r}}{D} \qquad (H_g/\bar{r} \ll 1), \qquad (9)$$

where h_g and $\bar{r}$ are the thickness of the gaseous disk and the cosmic-ray halo, respectively. This equation shows that the necessary value of $x \simeq 7$ gr/cm^2 can be obtained for different values of D and $\bar{r}$. The only condition to be satisfied is that the relation between D and $\bar{r}$ must be definite. This condition determines the function $D(\bar{r})$, which is shown in Figure 2 by curve 2. However, in combination with the analysis of the diffuse radio emission, we *can* get unique model parameters. From Figure 2 we see that the two curves describing the cosmic ray chemical composition and the diffuse radio emission intersect at the point: $\bar{r} \approx 10$ kpc and $D \approx 10^{29}$ cm^2/s. So, we immediately come to the conclusion that the cosmic-ray halo is very extended.

Using these parameters we can estimate the cosmic-ray lifetime T in the Galaxy

$$T \sim \bar{r}^2/D \simeq 10^8 \quad \text{years}. \qquad (10)$$

At first glance, this estimate is in contradiction with analyses of the secondary radioactive nuclei flux, which give a much smaller value of $T^* \simeq 10^7$ years (Garcia-Munoz et al., 1977). Consequently, the scale height of the cosmic-ray halo cannot be more than 1 kpc. What is the reason for this discrepancy?

The value of T^* is not measured in experiments. It is calculated in the framework of a model of cosmic-ray propagation from measurements of the fraction of surviving radioactive nuclei. The indicated value of $T^* \simeq 10^7$ years has been got from the leaky-box model, which assumes a uniform cosmic-ray distribution everywhere in the Galaxy. In this case the cosmic-ray lifetime in the Galaxy T^* is given by

$$T^* \simeq (1-f)\tau/f \quad (\tau \ll T^*), \tag{11}$$

where τ is the lifetime of the nucleus with respect to radioactive decay, and f is the surviving fraction of radioactive isotopes.

Let us take as an example the radioactive isotope ^{10}Be measured in the cosmic-ray flux. Its radioactive lifetime $\tau \simeq 2.2 \times 10^6$ years and the fraction f is about 0.15 (Garcia-Munoz et al., 1977). The estimate of T^* from eq. (11) gives the value $T^* \approx 10^7$ years. However, in the case of diffuse propagation in the Galaxy the radioactive nuclei travel during their lifetime a distance of the order $\sqrt{D\tau}$ which is much smaller than the halo scale height $\bar{r}$ if $\tau \ll T$. So, the distribution of radioactive nuclei is essentially nonuniform in the Galaxy and the leaky-box model cannot be used to estimate the cosmic-ray lifetime. In the case of this nonuniform distribution of radioactive nuclei, we do not calculate from eq. (11) the true lifetime of stable cosmic rays $T \simeq \bar{r}^2/D$, but a combination of T and τ (Prischchep and Ptuskin, 1975)

$$T^* \approx \sqrt{\tau T}. \tag{12}$$

So, we cannot exclude that really $T^* < T$. The measurements of radioactive nuclei are, therefore, not in contradiction with the extensive halo for the stable nuclei.

5. ULTRA HIGH ENERGY COSMIC RAYS

Let us see now whether the extensive galactic halo is of essential importance for the ultra high energy cosmic rays. The propagation of ultra high energy cosmic rays in the Galaxy was investigated in a paper by Syrovatskii, (1971). It was shown that galactic sources can provide the observed cosmic-ray flux up to energies $E \leq 10^{17}$ eV, if the confinement region is restricted to the galactic disk. Cosmic rays with higher energies are produced by extragalactic sources. However, this analysis did not include the influence of the cosmic-ray halo on cosmic-ray propagation.

The confinement of ultra high energy cosmic rays ($E = 10^{17} - 10^{19}$ eV) by an extended cosmic-ray halo was investigated by Berezinsky et al. (1979). It was shown that even a relatively small flattened halo with a half thickness $\bar{r} \simeq 3$ kpc retains the particles with energies up to 10^{19} eV for a rather long time if there is a sufficiently strong regular magnetic field component ($\sim 1\mu$G) in the halo. The explanation of this fact is very simple. If the Larmor radius of a particle is smaller than the scale height of the halo, this particle can be returned from the halo to the disk under the influence of large-scale magnetic fields. Figure 3 shows the residence time of cosmic rays in the Galaxy, with (+) and without (o) a cosmic-ray halo. One can see that the residence time of particles with energies $E = 10^{17} - 10^{19}$ eV is significantly larger for the model with a halo. For energies $E > 10^{19}$ eV, the residence time is the same for both cases, i.e. the large-scale magnetic field in the halo does not return the particles with these energies to the disk.

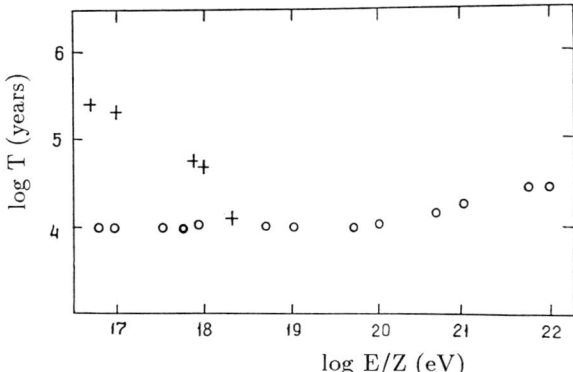

Figure 3. The residence time of ultra high energy cosmic rays in the Galaxy, if a halo exists (+) and if a halo is absent (o).

Due to the larger residence time in case a halo is present, galactic sources can provide the observed cosmic-ray flux at energies $E = 10^{17} - 10^{19}$ eV, which is impossible if a halo is absent.

In the case a halo is present, it is also possible to explain the observed values of anisotropy ($\sim 10^{-4}$) and the intensity of cosmic rays at these energies. The metagalactic models meet serious difficulties in explaining these data, because they give a value for the cosmic-ray anisotropy of about 10^{-5}.

We notice here that the crucial point of this model is the presence of the strong regular magnetic field in the halo. Recently, such regular magnetic fields were discovered in the halo of the galaxy NGC 4631 (Hummel et al., 1988).

6. THE COSMIC-RAY GRADIENT IN THE GALAXY

Gamma-ray astronomy provides methods to address the problem of cosmic-ray propagation in the Galaxy. The point is that the gamma-ray emission from the galactic disk at energies $E > 300$ MeV is mainly due to collisions of cosmic-ray protons and nuclei with the ambient gas (Stephens and Badhwar, 1981)

$$I_\gamma \propto n_H I_{cr} L, \qquad (13)$$

where n_H is the average density of the gas along the line of sight, I_{cr} is the intensity of cosmic rays, and L is the length of the disk filled with gas and cosmic rays. It can be seen that gamma-ray astronomy is as important in investigating the proton-nuclear component of cosmic rays as radio astronomy is for understanding the electron component (compare eqs. (6) and (13)), because it provides estimates of the density distribution of this proton-nuclear component in various regions of the disk.

The sources of cosmic rays (supernovae, pulsars and other active stars) are mainly concentrated close to the central region of the disk. Therefore, we can

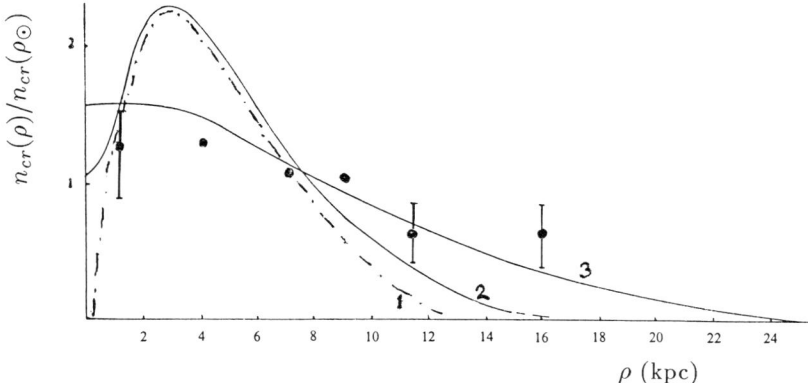

Figure 4. Distribution of the density of cosmic rays in the galactic disk, normalized to the density in the solar neighbourhood, $n_{cr}(\rho_\odot)$. The points indicate the cosmic-ray distribution extracted from the COS-B data (Bloemen, 1989).
1. density distribution of supernovae in the galactic disk.
2. cosmic-ray density distribution for a halo extent $\bar{r} = 600$ pc.
3. cosmic-ray density distribution for a halo extent $\bar{r} = 15$ kpc.

expect that the cosmic-ray density also decreases towards the periphery of the Galaxy. However, analyses of disk gamma-ray emission have shown that the gradient of cosmic rays is very small (see e.g. the review by Bloemen, 1989). The gradient of the cosmic-ray density is much less than the gradient of the density of the potential sources (see Figure 4). This fact imposes a strong constraint on models of cosmic-ray propagation in the Galaxy. A similar conclusion was presented by Jones and Stecker (1975). They used a somewhat steeper cosmic-ray gradient, obtained from the SAS-2 gamma-ray data, and derived a halo scale height of about 3 kpc.

In Figure 4, the calculations of the cosmic-ray distribution in the disk (see reviews by Bloemen, 1989; Dogiel and Ginzburg, 1989) are shown for two scales of the cosmic-ray halo: 600 pc (curve 2) and 15 kpc (curve 3). One can see that if the halo is small, the cosmic-ray density distribution becomes similar to the distribution of the sources. This is due to the fact that in the case of a small halo there is little mixing of cosmic rays within the Galaxy. Only in the case of a large halo the results of calculations agree with the distribution of cosmic rays in the disk obtained from the gamma-ray data. From this the conclusion follows that the experimental data point to an effective mixing of cosmic rays in the Galaxy, which is only possible if the Galaxy has an extensive halo. This situation can be understood in the framework of the static halo model. A "very strong" wind suppresses cosmic-ray mixture and the cosmic-ray distribution will in this case be similar to the source distribution, as in the case of a small static halo.

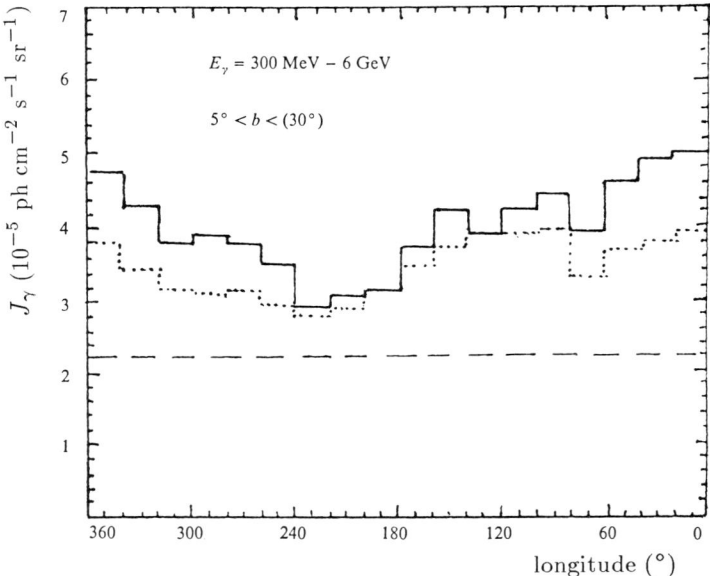

Figure 5. Longitude distribution of the gamma-ray intensity away from the galactic plane. The solid curve is the observed radiation and the dotted curve is the estimated contribution of the gaseous disk.

Another way to investigate this problem was suggested by Melisse and Bloemen (1990). They tried to distinguish between two models of cosmic-ray distribution: the "coupling model" and the "gradient model". In the first model, cosmic rays occupy only the thin gaseous disk and their density is proportional to the gas density. In the "gradient model", cosmic rays have a smoother distribution than the gas distribution and the volume filled by cosmic rays is much larger than the gaseous disk. Their analysis showed that the "gradient model" fits the distribution of the diffuse gamma-ray emission in the disk much better than the "coupling model". This supports the reality of the extended cosmic-ray halo.

7. THE GAMMA-RAY HALO

A noticable flux of radio emission from the halo means that the halo is filled with high-energy electrons (see Section 2). These electrons are also able to produce a gamma-ray flux from the halo through their inverse-Compton scattering on low-energy photons (relic, IR, optical, UV). Let us see if the gamma-ray flux from the halo can be observed and if there are any indications for the existence of this gamma-ray halo.

The estimates by Dogiel and Uryson (1988) show that the halo can account for a substantial part of the observed high-latitude gamma-ray intensity. The estimated gamma-ray flux of the halo is about 3×10^{38} erg/s at energies $E_\gamma > 100$ MeV, which is about 30% of the total gamma-ray flux of the Galaxy.

TABLE 1. Estimates of the extent of the cosmic-ray halo.

Observations	evidence for a halo?	extent
1. Diffuse galactic radio emission	yes	> 5 kpc
2. Edge-on galaxies	yes (for some of them)	$1 - 5$ kpc
3. Chemical composition	model dependent (yes, in combination with radio data)	—
4. Ultra high energy cosmic-ray anisotropy	seems to be yes	$\gtrsim 3$ kpc
5. Cosmic-ray gradient		
a. SAS-2 data	yes	~ 3 kpc
b. COS-B data	yes	~ 15 kpc
6. High-latitude gamma-ray excess		
a. p-p collision origin	yes	$\gtrsim 1$ kpc
b. Inverse-Compton origin	yes	~ 10 kpc

A substantial gamma-ray flux from the halo should show up as an anisotropic gamma-ray component, which is not correlated with the gas column density, i.e. which is not generated in the gaseous disk. An anisotropic excess of gamma-ray emission of this kind was indeed discovered in analyses of the diffuse gamma-ray emission (Bhat et al., 1985; Lebrun and Paul, 1985; Bloemen, 1989). The longitudinal distribution of this excess for the latitude region $5° - 30°$ at energies 300 MeV – 5 GeV is shown in Figure 5, taken from the review by Bloemen (1989). The inverse-Compton scattering of relativistic electrons in a large halo can easily provide the required flux of the gamma-ray excess (see the reviews by Bloemen, 1989, and Dogiel and Ginzburg, 1989). However, this excess could also be due to collisions of cosmic-ray nuclei with the diffuse warm ionized gas with a density of about 0.03 cm^{-3} and a scale height ~ 1 kpc (see Bloemen et al., 1988). Anyway, in both cases the observed gamma-ray excess is generated by the cosmic-ray halo.

8. CONCLUSIONS

Theoretical treatments of the stability of the galactic disk show that it should be surrounded by a region filled with cosmic rays and magnetic fields (i.e., the cosmic halo). The cosmic-ray scattering on magnetic field inhomogeneties increases the life time of cosmic rays in the Galaxy. We define the cosmic-ray halo as a region in which the scattering efficiency is high. The halo extent is different for different components of cosmic rays, but for stable cosmic ray nuclei it is more than 10 kpc. From analyses of observational data, we conclude that the halo existence is doubtless (see Table 1).

REFERENCES

Beck R.J., Bierman P., Emerson D.T., Wielebinski R. 1979, *Astron. Astrophys.* **77**, 25.
Berezinsky V.S., Mikhailov A.A., Syrovatskii S.I. 1979, *Proc. 18th Int. Cosmic Ray Conf.***2**, 86.
Berezinsky V.S., Bulanov S.V., Ginzburg V.L., Dogiel V.A., Ptuskin V.S. 1990, *Cosmic Ray Astrophysics*, ed. V.L.Ginzburg, North Holland.
Bhat C.L., Mayer C.J., Wolfendale A.W. 1984, *Astron. Astrophys.*, **140**, 284.
Bloemen J.B.G.M. 1989, *Ann. Rev. Astron. Astrophys.* **27**, 489.
Bloemen J.B.G.M., Reich P., Reich W., Schlickeiser R. 1988, *Astron. Astrophys.* **204**, 88.
Breitschwerdt D., McKenzie J.F., Völk H.J. 1987, in *Interstellar Magnetic Fields*, eds. R. Beck and R. Gräve, Springer Verlag, p.131.
Bulanov S.V., Dogiel V.A., Syrovatskii S.I. 1972, *Kosmich. Issled.* (Space Research, in Russian) **10**, 532, 721.
Bulanov S.V., Dogiel V.A., Syrovatskii S.I. 1975, *Kosmich. Issled.* (Space Research, in Russian) **15**, 787.
Cesarsky C.J. 1980, *Ann. Rev. Astron. Astrophys.* **18**, 289.
Dogiel V.A., Ginzburg V.L. 1989, *Space Sci. Rev.* **49**, 311.
Dogiel V.A., Uryson A.V. 1988, *Astron. Astrophys.* **197**, 335.
Dogiel V.A., Gurevich A.V., Zybin K.P. 1990, submitted to Astron. Astrophys.
Ekers R.D., Sancisi R. 1977, *Astron. Astrophys.* **54**, L973.
Garcia-Munoz M., Simpson J.A., Wefel J.P. 1977, *Proc. 17th Int. Cosmic Ray Conf.* **2**, 72.
Ginzburg V.L. 1953, *UFN* **51**, 343 (in Russian).
Ginzburg V.L. 1989, in *Essays on particles and fields*. MGK Menon Festschrift, Indian Academy of Sciences, Bangalore, p.103.
Ginzburg V.L., Ptuskin V.S. 1976, *Rev. Mod. Phys.* **48**, 161.
Hummel E., Lesch H., Wielebinski R., Schlickeiser R. 1988, *Astron. Astrophys.* **197**, L29.
Kuznetsov V.D., Ptuskin V.S. 1983, *Astrophys. Space Sci.*, **94**, 5.
Lerche I., Schlickeiser R. 1982, *Astron. Astrophys.* **107**, 148.
Lebrun F., Paul J.A. 1983, *Astrophys. J.* **266**, 276.
Melisse J., Bloemen J.B.G.M. 1990, *Proc. 21st Int. Cosmic Ray Conf.* **1**, 137.
Parker E.N. 1966, *Astrophys. J.* **145**, 811.
Prishchep V.L., Ptuskin V.S. 1975, *Astrophys. Space Sci.*, **32**, 265.
Shklovskii I.S. 1952, *Astron. Zh.* **29**, 418 (in Russian).
Stephens S.A., Badhwar G.D. 1981, *Astrophys. Space Sci.* **76**, 213.
Syrovatskii S.I. 1971, *Comments Ap. Sp. Phys.* **3**, 155.
van Woerden (Ed.), 1967, *Radio Astronomy and the Galactic System*, Academic Press.

RADIO STUDIES OF COSMIC RAYS IN THE GALAXY

W. REICH
Max-Planck-Institut für Radioastronomie
Auf dem Hügel 69
D-5300 Bonn 1, FRG

ABSTRACT. Changes of the cosmic ray electron spectrum throughout the Galaxy have been found, based on the comparison of large-scale radio continuum surveys. These observations are not compatible with the assumption of a static Galactic halo, but indicate the existence of a Galactic wind. Galactic plane surveys reveal sources of cosmic ray electrons in the Galactic disk. Recent studies of the population of radio sources show no evidence for a large number of compact Galactic non-thermal sources. Most of the extended sources are probably HII-regions. Relatively few new supernova remnants (SNRs) with low surface brightness could be identified. Most of the non-thermal emission in the disk-halo interface seems diffuse or unresolved, even at arcmin angular resolution.

1. INTRODUCTION

The radio continuum emission observed from the Galaxy for frequencies up to a few GHz is dominated by the synchrotron emission process. Thermal emission is seen within a few degrees of the Galactic plane. The intensity of synchrotron emission depends on the number of cosmic ray electrons and the magnetic field component perpendicular to the line of sight. It decreases with frequency for a power law electron energy spectrum $N \sim E^{-\gamma}$, with a direct relation of the energy spectral index γ to the observed brightness temperature spectral index $\beta = (1-\gamma)/2 + 2$. β is calculated from the measured brightness temperatures $T_\nu \sim \nu^{-\beta}$ at two frequencies ν.

Information on the radio emission from the Galaxy is based on two types of surveys. All-sky or large-scale surveys at medium angular resolutions have been made to study the disk and the high latitude emission, but also show large local structures like the giant radio loops. Because of our position inside in the Galaxy these surveys need absolute calibration. Model fits are required to get an idea of the three-dimensional distribution of the Galactic emissivity. A comparison of surveys at different frequencies results in spectral indices of the various Galactic components.

Sources for cosmic ray electrons are supposed to be highly concentrated in the Galactic plane. To resolve the mixture of diffuse emission and the large concentration of discrete emission complexes in the Galactic plane higher angular resolution surveys are needed. Galactic plane surveys are the basis for the identification of sources.

2. LARGE SCALE GALACTIC RADIO CONTINUUM SURVEYS

A list of absolutely calibrated large-scale surveys is given in Table 1. At present the all-sky survey at 408 MHz has the highest angular resolution and sensitivity. The 1420 MHz survey of the northern sky has a somewhat higher angular resolution and matches the 408 MHz survey in sensitivity. Observational problems increase with frequency and the listed 2.3 GHz and 2.7 GHz surveys are likely to be the limit for ground-based sensitive large-scale continuum surveys.

The 408 MHz survey has been used by Phillipps et al. (1981) and Beuermann et al. (1985) to obtain three-dimensional models of the Galactic emissivity. Their results are somewhat different, especially in respect to the z-extent and strength of the thick disk or halo component of the Galaxy. Thick disks or halos around galaxies are certainly better studied on the basis of sensitive radio observations of edge-on galaxies, which have become available during the last years (Hummel, this volume).

TABLE 1.
Large-Scale Surveys (HPBW < 11°)

Frequency [MHz]	HPBW [°]	Sensitivity [K]	Declination Coverage	Authors
10	2.6 ×1.9secz	20000	-5° to 74°*	Caswell 1976
30	11	1000	all-sky	Cane 1978
38	7.25×8.25	400	> -25°	Milogradov-Turin et al. 73,84
85	3.8 ×3.5	100	all-sky	Yates 1968
150	2.2	25	all-sky	Landecker, Wielebinski 1970
178	0.22×4.6	20	-5° to 88°	Turtle, Baldwin 1962
404	8.5 ×6.5	3	> -20°	Pauliny-Toth, Shakeshaft 1962
408	0.85	2	all-sky	Haslam et al. 1982
820	1.2	0.5	-7° to 85°	Berkhuijsen 1972
1420	0.59	0.05	> -19°	Reich 1982, Reich, Reich 1986
2300	0.33	0.04	southern sky*	Jonas et al. 1985
2700	0.3	0.03	northern sky* *incomplete	Reif et al. 1987

References: Berkhuijsen 1972, Astr. Astrophys. Suppl. 5, 263; Cane 1978, Aust. J. Phys. 31, 561; Caswell 1976, Mon. Not. R. Astr. Soc. 177, 601; Haslam et al. 1982, Astr. Astrophys. Suppl. 47, 1; Jonas et al. 1985, Astr. Astrophys. Suppl. 62, 105; Landecker, Wielebinski 1970, Aust. J. Phys. Astr. Suppl. 16, 1; Milogradov-Turin, Smith 1973, Mon. Not. R. Astr. Soc. 161, 269; Milogradov-Turin 1984, Mon. Not. R. Astr. Soc. 208, 379; Pauliny-Toth, Shakeshaft 1962, Mon. Not. R. Astr. Soc. 124, 61; Reich, W. 1982, Astr. Astrophys. Suppl. 48, 219; Reich, P., Reich, W. 1986, Astr. Astrophys. Suppl. 63, 205; Reif et al. 1987, Mitt. Astr. Ges. 70, 419; Turtle, Baldwin 1962, Mon. Not. R. Astr. Soc. 124, 36; Yates 1968, Aust. J. Phys. 21, 167

3. SPECTRAL INDEX VARIATIONS IN THE GALAXY

Variations of the radio spectral index with Galactic latitude have been interpreted in terms of a disk-halo model of the Galaxy (Webster, 1975; Sironi, 1976; Strong, 1977; Webster, 1978). These investigations are mainly based on low resolution drift scans at low frequencies. Since the derived spectra indicate some steepening with Galactic latitude, the data have been interpreted in terms of a flat spectrum disk and a steep spectrum halo component. However, the quality of the data used seems poor if compared with more recent observations (Lawson et al., 1987; Reich and Reich, 1988a,b) and in addition the spectra suffer from confusion of the Galactic background emission with emission from the giant radio loops. In fact, it has been shown in a spectral analysis by Milogradov-Turin (1985, 1987) that a steep spectrum halo is not in agreement with the data, if the influence of the steep spectrum North Polar Spur is taken into account. Lawson et al. (1987) presented a newly calculated spectral index map between 38 MHz and 408 MHz, which shows slightly steeper spectra at higher latitudes than in the Galactic plane ($\Delta\beta = 0.1$) mostly in the regions of the giant radio loops. Globally, flatter Galactic spectra are observed at low frequencies rather than at higher frequencies. This reflects the steepening of the electron energy spectrum with increasing energy (Webber, 1983). Even if the electron spectrum at high z is the same as in the Galactic plane, steeper radio spectra at high z are expected. It is evident from the characteristic frequency $\nu_c \sim B\ E^2$ that a decrease of the magnetic field strength B with z requires radiating electrons with higher energies E than in the Galactic plane.

The most detailed map of Galactic spectral indices across the northern sky has been calculated by Reich and Reich (1988a) based on the 408 MHz and the 1420 MHz surveys. Figure 1 shows the 1420 MHz data and the distribution of temperature spectral indices. The map of spectral indices shows significant variations along the Galactic plane, and from the plane towards higher Galactic latitudes. It has been discussed in detail by Reich and Reich (1988b). Besides some local structures steep spectra are seen in the first Galactic quadrant ($l < 55°$, $|b| < 20°$). Correcting for thermal and foreground emission a spectral index of $\beta = 3.1$ was found for the non-thermal emission in the inner part of the Galactic disk. This result is in close agreement with that of Hirabayashi (1974), who found $\beta = 3.08\pm0.11$ for the non-thermal spectrum between 1.4 GHz and 10 GHz. The non-thermal emission in the second and third quadrant has a spectral index close to $\beta = 2.85$ in the plane and the spectra flatten continously with increasing Galactic latitude. The flattest spectra are seen at $l \sim 230°$, $b \sim 30°$ and $l \sim 200°$, $b \sim -45°$. As concluded by Reich and Reich (1988b), these results are unexpected for a static or purely convective Galactic halo, but agree to some extent with the cooling-convection halo models as proposed by Lerche and Schlickeiser (1982). These models are based on the existence of a Galactic wind.

It is of interest that a spectral analysis by Bloemen et al. (1988) of the γ-ray emssion as observed by the COS-B satellite, shows a similar spectral flattening with increasing z in the second and third quadrant as found for the radio data. This suggests a hardening of both the cosmic ray proton and electron spectrum with increasing z at GeV energies in

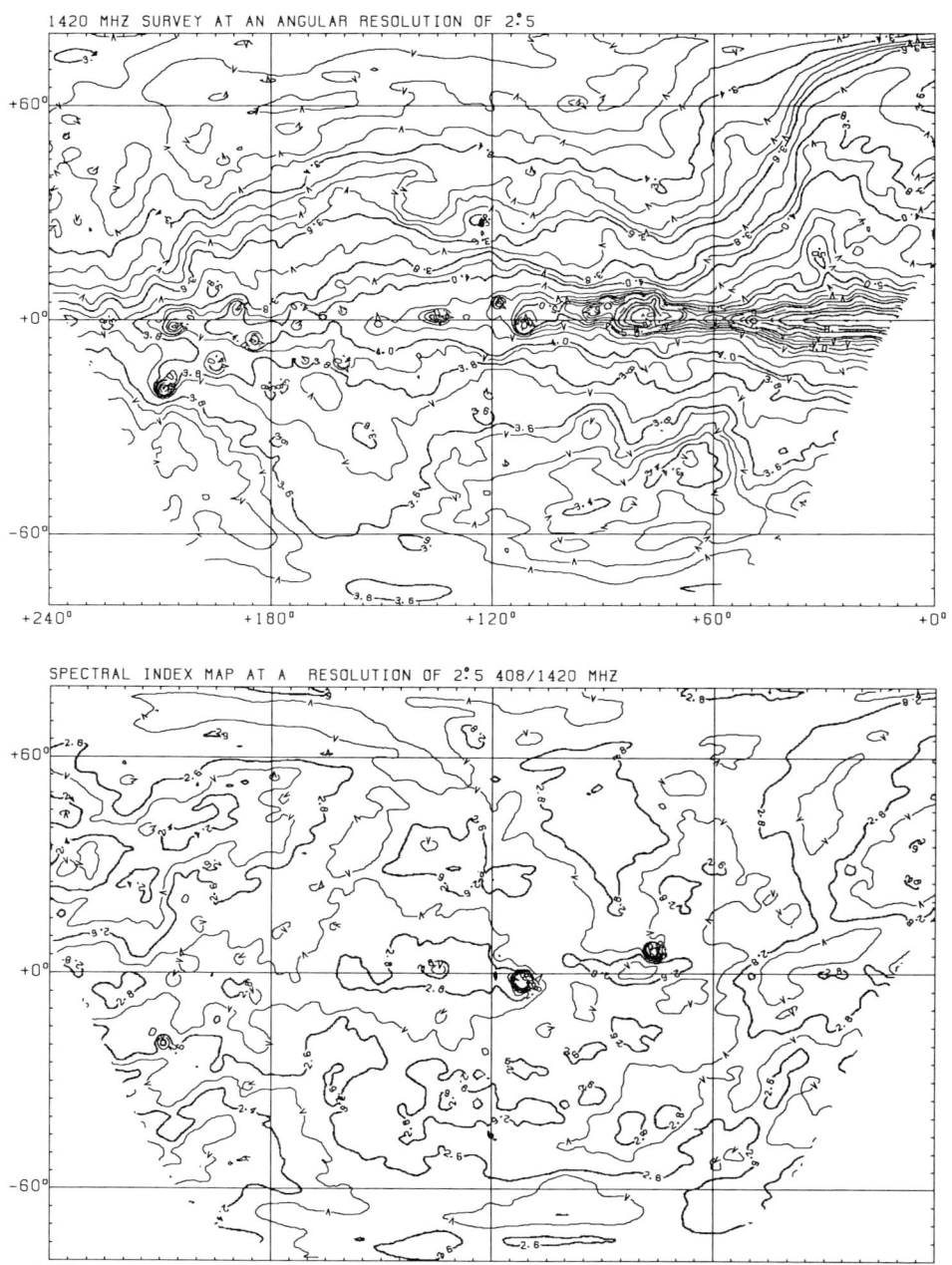

Figure 1. Coordinates are Galactic. Contours are in K T_B (top) and represent temperature spectral indices (bottom) (see Reich and Reich, 1988a,b).

the outer Galaxy, which further supports the asumption of the existence of a Galactic wind.

Reich and Reich (1988b) have used the edge-on view of the Galaxy as modelled by Beuermann et al. (1985) from the 408 MHz survey and decomposed the brightness temperatures along the plane into an inner and outer non-thermal component with spectral indices of $\beta = 3.1$ and $\beta = 2.85$, and a thermal component for the inner disk. Figure 2 shows the predicted spectral index variations. The rather small spectral differences result from the superposition of the three components in the line of sight. A very wide frequency coverage is necessary to sort out the three components even for our simple model. It is obvious that Galactic wind effects in edge-on galaxies must be much more pronounced than in our Galaxy if they are to be observable.

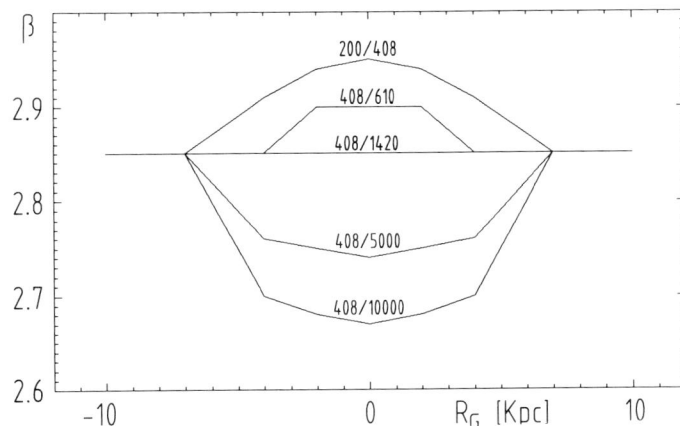

Figure 2. Variation of the temperature spectral index β of the Galactic disk emission when seen edge-on. The frequency pairs are indicated.

4. GALACTIC PLANE SURVEYS

Galactic plane surveys with arcminute resolution are listed in Table 2. Highest angular resolution has been obtained at 4.9 and 10 GHz, which is required to see details within complex structures. Even higher resolution data come from synthesis telescopes, where some source surveys are currently in progress. The maps with highest sensitivity are from the Effelsberg 1.4 GHz and 2.7 GHz surveys. A large number of sources has been listed. The identification of these sources is a long- term project, but global statistical results can be given.

5. COMPONENTS OF THE GALACTIC DISK EMISSON

Figure 3 shows the number distribution of compact sources with a peak flux density larger than 40 mJy at 2.7 GHz as a function of Galactic latitude. For $l < 100°$ there is an excess of about 900 sources at low latitudes with intrinsic sizes between 2'.5 to 11'. The scale height is about 0°.6. A weak excess is seen for sources with sizes between 1'.3 and 2'.5 for latitudes below 0°.5. Unresolved sources show no excess, although

TABLE 2.
Selected Single Dish Galactic Plane Surveys (HPBW < 10')

Frequency [GHz]	HPBW [']	Sensitivity [K]	Coverage l	b	Authors / Telescope
1.4	9	0.2	93° - 162° 236 sources	4°	Kallas, Reich 1980 Effelsberg 100-m
1.4	9.4	0.125	357° - 95.5° 884 sources	4°	Reich et al 1990a Effelsberg 100-m
2.7	8.2	0.2	190° - 61° 890 sources	2°	various see Day et al 1972 Parkes 64-m
2.7	4.3	0.05	357.4° -240° 6483 sources	5°	Reich et al 1984, 1990b Fürst et al 1990a,b Effelsberg 100-m
4.9	2.6	0.2	357.5° - 60° 1186 sources	1°	Altenhoff et al 1978 Effelsberg 100-m
5	4.1	0.2	190° - 40° 915 sources	2°	Haynes et al 1978 Parkes 64-m
10	2.7	0.1	356° - 56° 144 sources	1.5°	Handa et al 1987 Nobeyama 45-m

References: Altenhoff et al. 1978, Astr. Astrophys. Suppl. 35, 23; Day et al. 1972, Austr. J. Phys. Astr. Suppl. 25, 1; Fürst et al., 1990a,b Astr. Astrophys. Suppl. in press (October vol.); Handa et al. 1987, Publ. Astr. Soc. Japan 39, 709; Haynes et al. 1978, Austr. J. Phys. Astr. Suppl. 45, 1; Kallas, Reich 1980, Astr. Astrophys. Suppl. 42, 227; Reich et al. 1984, Astr. Astrophys. Suppl. 58, 197; Reich et al. 1990a, Astr. Astrophys. Suppl. 83, 539; Reich et al. 1990b, Astr. Astrophys. Suppl. in press (October vol.)

some weaker sources are likely to be missed in some highly confused areas of the Galactic plane. Garwood et al. (1988) have used the VLA to make a source survey at 1.5 GHz. They found an excess of about 20% of sources for $l < 40°$ above the extragalactic source counts. This excess is due to compact structures towards identified thermal sources. Figure 3 shows for $l > 100°$ a quite uniform source distribution with latitude. Source counts for small diameter sources with sizes less than 2.5 agree within a few percent with extragalactic source counts (Fürst et al., 1990). High resolution VLA observations of 135 sources smaller than 2' in the anticentre direction by Fich (1986) resulted in the identification of 11 HII-regions, but no compact SNR. These statistical results indicate a paucity of compact non-thermal Galactic sources, which might be young far distant SNRs, or more exotic objects like SS433 or Sco X-1.

The Effelsberg surveys have been used to identify extended low surface brightness SNRs. Confusion is lowest in the anticentre direction and SNRs with a surface brightness as low as $\Sigma_{1GHz} \sim 2 \cdot 10^{-22}$ W m^{-2} Hz^{-1} sr^{-1} are clearly visible, but only three sources are seen within $90° \leq l \leq 240°$. The surface brightness of the Galactic background emission is more than ten times higher and the few observable low surface brightness SNRs imply a minor contribution of this class of sources to the unresolved background. A number of new SNRs could be identified in the

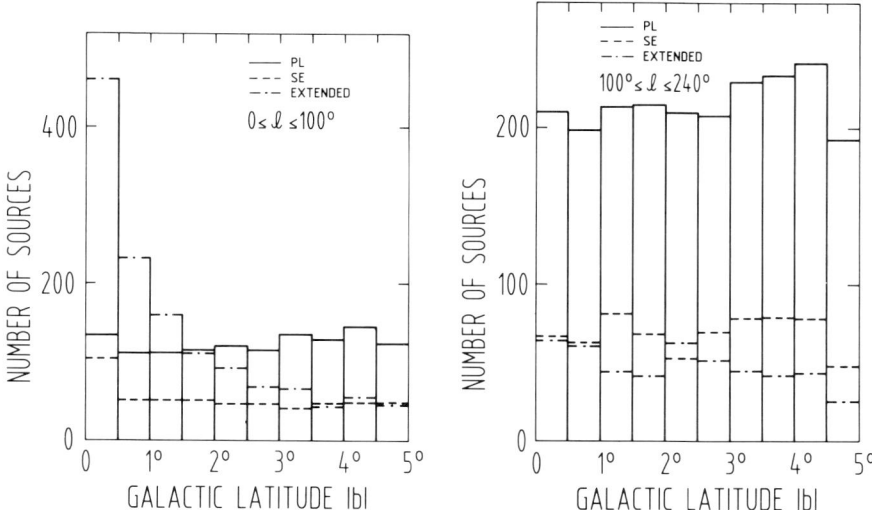

Figure 3. The number of radio sources at 2.7 GHz versus absolute Galactic latitude. Point-like (PL) sources are unresolved with intrinsic sizes less than 1!3. Slightly extended (SE) sources have intrinsic sizes between 1!3 and 2!5. Extended sources have sizes between 2!5 and 11'.

first galactic quadrant (Reich et al., 1988). Omitting the highly confused regions with $|b| \leq 0°.5$ the distribution of Σ for "known" sources (Green, 1988) and "new" ones is given in Table 3.

TABLE 3.
Distribution of SNR Surface Brightness
$357° < l < 76°$, $0°.5 \leq |b| \leq 5°$

$-\log(\Sigma)$	<19	19-20	20-21	21-21.7
"known"	1	5	8	3
"new"	1	1	13	13
all SNRs	2	6	21	16

The "new" SNRs are much weaker, on average, than those previously detected. No distances are known for most of the sources. Diameters may be obtained by applying a $\Sigma-D$ relation for shell-type sources. Using the relation given by Milne (1979) cumulative counts $N(<D)$ can be made. For adiabatic expansion $N(<D) \sim D^{5/2}$ is expected, which is seen for diameters up to 40 pc. Based on the $\Sigma-D$ relation only about five sources from Table 3 have larger diameters than 40 pc. Again there is no indication for the existence of numerous faint large diameter SNRs, which contribute significantly to the non-thermal disk emission.

Figure 4 shows the latitude dependence of the integrated 1.4 GHz, 2.7 GHz and IRAS 60 μm emission for $4° \leq l \leq 36°$. Data are at 9.4 angular resolution and have been scaled relative to 1.4 GHz for optically thin thermal emission. A flux density ratio of $S_{60\,\mu m}/S_{2.7\,GHz} = 1000$, which is typical for thermal emission in this longitude range, has been assumed (Fürst et al., 1987). Also shown is the integrated source emission obtained after subtraction of the smooth large-scale emission. About half of the total emission at $b = 0°$ is from discrete sources. Their thermal fraction is close to 90%. HII-regions are the major source contri-

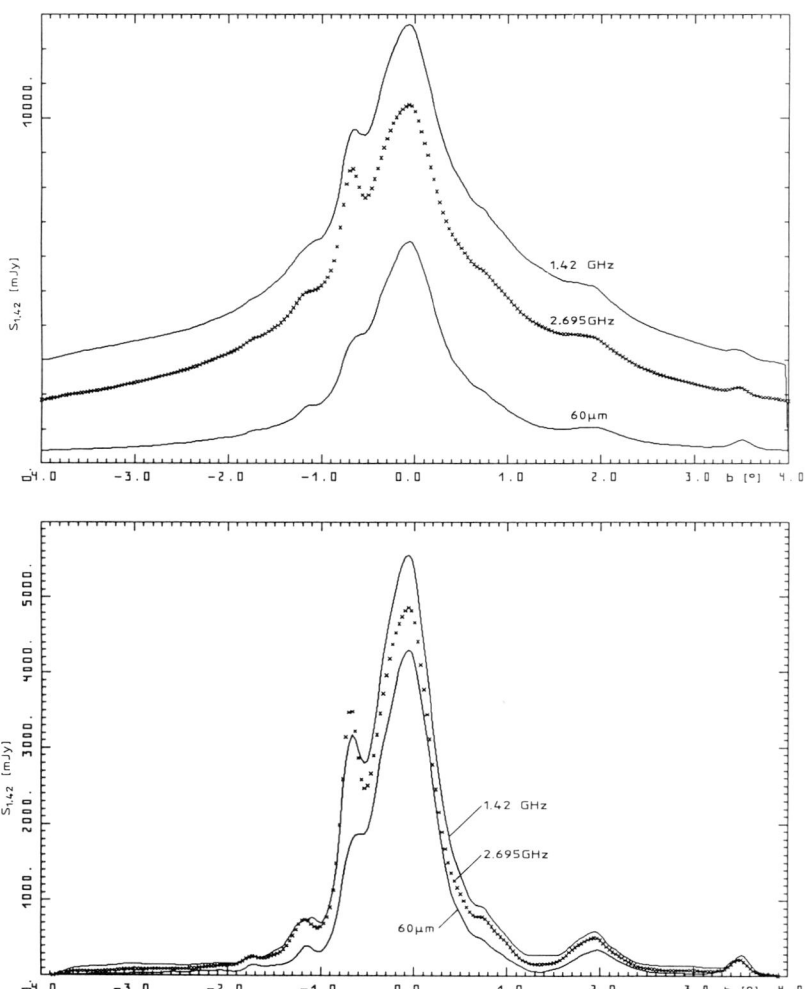

Figure 4. Top: Integrated Galactic emission (see text). $(T_B/S)_{1.4\,GHz} = 2$. Bottom: Integrated emission from discrete sources.

bution in this frequency range. The discrete source component contributes only a few percent to the total emission for $|b| > 1°$.

In Figure 5 the 1.4 GHz and 2.7 GHz data at $b = -2°.5$ are shown, with separate plots for the small-scale and the extended emission. The data are scaled for thermal emission. The non-thermal nature of most of the small-scale and the diffuse emission is evident. The survey maps show that the small-scale emission at a few degrees of Galactic latitude

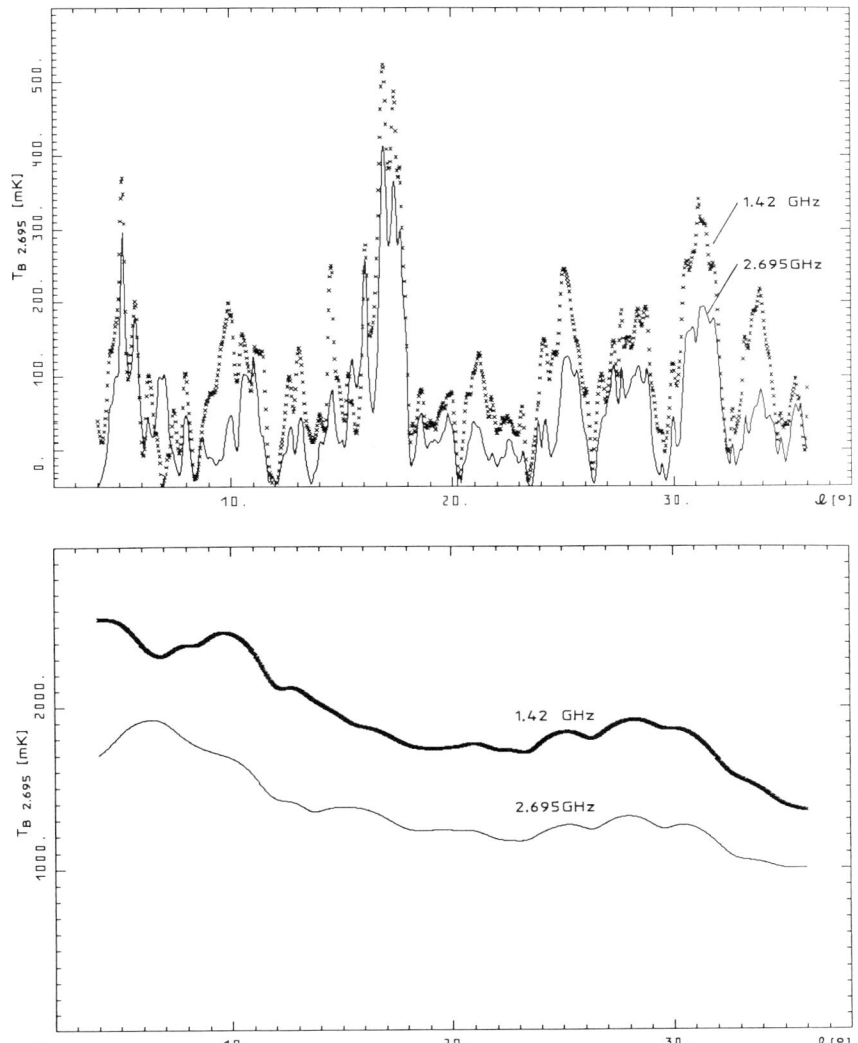

Figure 5. Small-scale (top) and large-scale emission (bottom) at $b = -2°.5$. The 1.4 GHz data have been scaled relative to 2.7 GHz for optically thin thermal emission.

is rather complex and can not be classified in a simple scheme. Its average fraction is of the order of 5% of the diffuse emission.

To summarize: The observed synchrotron emission a few degrees outside of the Galactic plane remains basically unresolved, even when observed with high sensitivity at arcmin angular resolution. This indicates rather small fluctuations in the magnetic field strength and the cosmic ray electron density in the disk-halo interface. A mainly uniform outflow of cosmic ray electrons from the disk sources into the halo is consistent with the observations. Alternatively, a convection of cosmic ray electrons from clustered supernova explosions in OB-associations via chimneys into the halo has been recently proposed (e.g. Norman and Ikeuchi, 1989). In view of the smooth radio emission these chimneys have to be sufficiently numerous and/or large in size with small internal fluctuations in the magnetic field strength and electron distribution to agree with the observations.

REFERENCES

Beuermann, K., Kanbach, G., Berkhuijsen, E.M. (1985) *Astron. Astrophys.* **77**, 25
Bloemen, J.B.G.M., Reich, P., Reich, W., Schlickeiser, R. (1988) *Astron. Astrophys.* **204**, 88
Fich, M. (1986) *Astron. J.* **92**, 787
Fürst, E., Reich, W., Sofue, Y. (1987) *Astron. Astrophys. Suppl.* **71**, 63
Fürst, E., Reich, W., Reich, P., Reif, K. (1990) *Astron. Astrophys. Suppl.*, in press (October volume)
Garwood, R.W., Perley, R.A., Dickey, J.M., Murray, M.A. (1988) *Astron. J.* **96**, 1655
Green, D.A. (1988) *Astrophys. Space Sci.* **148**, 3
Hirabayashi, H. (1974) *Publ. Astron. Soc. Japan*, **26**, 263
Lawson, K.D., Mayer, C.J., Osborne, J.L., Parkinson, M.L. (1987) *Monthly Not. Roy. Astron. Soc.* **225**, 307
Lerche, I., Schlickeiser, R. (1982) *Astron. Astrophys.* **107**, 148
Milne, D. (1979) *Australian J. Phys.* **32**, 83
Milogradov-Turin, J. (1985) in *Proc. IAU Symposium No. 106*, eds. H. van Woerden, R.J. Allen, W.B. Burton, Reidel, Dordrecht, p. 245
Milogradov-Turin, J. (1987) in *Proc. of the 10th ERAM*, ed. J. Palous, Czechoslovak Acad. Sc., Praha, p. 225
Norman, C.A., Ikeuchi, S. (1989) *Astrophys. J.* **345**, 372
Phillipps, S., Kearsey, S., Osborne, J.L., Haslam, C.G.T., Stoffel, H. (1981) *Astron. Astrophys.* **103**, 405
Reich, P., Reich, W. (1988a) *Astron. Astrophys. Suppl.* **74**, 7
Reich, P., Reich, W. (1988b) *Astron. Astrophys.* **196**, 211
Reich, W., Fürst, E., Peich, P., Junkes, N. (1988) in *Proc. IAU Coll. No. 101*, eds. R.S. Roger, T.L. Landecker, Cambridge Univ. Press, p. 293
Sironi, G. (1976) *Astrophys. Space Sci.* **44**, 159
Strong, A. (1977) *Monthly Not. Roy. Astron. Soc.* **181**, 311
Webber, W.R. (1983) in *Composition and Origin of Cosmic Rays*, ed. M.M. Shapiro, Reidel, Dordrecht, p. 83
Webster, A. (1975) *Monthly Not. Roy. Astron. Soc.* **171**, 243
Webster, A. (1978) *Monthly Not. Roy. Astron. Soc.* **185**, 507

IONIZATION IN THE INTERSTELLAR HI REGION BY LOW ENERGY COSMIC RAY ELECTRONS

X. CHI AND A.W. WOLFENDALE
Physics department, University of Durham,
Durham City, DH1 3LE, UK

Recent observations of the pulsar dispersion measures show a large scale height and relatively high density for the diffuse, ionized gas in HI regions. The maintenance of the ionization requires a pervasive ionizing agent. This agent could hardly be the UV photons from O stars or the extragalactic background UV photons due to their large optical depth; nor could it be the soft X–rays from the hot, ionized gas due to a low flux indicated by observations. Studies of the Galactic diffuse low energy γ–rays provides a clue to this problem. The extraordinarily high intensity of MeV electrons, which is derived from the γ–ray flux, could account for the ionization. In this scenario the electrons do both ionization and heating in HI regions, accordingly the atomic hydrogen gas is *warm* ($\sim 10^4$ K). The ionized gas density has a radial gradient with a scale length 4 kpc, there being more ionized gas in the inner Galaxy, due to a gradient in the electron intensity.

The MeV cosmic ray electrons are usually referred to as the 'seed' population, to distinguish them from the high energy population of cosmic rays in the common sense. The generation mechanism of these low energy electrons, in view of their excess over a single power–law injection spectrum extrapolated from the high part, is thought to be different from the standard diffusive shock accleration by young supernova remnants. A likely mechanism is the low–frequency turbulence due to the modified two stream instability in a magnetized plasma, such as it works for particle acceleration in solar flares and the Earth's bow shock. Here, we suggest that this mechanism may be operated effectively by interstellar shocks which are initiated by massive stars and old supernova remnants. Also, unlike the high energy electrons, these low energy electrons propagate slowly, thus they are relatively 'local' and have a large radial gradient, say, 4 kpc for the scale length, a value comparable to the stellar scale length. The diffusion coefficient, we suggest, should lie in the range $10^{27} \sim 10^{28}$ cm^2 s^{-1} and the electrons are able to travel a few hundred pc before losing most of their energy. Nevertheless, they could be convected into the halo by the galactic wind and ionize the gas there. The total energy input for these electrons is estimated to be $\sim 10^{41}$ erg s^{-1} and the ambient energy density is 0.05 eV cm^{-3}.

It can be remarked, finally, that it is unnecessary to postulate the decay products of dark matter particles as the source of the observed ionization. It is also unnecessary to invoke the diffuse ionized gas being *cool* as the interpretation of the inclination dependence of radio power at low frequency (57.5 MHz) observed in external spiral galaxies. This effect, we think, is caused by the spectral flattening of GeV cosmic ray electrons and by the absorption of HII regions.

OPTICAL RADIATION FIELD IN THE DISK AND HALO

X. CHI AND A.W. WOLFENDALE
Physics department, University of Durham,
Durham City, DH1 3LE, UK

The imminent launch of NASA's Gamma Ray Observatory (GRO) has focussed on the need to provide up to date estimates of the energy density of interstellar radiation away from the Galactic Plane, this parameter being a prerequisite for calculation of the flux of gamma rays coming from the inverse Compton scatterings of cosmic ray electrons. In this work we use recent information on the stellar distributions and on the extinction properties of dust in the interstellar medium in the calculation. An important feature of the radiation field is that the field energy density is still substantial in the halo, as was found in our previous work. The differences between groups are due to the total energy input and the dust distribution. Our ensuing energy densities are probably accurate to about 30%, the uncertainty being due to lack of precise knowledge of the input parameters rather than approximations used in the calculation. The result is given by a contour map in Figure 1.

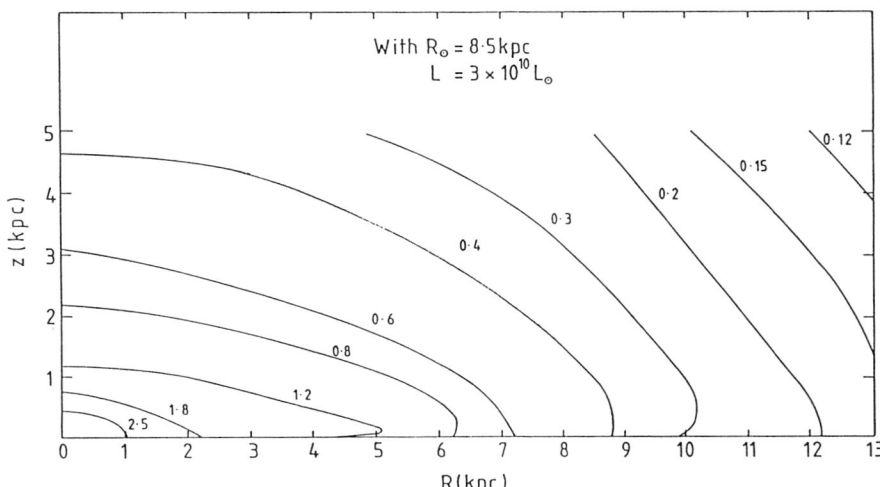

Figure 1. The distribution of the OPT energy density in the Galaxy, with $R_\odot = 8$ kpc and $L_{OPT} = 3 \times 10^{10} L_\odot$. The wavelength range of the OPT is defined as 0.1 μm $< \lambda < 8$ μm. The units in the contour map are eV cm^{-3}.

II

The Disk-Halo Interface in other Galaxies

LARGE SCALE STRUCTURE OF HI IN OTHER GALAXIES

THIJS VAN DER HULST AND JURJEN KAMPHUIS
Kapteyn Astronomical Institute, Postbus 800,
NL-9700 AV Groningen, The Netherlands

ABSTRACT. In this paper we discuss the large-scale structures (> 500 pc) of the neutral atomic hydrogen in the disk of nearby galaxies, with an emphasis on finding evidence for neutral gas features in the disk - halo interface. Most nearby galaxies studied in detail appear to have hole and shell type structures in their HI disks and several show evidence for gas at peculiar velocities, probably examples of gas streaming out of the disk into the halo. Not all of the observed phenomena can be explained by stellar winds and supernova explosions and other explanations will briefly be discussed.

INTRODUCTION

In the last decade many observational and theoretical studies of the interstellar medium in our Galaxy have uncovered a large variety of phenomena, features and processes which play a role in shaping the HI layer. Good examples are the shell- and worm-like structures found by Heiles (1979, 1984). It is believed that these structures are located in the disk and probably have extensions perpendicular to the plane of the disk. The origin of these structures is uncertain. Heiles searched for correlations between shells and OB associations or HII regions and concluded that there was no one-to-one correspondence between the shells and supershells and any other known population of astronomical objects.

In particular the supershells, all located in the outer Galaxy, show a complete lack of association with young stellar populations. The diameters and the kinetic energies of the largest shells are a few kpc and 10^{53} - 10^{54} ergs. Collective stellar winds/supernova explosions are barely sufficient to create these large HI cavities. An alternative explanation is the infall of gas clouds (Tenorio-Tagle et al. 1987, Tenorio-Tagle and Bodenheimer 1988), which more efficiently produce the mechanical energy necessary to drive out the huge HI shells. Evidence for collisions of gas clouds with the Galactic disk has been provided by Heiles (1984), Mirabel and Morras (1990) and Mirabel (cf. this volume). An excellent review of the different properties and the different physical models for the worms and shells can be found in e.g. Heiles (1990). The presence of shells and worm-like features suggests that neutral gas can be brought into the halo of the galaxy. The idea that neutral gas exists in the Galactic halo is of course not new: the since long known High Velocity Clouds (HVCs), HI with peculiar velocities ($V_{LSR} > 70$ km sec^{-1}) at high latitudes, are such a 'halo' population of gas clouds. These clouds are now known to cover a large part of the sky and consist of complexes in a variety of sizes, up to several kpc (for review see cf. van Woerden et al. 1985, Wakker 1990 and this volume). Since the distances to these clouds are unknown, masses, energies and origins are uncertain. The structure of a large fraction of the HVCs can be explained in terms of a galactic fountain model (Bregman 1980, Wakker 1990). The very high velocity clouds, however, are more likely associated with the Magellanic Stream and originate from a tidal interaction with the Magellanic Clouds (Wakker 1990).

In this review we will examine the situation in other galaxies with the question in mind: do other galaxies show the same type of phenomena and if so what can we learn about their origin? Wakker et al. (1989) made a first attempt to search for high velocity gas in existing HI databases of three nearby galaxies, but found no clear examples. As a result of the improved sensitivity of modern synthesis radio telescopes such as the Very Large Array (VLA) and the Westerbork Synthesis Radio Telescope (WSRT) it is now possible to detect HI masses of about 10^6 - 10^7 M$_\odot$ in nearby (D < 10 Mpc) galaxies. For comparison, if we assume that the Galactic HVCs are at a distance of 10 kpc above the plane, the HI masses of the largest HVC complexes are of the order 10^7 M$_\odot$ (Wakker 1990). In addition, the 'chimney' theory (Mac Low, McCray and Norman 1989) predicts that about 5 - 10% of the swept up material in a superbubble will be cold gas blown out into the halo. This implies that we should be able to detect objects like the large HVC complexes and the upper extreme of the superbubble range in other nearby galaxies from such sensitive synthesis observations.

The most suitable objects to search for HI outside the main disk are face-on and edge-on galaxies. The advantage of the edge-on galaxies is that one can directly observe gas high above the disk. Observing face-on galaxies on the other hand does yield information about the velocities of the gas in a direction perpendicular to the disk.

Radio telescopes such as the VLA and the WSRT have spatial resolutions up to $\approx 10''$ corresponding to about 500 pc at a distance of 10 Mpc. This implies that one could distinguish objects like the Heiles' supershells out to such a distance, but that the smaller HI features could only be well resolved in Local Group galaxies. Another consequence of the limited resolution is that beyond 5 Mpc one expects to only detect break-through or blow-out holes assuming that the thickness of the cold neutral medium is about 350 pc like in our Galaxy.

We will first review the evidence for HI at high distances above the plane from observations of edge-on galaxies and then discuss the presence of large holes and HI at peculiar velocities in face-on galaxies.

HI AT HIGH DISTANCES ABOVE THE PLANE

Edge-on galaxies offer the best perspective for finding HI cloud complexes at large distances from the plane and for finding features which extend from the plane to high z distances. A few galaxies have been observed in detail recently. These are NGC 3079 (Irwin and Seaquist 1990), NGC 891 (Broeils, Sancisi and van Albada, in preparation; Rupen, in preparation) and NGC 4565 (Rupen, in preparation).

NGC 3079 is a galaxy (D = 10 Mpc) with an active nucleus and has a peculiar distribution of non-thermal radio emission showing two radio lobes along the minor axis (Duric et al. 1983 and references therein). Irwin and Seaquist (1990) observed the galaxy in HI and found 15 extensions, which may be the equivalent of the worms and shells found in our Galaxy by Heiles (1979, 1984). These extensions are randomly distributed over the HI disk and not concentrated toward the nuclear region, so that a relation with the non-thermal nuclear activity is unlikely. In addition they found 5 arcs, 2 to 6 kpc in size, in the outer regions of the galaxy. Here, however, the situation becomes confused because of the warp of the galaxy.

Other edge-on galaxies observed in detail are NGC 4565 and NGC 891. Both show extensions to moderate distances from the plane ($\approx$ a few kpc) at marginal column density levels. NGC 891 has been observed in Hα by Rand et al. (1990) and Dettmar (1990), and at very low level they detect spurs of ionized gas extending to $\approx$ 2 - 4.5 kpc in z. No clear correlation exists between the Hα filaments and extensions seen in the HI. Perhaps we need to look deeper, at lower column density levels, and work on a new, very sensitive WSRT observation is in progress. The existing data of Broeils et al. do, however, reveal one promising case of a gas feature extending out to 5 kpc in z.

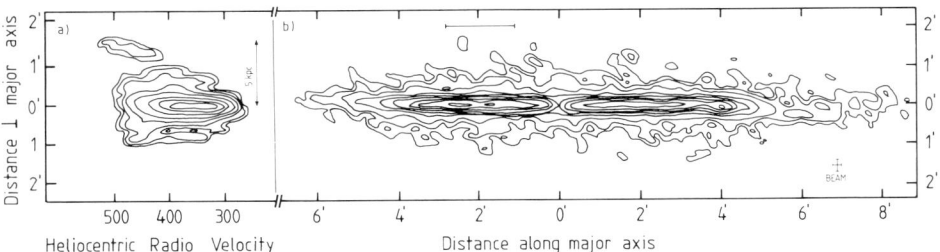

Figure 1. Contour map (b) of the total column density distribution of NGC 891 (contour levels are 1.6, 3.2, 6.4, 12.8, 19.2, 25.6, 32.0, 38.4, 44.8$\times 10^{20}$ cm^{-2}) at 13″×19″ resolution and, at the left, the position - velocity diagram (a) integrated over the high z gas (contour levels are 2.5, 5, 10, 20, 40, 80, 120, 160 mJy/beam). The integration interval is given by the bar in (b).(Image courtesy: A. H. Broeils)

Figure 1 shows the integrated HI map of NGC 891 (b) and a position-velocity map (a) integrated along the high-z feature indicated on Figure 1a. The position-velocity map does show a clear continuity in velocity, suggesting strongly that this feature is real and that the high z gas may be related to phenomena in the disk. The precise connection with the disk is, however, not yet clear and until more examples of this kind are found one could only speculate.

HOLES

If galaxies have a general flow of material from the disk into the halo produced by winds and supernova explosions, one would expect to find holes and cavities in the HI layer. Face-on galaxies are ideal objects to search for such effects. The first examples of holes are the expanding HI shells found in the SMC by Hindman (1967). Allen and Goss (1979) noted the existence of large holes with diameters of about 1 to 5 kpc in the giant Sc galaxy M 101. Since then more information has become available on nearby galaxies and the existence of holes, near-holes and shells in the HI distribution of galaxies appears to be more common than originally thought.

HI observations of the Large Magellanic Cloud (McGee and Milton 1966, Rohlfs et al. 1984) show several large holes. In the LMC at least 5 out of the 9 Hα supershells (Meaburn 1980, Meaburn et al. 1987) are associated with HI holes or loop-like features (Dopita, Mathewson and Ford 1985). The HI masses in the shells are in the range 6 - 25 10^6 M$_\odot$.

Brinks and Bajaja (1986) searched the HI observations of the Sb galaxy M 31 for holes and found 141 examples. Most of these are roughly elliptical in shape with sizes ranging from 100 pc (the resolution limit) to $\approx$ 1 kpc. The missing HI mass in the largest holes is of the order of a few 10^6 M$_\odot$. The estimated energies and ages of these largest holes determined, assuming that the surrounding HI gas is in the snowplough phase, are 10^{52} - 10^{53} ergs and about 10^7 years, respectively. None of these holes should be classified as a supershell, especially because the largest holes are probably inter-arm regions. Only the smaller ($<$ 300 pc) holes are associated with OB associations. The larger holes do not clearly show this association with star forming regions, a fine exception being the hole around the OB association NGC 206 (Brinks 1981). One should keep in mind, however, that the inclination of M 31 is not very favourable and the spiral arms are mostly seen superposed on one another.

In M 33, considerably more face-on than M 31 though still inclined by 59 degrees, Deul and den Hartog (1990) found 148 holes in the HI distribution. The sizes of these holes range from 40 pc (the resolution limit) to 1 kpc, with a tendency for the largest holes (probably all inter-arm regions) to be located at larger galactocentric distances. Although the starformation rate in M 33 is higher than in M 31, the distribution, sizes, energies and missing masses of the holes are very similar. Deul and den Hartog found a correlation between the holes and the distribution of OB

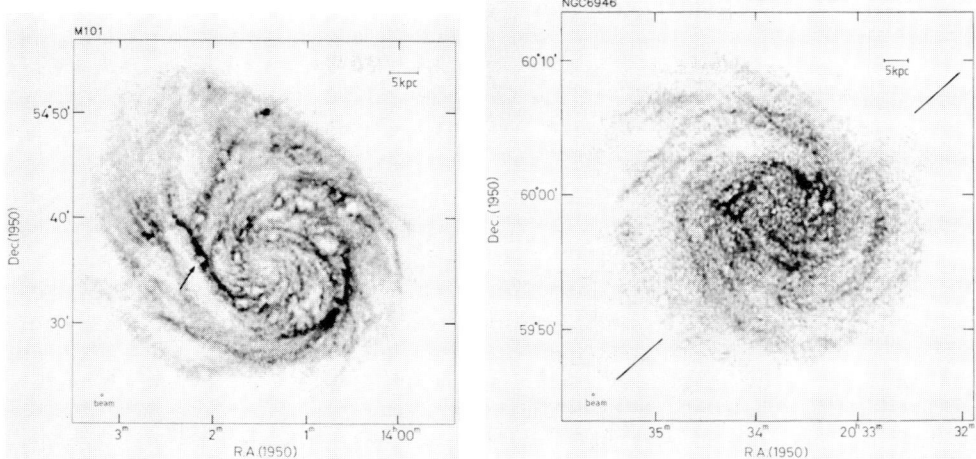

Figure 2. Grey-scale picture of the total HI column density distribution of M 101 (13″×16″ resolution, left panel) and of NGC 6946 (13″×16″ resolution, right panel). The arrow in the left panel gives the position of the superbubble near NGC 5462 (see Figure 3) and the line in the right panel indicates the position angle of Figure 4 and crosses the high velocity gas and elongated hole at $\alpha = 20^h\ 33.4^m\ \delta = 59°\ 59.4'$.

associations and some correlation with the distribution of Hα emission. The larger holes, however, anti-correlate with the distribution of recent star formation.

Another example of a galaxy with large holes in the HI is the nearby dwarf irregular galaxy IC 10, where Shostak and Skillman (1989) found seven holes. Two of these holes have diameters larger than 500 pc. The energy input derived is $5\ 10^{52}$ - $3\ 10^{53}$ ergs. The largest holes correspond to the supershells found in our Galaxy (Heiles 1979).

Two nearly face-on, large galaxies which have recently been observed in HI and are excellent candidates for a study of the structure of the HI, are the ScI galaxies M 101 and NGC 6946 (Kamphuis, Sancisi and van der Hulst, in preparation). Both have several very large star forming regions along well defined, massive spiral arms. The HI extends farther than the Holmberg radius in both galaxies and shows spiral structure even outside the optical image. The HI column density distributions of these galaxies are shown in Figure 2. The resolution is 15″ or 500 - 700 pc at the distances of these galaxies.

It is obvious from Figure 2 that the HI disks of M 101 and NGC 6946 have tens of large holes with sizes of 1 to more than 5 kpc. Some of the large holes are ambiguous and could equally well be considered as inter-arm regions. Most of the holes are located near regions of high column density, which probably is a selection effect. The HI disks give the impression that there is a network of holes, and it is quite conceivable that the general structure of the ISM in galaxies like M 101 and NGC 6946 is very filamentary, quickly giving the impression of being full of holes and shells. The large, well defined holes occupy about 10% of the disk and are located throughout the whole HI disk, even in the outer regions where the spiral arms are barely visible in the optical. There is no strong correlation between large holes and OB associations and HII regions.

HIGH VELOCITY GAS

Nearly face-on galaxies are in addition ideal objects to search for gas motions perpendicular to the plane. Radial velocities exceeding the general rotation of the disk could be associated with random motions of the gas in the disk, with expansion (or contraction) of features such as a superbubble or with a collection of high velocity clouds as observed in our Galaxy.

In several face-on galaxies the velocity dispersion of the HI has been measured (cf. van der Kruit and Shostak 1984, Murray, Helou and Dickey 1990). In these galaxies the velocity dispersion appears to be very constant throughout the whole disk with values of 6 - 12 km sec^{-1}.

Superposed on this small, constant velocity dispersion, gas with peculiar velocities has now been found in several nearby galaxies. The most extreme example is M 101 where van der Hulst and Sancisi (1988) discovered two large, high velocity regions within the Holmberg radius. The velocities of these gas complexes range up to 160 km sec^{-1} redshifted with respect to the rotation of the disk. The HI masses involved are 2 10^8 M$_\odot$ and the kinetic energies are of the order 10^{55} ergs. The velocity structure of these regions suggests a connection with the underlying holes in the disk. New, more sensitive observations reveal that these two regions are probably one large (40 × 20 kpc) complex covering almost half of the optical disk (Sancisi et al. 1990). This feature is simply too massive to be caused by winds and supernova blouw-outs. The most likely explanation is that the high velocity gas is caused by a gas-rich object, perhaps a small galaxy, which punched through the disk of M 101. The result is now visible as an outflow. The fate of the outflowing gas could be that it breaks up into small fragments which rain back into the disk forming a population of high velocity clouds in M 101. Evidence for a general population of high velocity clouds similar to the HVCs in our Galaxy has, however, not been found in M 101.

There is no systematic correspondence between the gas with excess velocities and star forming regions. The majority of the holes in the HI distribution of M 101 show at most moderate (10 - 30 km sec^{-1}) velocity deviations with respect to the bulk HI in the disk. So far there is one glaring exception (Kamphuis, Sancisi and van der Hulst, in preparation): the hole at the edge of the giant HII region NGC 5462, marked in Figure by a small arrow, shows clear evidence for systematic blue- and redshifted gas indicating systematic expansion with velocities up to 80 km sec^{-1}. Figure 3 shows a position-velocity map across this hole at a position angle of 27 degrees. The expanding shell structure is immediately obvious from this Figure. Supernova SN 1951h is located inside the superbubble. The total HI mass and kinetic energy involved are of the order 10^7 M$_\odot$ and a few 10^{53} ergs, respectively, implying that at least several hunderd supernova events would be required to produce it. This is not inconceivable: the neighbouring HII region, for comparison, is powered by the equivalent of 630 O5 stars (Israel et al. 1975).

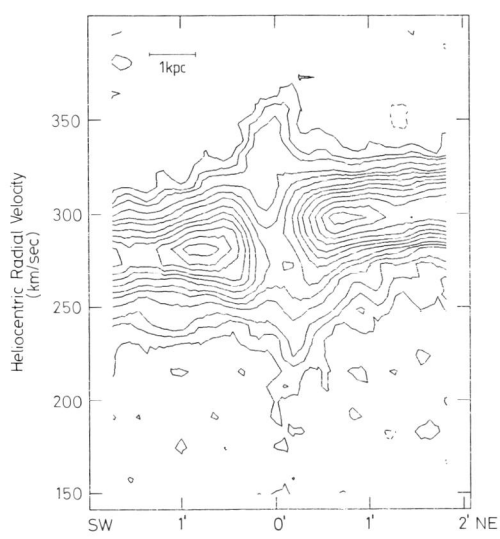

Figure 3. Position - velocity diagram centered on the hole near the giant HII region NGC 5462 in M 101, indicated by the arrow in the left panel of Figure 2. Contour levels are -5, 5, 10, 15, 25, 35, 45, 60, 75, 90, 110, 130, 150, 170 mJy/beam.

The other ScI galaxy observed in detail, NGC 6946, contains at least 9 regions with gas at velocities clearly deviating from the general rotation (Kamphuis, van der Hulst and Sancisi, in preparation). Most of the velocity deviations are associated with holes and range from 30 to 80 km sec^{-1}. The peculiar velocities are either red- or blue shifted, but not both at the same location. The HI masses and kinetic energies involved are of the order of 10^7 M$_\odot$ and 10^{53} ergs, respectively. There is one HI feature covering a large area and showing no clear connection with underlying structures of the disk. An example of high velocity gas is shown in Figure 4, which is a position-velocity diagram at the location indicated in the right panel of Figure 2.

A third example of a galaxy with high velocity gas is NGC 628 (Kamphuis and Briggs, in preparation). The outer parts of the extended gas disk in this object are heavily distorted. The inner part, however, behaves normally as a flat, rotating disk. In this inner region Kamphuis and Briggs discovered at least three gas

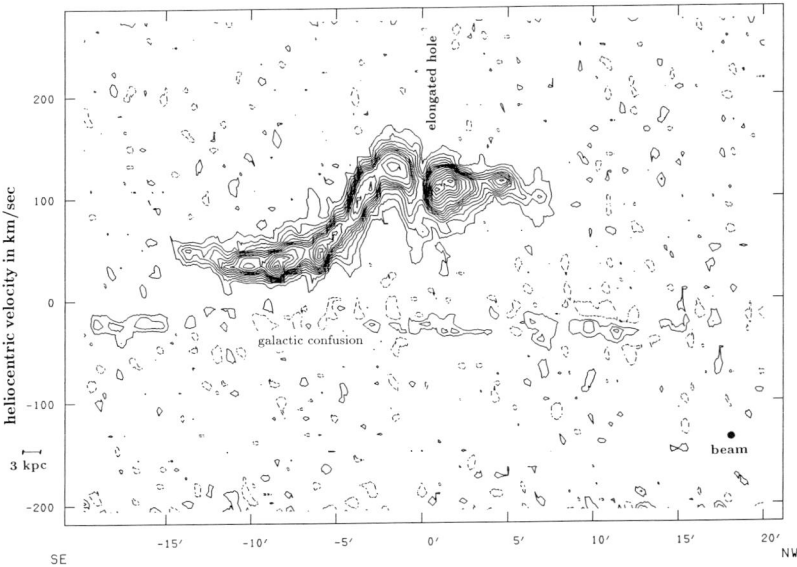

Figure 4. Position - velocity diagram (position angle = -48°, contour levels are -2 to 16 mJy/beam in steps of 2 mJy/beam, followed by 19 to 43 mJy/beam in steps of 4 mJy/beam) centered on the elongated hole in NGC 6946 at $\alpha = 20^h\ 33.4^m\ \delta = 59°\ 59.4'$ (see the right panel of Figure 2).

complexes with anomalous velocities up to 50 - 70 km sec^{-1}. The association with the underlying disk is not clear, due to the relative low angular resolution of the observations. Outside the Holmberg radius, two high velocity gas complexes have been found at the eastern and western edge of the galaxy. Each of these regions have a few 10^7 M$_\odot$ of HI and velocity excesses of 100 km sec^{-1}. The outer regions of NGC 628 are in addition heavily warped (see also Shostak and van der Kruit 1984). The symmetric placement of the two outer high velocity gas complexes suggests an explanation in terms of a tidal model, perhaps resulting from capture of a small companion.

The large distance to galaxies such as M 101, NGC 6946 and NGC 628 limits us to finding only the larger scale high velocity features. A closer look at the HI synthesis data of M 33 (Deul and van der Hulst 1987) now reveals that fainter, smaller features can be found in addition to the expanding gas associated with the holes discussed by Deul and den Hartog (1990). About a dozen regions have been located in a preliminary analysis of the data. These are not clearly associated with star forming regions. An example is shown in Figure 5, which is a right ascension-velocity map centered in the south-east of M 33. This map shows both a hole (at $\alpha = 1^h\ 30^m\ 30^s$) and a high velocity gas cloud (at $\alpha = 1^h\ 30^m\ 40^s$). The radial velocity of the gas cloud (uncorrected for projection effects) is 80 km s^{-1} with

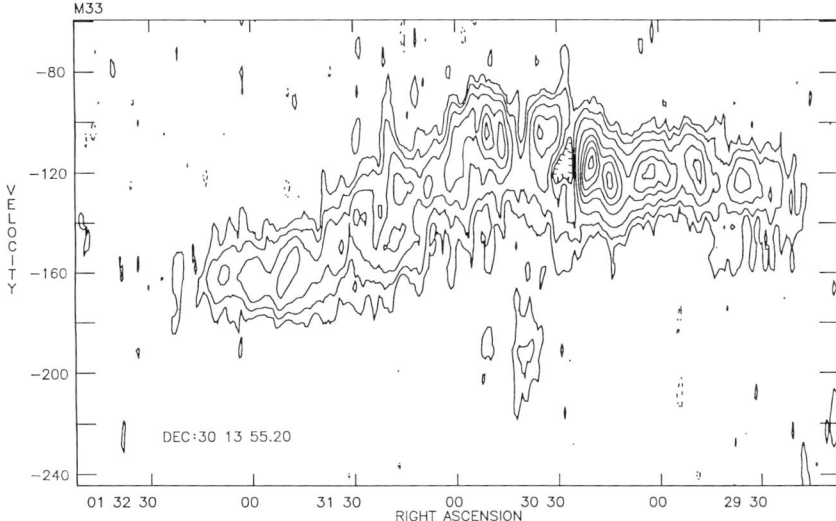

Figure 5. Right Ascension - velocity diagram at a Declination of 30° 24' 11.23" showing high velocity gas in M 33. Contour levels are -21.3, 21.3, 42.6, 85.2, 170.4, 255.6, 340.8, 426.0, 511.2 mJy/beam. Relative minima are hatched.

respect to the bulk rotation. The HI mass involved is $5.6\ 10^5$ $M_\odot$. Such a feature is much more in the range of parameters found for the HVCs in our galaxy.

DISCUSSION

The holes

It is clear from HI observations of nearby spiral and irregular galaxies, that large holes in the HI distribution are not uncommon. The number found in galaxies varies from a few to dozens and more (Table 1). In some galaxies the large holes are concentrated at large galactocentric distances (e.g. our Galaxy, M31 and M 33), but in other galaxies, they are randomly distributed throughout the disk (e.g. M 101, NGC 6946). Some of the larger "holes" are not real holes but rather inter-arm regions and should be excluded from the population of structures which is created by explosive events. The area covered by the (well defined) large holes is of the order of 15% of the disk or less. It is quite conceivable that the galaxies described in the previous section also have a collection of smaller holes, such as seen in our Galaxy, M 31 and M 33, covering a larger fraction of the disk. The limited angular resolution unfortunately prevents us from detecting those.

Assuming that all of these large holes (diameter > 500 pc) are caused by stellar winds and/or supernova explosions, the energies required are such that most of them are classified as supershells (E > 3 10^{52} ergs, Heiles 1979). The missing (partly driven out) HI masses are of the order of 10^6 - 10^8 M$_\odot$. The lifetimes of the large holes, derived from their linear sizes and expansion velocities, vary from about 10^7 to a few 10^8 years. Since most of the large holes show a lack of correlation with present star formation, another lower limit to the age for these holes is the lifetime of an HII region ($\approx$ 2 10^7 years). The upper limit comes from the time needed for the surrounding gas to fill in the holes. Approximate values for some of the parameters of the large holes are given in Table 1.

Table 1. Parameters of large holes (> 500 pc) in nearby galaxies

Galaxy	Number[a]	Max. sizes (kpc)	Missing HI masses[b] (10^6 M$_\odot$)	Energies[c] (10^{52} ergs)	Ages[d] (10^7 yr)
M 101	> 30	> 3	2 - 100	4 - 1100	2 - 15
NGC 6946	> 20	5	5 - 200	14- 5300	2 -25
IC 10	2	0.8	1.7 - 3	5 - 30	2.5 - 2.8
M 31	< 15	1.2	0.9 - 6.6	2 - 19	1 - 4
M 33	< 16	1.2	0.3 - 11	0.7 - 50	1 - 10
SMC	3	1.6	2 - 12	20 - 600	1.5 - 2.5
Our Galaxy	34	2.6	2 - 50	0.5 - 160	2 - 13

Notes: a Including possible inter-arm regions
b Calculated as $\pi R^2_{hole} \times$ thickness disk $\times$ average ambient HI density
c Based on elliptical shaped holes, following calculation of Heiles (1979)
d Estimated as $R_{hole}/V_{expansion}$

If holes like the ones discussed above result from explosive events which break out of the disk one does not expect to find sizes much in excess of 5 - 6 times the scale height of the HI gas layer (Mac Low, McCray and Norman 1989, Heiles 1990). The sizes found are, however, a few kpc and much larger than the generally assumed thickness of the cool HI disk.

The origin of such large holes is not clear. Superbubbles in the outer parts may have larger sizes due to flaring of the HI layer. The large holes in the inner parts require another explanation. One possibility is that the thickness of the HI layer is larger than a few hundred parsecs such as suggested for our Galaxy (Lockman 1984 and this volume). Another possibility is that the higher star formation rate in the inner parts causes superbubbles to overlap so they show up as very large cavities. A third possibility is that another mechanism, such as a dominant magnetic field (cf. Cox 1990), prevents the hot gas from escaping into the halo and tends to keep the gas down into the disk.

Not all of the holes show evidence for expansion. This could imply that the gas with anomalous velocities is diffuse so that the observations do not pick it up, or that the holes are old and have stopped expanding. The maximum velocity excesses found associated with a minor fraction of the holes are 50 - 80 km sec^{-1}. None (with the exception the hole near NGC 5462) of the observed velocity deviations are symmetric: they are either blue or redshifted. These asymmetries may imply that the supernova explosions driving the gas out are not in the midplane of the disk or that the ambient HI (and the molecular cloud distribution) is very clumpy.

The High Velocity Gas

High velocity gas features have been found in other galaxies. The edge-on galaxies examined here do reveal some large HI features extending up to 5 kpc above the plane and the face-on galaxies do occasionally show large high velocity complexes. These complexes are, however, larger and more massive than the HVC complexes in our Galaxy, except for the feature found in M33. The reason for not finding a wide-spread population of gas clouds in other galaxies similar to the HVCs in our Galaxy may be a result of the limited sensitivity and resolution which restricts us to only finding HI complexes of 10^6 M$_\odot$ or more. This would imply that the high velocity complexes found occasionally in other galaxies represent the upper extreme of the HVC mass spectrum. If on the other hand the Galactic HVCs are not as distant as assumed one would not expect to find similar features except in the nearest galaxies such as M31 and M33. A third possibility is that high velocity clouds are not a general phenomenon in spiral galaxies.

The larger high velocity gas complexes such as observed in M 101 and NGC 628 suggest that galaxies may suffer episodic accretion of large gas clouds or gas rich, dwarf galaxies (Sancisi 1990, Sancisi et al. 1990). Such events do throw gas out of the disk into the halo and provide a mechanism for occasional feeding of the gas halo of galaxies.

We are entering an era of very fruitful research focussing on problems related to the interstellar medium in nearby galaxies. This review did not address all possible issues but is intended to outline the important first results which have emerged from detailed observations of the HI in a small number of nearby galaxies. These first results discussed in this review clearly illustrate that further work in this field is important and fruitful, and is expected to broaden our insights in the years to come.

ACKNOWLEDGEMENTS

We thank A. H. Broeils and M. Rupen for the use of unpublished data and thank R. Sancisi for his comments on an earlier version of this manuscript. We appreciate the help of G. Comello, W. Haaima and G. Tamminga for preparing the figures.

REFERENCES

Allen, R.J. and Goss, W.M.: 1979, Astron. Astrophys. Suppl. **36**, 135.
Bregman, J.N.: 1980, Astrophys. J. **236**, 577.
Brinks, E.: 1981, Astron. Astrophys. (Letters) **95**, L1.
Brinks, E. and Bajaja, E.: 1986, Astron. Astrophys. **169**, 14.
Cox, D.P.: 1990, in *The Interstellar Medium in Galaxies*, ed. H. A. Thronson and J. M. Shull, Kluwer Ac. Publ, p. 181.
Dettmar, R.J.: 1990, Astron. Astrophys. (Letters) **232**, L15.
Deul, E.R. and van der Hulst, J.M.: 1987, Astron. Astrophys. Suppl. **67**, 509.
Deul, E.R. and den Hartog, R.H.: 1990, Astron. Astrophys. **229**, 362.
Dickey, J.M., Murray Hanson, M. and Helou G.: 1990, Astrophys. J. **352**, 522.
Dopita, M.A., Mathewson, D.S. and Ford, V.L.: 1985, Astrophys. J. **297**, 599.
Duric, N., Seaquist, E.R., Crane, P.C., Bignell, R.C. and Davis, L.E.: 1983, Astrophys. J. (Letters) **273**, 574.
Heiles, C.: 1979, Astrophys. J. **229**, 533.
Heiles, C.: 1984, Astrophys. J. Suppl. **55**, 585.
Heiles, C.: 1990, Astrophys. J. **354**, 483.
Hindman, J.V.: 1967, Aust. J. Phys. **20**, 147.
Irwin, J.A. and Seaquist, E.R.: 1990, Astrophys. J. **353**, 469.
Israel, F.P., Goss, W.M. and Allen, R.J.: 1975, Astron. Astrophys. **40**, 421.
Lockman, F.J.: 1984, Astrophys. J. **283**, 90.
Mac Low, M., McCray, R. and Norman, M.L.: 1989, Astrophys. J. **337**, 141.
McGee, R.X. and Milton, J.A.: 1966, Aust. J. Phys. **19**, 343.
Meaburn, J.: 1980, Mon. Not. R. Astr. Soc. **192**, 365.
Meaburn, J., Marston, A.P., McGee, R.X. and Newton, L.M.: 1987, Mon. Not. R. Astr. Soc. **225**, 591.
Mirabel, I.F. and Morras, R.: 1990, Astrophys. J. **356**, 130.
Rand R. J., Kulkarni, S. R. and Hester J. J.: 1990, Astrophys. J. **352**, L1.
Rohlfs, K., Kreitschmann, J., Siegman, B.C. and Feitzinger, J.V.: 1984, Astron. Astrophys. **137**, 343.
Sancisi, R.: 1990, in Proceedings Erice Workshop *Windows on Galaxies*, ed. Fabbiano, G., Gallagher, J. and Renzini, A., Kluwer Ac. Publ., in press.
Sancisi, R., Broeils, A. H., Kamphuis, J. and van der Hulst, J.M.: 1990, Proceedings of the Heidelberg Conference on *Dynamics and Interactions of Galaxies*, ed. Wielen, R., Springer., p. 304.
Shostak, G.S. and van der Kruit, P.C.: 1984, Astron. Astrophys. **132**, 20.
Shostak, G.S. and Skillman, E.D.: 1989, Astron. Astrophys. **214**, 33.
Tenorio-Tagle, G. and Bodenheimer, P.: 1988, Ann. Rev. Astron. Astrophys. **26**, 145.
Tenorio-Tagle, G., Franco, J., Bodenheimer, P. and Rozyczka, M.: 1987, Astron. Astrophys. **179**, 219.
van der Hulst, J.M. and Sancisi R.: 1988, Astron. J. **95**, 1354.
van der Kruit, P.C. and Shostak, G.S.: 1984, Astron. Astrophys. **134**, 258.
van Woerden, H., Schwarz, U.J. and Hulsbosch, A.N.M.: 1985, in *The Milky Way Galaxy, IAU Symp* **106**, eds. H. van Woerden, R.J. Allen, W.B. Burton, p. 387.
Wakker, B.P.: 1990, Ph.D. thesis Groningen University.
Wakker B.P., Broeils, A.H., Tilanus, R.P.J. and Sancisi, R.: 1989, Astron. Astrophys. **226**, 57.

SMALL-SCALE PROPERTIES OF HI IN NEARBY GALAXIES

ROBERT BRAUN
NFRA, P.O. Box 2
7990 AA Dwingeloo
The Netherlands

ABSTRACT. The instrumental requirements and current observational results are outlined for the study of resolved neutral hydrogen structures in both emission and absorption in external galaxies. Neutral super-shell structures exterior to ionized shells around OB associations seem to be a common phenomenon. Further analysis of existing data should allow quantification of the degree and rate of energy deposition in the ISM by massive stars. HI absorption studies offer great promise in constraining the physical properties of the neutral ISM, including an estimate of the gas pressure. Together these data will provide hard constraints on the existence and prevalence of chimney formation, and hence disk – halo interactions in a variety of galactic environments.

1. INTRODUCTION

Within the last few years a new phase in the study of the neutral, interstellar medium (ISM) has begun. Instrumental capabilities have now advanced to the point where observations of the ISM in the closest external galaxies can probe physical scales and sensitivities comparable to those accessible to studies of the Galaxy. The opportunity to study external systems at known distances, at a variety of inclinations and over a range of galaxy types opens a wide range of possibilities for better understanding the physical conditions and processes in the neutral ISM which have remained elusive up to this point.

A basic requirement of an observation which is intended to address physical conditions of the ISM is that the source structure be resolved both spatially and, in the case of a spectral line like that of HI, in velocity. In this paper we will consider to what extent these requirements can be satisfied in the context of current instrumentation, which will lead to definitions of the terms "small" and "nearby" which appear in the title. Since observations of resolved HI emission and absorption have only been fully reduced for one external galaxy to date (M31) we will consider the emerging results for this system in some detail.

2. OBSERVATIONAL REQUIREMENTS

HI emission structures observed in the Galaxy have a large range of physical sizes, from some 100's of pc down to only a few pc. Most of the power appears to be concentrated in the larger spatial scales, as evidenced both by power spectrum analyses (e.g. Crovisier and Dickey 1983) and the simple observational result that the highest brightness temperature seen in emission of about 125 K was observed with the 35 arcmin angular resolution of the early surveys done with 25 m telescopes, and this value has not increased significantly with the much higher angular resolution of more recent work. The corresponding spatial resolution is about 100 pc at a typical inner galaxy distance of about 10 kpc. Similarly, velocity resolutions in excess of about 5 km/s have not resulted in higher observed brightnesses. Observations of HI absorption, which are sensitive to the high opacity component of the neutral gas, reveal somewhat narrower linewidths than those seen in emission, although even here the velocity dispersions, σ, are usually greater than 2 km/s (e.g. Dickey, Salpeter and Terzian 1978) so that the linewidth, FWHM = 2.36 σ, is typically greater than about 5 km/s.

The range of brightness of HI emission, just as the range of structural scales, spans about two orders of magnitude from a few to a few hunderd degrees K. At low brightnesses, there appears to be a threshold column density of a few times 10^{19} cm^{-2} below which neutral hydrogen is not detected independant of sensitivity (e.g. Sancisi et.al. 1990). At high brightnesses, as we will see below, the limit is primarily determined by the kinetic temperature of high opacity HI which is expected to reach between some ten's and a few hunderd K depending on the physical conditions (e.g. Draine 1978). A practical criterion for the study of resolved HI emission is then a sensitivity better than about 20 K, corresponding to a column density less than about 2 $\times 10^{20}$ cm^{-2}. Together with the 100 pc spatial resolution this implies a sensitivity to an HI mass less than about 2 $\times 10^4$ M$_\odot$.

These observational constraints are summarized in Table 1 for the study of resolved HI emission at a variety of distances. The column density noted above, together with a kinetic temperature of 200 K, lead to a limiting opacity in HI absorption, $\tau = 0.1$. Twelve hours of integration with an extended VLA configuration give sensitivity to this opacity against background sources brighter than about 5 mJy independant of the absorber's distance. Integrating background source counts at 20 cm wavelength (e.g. Windhorst, Van Heerde and Katgert 1984) down to 5 mJy leads to an expected source density of 26 deg^{-2} or about 1 per 10 by 10 arcmin, the angular size of a 15 kpc galactic disk seen at a distance of 5 Mpc. Together with the integration times listed in Table 1, it becomes clear that only galaxies within about this distance are currently accessible to studies of resolved HI emission, and have a reasonable *a priori* probability of detectable absorption against a background source. It is somewhat frustrating to note that the volume of space we can effectively probe will not extend to the Virgo cluster or various well known "nearby" galaxies until a new instrument of 40 km extent and about 10 times the collecting area of the VLA is built. Such an instrument is not yet being seriously discussed.

Table 1.
Instrumental Requirements for the Study of Resolved HI Emission

Source (1)	Distance (2)	Baseline (3)	Instrument (4)	Integration Time (5)
The Galaxy	10 kpc	25 m	many	$\sim 1^m$ / position
LMC/SMC	50 kpc	125 m	compact AT	$\sim 2^h$ / field
eg. M81, M101 groups	5 Mpc	12 km	VLA B config.	$\sim 12^h$
			GMRT	$\sim 2^h$??
eg. Virgo cluster	20 Mpc	40 km	VLA A config.	$\sim 1/2$ year
			GMRT	~ 1 month ??

3. STRUCTURES IN EMISSION

High resolution observations of HI in nearby galaxies have revealed a wealth of structural information, even though almost none have had the requisite resolution (better than about 100 pc and 5 km/s) to resolve the emission structures. The most noted form of structure, up to this point, has been local minima in the distribution of HI. Hindman (1967) detected three such minima in the SMC, Shostak and Skillman (1989) report seven in IC10, while extensive catalogs have been published by Brinks and Bajaja (1986) for M31 (140 entries) and by Deul and Den Hartog (1990) for M33 (150 entries). It should be noted that less than 10 % of the HI "holes" in the M31 and M33 catalogs were considered by these authors to be of "high quality" in the sense of possessing a relatively elliptical boundary and reasonably high contrast.

While some fraction of the HI "holes" detected in nearby galaxies almost certainly represent physical entities, the question which naturally arises is how many of these local minima are really connected structures rather than simply a consequence of the projected topology. This distinction is aggravated by the obviously filamentery, frothy HI structure which is apparently left behind by multiple generations of massive star formation. In cases of moderate signal-to-noise, where a clear limb-brightening and possibly an expansion signature are not detectable it must be concluded that it remains impossible to identify physically connected structures on the basis of the distribution of HI emission alone.

The intrinsic ambiguity of moderate signal-to-noise local HI minima, suggests an alternate approach in the analysis of small-scale HI emission structures in which a search for HI counterparts is undertaken relative to positions of a less confused population such as molecular clouds, HII regions, OB associations or X-ray bubbles. Specifically, HII super- or super-giant shells offer an obvious parent population to search for associated, expanding HI shells.

3.1 The Magellanic Clouds

The type of directed search that was just outlined has as yet been appplied in only a limited way to the LMC/SMC, although the possibilities are very graphically

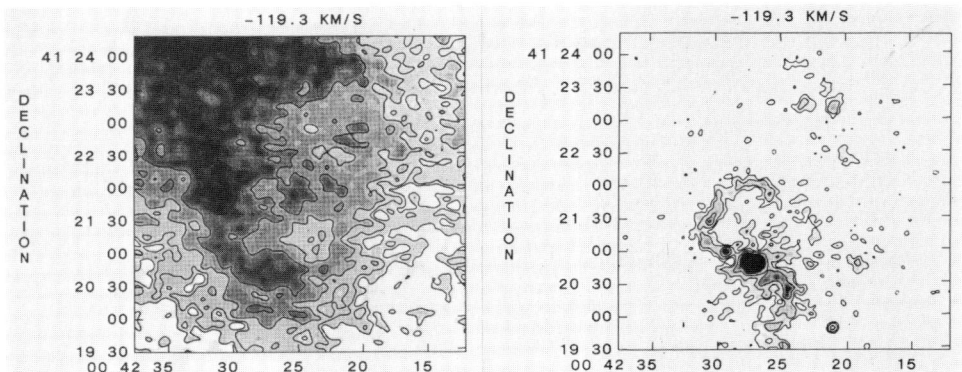

Figure 1. Narrow-band images in neutral hydrogen (left) and ionized hydrogen (right) at a heliocentric velocity of −119.3 km/s of a region of about 1 kpc on a side within Messier 31. Neutral counterparts are seen exterior to many of the ionized shells surrounding OB associations at their systemic velocities. Expansion signatures are sometimes apparent at outlying velocities.

illustrated by the narrow-band image in Hα + [NII] of the LMC made by Davies, Elliott and Meaburn (1976). The tabulated properties of the many observed nebulae (Figure 3 of Meaburn 1980) indicate a population of super-shells with diameters less than about 300 pc associated with individual OB associations. These in turn are sometimes organized into super-giant shells with diameters between 600 and 1200 pc. Existing HI data for the LMC and SMC are severely limited by the 220 pc spatial resolution obtainable with the Parkes 64 m telescope. Comparative studies, like that of Dopita, Mathewson and Ford (1985), have therefore only been able to access the super-giant shells, which are in fact found to be clearly associated with expanding neutral shells. Major advances in this study are bound to follow from the availability of the Australia Telescope (AT) during the coming years.

3.2 Messier 33

Comparable opportunities exist in M33, where the photographic imagery of Courtes et al. (1987) delineates the extensive population of Hα super-shells. The recent HI survey of Deul and Van der Hulst (1987), with 65 pc by 8 km/s resolution and 3000 M$_\odot$ sensitivity may partially resolve emission structures, although another factor of two in velocity resolution would clearly be desirable. A systematic analysis of the HI database from the perspective of super-shell associations is now being planned (Van der Hulst, private communication).

3.3 Messier 31

The only external galaxy for which reduced observations now exist that satisfy the criterion derived in §2 for resolved detection of HI emission structures is M31. The data covering the north-east half of M31 at 35 pc by 5 km/s resolution and 500

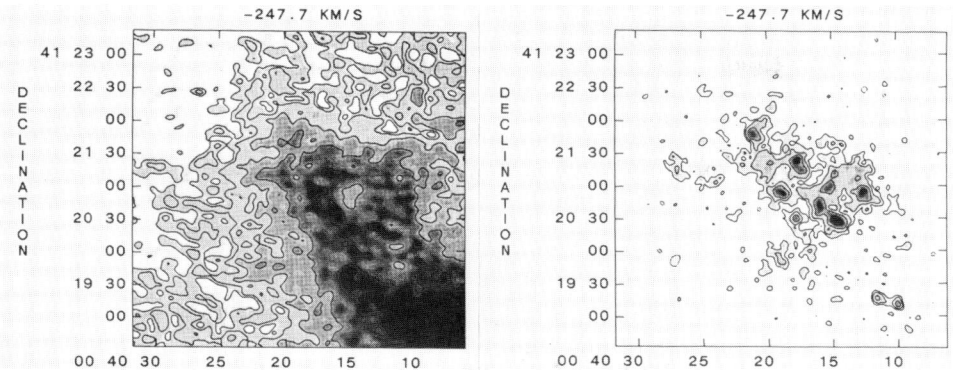

Figure 2. As for Figure 1, but for another region in Messier 31 and at a velocity of -247.7 km/s. The major shell structure in this example shows a clear expansion at outlying velocities in both HI and Hα at about 20 and 30 km/s respectively.

M$_\odot$ sensitivity were recently published (Braun 1990). Analysis of limited portions of this database has already taken place in conjunction with Hα kinematic data (Brinks, Braun and Unger 1989, 1990), while a systematic analysis in conjunction with narrow-band imagery in Hα and [SII] (Walterbos and Braun 1990) is still underway. Some of the correspondences between Hα and HI super-shells are illustrated in Figures 1 and 2, where corresponding images are shown at the indicated velocities for two regions of about 1 kpc on a side.

Virtually all of the Hα shells seen in these figures have at least low contrast (> 50 % of the background) HI counterparts at the systemic velocities. The best developed examples show neutral shells which are clearly exterior to the ionized shells. A detectable expansion signature in HI is less common, and when present is always slower than in Hα. In particular, the major shell structure in Figure 2 (at $\alpha_{50} = 0^H 40^M 15^S$, $\delta_{50} = 41°20'50"$) has a clear expansion signature in both Hα and HI with expansion velocities of about 30 and 20 km/s respectively. The kinetic energy associated with the expanding HI shell in this case is a few times 10^{51} erg. In is interesting to note that none of the HI structures in these figures has been previously identified as an HI "hole".

No obvious cases of disk–halo communication have yet come to light in the analysis of HI and Hα counterparts. More detailed study of individual regions as well as statistical analysis of counterparts to stellar associations and a sample of more than 900 emission nebulae cataloged by Walterbos and Braun (1990) will be carried out in the coming year. This analysis should allow assessment of the degree and rate of kinetic energy deposition in the ISM and hence hard observational data pertaining to the issue of chimney formation.

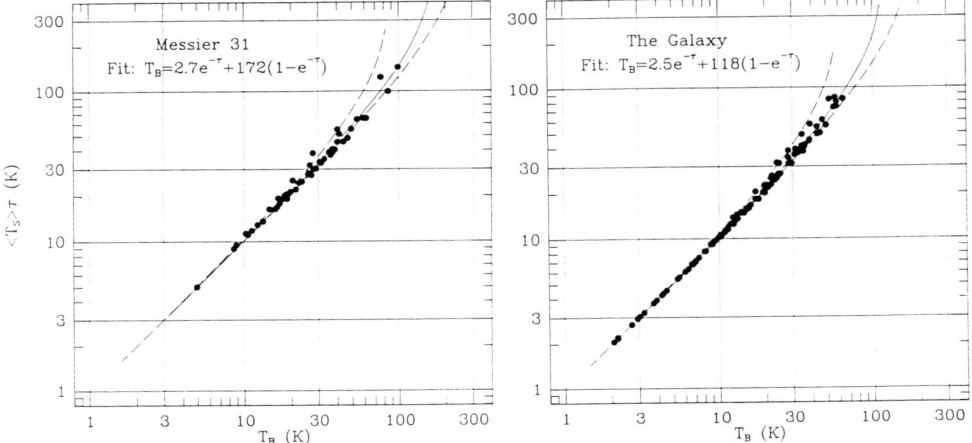

Figure 3. Implied column density as a function of brightness temperature of emission for M31 and the Galaxy. This product isolates the asymptotic behaviour at large T_B. Asymptotic temperatures derived here are consistent with those obtained directly from HI emission studies, and are almost 50 % higher in M31 than in the Galaxy.

4. ABSORPTION STUDIES

There have, as yet, been few detections of HI absorption outside of the Galaxy. Besides a handful of detections against bright nuclear sources (Van Gorkom et al. 1989) single lines-of-sight have been detected against a few systems including NGC891 (Rupen et al. 1987) and M31 (Dickey and Brinks 1988). The recent high sensitivity survey of M31 (Braun 1990 (B90)) has made possible the detection of seven lines-of-sight through that galaxy's disk, with a total equivalent width of absorption and sampled path length comparable to that obtained in the Arecibo surveys of HI absorption in the Galaxy (Dickey, Salpeter and Terzian 1978, (DST78) Payne, Salpeter and Terzian 1982 (PST82)). The analysis of these data (Braun and Walterbos 1990) has been very instructive in delineating the physical properties of the neutral gas in M31.

In analyzing the absorption properties of HI, the implied spin temperature, $<T_S>$ defined by,

$$< T_S > = \frac{T_B}{(1 - e^{-\tau})} \quad (1)$$

has often been plotted as a function of the measured opacity, τ, since Lazereff (1975) sugggested there might be some correlation between these quantities. In fact, these quantities are clearly not independant of eachother and their relation offers little insight into the gas properties. A more revealing relationship turns out to be that between implied column density, $< T_S > \tau$ and emission brightness, T_B. These quantities are plotted in Figure 3 using the data of DST78 and PST82

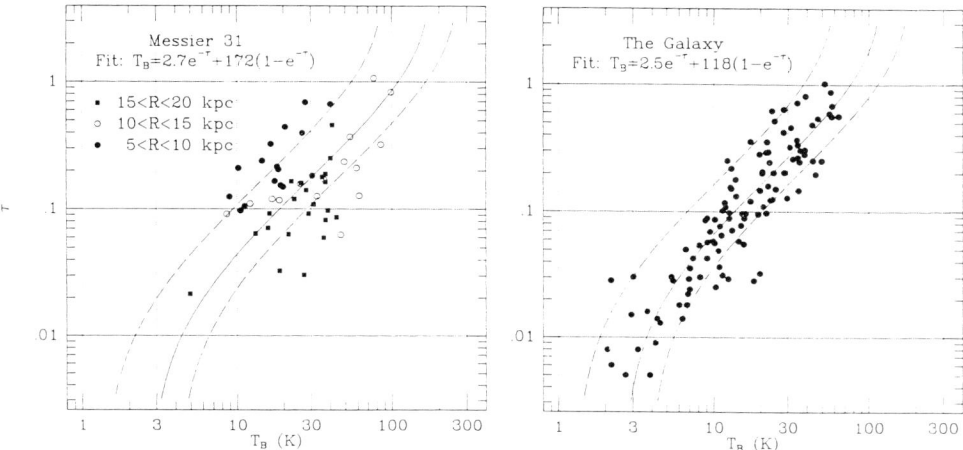

Figure 4. HI opacity against emission brightness for M31 and the Galaxy. This form is sensitive to the asymtotic behaviour at small T_B. The values for M31 are binned by radius in that system, while the galactic data are for gas within about 1 kpc of the sun. The plotted fit and envelope at ± 50 % of the variable values corresponds to a simple two component model. The inner disk of M31 has a systematically lower temperature than the outer disk.

for the Galaxy and B90 for M31. A very tight correlation of the implied column density is seen with the emission brightness. This relation follows the one-to-one correspondence expected for low optical depth at low T_B and then becomes increasingly non-linear for high T_B. It allows a sensitive determination to be made of the asymptotic brightness, T_∞ from,

$$T_B = T_\infty (1 - e^{-<T_s>\tau/T_\infty}) \qquad (2)$$

yielding $T_\infty = 118 \pm 2$ K for the Galaxy and $T_\infty = 172 \pm 3$ K for M31. These values of asymptotic brightness are in good agreement with the highest values of T_B observed directly in emission, indicating that the "single component" opacity corrections that have sometimes been employed in determining HI masses and dust-to-gas ratios are a good approximation to the actual dependance of opacity on brightness.

The different linewidths and detailed line shapes of HI absorption and emission spectra have made it clear from early on that HI emission is due to at least two distinct physical components; a cool component of significant opacity and a warm component with negligible opacity. This need is also obviated by the relationship between the primary observables shown in Figure 4; the opacity and the emission brightness. Not only is there an asymptotic brightness at high opacity, but also at low opacity; a relationship incompatible with a single temperature gas. The simplest physical model which can accomodate these trends is one composed of two temperatures, T_c and T_w for which $\tau_w << 1$ and $T_w \tau_w =$ constant. Placing one half of the warm gas in front and one half behind the cooler gas yields an emission brightness,

$$T_B = \frac{T_w \tau_w}{2} + T_c(1 - e^{-\tau_c}) + \frac{T_w \tau_w}{2} e^{-\tau_c} \qquad (3)$$

which yields the asymptotic brightness, $T_\infty = T_c + T_w \tau_w/2$. Rewriting eqn. 3 in terms of T_∞ yields,

$$T_B = T_w \tau_w e^{-\tau_c} + T_\infty (1 - e^{-\tau_c}) \qquad (4)$$

Having already fixed the value of asymptotic brightness from the $<T_S> \tau - T_B$ relation above, permits a single parameter fit for the column of warm gas, $T_w \tau_w$. This yields $T_w \tau_w = 2.5 \pm 0.1$ K for the Galaxy and $T_w \tau_w = 2.7 \pm 0.2$ K for M31, which implies that the cool HI component temperatures are, $T_c = 117 \pm 2$ K for the Galaxy and $T_c = 171 \pm 3$ for M31. The fits to eqn. 4 are overlaid on the data in Figures 3 and 4 together with an envelope at plus and minus 50 % of the variable values to illustrate the functional dependance and quantify the degree of scatter.

While the contribution of warm HI is not significantly different in M31 and the (solar neighbourhood of the) Galaxy, and corresponds to about 2×10^{19} cm^{-2} per 5 km/s interval, the cool component temperatures are significantly different. Furthermore, there is evidence for a gradient in the temperature of cool HI in M31, as illustrated by the binning by radius in Figure 4. Cool component temperatures vary from about 80 K in the inner galaxy to about 200 K beyond about 10 kpc radius. The kinetic temperature of neutral hydrogen is regulated by photoelectric heating from dust and radiative cooling by gas-phase heavy elements (e.g. Draine 1978). The relevant physical parameters for these processes are the ionization rate due to cosmic rays and soft X-rays, the gas-to-dust ratio, the gas-phase metal abundance and the gas pressure. The only plausible mechanism for producing the higher cool HI temperatures (at R > 10 kpc) in M31 than in the Galaxy appears to be a lower gas phase pressure by about a factor of 2 (Braun and Walterbos 1990).

With knowledge of the gas temperature, it becomes possible to derive line-of-sight filling factors for the various components, by assuming values of the gas pressure. Since the HI opacity is given by,

$$\tau = \frac{1.7ns}{T \Delta V} \qquad \left[\frac{cm^{-3} pc}{K\, km/s} \right] \qquad (5)$$

in terms of a path length, s, the line-of-sight filling factor can be written as,

$$f = \frac{s}{S} = \frac{T^2 \tau \Delta V}{1.7 p S} \qquad (6)$$

in terms of the component scale height, S and the gas pressure, p. Various lines of evidence suggest that the warm HI component has a temperature of 8000 K, while the exponential scale heights of the dense and tenuous components appear to be

Table 2.
Global Properties of Neutral Gas in M31 and The Galaxy

Quantity			Messier 31			The Galaxy[a]		
Comment (1)	Symbol (2)	Unit (3)	Cool (4)	Warm (5)	Ref.[b] (6)	Cool (7)	Warm (8)	Ref.[b] (9)
Equivalent width to midplane	$<\tau\Delta V>_\perp$	km s^{-1}	0.936	...	1	0.706	...	2,3
Velocity width to midplane (at $T_B > 5$ K)	$<\Delta V>_\perp$	km s^{-1}	...	6.62	1	...	7.03	2,3
Brightness temperature	$<T\tau>$	K	...	2.70 ± 0.16	1	...	2.54 ± 0.12	1
Temperature	$<T>$	K	171 ± 3	8000	1,4	117 ± 2	8000	1,4
Column to midplane	$<T\tau\Delta V>_\perp$	K-km s^{-1}	160.	17.9	1	82.6	17.9	1
Scale height	S	pc	150	400	1,5	150	400	4
Thermal pressure	$<p>$	cm^{-3}K	1500	1000	1	2960	2070	4
Density	$<n>$	cm^{-3}	9	0.13	1	25	0.26	1
Line-of-sight filling factor	$<f>$	%	7	20	1	1.3	9.5	1
Opacity-corrected HI mass	M_{HI}	10^9 M$_\odot$	4.6		1,6	3.5		1,7
Gas-to-dust ratio	N_H/E_{B-V}	10^{21}cm^{-2}mag^{-1}	4.4–5.6		1	4.8–5.3		8
Gas-phase "metallicity"	$<A>$	[O/H] 10^{-4}	5.0 ± 1.0		9,10	2.5 ± 0.5		10,11

[a]Galactic values refer to the extended (1 kpc) solar neighbourhood.
[b]References.—(1) Braun and Walterbos (1990), (2) Dickey, Salpeter and Terzian (1978), (3) Payne. Salpeter and Terzian (1982), (4) Kulkarni and Heiles (1988), (5) Braun (1990), (6) Cram et al. (1980), (7) Henderson, Jackson and Kerr (1982), (8) Savage and Mathis (1979), (9) Blair, Kirshner and Chevalier (1982), (10) Dopita et al. (1984), (11) Talent and Dufour (1979).

about 150 and 400 pc respectively (e.g. Lockman, Hobbs and Shull 1986). Midplane gas pressures in the Galaxy are approximately 4000 cm^{-3}K (Kulkarni and Heiles 1988). Using these values together with the mean column to the midplane $<T\tau\Delta V>_\perp$ and eqn. 6 yields the values of Table 2.

One of the noteworthy results of this analysis of HI absorption and emission is the small line-of-sight filling factors derived for the neutral gas. Taken together with the very large surface covering factor of neutral gas in both the Galaxy and M31, it becomes possible to constrain the three-dimensional topology of this component of the ISM. The most obvious topology which reproduces such filling factors is the one suggested by the projected distribution itself; an extensive, frothy network of bubbles like that detected in §3.

5. FURTHER PLANS AND DEVELOPMENTS

As indicated at the outset, a fascinating phase in our study of the neutral ISM has begun. Our first relatively unconfused glimpses of resolved HI structures in external galaxies are in hand. Neutral shells such as discussed in §3.3 appear to be a common phenomenon external to the ionized shells surrounding evolved OB associations. More extensive analysis of the M31 and M33 databases will be vital in refining our understanding of the physical properties, phase and energy balance of the ISM in normal galaxies. At the same time, our curiosity has been arroused by the fact that the neutral ISM of M31 appears to be so different than that of the Galaxy in terms of average pressure, temperature and density. What is the actual range of physical conditions which can occur and how do they depend on the galaxy type and position within the galaxy?

In an effort to address these questions, a sample of eleven major galaxies within about 5 Mpc (NGC 55, 247, 2366, 2403, 3031, 4236, 4736, 4826, 5457 and 7793) is being observed by an extended group (Braun, Van Gorkom, Walterbos, Kennicutt, Norman, Tacconi-Garman) utilizing not only resolved neutral hydrogen but all accessible ISM tracers. The coming years will likely yield important insights in our understanding of the ISM in general and the degree and prevalence of communication between the disk and halo in particular.

REFERENCES

Blair, W.P., Kirshner, R.P., Chevalier, R.A. (1982) *Ap.J.* **254**, 17
Braun, R. (1990) *Ap.J.Suppl.* **72**, 755
Braun, R., Walterbos, R.A.M. (1990) *Ap.J.* submitted
Brinks, E., Bajaja, E. (1986) *Astr.Ap.* **169**, 14
Brinks, E., Braun, R., Unger, S.W. (1989) in *IAU Col. 120, Structure and Dynamics of the Interstellar Medium*, eds. G. Tenorio-Tagle, M. Moles, J. Melnick, Springer-Verlag, New York
Brinks, E., Braun, R., Unger, S.W. (1990) in prep.
Courtes, G., Petit, H., Sivan, J.-P., Dodonov, S., Petit, M. (1987) *Astr.Ap.* **174**, 28
Cram, T.R., Roberts, M.S., Whitehurst, R.N. (1980) *Astr.Ap.* **40**, 215
Crovisier, J., Dickey, J.M. (1983) *Astr.Ap.* **122**, 282
Davies, R.D., Elliott, K.H., Meaburn, J. (1976) *Mem.R.Astr.Soc.* **81** 89
Deul, E.R., Den Hartog, R.H. (1990) *Astr.Ap.* **229**, 362
Deul, E.R., Van der Hulst, J.M. (1987) *Astr.Ap.Suppl.* **67**, 509
Dickey, J.M., Brinks, E. (1988) *M.N.R.A.S.* **233**, 781
Dickey, J.M., Salpeter, E.E., Terzian, Y. (1978) *Ap.J.Suppl.* **36**, 77
Dopita, M.A., Binette, L., D'Odorico, S., Benvenuti, P. (1984) *Ap.J.* **276**, 653
Dopita, M.A., Mathewson, D.S., Ford, V.L. (1985) *Ap.J.* **297**, 599
Draine, B.T. (1978) *Ap.J.* **36**, 595
Henderson, A.P., Jackson, P.D., Kerr, F.J. (1982) *Ap.J.* **263**, 116
Hindman, J.V. (1967) *Aust.J.Phys.* **20**, 147
Kulkarni, S.R., Heiles, C. (1988) in Galactic and Extragalactic Radio Astronomy, eds. K. Kellerman, G. Verschuur, Springer-Verlag, Heidelberg, p. 95
Lockman, F.J., Hobbs, L.M., Shull, J.M. (1986) *Ap.J.* **301**, 380
Meaburn, J. (1980) *M.N.R.A.S.* **192**, 365
Payne, H.E., Salpeter, E.E., Terzian, Y. (1982) *Ap.J.Suppl.* **48**, 199
Rupen, M.P., Van Gorkom, J.H., Knapp, G.R., Gunn, J.E., Schneider, D.P. (1987) *A.J.* **94**, 61
Sancisi, R., Van Gorkom, J.H., Cornwell, T.J., Van Albada, J. (1990) in prep.
Savage, B.D., Mathis, J.S. (1979) *A.R.A.A.* **17**, 73
Shostak, G.S., Skillman, E.D. (1989) *Astr.Ap.* **214**, 33
Talent, D.L., Dufour, R.J. (1979) *Ap.J.* **233**, 888
Van Gorkom, J.H., Knapp, G.R., Ekers, R.D., Ekers, D.D., Laing, R.A., Polk, K.S. (1989) *A.J.* **97**, 708
Walterbos, R.A.M., Braun, R. (1990) in prep.
Windhorst, R.A., Van Heerde, G., Katgert, P. (1984) *Astr.Ap.Suppl.* **58**, 1

DIFFUSE IONIZED GAS IN NEARBY GALAXIES

R.A.M. WALTERBOS
Astronomy Department, University of California
Berkeley, CA 94720, USA

ABSTRACT. We discuss the distribution and spectral characteristics of diffuse ionized gas in nearby galaxies. The existence of this elusive component of the interstellar medium (ISM), also referred to as the Warm Ionized Medium, is by now well established from deep imaging and spectroscopic surveys in several emission lines in external galaxies. Diffuse ionized gas is characterized by a relatively high ratio of [SII] over Hα intensities, typically twice as high as for discrete HII regions. The diffuse gas has been mapped in both edge-on and more face-on galaxies providing information on the radial and vertical distributions. Emission from diffuse ionized gas is strongest around star forming regions. The vertical distribution appears related to the radio continuum thick-disk emission. We also briefly discuss ionization mechanisms, and the connection between star formation characteristics and morphology of the interstellar medium.

1. INTRODUCTION

The existence of diffuse ionized gas in the interstellar medium of the Galaxy has been established in several independent ways. Evidence for this widely distributed, low-density ($n_e \sim 0.2$ cm^{-3}), relatively cool ($< 10,000K$) ionized gas follows from imagery and spectroscopy in emission lines, especially the Hα line, from pulsar dispersion measurements, observations of scintillation of compact radio sources, absorption of non-thermal radio emission, Faraday rotation, and thermal radio continuum emission. The first two probes have provided most information on the distribution and characterisctics of this gas in the general solar neighborhood (see Reynolds, this volume). The gas is referred to as Diffuse Ionized Gas (DIG) or Warm Ionized Medium (WIM). We will use the latter acronym in this paper. With its large vertical extent, more than a kpc above the disk in our Galaxy, and significant filling factor, ≥ 0.2, the WIM evidently plays an important role in the disk-halo interface.

There are several reasons why it is necessary to study this component of the ISM in galaxies other than our own. First, the global distribution of the WIM in our Galaxy is not known because of our location in the disk. Second, the connection between the young stellar population, presumably responsible for ionizing the gas, and the WIM is difficult to derive for the same reason. Thirdly, if OB stars power the ISM, then ISM properties are likely to depend on galaxy type and on position in the disk, since it is well established that star formation characteristics vary strongly both with galaxy type and galactic radius (*e.g.* Kennicutt, 1983, 1989). Hence it is important to study the WIM in a variety of galaxies. By observing both edge-on and more face-on systems, all these issues can be addressed.

2. OBSERVATIONS

Of the above-mentioned techniques that are in principle able to detect the WIM in external galaxies, direct imagery and spectroscopy of (optical) emission lines are most promising with current observational capabilities. The thermal radio continuum emission of the WIM is very weak and difficult to separate from the generally much stronger non-thermal synchrotron emission and from the contribution of discrete HII regions, especially at the low resolution of most radio surveys. To detect the turnover in the non-thermal radio spectrum due to absorption by thermal electrons requires observations at very low frequencies at which not many radio telescopes operate (see Israel, this volume). Finally, with the exception of the Magellanic Clouds, galaxies are too far away to detect radio pulsars. Thus we are left with optical imagery and spectroscopy.

The study of the *morphology* of the WIM in nearby galaxies is not new, of course. Photographic surveys of the Magellanic Clouds (Davies *et al.*, 1976, Meaburn, 1980), M31 (Pellet *et al.*, 1978), M33 (*e.g.* Courtès *et al.*, 1988), and other galaxies (*e.g.* Monnet, 1971, Sabbadin and Bianchi, 1979, Hunter, 1982, Hodge and Kennicutt, 1983) have shown filaments, shells, and bubbles of ionized gas, up to more than a kpc in diameter. Some of these surveys also pointed out the existence of low-intensity *diffuse* ionized gas. Strictly speaking, filaments or bubbles are not diffuse gas, in the sense of having a homogeneous distribution. One of the characteristic optical signatures of the WIM is a relatively high ratio of forbidden sulfur, [SII], over Hα line intensities, typically twice as high as discrete HII regions, but not as high as is found in supernova remnants. If we adopt that characteristic of the WIM as a diagnostic, the faint loops and shells can actually be seen as part of the WIM, since they share this property. This high line ratio is characteristic of gas ionized by a strongly diluted photon field (or shocks) and as such both diffuse gas and loops may be part of the same phenomenon. Lasker (1977,1979) obtained Hα and [SII] photographs of several loops in the LMC and drew attention to their relatively high ratio of [SII]/Hα emission.

The earlier photographic surveys typically reached emission measures, $EM = \int n_e^2 dl$, where n_e is the electron density and the integral is along the line of sight, down to 40 pc cm^{-6}, and suffered from variations in the plate background, which limited quantitative work. Observations of the WIM in the Galaxy by Reynolds (1987, 1988) show emission measures down to a few pc cm^{-6}, well below the photographic limit. However, such faint limits can be reached with CCD detectors on even rather modest telescopes. CCD observations of the WIM in other galaxies have only been obtained over the past few years, and allow the first detailed studies of this component of the ISM.

3. DISTRIBUTION AND SPECTRAL CHARACTERISTICS

3.1 The WIM in Nearby Spiral and Irregular Galaxies

From a study of the Hα luminosity functions in several spiral galaxies, Kennicutt et al. (1989) concluded that a substantial fraction of the total Hα luminosity from a galaxy is contributed by diffuse ionized gas, rather than discrete HII regions. For example, for the Large Magellanic Cloud, at least 20-30% of the total Hα emission comes from diffuse gas (Kennicutt and Hodge, 1986). Diffuse gas shows up clearly in two recent very deep CCD surveys of M33 (Hester and Kulkarni, 1990) and M31 (Walterbos and Braun, 1990a, 1990b), and it has also been found in several edge-on systems, which will be discussed below. Fig. 1 shows a field in the NE arm of M31, which was obtained with the No-1 36" telescope at Kitt Peak. The WIM shows up both as faint shells and loops, and as diffuse gas with no apparent structure. That the gas has truely different characteristics from the normal, discreet HII regions is apparent from Fig. 2, which shows cross-cuts through a discrete HII region and through a region with mainly diffuse gas. The ratio of [SII]/Hα intensities is consistently higher for the WIM than for the discreet HII regions, just as is found in our Galaxy (Reynolds, this volume). The diffuse gas can be traced to emission measures well below 5 pc cm^{-6}.

Our survey consisted of 19 fields, imaged in both Hα and [SII], which cover large sections of the spiral arms in the Northern half of M31, over a range in radius from 4 to 15 kpc. The ratio of [SII]/Hα intensities in the WIM is 0.5, and does not seem to vary with radius. A systematic decrease in that ratio with distance from the center is observed for discreet HII regions. The emission from the WIM is strongest closest to the major star forming regions. This is apparent in Fig. 1 and is generally true across the galaxy.

Intensities for the gas with high [SII]/Hα ratio reach up to $EM = 80$ pc cm^{-6}, which is a few times higher than the maximum intensities that Reynolds (1985) finds for our Galaxy. However, this may not be surprising in view of the fact that our fields are located in bright spiral arms. Direct summation of the flux of the WIM indeed shows that it contributes between 20 to 40% of the total Hα flux in M31. The exact value depends on where the transition from WIM to discreet HII region is assumed to take place. Also, our frames do not cover the interarm regions, and intensities were assumed to be zero in the corners of the frames.

The survey of M33 by Hester and Kulkarni shows very similar results. No [SII] maps have been published yet, so it is not yet known if a similar change in ratio of [SII]/Hα intensities occurs, but the presence of the WIM is evident from their high quality image. Their frame covers both arm and interarm regions. Weak interarm

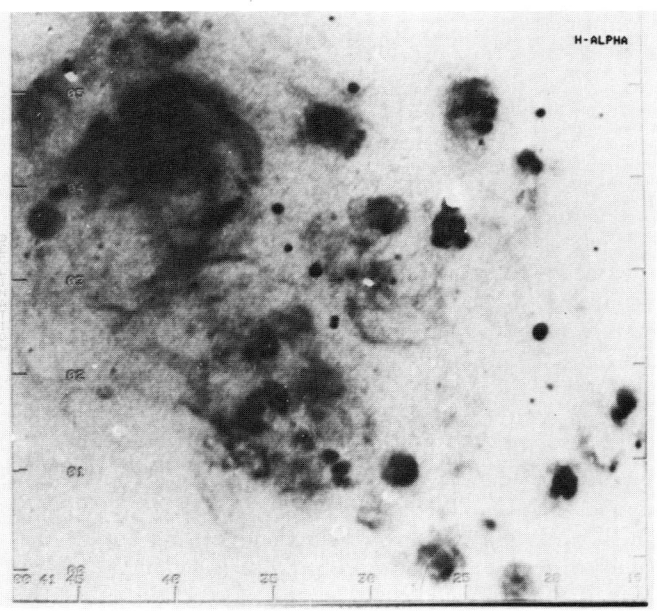

Figure 1. Continuum-subtracted CCD frame in Hα of a region in the NE arm at 10 kpc distance from the center of M31. The frame measures 6.6 arcmin or 1.3 kpc on the side. North is up. The bright HII region in the top-left corner is Pellet 550, one of the largest HII regions in M31. The WIM shows up as faint diffuse emission surrounding the discrete HII regions; also many of the faint loops and shells have spectral characteristics similar to that of the WIM (see text).

emission is seen, but the intensities are higher in the spiral arms. The average emission measure of the WIM is about 40 pc cm^{-6}. Various loops and shells are seen, more homogeneous diffuse gas and possibly sheets of ionized gas (Hester and Kulkarni, 1990). The presence of the WIM in the interarm regions in M33 is not surprising, since there are massive stars there as well (Freedman, 1984).

The WIM has also been detected in irregular galaxies, other than the LMC. In particular, Hunter and Gallagher (1989) drew attention to various loops and filaments in their survey of several irregular galaxies, which have the characteristic line intensity ratios of the WIM. In this case, however, some of the loops were far displaced from O stars so the source of ionization may be a problem. Also, the emission measures are much higher, reaching values of several hundred pc cm^{-6}. Still, in view of the spectral characteristics, this material, which they refer to as *froth*, should be considered as part of the WIM.

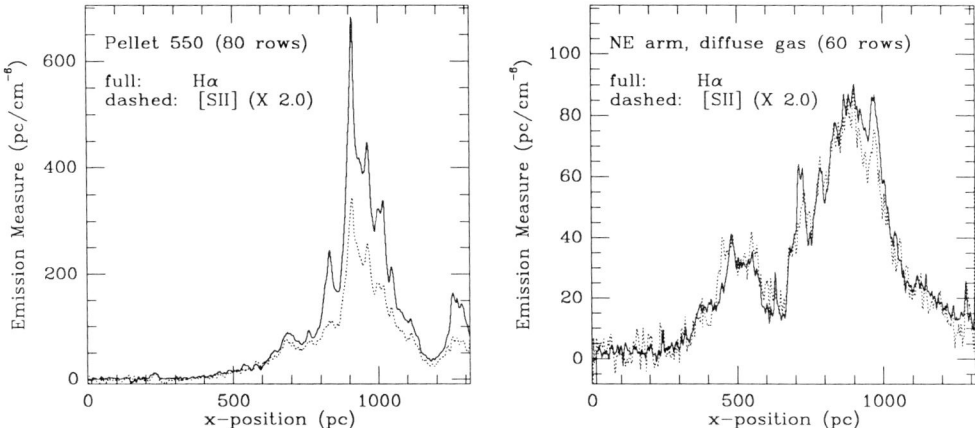

Figure 2. Two cross-cuts in Hα and [SII] emission through different regions in the NE spiral arm in M31. The [SII]-intensities have been multiplied by two. The cuts are averages of the indicated number of rows in a CCD frame. The left plot contains the bright HII region Pellet 550, while the right plot corresponds to a region with mainly diffuse gas.

3.2 The Vertical Distribution

Some edge-on galaxies have been mapped recently, to determine the distribution of the WIM above the disk. In the Milky Way, the WIM has a scaleheight of about 1 kpc, significantly larger than that of atomic hydrogen (see Reynolds, this volume). The best example to date of a galaxy with a prominent Hα disk is NGC 891, an Sb galaxy that is a close twin of our Galaxy in its optical structure (van der Kruit, 1984). Rand et al. (1990) and Dettmar (1990) found a very extended Hα disk, out to some 4 kpc above the plane. This galaxy may also show evidence for outflow of material from the disk through chimneys (Norman and Ikeuchi, 1989), the walls of which show up as ionized filaments protruding from the disk. More details about these and other observations of NGC 891 can be found elsewhere in this volume. Hester et al. also present some results for NGC3079 in these proceedings.

A somewhat different case than NGC891 is presented by NGC4244, a nearby Sc edge-on system. We recently obtained Hα imagery with the No-1 36" telescope at Kitt Peak of this and other galaxies (Walterbos, Braun and Kennicutt, in prep.). Fig. 3 shows a grey scale plot of the continuum-subtracted image. No thick Hα disk is evident in our images. Our observational parameters were quite similar to those of Dettmar in his study of NGC891, so if a comparable disk were present, we ought to have seen it. The full width at $EM = 10$ pc cm^{-6} is about 800 pc, while it is 2000 pc in NGC891 (Rand et al., 1990). Our current data do not rule out the presence of a weaker disk and deeper exposures would be useful. What is noteworthy is that the distribution of HII regions is remarkably wide in the z-direction. HII regions are generally found out to 300 pc above the plane of the disk. The inclination of this galaxy is 88 degrees so some of the apparent width results from projection effects, but this is not the case near the outer edges. One remarkable HII region is found at an apparent height of 750 pc above the disk. Its luminosity is comparable to that of Orion in the Milky Way, so it is a small star forming complex. It might be an outlying HII region in the warped part of the disk. It will be interesting to compare the ionized gas distribution with high resolution HI data. We have recently observed this galaxy at the VLA (Braun et al., in prep.).

A trend appears which suggests that thick Hα disks occur in galaxies that have thick radio continuum disks. NGC891 is known to have a prominent nonthermal radio continuum disk (Allen, these proceedings), as was also pointed out in the papers by Rand et al. and Dettmar. Also our Galaxy has a thick radio disk (Beuermann et al., 1985). The extent of the radio disk is very similar to that of the Hα emission in both cases. NGC4244, on the other hand, has not been detected in radio continuum (Hummel et al., 1984). The underlying mechanism that may explain both the thick radio and Hα disks is most likely the level of star formation. Galaxies with low star formation activity do not manage to vent material into the halo and thus there are no channels for either the gas or the ionizing photons to reach high z-distances. To illustrate this, we can look at the rate of star formation per unit area for these systems. Although not all the far-infrared emission is directly related to star formation activity, it may be the best indicator for edge-on systems where optical tracers suffer too much from extinction. Using data from Rice et al. (1988), the average far-infrared luminisity per unit area, in solar luminosities per square pc, is 4.1 for NGC891 and 0.4 for NGC4244, which is even lower than a value of 0.8 for M31, a galaxy known to have very low star formation activity (Walterbos, 1988). Evidently this correlation can easily be tested by observations of more galaxies.

4. SOURCES OF IONIZATION

Originally, the relatively high ratios of [SII]/Hα intensities that were observed for the WIM seemed to point to shocks as the most likely source of ionization, since spectra of supernova remnants, where shocks are clearly important, show high values for this ratio, between 0.5 and 1.2 (e.g. Blair et al., 1981). Models of shock ionization (e.g. Brand and Mathis, 1978) indeed produce high ratios, but the problem is that the observed values in the WIM in M31 and also in NGC891 are extremely constant, whereas the ratio is critically dependent on shock velocity. A problem is also that all the energy from supernovae would be required to ionize the WIM in the Galaxy (Reynolds, 1988). Photo ionization models generally do not reproduce the high ratios of [SII]/Hα emission. However, Mathis (1986) calculated the effects of a very dilute photon field in a tenuous medium, as is the case for the WIM, and did reproduce the observed line ratios. Thus, the current data seem to be more in agreement with OB stars as the major ionizing source for the WIM. Also energetically, this seems to be feasible (Reynolds, 1988). There are

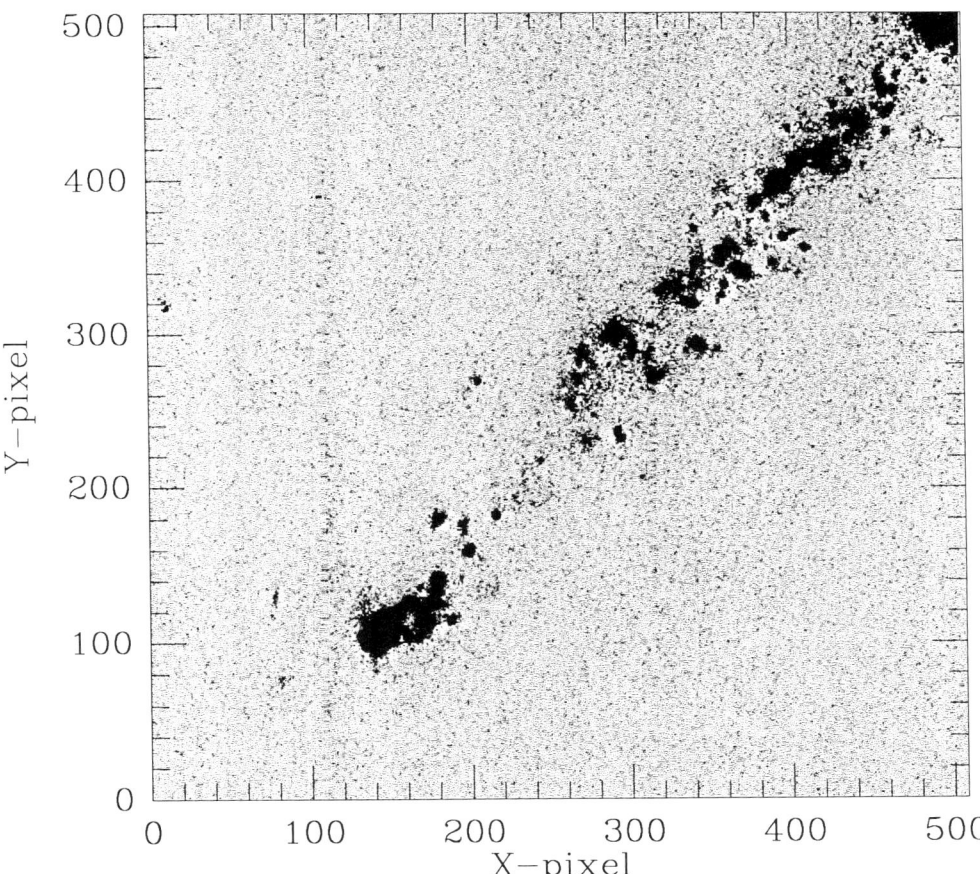

Figure 3. Hα-image of the SW part of the edge-on galaxy NGC4244, obtained with the No1 36" telescope at Kitt Peak National Observatory. Scale: 100 pixels equals about 1100 pc. Note the wide distribution of HII regions; the one at $X = 205$, $Y = 270$ is about 750 pc above the plane. This galaxy does not seem to have a prominent thick Hα disk, however.

two potential problems with the photo ionization model though. First, it is not clear how the ionizing photons can reach such large distances above the plane; this may put strong restrictions on the morphology of the interstellar medium (see also Norman, these proceedings). Second, as was pointed out by Mathis (1986), the timescale for establishing the ionization equilibrium is so long that supernovae are likely to occur in the meantime, so shocks must happen. Clearly, further modelling in this area would be useful. Also the role of runaway O stars may be important and deserves further study.

5. MORPHOLOGY OF THE INTERSTELLAR MEDIUM AND GALAXY TYPE

An important issue, both for the explanation of the thick Hα disks and the general disk-halo interaction, is the question of which galaxies are likely to contain star forming regions that are large enough to produce breakout of the disk and manage to vent material into the halo. One crucial parameter is the number of HII regions in a given galaxy that produces a large enough number of supernovae so that this may occur. An HII region with an observed Hα luminosity of 10^{38} $ergs/s$ (L_{38}), corresponds to the equivalent of about 4 OV stars (Kennicutt, 1988), assuming 1 mag of extinction in Hα. Depending on the lower mass cutoff and Initial Mass Function, this translates to about 40-60 supernovae, roughly the minimum amount required to produce superbubbles, according to theoretical models (see Norman, these proceedings). Blowout will require significantly higher numbers, especially since the thickness of the disk of interstellar material is much larger than assumed in most model calculations, which in general do not include a thick Hα disk.

From the study of the HII region luminosity functions by Kennicutt *et al.* (1989), it is possible to estimate the number of HII regions in different galaxies, brighter than this characteristic luminosity (*cf.* Heiles, 1990). This number does not directly translate into the expected number of superbubbles or chimneys, because the timescales involved in creating the bubble are somewhat longer than the HII region phase, but the numbers will not be different by an order of magnitude. The important result of the study by Kennicutt *et al.* is the strong dependence of the luminosity function, in particular the number of bright HII regions, on galaxy type. Thus, Sa and Sab galaxies do not have HII regions brighter than 5 L_{38} so they may never be able to produce blowout. Relatively quiescent galaxies such as M31 and M81 have of order 50 HII regions brighter than L_{38}, and again few blowout regions are expected. Luminous Sc galaxies, such as M51 and M101 have some 300-500 regions brighter than L_{38} and blowout is certainly likely to occur here, although not in hundreds of cases. The main point is that although blowout and chimney phenomena are likely to occur in galaxies, most of the mass transfer between disk and halo may occur in relatively few, large events rather than in hundreds of small regions. Further observational studies can certainly address this issue. For example, it is striking that most of the Hα shells in the highly inclined galaxy M31 are quite circular in shape, suggesting they are still intrinsically spherical, hence have not reached blowout. The HI holes in M31, cataloged by Brinks and Bajaja (1986), do show some departures from sphericity, but holes larger than 300 pc show only moderate flattening, the average axial ratio being about 0.75.

6. FUTURE DEVELOPMENTS

Further observations will greatly improve our understanding of the distribution of the WIM in galaxies of various types. Observations in other emission lines and correlation of the distribution of the WIM with the young stellar population will allow us to test the photo ionization model. More work needs to be done on trying to derive the actual morphology of the WIM; is it mainly filamentary or sheet-like? How often does blowout occur, and can we find clear regions where it is happening? What is the exact role of the WIM in the chimney or fountain models? One parameter that will be hard to get a handle on, unfortunately, is the electron density in the WIM, since the canonical methods using forbidden line ratios do not discriminate at the low electron densities in the WIM. Yet, that parameter is crucial for deriving filling factors. The connection between the WIM and the hotter ionized gas in the halo, apparent in UV absorption lines (Savage, these proceedings), needs to be studied. Further theoretical work on the modelling of the ionization would be very useful to see if we really understand the ionization mechanism. Finally, current estimates for the filling factor of the WIM seem to indicate that it fills a substantial fraction of the volume in interstellar space; models for the ISM are required that can account for this.

ACKNOWLEDGEMENTS

My work in this area has greatly benefitted from discussions with R. Braun, R.J. Dettmar, J. Hester, R.C. Kennicutt, and C. Norman. I received support from NASA grant NAS5-28086. I am grateful to the IAU for partial financial support to come to this meeting.

REFERENCES

Beuermann, K., Kanbach, G., Berkhuijsen, E.M. (1985) *Astr. Ap.* **153**, 17
Blair, W.P., Kirshner, R.P., Chevalier, R.A. (1981) *Ap.J.* **247**, 879
Brand, P.W.J.L., Mathis, J.S. (1978) *Ap.J.* **223**, 161
Brinks, E., Bajaja, E. (1986) *Astr. Ap.* **169**, 14
Courtès, G., Petit, H., Sivan, J.-P., Dodonov, S., Petit, M. (1988) *Astr. Ap.* **174**, 28
Davies, R.D., Elliott, K.H., Meaburn, J. (1976) *Mem. R. Astr. Soc.* **81**, 89
Dettmar, R.-J. (1990) *Astr. Ap. (Lett.)* **232**, L15
Freedman, W.L. (1984) Ph.D. thesis, University of Toronto
Heiles, C. (1990) *Ap.J.* **354**, 483
Hester, J.J., Kulkarni, S.R. (1990) to be published in *The Interstellar Medium in External Galaxies*, Poster Proc., ed. D. Hollenbach, NASA, in press
Hodge, P.W., Kennicutt, R.C. (1983) *A.J.* **88**, 296
Hummel, E., Sancisi, R., Ekers, R.D. (1984) *Astr. Ap.* **133**, 1
Hunter, D.A. (1982) *Ap. J.* **260**, 81
Hunter, D.A., Gallagher, J.S. (1990) *Ap.J.*, in press
Kennicutt, R.C. (1983) *Ap.J.* **272**, 54
Kennicutt, R.C. (1988) *Ap.J.* **334**, 144

Kennicutt, R.C. (1989) *Ap.J.* **344**, 685
Kennicutt, R.C., Edgar, B.K, Hodge, P.W. (1989) *Ap. J.* **337**, 761
Kennicutt, R.C., Hodge, P.W. (1986) *Ap. J.* **306**, 130
Lasker, B.M. (1977) *Ap. J.* **212**, 390
Lasker, B.M. (1979) *Pub. A.S.P.* **91**, 153
Mathis, J.S. (1986) *Ap.J.* **301**, 423
Meaburn, J. (1980) *M.N.R.A.S.* **192**, 365
Monnet, G. (1971) *Astr. Ap.* **12**, 379
Norman, C.A., Ikeuchi, S. (1989) *Ap.J.* **345**, 372
Pellet, A., Astier, N., Viale, A., Courtès, G., Maucherat, A., Monnet, G., Simien, F. (1978) *Astr. Ap. Suppl.* **31**, 439
Rand, R.J., Kulkarni, S.R., Hester, J.J. (1990) *Ap.J. (Lett.)* **352**, L1
Reynolds, R.J. (1985) *Ap.J.* **294**, 256
Reynolds, R.J. (1987) *Ap.J.* **323**, 118
Reynolds, R.J. (1988) *Ap.J.* **333**, 341
Rice, W.A., Lonsdale, C.J., Soifer, B.T., Neugebauer, G., Koplan, E.L., Lloyd, L.A., de Jong, T., Habing, H.J. (1988) *Ap.J.Suppl.* **68**, 91
Sabbadin, F., Bianchi, A. (1979) *Pub. A.S.P.* **91**, 281
van der Kruit, P.C. (1984) *Astr. Ap.* **140**, 470
Walterbos, R.A.M. (1988) in *Galactic and Extragalactic Star Formation*, eds. R.E. Pudritz and M. Fich, NATO ASI Series, Kluwer, 361
Walterbos, R.A.M., Braun, R. (1990a) to be published in *The Interstellar Medium in External Galaxies*, Poster Proc., ed. D. Hollenbach, NASA, in press
Walterbos, R.A.M., Braun, R. (1990b), in prep.

IS THERE EVIDENCE FOR DISK-HALO CONNECTIONS IN M31 ?

ELLY M. BERKHUIJSEN, GÖTZ GOLLA, RAINER BECK
Max-Planck-Institut für Radioastronomie
Auf dem Hügel 69
D-5300 Bonn 1
Federal Republic of Germany

ABSTRACT. There is some evidence for disk-halo connections in M31, i.e.: (1) The central area seems inclined w.r.t. the main disk. (2) In several regions in the southern half the magnetic field has a significant component perpendicular to the disk. However, (3) any general emission from a thick disk at $\lambda 75$ cm is ~100x weaker than that from our Galaxy. The uniform disk field seems to inhibit cosmic-ray diffusion perpendicular to the plane, and/or the field at high z is exceptionally weak.

1. CENTRAL AREA

Beck et al. (1989) showed that the projected uniform magnetic field ($B_\perp$) runs along the Hα-arm at about 2' (400 pc) south and east of the nucleus (Ciardullo et al., 1988). If this arm is seen from an angle of ~20° it would be ~55° inclined to the main disk of M31. North of the nucleus many thin Hα filaments are seen, typically 0.5 kpc long. These filaments possibly stick out of the main disk with inclinations of ~30°. If the magnetic field runs along the filaments, the complicated structure in the radio beam could account for the non-detection of polarized emission at $\lambda 20$ cm in this area.

2. DISK

In Fig. 1 the magnetic field structure within 30' from the minor axis (Berkhuijsen et al., 1987) is shown superimposed onto the distribution of integrated HI of Brinks and Shane (1984). The field is generally aligned with the spiral arms, but some deviations are noticeable.

Near the edge of the map in the southwestern quadrant part of the field lines is orientated along an HI bridge connecting the brightest inner arm with an outer arm. This quadrant was also observed at $\lambda 20$ cm with the VLA; the field structure at a resolution of 3' derived from a comparison with the $\lambda 6.3$ cm data taken at Effelsberg (Berkhuijsen et al., 1987) is in good agreement with that shown in Fig. 1. Beck et al. (1989)

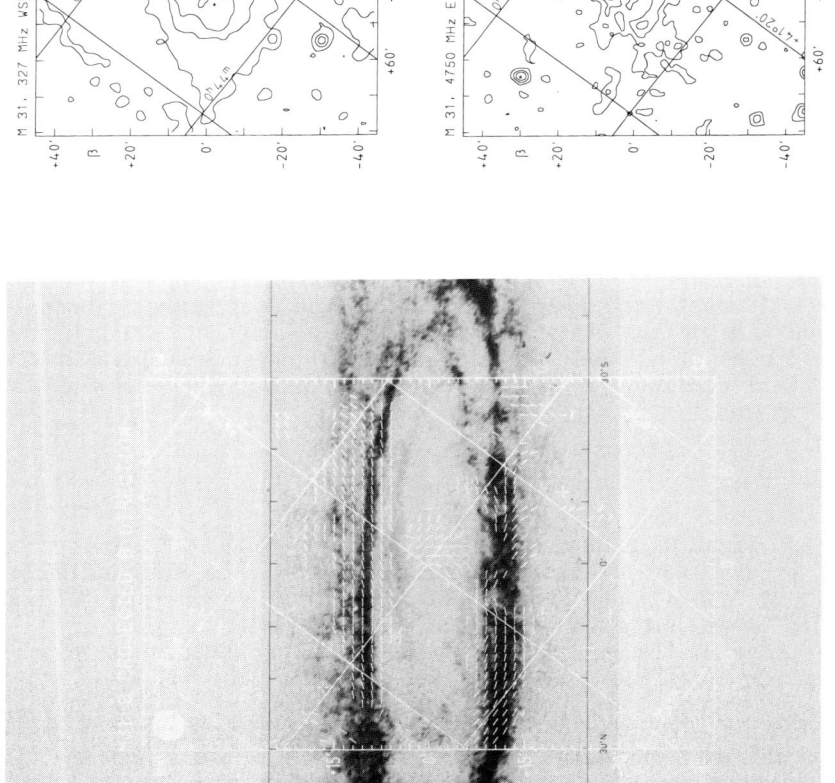

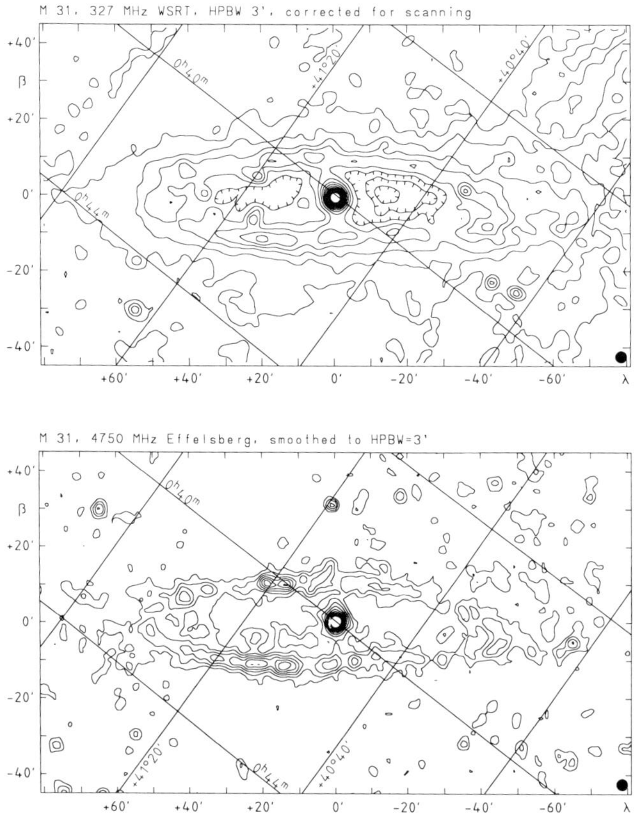

Figure 2. Radio continuum map of M31 (HPBW = 3') at: (a) λ92 cm, contour levels 5 (= 1.4σ), 20, ..., 215 mJy/beam, hatched contours enclose minima; (b) λ6.3 cm, contour levels 2.5 (= 2.2σ), 4.5, ..., 22.5 mJy/beam.

Figure 1. Distribution of B⊥ superimposed onto the integrated HI distribution.

could explain the distributions of $B_\perp$ and of the rotation measures (RM) by a magnetic loop coming out of the plane, a structure reminiscent of a Parker-Jeans instability.

Field lines strongly deviating from the arms are also seen in the southeastern quadrant of Fig. 1. Interestingly, spurs perpendicular to the arms in the distributions of cool dust (Walterbos and Schwering, 1987) and blue light (Walterbos and Kennicutt, 1987) have been detected, which are coincident with the magnetic field deviations near $(\lambda,\beta) = (-9',-20')$ and $(-27',-15')$. The authors note, however, that these spurs could be foreground structures in the Milky Way. Another remarkable fact is the large 'hole' on the spiral arms between $\lambda = -15'$ and $-35'$ in the $^{12}CO(1-0)$ distribution of Dame et al. (1991). In this area the number of massive molecular clouds may be too small to keep the magnetic field in the disk (Beck et al., 1991; Beck, 1991).

3. THICK DISK AND EXTENDED HALO

Recently, Golla and Walterbos (in prep.) completed a map at $\lambda 92$ cm of M31 with the Westerbork interferometer. After subtraction of unrelated point sources this map was smoothed to a resolution of 3' for comparison with the Effelsberg map at $\lambda 6.3$ cm (Fig. 2a and 2b). Both maps show extended minima on either side of the nucleus. Although these minima contain dust and stars they emit very little nonthermal emission. An upper limit to the thick-disk emission was estimated by taking a 10' wide cut through the southern minimum perpendicular to the major axis (Fig. 3). At $\lambda 92$ cm the emission from the minimum is < 10 mJy/b (= 5 K), whereas that at $\lambda 6.3$ cm is < 0.5 mJy/b (=1.1 mK = 1σ). Hence, the flux density spectrum has a slope steeper than 1.1. These values are consistent with the estimate of < 2 K at 408 MHz at $\lambda = -15'$ with a 4!5 beam by Gräve et al. (1981). This upper limit for the emission from a thick disk in M31 is ~100x less than the emission from the thick disk of our Galaxy (Beuermann et al., 1985, Fig. 9) and > 200x less than that of NGC 891 (Hummel et al., 1990) at 408 MHz.

Weak evidence for an extended halo of < 4.2 K at 408 MHz and a diameter of $2°.4\pm0°.1$ has been given by Gräve et al. (1981). Such a halo would not be detectable in our maps.

The distribution of the emission at $\lambda 92$ cm is much broader than that at $\lambda 6.3$ cm (Figs. 2

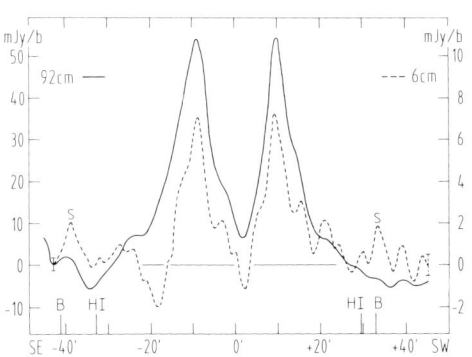

Figure 3. Cuts through the southern minimum perpendicular to the major axis averaged between $\lambda = -8'$ to $-17'$. At $\lambda 92$ cm the emission at $\lambda > 20'$ is influenced by a ripple in the background. At $\lambda 6.3$ cm s = unsubtracted point source.

and 3) suggesting that the high-energy electrons have suffered diffusion losses. This is supported by the spectral distribution along the cuts in Fig. 3 with typical values of $\alpha = 0.73$ on the emission peaks steepening to $\alpha = 1.0$ to 1.1 on the relative minima in the emission ring at $\lambda 6.3$ cm. If the nonthermal spectral index $\alpha_{nth} = 1.0$ the fraction of nonthermal emission at $\lambda 92$ cm is $> 90\%$ and at $\lambda 6.3$ cm $> 40\%$. Assuming that the distribution at $\lambda 6.3$ cm reflects the distribution of the sources of relativistic electrons the observed differences in width may be used to estimate the diffusion length of the electrons. The beam-corrected half-widths in Fig. 3 are 9!5 and 6!1 in the SE and 6!9 and 4!0 in the SW at $\lambda 92$ cm and $\lambda 6.3$ cm, respectively. If the diffusion were entirely *in* the plane of M31 (distance = 690 kpc, inclination of ring = 74°) the full half-width is 2.1 to 2.5 kpc larger at $\lambda 92$ cm than at $\lambda 6.3$ cm, yielding diffusion lengths of ~1 kpc. Similarly, if the diffusion were entirely *perpendicular* to the plane the diffusion length is ~0.3 kpc. In this case the full thickness of the nonthermal emission in z at half intensity is 2 kpc in the SE and 1.4 kpc in the SW.

The exceptionally weak emission at high z of M31 may be a consequence of the high uniformity of the magnetic field parallel to the spiral arms, which hampers cosmic-ray diffusion perpendicular to the arms, and/or of a weak magnetic field in the thick disk (Beck, 1991).

REFERENCES

Beck, R. (1991) this volume
Beck, R., Loiseau, N., Hummel, E., Berkhuijsen, E.M., Gräve, R., Wielebinski, R. (1989) *Astron. Astrophys.* **222**, 58
Beck, R., Berkhuijsen, E.M., Bajaja, E. (1991) in *Dynamics of Galaxies and Molecular Cloud Distribution*, IAU Symp. 146, eds. F. Combes and F. Casoli, Kluwer, Dordrecht
Berkhuijsen, E.M., Beck, R., Gräve, R. (1987) in *Interstellar Magnetic Fields*, eds. R. Beck and R. Gräve, Springer, Heidelberg, p. 38
Berkhuijsen, E.M., Wielebinski, R., Beck, R. (1983) *Astron. Astrophys.* **117**, 141
Beuermann, K., Kanbach, G., Berkhuijsen, E.M. (1985) *Astron. Astrophys.* **153**, 17
Brinks, E., Shane, W.W. (1984) *Astron. Astrophys. Suppl.* **55**, 179
Ciardullo, R., Rubin, V.C., Jacoby, G.H., Ford, H.C., Ford, W.K. (1988) *Astron. J.* **95**, 438
Dame, T.M., Thaddeus, P., Koper, E. (1991) in *Dynamics of Galaxies and Molecular Cloud Distribution*, IAU Symp. 146, eds. F. Combes and F. Casoli, Kluwer, Dordrecht
Gräve, R., Emerson, D.T., Wielebinski, R. (1981) *Astron. Astrophys.* **98**, 260
Hummel, E., Dahlem, M., van der Hulst, J.M., Sukumar, S. (1990) *Astron. Astrophys.* (in press)
Walterbos, R.A.M., Kennicutt, R.C. Jr. (1987) *Astron. Astrophys. Suppl.* **69**, 311
Walterbos, R.A.M., Schwering, P.B.W. (1987) *Astron. Astrophys.* **180**, 27

X-RAY HALOES AND COOLING FLOWS

A.C. FABIAN
Institute of Astronomy, Madingley Road,
Cambridge CB3 0HA, UK

ABSTRACT. The properties of hot gaseous haloes in massive early-type galaxies are briefly reviewed. Gas flows in such haloes are complex yet so large-scale that they may guide us in the understanding of flows around disk galaxies. The intracluster medium is discussed as a further illustration of the properties of diffuse hot gas trapped in a gravtational well. Finally, the possibility of the existence of a significant diffuse medium in the Local Group, and in groups in general, is revived. Such a medium would generate a substantial disk-halo interaction with our Galaxy.

1. INTRODUCTION

Hot gaseous haloes are common in massive early-type galaxies, in groups and in clusters of galaxies. They provide an observable guide to what any large gaseous haloes of disk galaxies may be like and in particular show the behaviour of diffuse hot gas trapped in a gravitational potential. X-ray, and other, observations of the hot gas in early-type galaxies are reviewed and models for the flows in these galaxies discussed. The gas is shown to multiphase and the flows complex. Some clues for understanding the operation of thermal conduction, magnetic fields, mixing and star formation in such gas are considered. The more extensive, and better observed, hot atmospheres in clusters are used to illustrate these processes further.

Much of the gas in a protogalaxy may exist in the form of a hot halo if gravitational collapse succeeds in heating it to the virial temperature of the galaxy. The study of such haloes may therefore be important for understanding galaxy formation. Large-scale gaseous haloes may persist now in all groups of galaxies, as well as in clusters. This leads to disk-halo interactions being common, in the sense that the gas in disk galaxies in groups (*i.e.* perhaps most disk galaxies) is interacting with an intra–group or –cluster medium. Such interactions, which are commonly ignored in groups, may explain some observed phenomena in our own and other galaxies.

Dramatic disk-halo interactions occur when a group merges with another group or with a cluster of galaxies. Then the pressure of the surrounding halo rises ten- to onehundred-fold and the cluster halo controls the behaviour of the disk gas.

2. THE HOT INTERSTELLAR MEDIUM OF EARLY-TYPE GALAXIES

The hot ($T \sim 10^7$ K) interstellar medium (ISM) of massive early-type galaxies was discovered by X-ray images from the *Einstein Observatory* (Forman et al. 1979). Prior to then, the apparent lack of any substantial ISM in elliptical galaxies was ascribed to galactic winds.

The X-ray data showed that several Virgo cluster ellipticals have substantial haloes of $10^9 - 10^{10}$ M$_\odot$ of hot gas. Of particular importance was the discovery of a 'plume' of X-ray emission to the galaxy M86, the radial velocity of which indicates it is falling through the intracluster medium (ICM) at about 1500 km s^{-1}. A lack of radio, or other, emission in the plume indicated that the X-ray emission was due to thermal bremsstrahlung and line radiation from diffuse hot gas. What X-ray spectral data there was, agreed with that conclusion. The plume is then due to ram-pressure stripping of the hot halo of M86 by the Virgo cluster ICM.

These results were added to by observations of haloes in galaxies in groups (Biermann & Kronberg 1983) and in relatively isolated ellipticals (Nulsen et al. 1984). Forman, Jones & Tucker (1985),and Canizares, Fabbiano & Trinchieri (1987) (for a review see Fabbiano 1989) have since scoured the *Einstein Observatory* database for detections and upper limits of the X-ray luminosity of early-type galaxies and find that $L_X \propto L_B^\beta$, where L_X and L_B are the X-ray and B-band luminosities of the galaxy and $\beta \sim 1.5 - 2.3$. The exact value obtained for β depends on what corrections are made for other sources of X-ray emission (*e.g.* binaries) in the galaxy and on what distances are adopted (redshifts are not the best guide to distance in the nearby, Virgo distance, flow where most of the detected galaxies lie). The most recent determinations (see *e.g.* Sarazin 1990) prefer a value of $\beta \sim 2.3$. Ellipticals in clusters with a massive ICM, such as the Coma cluster, are deficient in such individual haloes (Canizares 1988), presumably due to ram-pressure stripping.

No clear detections have yet been made of X-ray emitting gaseous haloes around disk galaxies (Bregman & Glassgold 1982; McCammon & Sanders 1984). Indeed, disks and extensive hot haloes appear to be mutually exclusive, at least in the *Einstein Observatory* data, since Bender et al. (1989) have shown that X-ray haloes are most commonly found around 'boxy' galaxies and rarely around 'disky' ones (the terms 'boxy' and 'disky' refer to the shape of the isophotes). Whether this is due to an absence of gas around disk-, or 'disky'–galaxies, or to a lower temperature putting the emission outside the *Einstein Observatory* waveband is not clear. The recently launched X-ray telescope on ROSAT has much more sensitivity at softer X-ray wavelengths and will resolve this issue, as well as finding hundreds more X-ray emitting elliptical galaxies.

The detection of such haloes, particularly in isolated galaxies, indicated that the gas is quasi-static – any flows must be subsonic. If they were not, then the gas would disperse in a flow time,

$$t_{flow} < R/c_s,$$

where R is a characteristic radius for the bulk of the observed gas (say 10 kpc) and c_s is the speed of sound in the gas ($\sim 10^2$ km s^{-1}, comparable with the velocity dispersion of stars within the galaxy). The mass loss rate of the galaxy would then be

$$\dot{M} \gtrsim \left(\frac{M_{gas}}{t_{flow}}\right) > 10 \, M_\odot \, \text{yr}^{-1}.$$

This is much greater than the rate at which the ISM of the galaxy can be replaced by stellar mass loss,

$$\dot{M}_\star \sim 1 \, M_\odot \, \text{yr}^{-1}.$$

It would therefore be very surprising to detect haloes commonly in such galaxies. The simplest conclusion from the observed hot gaseous haloes is that the halo gas is in hydrostatic equilibrium.

This led to several conclusions which are somewhat controversial. The first is that the high temperature of the gas would lead to a wind developing and a loss of the gas unless the potential well of the galaxy was large. Massive elliptical galaxies have massive dark haloes (Forman, Jones & Tucker 1975; Fabian et al. 1986; but see Trinchieri, Fabbiano & Canizares 1986). The second is that the supernova rate in elliptical galxies must be low, less than one per 400 yr, or too much heat would be injected into the halo and it would be heated to the temperature required for escape. The supernova rate in ellipticals has been revised downward in the past few years (Van den Bergh, McClure & Evans 1987) towards this value. One question that had worried me about supernovae in elliticals was the possible requirement of binaries if the supernovae were due to an old stellar population. Are binaries present in elliptical galaxies? This question in now answered to some degree by the discovery of novae in the radio galaxy Centaurus A, which does have a large spheroidal body of stars.

The third conclusion concerns the cooling rate of the hot halo. The radiative cooling time (due to the emission of X-rays such as those detected), t_{cool} is less than a Hubble time within the inner 30 kpc of many large ellipticals and is much shorter ($< 10^9$ yr) within the inner few kpc. This suggests that cooling flows operate in massive elliptical galaxies, where

$$\dot{M}_{cool} \sim 1 \, M_\odot \, \text{yr}^{-1} \sim \dot{M}_\star.$$

The accumulation of stellar mass loss in the hot halo is balanced by radiative cooling of the gas, presumably leading to some star formation. Whether 'normal' (disk-galaxy like) star formation continues to take place in early-type galaxies is debatable (Pickles 1986; Burstein et al. 1988). Perhaps the gas chiefly forms low-mass stars, as must be the case in cluster cooling flows (see next Section).

Detailed studies of the X-ray surface brightness profile and of the effects of cooling and of mass and energy injection from stars and supernovae show that gas flows in ellipticals are complex (Thomas 1986; Thomas et al. 1986; Sarazin & White 1988; Sarazin & Ashe 1989; Sarazin 1990). The gas must exist in many phases, some of which may flow out as some flow in. Gas cools out of the hotter phases over a wide range of radii, not just in the centre. Not all the stellar mass loss (predominantly from red-giant winds and planetary nebulae) may be heated to the X-ray emitting phases, but the denser parts may just be slowed down and remain intact. What happens then is unclear. The flows probably resemble a giant galactic fountain, such as discussed elsewhere in this Volume for disk galaxies, except that this time the geometry is more neary spherical. Further study of such behaviour probably requires further X-ray observations of higher spatial and spectral quality. Fortunately these should soon be forthcoming from ROSAT, ASTRO–D and later X-ray missions in the 1990s.

The evolution of such haloes has been discussed by Mathews (1989; see also references in Sarazin 1990).

3. THE INTRACLUSTER MEDIUM

The ultimate hot halo is of course the intracluster medium (ICM). All clusters of galaxies appear to have diffuse gas spread throughout them with a mass of about 10 to 30 per cent of the total mass of the cluster. The gas has a temperature of between 2×10^7 K for the poorest clusters to 10^8 K or more for the richest ones. Their X-ray properties have been well-reviewed by Sarazin (1988).

Such gas is not or particular relevance to the topic of the 'disk-halo connection', except that it may have stripped the disk gas from many member galaxies and its large scale means that some processes that must occur in hot halo gas have been clearly exposed for observation.

On the first point, it should be noted that if our galaxy could be placed in the core of a rich cluster, then the mean pressure of the ISM would rise by a factor of about 100. If the galaxy then fell in the gravitational well of the cluster it would experience large ram-pressure forces, first discussed by Gunn & Gott in 1972, and turbulent stripping process (Nulsen 1982). Such a 'disk-halo' interaction would destroy the disk gas, and so its reason for being thin. This could account for some S0 galaxies (Biermann & Shapiro 1979). Worth noting now is that clusters do not appear to have formed at some early time intact, but are evolving now through the merger of subclusters (Edge *et al.* 1990; see also Gioia *et al.* 1990). This means that whole groups are being subsumed into clusters now.

On the second point, we now see that the ICM is again complex, particularly in most cluster cores where the data are best. The radiative cooling time of the gas within about 100 kpc of the central galaxy in most ($\gtrsim$ 70per cent; Pesce *et al.* 1990) clusters is less than a Hubble time. The strongly-peaked X-ray surface brightness profiles within this region show that the cooling time continues to decrease inward, indicating the presence of a cooling flow. This just means that when the gas loses energy by radiation, its density must rise by inflow in order to maintain the pressure necessary to support the weight of the overlying gas. X-ray spectra of a few clusters shows that the gas does cool (Canizares, Markert & Donahue 1988; Mushotzky & Szymkowiak 1988). Instead of the one or less solar masses per year cooling in elliptical galaxies, here we have tens to hundreds of solar masses cooling out of the hotter phases per year. Optical colours and spectra of the central galaxies show that 'normal' star formation cannot be taking place and that whatever the cooled gas forms, it cannot be massive blue stars. Probably the gas just forms into low–mass stars, since the molecular clouds that give birth to massive stars in the Solar Neighbourhood cannot exist tens of kpc out from the central galaxy in the ICM. If one were to be placed there, it would be broken into pieces by the ICM on falling into the galaxy. Reviews of cooling flows and of how star formation might take place in them are given by Fabian, Nulsen & Canizares (1984; 1990), Sarazin (1988) and Fabian (1990).

Points of interest for the study of hot haloes elsewhere are the operation of thermal instability (Balbus & Soker 1989); the inference of turbulence in the gas (Heckman et al. 1989; Loewenstein & Fabian 1990); of tangled magnetic fields which reduce thermal conduction (see above reviews – the magnetic fields themselves are studied on scales of kpc by Faraday depolarization of central radio sources; Dreher et al. 1987; Fomalont et al. 1989; Owen et al. 1990); the occurrence of dust in the central galaxies, either in dust lanes such as in NGC 4696 (Sparks et al. 1989) or observed through infrared emission (Bregman et al. 1990), of warm clouds of gas embedded in the hot gas (see e.g. Heckman et al. 1989), of cold HI clouds observed through 21 cm absorption of a radio source (Jaffe et al. 1988; McNamara et al. 1990), or 21 cm emission (Bregman et al. 1988); of molecular gas observed through its CO emission (Lazareff et al. 1989; Mirabel et al. 1990). Jaffe's (1990) observation of extended HI absorption across the inner tens of kpc of the cooling flow in the Perseus cluster suggests that there is at least $10^{10}\,M_\odot$ of cold gas accumulated there. This could be a large underestimate of the total mass of cold gas in the flow if it is optically thick and extends (in clouds) out to radii of 100 – 200 kpc. We have been studying (Fabian et al. 1990) the implications of such large quantities of cold gas. Cloud-cloud collisions may be common in the inner parts of the flow and contribute to the observed optical line emission. Cold clouds distributed throughout the core of the ICM may be responsible for some of the narrow absorption line clouds observed in the spectra of distant quasars. Further studies of the ICM should allow us to understand how magnetic fields, turbulence, cloud and star formation take place in all hot haloes.

4. GAS FLOWS IN LOOSE GROUPS AND SPIRAL GALAXIES

The general existence of hot haloes in early-type galaxies and clusters of galaxies suggests that they could exist in groups of galaxies. After all, it is unlikely that galaxy formation was 100 per cent efficient anywhere. The most difficult aspect is possibly the depth of the potential well in a loose group, such as the Local Group. The early star-forming phases of most galaxies are likely to be rich in supernovae (otherwise galaxies would not be metal-rich). These can blast much of the rest of the gas out of the potential well of small galaxies and lead to a metal-enriched intergalactic medium. This is probably the source of the metals in the ICM (Larson & Dinerstein 1975; see discussion in Thomas & Fabian 1990). A problem with a loose group is that much of the metal-enriched gas may be ejected out altogether. However, if the velocity is less than $100\,\mathrm{km\,s^{-1}}$, the gas can travel no more than 1 Mpc in a Hubble time so will remain in the Group, if not bound to it. Consequently, we can expect that most groups, no matter how loose, contain an IntraGroup Medium, which is almost as metal-rich as that in clusters (say one-tenth Solar). The temperature of this gas will be between 10^6 K and 10^7 K if it does not all form a wind, and the total mass of gas may even exceed that of the component galaxies in the group (the scaling arguments in Thomas & Fabian 1990 are relevant here).

What limits can we place on the existence of such gas in the Local Group? For that we can use the soft X-ray Background limits of McCammon & Sanders (1990), which have not changed significantly since the 1970s, when Hunt & Sciama (1972) made a similar estimate. Assuming a scale size of 1 Mpc, slightly larger than the distance to the Andromeda galaxy, I find a gas density of $\sim 10^{-5}$ cm^{-3} if $T \sim 10^7$ K and 2×10^{-5} cm^{-3} if the temperature is about 3×10^6 K. The total X-ray luminosity is between $10^{41}-10^{42}$ erg s^{-1} and the mass of gas about 10^{12} M$_\odot$. The pressure of the gas is only a few percent of that in our own ISM, so there are no overpressure problems. The cooling time of the gas, if smoothly distributed, is between 1 and 100 Hubble times, so if the gas is clumpy, then cooling and cooled clouds may occur. The ram power due to the motion of our Galaxy through this medium is a few per cent of the total kinetic power due to supernovae. This is not much power compared with the total Galactic power budget, but the ram power is deposited at the outskirts of the galaxy, principally as turbulence (see Begelman & Fabian 1990 for a discussion of a turbulent mixing layer above the galactic disk). Cold clouds embedded in the intragroup medium (perhaps clouds that have never been heated up, or regions that were denser and have since cooled) then impact our galaxy from a particular direction, related to the vector of our velocity and that of the local intragroup medium. They are then a source of high-velocity clouds.

Some discussions of the behaviour of such a medium can be found in the 1970s (Kahn & Woltjer 1959; Hunt & Sciama 1972; Oort 1969) and its effect on the Magellanic Stream (Mathewson *et al.* 1977). It appears to me to have been largely ignored in the 1980s, except as a source for high-velocity clouds (see *e.g.* Mirabel & Morras 1990), but does require further study, either to rule it out or to establish its properties. Simple ways in which such gas may be detected directly or indirectly are through its X-ray emission (either in the Local Group or in other groups – ROSAT may be crucial here), through absorption studies of 'intragroup clouds' (perhaps some of the narrow metal absorption lines commonly found in the spectra of distant quasars, Bergeron (1988), are due to intragroup clouds) its impact on our and other galaxies - either in terms of energy deposited, mixing, turbulence, abrupt edges to the HI (see discussion in terms of photoionization for such edges by Kenney 1990) or HI warps, tails and streams. Some medium with roughly the properties described above must exist.

REFERENCES

Balbus, S. & Soker, N., 1989. *Astrophys. J.*, **341**, 611.
Begelman, M.C. & Fabian, A.C., 1990. *Mon. Not. R. astr. Soc.*, **244**, 26P.
Bender, R., Surma, P., Döbereiner, S., Möllenhoff, C. & Madejsky, R., 1989. *Astr. Astrophys.*, **217**, 35.
Bergeron, J. 1988. In *QSO Absorption Systems; Probing the Univers*, eds Blades, J.C., Norman, C.A. & Turnshek, D., C.U.P.
Biermann, P. & Shapiro, S.L, 1979. *Astrophys. J.*, **230**, L33.
Biermann, P. & Kronberg, P.P., 1983. *Astrophys. J.*, **268**, L69.
Bregman, J.N. & Glassgold, A.E., 1982. *Astrophys. J.*, **263**, 564.
Bregman, J.D., McNamara, B. & O'Connell, R., 1990. *Astrophys. J.*, in press.
Bregman, J.N., Roberts, M.S. & Giovanelli, R., 1988. *Astrophys. J.*, **330**, L93.
Burstein, D., Bertola, F., Buson, L.M., Faber, S.M. & Lauer, T.R., 1988. *Astrophys. J.*, **328**, 440.
Canizares, C.R., 1988. In *Cooling Flows in Clusters and Galaxies*, ed. A.C. Fabian, Kluwer, Dordrecht, 376.
Canizares, C.R., Markert, T.H. & Donahue, M.E., 1988. In *Cooling Flows in Clusters and Galaxies*, ed. A.C.Fabian, Kluwer, 63.
Canizares, C.R., Fabbiano, G. & Trinchieri, G., 1987. *Astrophys. J.*, **312**, 503.
Dreher, J.W., Carilli, C.L. & Perley, R.A., 1987. *Astrophys. J.*, **316**, 611.
Edge, A.C., Stewart, G.C., Fabian, A.C. & Arnaud, K.A., 1989. *Mon. Not. R. astr. Soc.*, **245**, 559.
Fabbiano, G., 1989. *Ann. Rev. Astr. Astrophys.*, **27**, 87.
Fabian, A.C., 1990. In *Baryonic Dark Matter*, eds. D. Lynden Bell & G. Gilmore, Kluwer, Dordrecht, 195.
Fabian, A.C., Nulsen, P.E.J. & Canizares, C.R., 1984. *Nature*, **311**, 733.
Fabian, A.C., Nulsen, P.E.J. & Canizares, C.R., 1990. *Astr. Astrophys. Rev.*, in press.
Fabian, A.C., Thomas, P.A., Fall, S.M. & White, R.A., 1986b. *Mon. Not. R. astr. Soc.*, **221**, 1049.
Fabian, A.C., Thomas, P.A., Daines, S.J. & Johnstone R.M., 1990. Preprint.
Fomalont, E.B., Ebnetter, K.A., van Breugel, W.J.M. & Ekers, R.D., 1989. *Astrophys. J.*, **346**, L17.
Forman, W., Schwarz, J., Jones, C., Liller, W. & Fabian, A., 1979. *Astrophys. J.*, **234**, L27.
Forman, W., Jones, C. & Tucker, W., 1985. *Astrophys. J.*, **293**, 102.
Gioia, I.M., Henry, J.P., Maccacaro, T., Morris, S.L., Stocke, J.T. & Wolter, A., 1990. *Astrophys. J.*, **356**, L35.
Gunn, J.E. & Gott, J.R., 1972. *Astrophys. J.*, **176**, 1.
Heckman, T.M., Baum, S.A., van Breugel, W.J.M. & McCarthy, P.,1989. *Astrophys. J.*, **338**, 48.
Hunt, R. & Sciama, D.W., 1972. *Mon. Not. R. astr. Soc.*, **157**, 335.
Jaffe, W., de Bruyn, A.G. & Sijbreng, D., 1988. In: *Cooling Flows in Clusters and Galaxies*, ed. Fabian, A. C., Kluwer, Dordrecht, Holland, p145.
Jaffe, W., 1990. *Astr. Astrophys.*, in press.
Kahn, F.D. & Woltjer, L., 1959. *Astrophys. J.*, **130**, 705.
Kenney, J.D.P., 1990. In *The Interstellar Medium in External Galaxies*, eds H.A. Thronson & J.M. Shull, Kluwer, Dordrecht, 151.
Larson, R.B. & Dinerstein, H.L., 1975. *Publ. astr. Soc. Pacific*, **87**, 911.
Lazareff, B., Castets, A., Kim, D-W. & Jura, M., 1989, **336**, L13.
Loewenstein, M. & Fabian, A.C., 1990. *Mon. Not. R. astr. Soc.*, **242**, 120.

Mathews, W., 1989. *Astr. J.*, **97**, 42.
Mathewson, D.S., Schwarz, M.P. & Murray, J.D., 1977. *Astrophys. J.*, **217**, L5.
McCammon, D. & Sanders, W.T., 1990. *Astrophys. J.*, **287**, 167.
McCammon, D. & Sanders, W.T., 1990. *Ann. Rev. Astr. Astrophys.*, **28**, 657.
McNamara, B.R., Bregman, J.N. & O'Connell, R.W., 1990. *Astrophys. J.*, **360**, 20.
Mirabel, I.F. & Morras, R., 1990. *Astrophys. J.*, **356**, 130.
Mirabel, I.F., Sanders, D.B. & Kazes, I., 1989. *Astrophys. J.*, **340**, L9.
Mushotzky, R.F. & Szymkowiak, A.E., 1987. In *Cooling Flows in Clusters and Galaxies*, ed. A.C.Fabian, Kluwer, 47.
Nulsen, P.E.J., 1982. *Mon. Not. R. astr. Soc.*, **198**, 1007.
Nulsen, P.E.J., Stewart, G.C. & Fabian, A.C., 1984. *Mon. Not. R. astr. Soc.*, **208**, 185.
Oort, J.H., 1969. *Nature*, **224**, 1158.
Owen, F., Eilek, J.A. & Keel, W.C., 1990. *Astrophys. J.*, **362**, 449.
Pesce, J.E., Edge, A.C., Fabian, A.C. & Johnstone, R.M., 1990. *Mon. Not. R. astr. Soc.*, **244**, 58.
Pickles, A.J., 1985. *Astrophys. J.*, **296**, 340.
Sarazin, C.L., 1988. *X-ray Emission from Clusters of Galaxies*, C.U.P.
Sarazin, C.L., 1990. In *The Interstellar Medium in External Galaxies*, eds H.A. Thronson & J.M. Shull, Kluwer, Dordrecht, 201.
Sarazin, C.L. & White, R.E., 1988. *Astrophys. J.*, **335**, 688.
Sarazin, C.L. & Ashe, G.A., 1989. *Astrophys. J.*, **345**, 22.
Sparks, W.B., Macchetto, F. & Golombek, D., 1989. *Astrophys. J.*, **345**, 153.
Thomas, P.A., 1986. *Mon. Not. R. astr. Soc.*, **220**, 949.
Thomas, P.A., Fabian, A.C., Arnaud, K.A., Forman, W. & Jones, C., 1986. *Mon. Not. R. astr. Soc.*, **222**, 655.
Thomas, P.A. & Fabian, A.C., 1990. *Mon. Not. R. astr. Soc.*, **246**, 156.
Trinchieri, G., Fabbiano, G. & Canizares, C.R., 1986. *Astrophys. J.*, **310**, 637.
Van den Bergh, S., McClure, R.D. & Evans, R., 1987. *Astrophys. J.*, **323**, 44.
Vedder, P.W., Trester, J.J. & Canizares, C.R., 1988. *Astrophys. J.*, **332**, 725.

OPAQUE SPIRAL DISKS: SOME EMPIRICAL FACTS AND CONSEQUENCES

EDWIN A. VALENTIJN
European Southern Observatory, Karl-Schwarzschild-Str. 2
D - 8046 Garching bei München

1. INTRODUCTION

A recent analysis of the surface brightness profiles of a complete sample of 9381 spiral galaxies extracted from ESO-LV (Lauberts and Valentijn, 1989) showed that many galaxy disks of especially Sb–Sc galaxies are opaque (Valentijn, 1990). This paper studied how the observed surface brightness μ^{obs} varies as a function of the observed axial ratio a/b, by fitting the data of samples of spirals with

$$\mu^{obs} = \mu^{face} - 2.5\, C\, log(a/b), \qquad (1)$$

assuming the a/b to give the inclination angle of the disks to the line of sight. A sample of galaxies that were fully transparent at the particular radius used for the test would have $C \simeq 1$, while $C <\sim 0.25$ signifies opaque conditions, the transition value being heavily dependent on the spatial distribution of the absorbing material and the effect of multiple scatterings (Bruzual et al. 1988). The most frequently-used C values range from 0.5–0.9 (Holmberg 1975, de Vaucouleurs et al. 1976, Sandage and Tammann 1981). However, Valentijn (1990) derived C values well below 0.25 for large samples of spirals throughout the galaxy disks; this result is not significantly affected by selection effects in either magnitude, angular diameter or axial ratio, nor by the presence of bulges (at least for types Sb and later).

In an interesting study, Disney et al. (1989) discussed the possibility of opaque disks on the basis of the IRAS infrared flux. However, the IRAS detectors were most sensitive to warm ($\sim 60K$) dust components, with mostly cirrus-like distributions. Applying their models to the *ordinary* ESO-LV disks, the detected IRAS fluxes or the upper limits are too low to make them opaque by this component, unless large bolometric corrections from the IRAS-band to the total FIR-band are applied. Such large bolometric corrections actually represent the introduction of a *second* cooler component, which *a priori* need not have any direct physical association with the warm component, neither to follow its spatial distribution. In fact, if the cirrus were to follow the spatial distribution of the stellar component, then the Disney et al. models would have great difficulties in explaining the observed exponential visual light profiles in otherwise opaque disks. A recent study of Chini et al. (1986) demonstrated a substantial sub-mm signal in 18 Sb-Sc's, with a typical black-body temperature of $16K$. Adopting a λ^{-2} dependency of the grain opacities (not applied by Disney et al.) an average L^{cold}/L^{warm} luminosity ratio of ~ 1.4 was obtained, emphasizing the potential importance of the cool component (L^{warm}

from IRAS data). Lacking detailed knowledge of both the grain opacities at sub-mm wavelengths (*e.g.* Draine 1989) and the actual temperature distribution of the cold material, it is yet not possible to predict from the observed sub-mm continuum flux the effective total visual light cross-section of the cool component.

Here, I review the results for the Sb's and the Sc's, as obtained from the analysis of the optical ESO-LV data, and discuss the implied constraints for the properties of the absorbing components in spiral disks, which leads to an alternative interpretation of flat rotation curves and a revised extinction model for the Galaxy. I will argue that the presently available data are best understood, when in addition to a cirrus-like dust component that causes extinction (*i.e.* absorption plus scattering) a second component is causing complete obscuration (occultation). This second component could be identified with compact opaque clouds that have a temperature well below the typical IRAS temperature of $\sim 60K$ and a spatial distribution described by an exponential with a scale length larger than that of the stars.

Some key observations of the analysis of the ESO-LV data are:
- *key 1:* The central Blue surface brightness (μ_0^B) has a very small dispersion of $0.6^m - 0.7^m$ around the mean value (Freeman, 1970).
- *key 2:* The mean μ_0^B increases from 20.5 for Sa's to 22.3 for the latest spiral sub-types (Sd-I's), which also have on average a lower luminosity, but μ_0^B is not correlated with the luminosity when considering a given sub-type.
- *key 3:* The central C values range from –0.18 to +0.10.
- *key 4:* Both the μ^B at the half total light radius r_e and a profile fit between $r = 5''$ and the $\mu^B = 25$ isophote indicate C values in the range 0.08 to 0.22.
- *key 5:* The diameter ratio D_{26}/D_e does not depend on a/b, which directly implies a C value at the $\mu^B = 26$ isophote similar to that at D_e.
- *key 6:* Key 3, 4 and 5 combine to indicate a modest radial variation of C. Typical Sb-c disks have $C = 0.0$ at the center and $C \simeq 0.25$ at the outer isophotes.

2. THE PROPERTIES OF THE ABSORBING BODIES

2.1. Single layer models

I will discuss the observations in terms of single layer models (called slabs in Disney et al. 1989); this particular choice will be justified in Section 2.3. A layer containing uniformly distributed stars with spatial density $\eta_\star$ and average luminosity per star $\epsilon_\star$, and uniformly distributed obscurers with spatial density η_d, each obscurer having an effective cross-section σ_d, has an intensity

$$I = \epsilon_\star \eta_\star \lambda (1 - e^{-\tau}) = \epsilon_\star \frac{\eta_\star}{\eta_d \sigma_d}(1 - e^{-T\eta_d \sigma_d}), \tag{2}$$

and optical depth :
$$\tau \equiv \frac{T}{\lambda} = T \eta_d \sigma_d, \tag{3}$$

with T the metric thickness of the layer and λ the mean free path; $\lambda = (\eta_d \sigma_d)^{-1}$.

For $\tau \gg 1$:
$$I = \epsilon_\star \lambda \eta_\star = \epsilon_\star \frac{\eta_\star}{\eta_d \sigma_d}. \tag{4}$$

For $\tau \ll 1$:
$$I = \epsilon_\star T \eta_\star. \tag{5}$$

The relation between optical depth and absorption in magnitudes, A, is

$$A = 2.5\,log\left(\frac{\tau}{1-e^{-\tau}}\right), \qquad (6)$$

while the face-on optical depth τ_0 can be derived from the observed C values by evaluating numerically:

$$C\,log(a/b) = log\left(\frac{1-e^{-(a/b)\tau_0}}{1-e^{-\tau_0}}\right). \qquad (7)$$

Since $\tau = (a/b)\tau_0$, the previous two equations can be applied to transform between C, A, A_0, τ and τ_0.

2.2. The central regions: $\tau \gg 1$

Low C values obtained for the very central regions should always be assessed most critically, because it is very difficult to quantify the contribution of nuclei or small central spheroidal components, whose effect will be to pull the C values towards lower numbers. However, for a sub-sample of pure exponential disk systems the low central C values were reproduced, and especially for Sb's and Sc's their average μ_0^B hardly differed from the total sample. Thus, there is strong evidence that the observed low central C values are associated with the central parts of the *disks*, indicating opaque conditions. Equation (4) then implies that the central flux is basically determined by the star-to-obscurer density ratio, which leads to
• *Concl. 1: the small dispersion of μ_0 (key 1) reflects a limited range in the star-to-obscurers density ratio in different galaxies, which ratio however decreases for the later types (key 2).* This is in opposition to previous suggestions that *key 1* results from a constant surface density of stars in transparent systems, which implicitly invokes a constant metric disk thickness, T (van der Kruit, 1987). This latter hypothesis would also require a constant T for disks that vary over a factor of 100 in intrinsic diameter. Detailed modeling indicates that the observed negative C's can occur when additional extinction by surrouding semi-transparent regions (along the line of sight) acts on the light emitted by fully opaque central regions.

2.3. $C = 0.2$ disks: no $\tau \gg 1$ layers

The low C values at the half total light radii (*key 4*) and the evidence for similar C values at the outer isophotes (*key 5*) constrain the spatial distribution of the obscuring component. A disk with $C = 0.2$ could originate from one of the following configurations:
i) A single layer of dust and stars as described in Section 2.1 with $\tau_0 \sim 1.3$,
ii) A single layer of dust and stars with $\tau \gg 1$, in which the effect of multiple scatterings brighten the surface brightness at larger inclinations (Bruzual *et al.* 1988),
iii) A two layer situation in which 80% of the flux originates from a $\tau \gg 1$ layer and the remaining part from a transparent layer.
However, in case *ii)*, the total dust masses involved would have to be very high (Section 2.3.2), while for both *ii)* and *iii)* the $\tau \gg 1$ layer will have great difficulties in producing an exponential light profile (2.3.1). Interestingly enough, after excluding $\tau \gg 1$, a C value around 0.2 implies that the *detected light* should essentially

originate from a single layer à la i), with < 20% of the detected flux arising from a surrounding transparent layer. This justifies the application of the single layer equations and leads to
• *Concl 2: key 6 implies through (7)that $\tau_0 = T\, \eta_d\, \sigma_d$ only modestly decreases with radius.*

2.3.1. Scale height and scale length of the obscuring component

If the scale height t of the obscuring component were *much smaller* than that of the stars, we would observe C values near unity and not even notice that we miss half of the system. The existence of small t layers can not be ruled out, but there is no observational evidence that requires them. Even if they exist, the observed C must be due to another component.
• *Concl. 3: The scale height of the obscuring component must be intermediate or equal to that of the stars.*
• *Concl. 4: For well-mixed stars and obscurers a $C = 0.2$ disk then corresponds to $\tau_0^B \simeq 1.3$ and face-on $A_0^B \simeq 0.63^m$, via (6) (for $C = 0.08$: $A_0^B \simeq 1.0^m$).* Formally, these values are lower limits, due to the unknown contribution of stars above the obscuring layer and the effect of multiple scatterings off cirrus-like distributed dust. As pointed out above, these contributions must be small.

The combination of exponential light profiles in otherwise $C = 0.2$ disks sets more constraints: equation (2) should now be read as a radial description, with the condition of the modest radial variation of τ in (3). For $\tau \gg 1$, models with $\eta_\star/\eta_d$ constant with radius will have constant surface brightness profiles. Although some spiral arms of ESO-LV spirals exhibit flat surface brightness profiles, many $C = 0.2$ disks are observed to have an exponential light profile. So both for $\tau \gg 1$ models and for the favoured $\tau = 1.3$ situation:
• *Concl. 5: the ratio of the density of the stars to that of the obscuring bodies (multiplied by their cross-section) should decrease with radius in $C = 0.2$ disks with constant or modestly varying τ in order to explain their exponential light profile.*
This can be formalised in terms of the scale length of the obscuring component, α_d, compared to that of the stars, $\alpha_\star$, and that of the observed light, α_{obs}. If the stars have $\eta_\star(r) = \eta_\star(0)\exp(-r/\alpha_\star)$ and the obscurers have $\eta_d(r) = \eta_d(0)\exp(-r/\alpha_d)$, while the observed light profile is $I(r) = I(0)\exp(-r/\alpha_{obs})$, then substituting this in (2) results in:

$$\alpha_{obs} = \frac{\alpha_\star}{1 - \alpha_\star/\alpha_d}, \qquad (8)$$

assuming the obscurer's cross-sections do not vary with radius.
• *Concl. 6: (8) formalises how the observations (i.e. $\alpha_{obs} > 0$) require the scale length of the obscuring component to be larger than that of the stars ($\alpha_d/\alpha_\star > 1$).*

2.3.2. The nature of the obscuring bodies

The observed nominal value of τ at the half total light radius corresponds to the local determination of the product: $\tau = T\, \eta_d\, \sigma_d \simeq 1.3$. This can be used to estimate the total mass of the obscurers if they were composed of cirrus-like dust. When describing the macroscopic properties of cirrus with an exponential radial density distribution, and the more uncertain microscopic properties of the grains with mass density s and a ratio Q between the effective cross-section σ_d and the geometrical cross-section, a rough total mass interior to a galactic radius r due to grains of radius R (in units of $1\mu m$) can be estimated according to: (9)

$$M(<r) = Tm_d \int_0^r 2\pi r\eta_d(r)dr = 4.2 \cdot 10^6 e^{1.67n\alpha_\star/\alpha_d} \tau_{ne} s_{g/cm^3} Q^{-1} R_{1\mu m} Y_{1kpc^2}(M_\odot),$$

with $Y(r)$ in units of kpc^2 : $\qquad Y = \alpha_d^2 - \alpha_d (r+\alpha_d)e^{-r/\alpha_d},$ (10)

and τ_{ne} the measured value of τ at a particular radius $n \times r_e$. If the radial density distribution of *cirrus* follows that of the stars, as the IRAS data suggest (*e.g.* Wainscoat et al., 1987 in N891, N4565 and N5907, and Bloemen et al. 1990 in the Galaxy), for instance with $\alpha_\star = \alpha_d = 4kpc$, and using $s = 2.5$, $Q = 2$, and the typical post-IRAS grain sizes in the range $R = 0.01 - 0.1\mu m$ (*e.g.* Rowan-Robinson 1986, Cox and Mezger 1989) we obtain $M(r < 10kpc) = 5.3^n \times 0.8 - 8 \cdot 10^6 M_\odot$ for the mass associated with the small grains in some typical ESO-LV disks. We observed, however (*key 5*), a modest radial variation of τ beyond r_e, for instance out to $r/r_e = n = 2$. This would then imply dust masses $M(r < 10kpc) = 0.2 - 2 \cdot 10^8 M_\odot$. Noting the well documented star-to-cirrus-dust mass ratio of 1000 to 1 (at least in the solar neigbourhood), we see that the deduced masses are very high and seem acceptable only for ultra-small grains. However, if the ultra-small grains were accompagnied by larger grains or extended to create $\tau \sim 1$ at $n = 2.5$ (about the 26th isophote for many ESO-LV disks), the predicted $M(r < 10kpc)$ would exceed plausible values. Evidently, even larger masses will be obtained when the main absorption job is performed by any larger solid bodies.

• *Concl. 7: There appear a number of reasons why we should face the possibility that the main absorption is due to another component than grains in a cirrus-like configuration: i) cirrus grains could lead to exceptional dust masses and the typical cirrus temperatures predict IRAS fluxes much higher than is actually observed (see also Introduction), ii) Concl. 5 ($\eta_\star/\eta_d$ decreasing with radius) or equivalently Concl. 6 ($\alpha_d > \alpha_\star$) are not consistent with the spatial distribution of cirrus as mapped by IRAS.* Although, at larger radii, where the interstellar radiation field will be less intensive, cirrus cooler than that seen by IRAS might produce $\alpha_d > \alpha_\star$ (solving *ii)*), *i)* still invites to consider alternatives.

Prominent candidates to contribute to the absorption are compact *opaque* molecular clouds. The fundamental reason why they can be more effective is that the clouds represent large bodies, most likely in a virialized state with virial mass $m_{cl} = 3 G^{-1} R v^2$, where R is the radius and v the one-dim. velocity dispersion. Solomon et al. (1986) measured $v = (S/1.01)^{0.5}$, and thus $m_{cl} = 700 f S^2(M_\odot)$; here S is the metric size dispersion in pc and f a projection factor (π). Thus, contrary to individual dust particles, the ratio (cloud-mass/cross-section) is independent of the size, and absorption models involving molecular clouds will be invariant for the chosen cloud dimensions and hence their mass. This geometric difference compared to cirrus dust particles give compact clouds a higher overall absorption efficiency. We can compute the total cloud mass required to give a particular absorption even without knowing the cloud dust content, by using the observed size–line-width relation. Issa et al. 1990 pointed out that spatial confusion might seriously affect the size–line-width relation and especially the mass determinations for the largest clouds. However, the size–line-width relation is also observed for the smallest clouds and recently Lee et al. 1990 presented a linear relation between cloud mass estimates from CO data and virial masses.

Assuming the compact clouds obscure the background out to their gravitational radius $R = qS$, with $q = 2.5$, Spitzer 1969, (note, S was defined as size *dispersion*) the exponential distribution of η_d and the fixing of the optical depth at $n \times r_e$ lead to the following expression for the total obscuring mass in the form of molecular clouds in a disk within radius r:

$$M(<r) = Tm_{cl}\int_0^r 2\pi\eta_d(r)dr = 2.2 \cdot 10^8 f q_{2.5}^{-2} \tau_{ne} e^{1.67n\alpha_\star/\alpha_d} Y_{kpc^2}(M_\odot), \quad (11)$$

with Y in units of kpc^2 as in (10). This expression for $M(<r)$ can be used to compute the required obscurers mass from the observed *photometric* parameters $\alpha_\star$ and τ_{ne} for a given α_d. The main uncertainty in computing $M(<r)$ using (11) relates to deviations from the assumed spherical symmetry of the clouds and corresponds to different values of the constants f and q. However, since both Solomon et al. and Lee et al. find the clouds to conform to a virialized state the choosen values of the constants might be quite correct. For $\alpha_\star = 4kpc$ and $\alpha_d = 6 - 16\,kpc$ a $\tau_e = 1.3$ disk would need to contain $3 - 4 \cdot 10^{10} M_\odot$ of compact clouds within $r = 8kpc$, if they were to do the main absorption job. This is a large mass, but since this is for the situation with $\alpha_d > \alpha_\star$, this mass will only modestly increase with increasing n, in contrast to the cirrus models with $\alpha_\star = \alpha_d$. Note that a major portion of the total cloud mass corresponds to the systematic difference between cloud masses from CO data and those resulting from the virial theorem, the latter being a factor 3.5 higher.

• *Concl. 8: Compact opaque clouds are good candidates to provide a substantial contribution to $\tau \simeq 1.3$ in ordinary Sb-Sc disks.*

3. FLAT ROTATION CURVES - NO DARK HALOES REQUIRED

The analysis of the rotation curves in a number of Sb-Sc spirals has led many authors to suggest the existence of massive dark haloes (see van Albada and Sancisi 1986, for a review). The observed radial luminosity profile with its associated α_{obs}, together with an assumed constant M/L is used to set the mass distribution of the disk. In the so-called *maximum disk fits* the maximum dynamical mass is put in the stellar disk component, while the remaining mass-discrepancy is assigned to a halo. Alternative solutions with a radially increasing M/L in the disks were discarded, lacking any physical or observational justification. The resultant M/L ratio of the disk is only a side product and not essential. It formerly appeared that the maximum disk fits were resistent against any effects due to extinction. Firstly, if there were more extinction L would go up and correspondingly the M/L would go down, changing only the side product of the fits and not the dynamical mass M itself; further, if there were more extinction it was assumed to be more concentrated towards the centre, and this would only increase the mass-discrepancy. However,
• *Concl. 9: Concl. 6 ($\alpha_d > \alpha_\star$), or equivalently concl. 5 ($\eta_\star/\eta_d$ is decreasing with radius) indicate that the fractional contribution of the obscurers to the projected*

mass density increases with radius.

Irrespective of the true nature of the obscuring bodies, (8) clearly indicates that the true scale length of the stellar distribution can be smaller than the observed value, while α_d can be larger, and if the obscuring bodies carry mass, then *the apparent total M/L ratio in the disks will increase with radius.*

Motivated by these results, González-Serrano and Valentijn (1990) fit the rotation curves of two Sc's previously thought to have very massive haloes. These fits were done with one of the standard codes used for maximum disk fits (Begeman 1987), but now dis-abling the halo component and substituting instead a second disk component. With surprisingly few iterations, combinations of α_* and α_d were found that reproduced both the observed α_{obs} through (8) and the rotation curves (Fig 1). The iterations essentially improved the fit to the steep central parts.

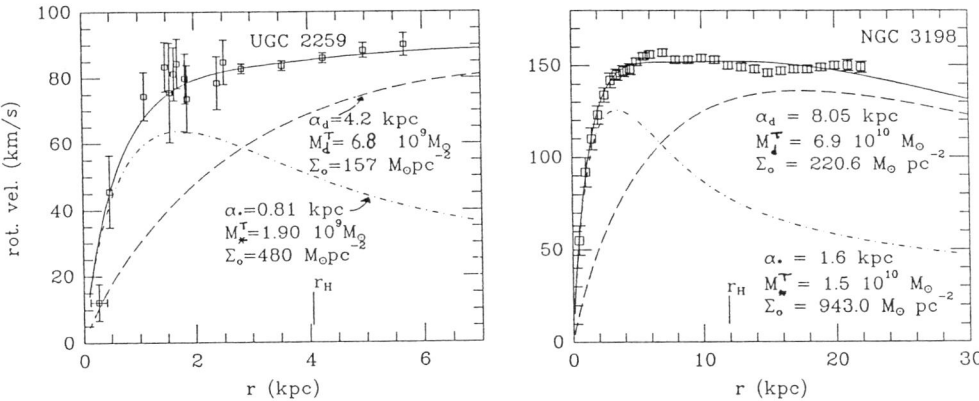

Fig. 1. Rotation curve fits with a two-disk model

The total mass out to the Holmberg radii, involved in the second (obscuring) component exceeds that of the first (stellar) component by a factor of 2–3, quite a large ratio; since here we deal with the most mass-discrepant Sc's with large HI fluxes, these factors do not necessarily represent a standard value.

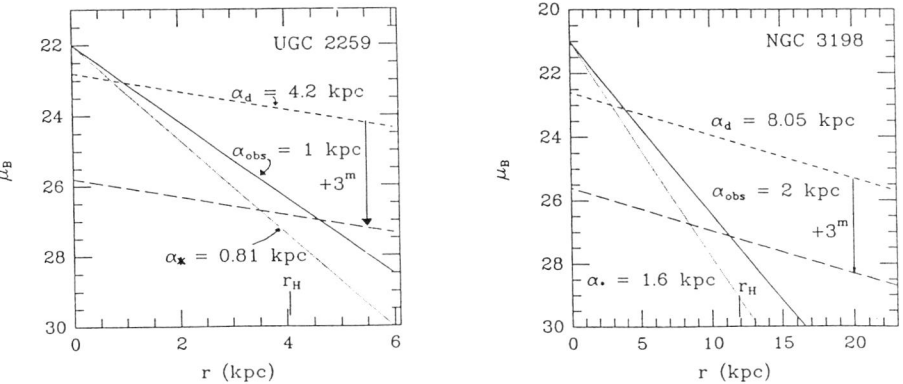

Fig. 2. Hypothetical surface brightness profiles of the two disk components that fitted the rotation curves of UGC 2259 and NGC 3198

• *Concl. 10: A two-disk model with one component representing the visible stellar mass, the second component representing the mass in the obscuring component, can reproduce the observed rotation curves, with the inter-relation between the two components in agreement with the above listed contraints on absorption in $C = 0.2$ disks.*

Fig. 2 gives an impression of the mass distribution corresponding to the solutions of the fits. This is displayed in units of surface brightness, arbitrarily taking a similar intrinsic M/L for both components. Obviously, the second obscuring component must have a much larger M/L than the stellar component and Fig. 2 demonstrates that, were the second component to contain stars that it self-obscures, an internal extinction of $\geq \sim 3^m$ would make it totally negligible in terms of surface brightness; simultaneously, in the case of a $C = 0.2$ disk, the effective absorption of the disk due to obscuration can be $\sim 0.63^m$.

The solutions for $\alpha_\star$ and α_d from the rotation curve fits can be substituted into the disk mass equation, (11), to evaluate the mass of the obscuring component, when it consists of compact clouds. Most strikingly, *the mass determinations from the dynamical studies (i.e. rotation curve fits) differ less than 10–20% from the completely independently derived values from photometric studies (i.e. the mass required to make the disks have $\tau_e = 1.3$ due to compact clouds)*. Also the central projected mass densities differ by less than 10–20% between the two methods. Note, that the *two* components used in the fit represent the prime emission and absorption parts of the *single* layer model, which should in reality be dynamically coupled and are thus not plagued by the well-known "disk-halo" conspiracy problem.

The strongest point that led to the introduction of dark haloes was the extent of flat rotation curves from HI data, which for UGC 2259 and NGC 3198 reached, respectively, ~ 1.5 and ~ 2 times the Holmberg radius. Several arguments can be made to explain these observations in terms of the two-disk solutions. First, the extrapolation of the surface brightness profiles in Fig. 2 shows that if these components extended to the HI radius, they would not be detected at optical wavelengths. Second, if the HI flux distribution of these galaxies is described with an exponential, the corresponding scale lengths are very close to that obtained for the obscuring components from the two-disk fits. Thus it appears that the two-disk solutions matches a configuration in which 'dark' matter is spatially associated with the HI gas. Physically, this match corresponds to a limited range of the mass ratio of HI-to-obscuring bodies; if this limited range were universal, it would not be surprising that the "classical" flat rotation curve spirals, which come from *HI selected* surveys, also contain the most "missing" mass.

4. THE GALAXY

4.1. A special Galaxy, a local hole in the ISM, or a typical $C = 0.2$ disk?

If the Galaxy were like ESO-LV Sbc disks with $\tau \simeq 1.3$, then at the location of the Sun, about two scale lengths from the center, we would expect a total face-on extinction of $A_B \sim 0.6^m$. Disney *et al.* 1989 summarized that a value of $\tau \simeq 1$ (corresponding to $\sim 0.5^m$) would be allowed by observational data taken at *low* latitudes. The actual extinction to the Poles is subject to debate, with estimates of the total disk extinction ranging from $0.0 - 0.3^m$, which values do not really match the 0.6^m suggested by ESO-LV data. A local hole in the ISM could be the explanation (*c.f.* Cox). Alternatively, the Galaxy could be special and contain less obscurers than $C = 0.2$ disks. However, all the Galactic extinction studies trace the effect of optically thin cirrus, either by using IRAS data (Boulanger and Pérault 1988) or by measuring the reddening of *detected* objects. The present work suggests a substantial contribution by occultation due to cool compact *opaque* clouds, which would be essentially un-noticed by such studies. It seems interesting to explore in more detail whether compact clouds could, at the location of the Sun, result in face-on extinctions similar to those seen in typical ESO-LV Sbc's. Most Galactic extinction studies, either based on reddening (Burstein and Heiles 1978) or on galaxy counts (Shane and Wirtanen 1967), had severe problems matching the Northern and Southern Galactic hemispheres, essentially because of the different origins of the data. The ESO-LV survey has the advantage of covering latitudes between $b = -90^0$ and $b = +45^0$ with homogenously acquired data.

4.2. A new extinction map

In Fig 3, an extinction map determined from the average reddening of ESO-LV galaxies is reproduced.

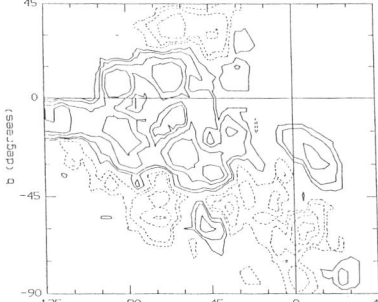

Fig. 3. Reddening map constructed from the central $(B - R)_{10''}$ of 11847 galaxies. For each object, the deviation from the mean value of objects with similar morphological type was computed. Then, with a typical 5^0 resolution the average relative reddening was derived. Contour values are $\Delta(B - R) = \pm 0.025, 0.05, 0.1, 0.15$; negative contours are dashed. Note, $A_B = 2.30 E(B - R)$.

Although the reddening map has a coarse resolution, it is clear that it exhibits much more structure than could be described by a simple cosecant law.
- *Key 7:* Structures in the reddening map perpendicular to the plane are evident.
- *Key 8:* The reddening map is asymmetric, with more reddening in the South.
- *Concl 11: This demonstrates that the Sun is located slightly above the plane of the main dust layer.* This is consistent with a study of the Galactic background light, using Pioneer 10 data, in which Toller (1981) deduces that the Sun is $12.2 \pm 2.1 pc$ above the Galactic plane.

4.3. Counts of ESO-LV galaxies

The ESO-LV data base also provides the projected surface density of galaxies.
- *Key 9:* When extracting sub-samples with selection criteria on either B, μ, type, galaxy density in the environment (clusters, groups, free field), or position on the sky (in/out the supergalactic plane), the surface density in the North ($b > 0^0$) is found to be a factor ~ 1.6 higher than in the South ($b < 0^0$).

This density difference was also found in the Shane-Wirtanen counts, but was not taken seriously because of the differences between the data taken in the North and the South and the potentially dramatic effect of the sky-brightness on the cut-off levels of the counts. ESO-LV avoids these problems; its accurate magnitudes allow the construction of a differential count with a Euclidian normalization (Fig 4).

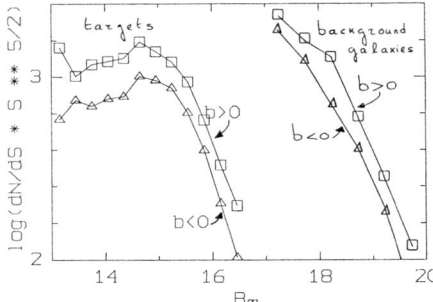

Fig. 4. Differential counts (bin 0.3^m) of ESO-LV target and background galaxies normalized with a Euclidian world model. The horizontal parts of the counts of target galaxies indicate that the counts are not hampered by selection effects, which start to operate at about $B = 15.5$ for this particular subsample (all types, $\mu_0^B > 21$). The surface density difference North - South is evidently a factor 1.6 over the whole horizontal range

Although the North-South density difference is very persistent, no significant galactic latitude dependency was found. Both the galaxy distribution on the sky (e.g. the super-galactic plane) and the reddening map (key 7) show substantial structure perpendicular to the plane, possibly masking any true galactic latitude dependency. The fact is that such a latitude dependency is not observed, and the projected galaxy density has a discrete jump when passing the plane. We are interested in the relative difference in surface density between North and South, and by correlating the relative reddening map of Fig 3 with a map of galaxy counts, we can evaluate how the component that causes the reddening affects the counts. The answer of this exercise is clear:

• *Concl 12: The difference in reddening North-South would explain at most a 0.2^m shift in the un-normalized differential counts, while the observed shift amounts to $\sim 0.8^m$ towards fainter magnitudes for the South.* This is also illustrated in Fig 5.

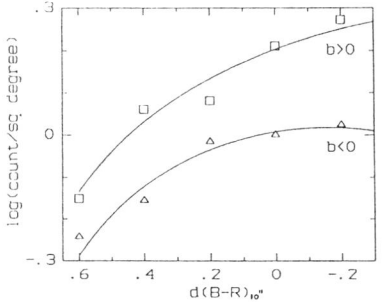

Fig. 5. Galaxy counts versus reddening for North and South. For each field a single count and a single average reddening was computed. The curves show the polynomial regression of the 92 Northern fields and the 313 Southern fields. The points represent the same data binned in $d(B - R) = 0.2$. The regressions of the individual curves match the expected effect of extinction via reddening on the counts. However, the offset between the two curves demonstrates that the difference North-South can not be explained with reddening effects (i.e. Concl 12)

• *Concl 13: The North-South offset in Fig. 5 also implies that, if this density difference is caused by another obscuring component, then that component should have a very different spatial distribution than the dust that causes the reddening.*

Hence, the count difference North-South cannot be caused by reddening due to cirrus-type dust. The possibility remains that it reflects an intrinsic anisotropy. Claims of this nature were presented by Lahav (1987) based on counts of optical galaxies and by Yahil et al.(1986) and Londsdale et al.(1990, preprint) based on counts of IRAS galaxies. ESO-LV has a typical depth of ~ 30 Mpc (with $H_0 = 100$) and a local anisotropy on such a scale is momentarily not perceived as unphysical. However, the present notion of the discrete jump when passing through the plane, together with the location of the Sun above the plane and the suggested presence of occulting clouds, should make us most suspicious. In addition, by constructing a similar Euclidian count for a sub-sample of 3000 well-classified background galaxies (total of 60,000), present on the edges of the scans of ESO-LV objects, the North-South difference was extended to $B \sim 19.5^m$ (Fig 4 - Jörsäter and Valentijn *in preparation*), corresponding to a depth of $\sim 200 Mpc$. This further points to a Galactic origin for the count difference.

Discarding a high content of sub-mm sized particles that would make the compact clouds opaque at IR wavelength, a discordant note for this interpretation might be the further confirmation of the IRAS galaxies anisotropy at $|b| > 50^0$ (Lonsdale *et al.*). However, it remains to be proven that this result is not affected by the suggested location of the Sun above the plane (*key 8*), which if not corrected for in the substraction of the forground emission, with different amplitudes in different directions, might result in an artificial count asymmetry in a similar way as hampered the interpretation of the Shane-Wirtanen counts.

4.4. *A model - The Galaxy as a typical $C = 0.2$ spiral disk*

In this section, an extinction plus occultation model for the Galaxy will be explored that obeys the above-labeled *conclusions* and also reproduces the North-South count difference. If the cirrus-type dust contributes 0.25^m to a total disk face-on extinction of 0.63^m (*concl. 4*), the opaque clouds must contribute the remaining $\sim 0.4^m$ or $\tau = 0.8$, if they were to resolve the discrepancy. To test this hypothesis, a computer programme was made in which obscuring bodies were randomly distributed in a disk. The apparent face-on τ_0 and the fractions ($f(b)$ in percents) of the sky obscured to observers in the disk at $T = 0$ were computed, carefully correcting for the effects of overlapping clouds. In Table 1, an abstract is presented of the results of a typical run with $R = 5pc$ spherical clouds, but since the effective obscuration is invariant for cloud dimensions (Section 2.3.2), as long as τ remains fixed the data in Table 1 could as well represent other or a mix of cloud dimensions. Different diameters only affect the gamble at high galactic latitudes. For $R = 12.5pc$, the model corresponds to a spatial density of $1.6 \cdot 10^{-5}$ pc^{-3}, *i.e.* filling 13% of the volume with an average cloud separation of $40pc$. This is a relatively high density, but for instance, correcting the Lee *et al.* CO survey for distance dependent selection effects, their survey indicates a cloud density $\sim 10^{-6}$ pc^{-3} above a treshold temperature of $4K$ for a minimum cloud diameter of $\sim 7pc$. The authors mention that 68% of the detected flux is below that treshold temperature and thus the value of $\sim 10^{-6}$ pc^{-3} is a strong lower limit on the actual cloud density.

Table 1. Model of fractions of the sky $f(range\ b^0)$ obscured by compact clouds

T(pc)	τ_0	$f(10-20)$	$f(20-30)$	$f(30-40)$	$f(40-50)$	$f(50-60)$	$f(60-70)$	$f(70-90)$
30	0.22	55	23	26	27	0	0	0
35	0.26	66	28	33	33	0	0	0
40	0.30	71	31	35	36	14	20	0
50	0.38	82	41	48	39	28	20	0
60	0.45	89	49	59	44	31	24	7

Searching for conditions that produce the observed North-South count difference and adopting a location of the Sun 12.5 pc above the plane (*concl. 11*), the data in Table 1 suggest a solution with a total thickness T of 95 pc for the obscuring bodies, corresponding to a 35 pc layer to the North and a 60 pc layer to the South, with total $\tau = 0.71$ and a ratio of obscured sky $f(10^0 - 40^0)$ North/South = 1/1.54, (*i.e. key 9*). The low obscured fractions of the sky at high latitudes (see Table 1), also demonstrate that at these areas the effect of obscuring clouds need not be noticable by the terrestrial observer, but this is subject to strong chance occurences.

• *Concl 14:* *A model as tabulated in Table 1 could reproduce the observed North-South count difference, obeying at the same time all the labeled conclusions in this paper. This would make the Galaxy, with respect to the aspects discussed here, a typical* $C = 0.2$ *Sbc.* Applying (11), the total mass of the clouds in this model within the solar circle is $\sim 2 \cdot 10^{10} M_\odot$ for $\alpha_\star = 4 kpc$ and α_d in the range $6-20$ kpc; the projected cloud mass density near the Sun is 75 $M_\odot pc^{-2}$.

ACKNOWLEDGEMENTS. I thank H. van der Laan for his continous encouragement and S. Casertano and M. Rupen for very useful comments on the manuscript.

Albada, van T. S., Sancisi, R. 1986: Phil. Tranc. Roy. Soc., London **A320** 447-465
Begeman, K. 1987: Ph.D Thesis, Kapteyn Laboratorium, Groningen
Bloemen, J.B.G.M., Deul, E.R., Thaddeus, P. 1990: *Astron. Astrophys.*, **233**, 437
Boulanger, F., Pérault, M. 1988: *Astrophys. J.*, **330**, 964
Bruzual, G.A., Magris, G.C., Calvet, N. 1988: *Astrophys. J.*, **333**, 673-688
Chini, R., Kreysa, E., Krügel, Mezger, P. G. 1986: *Astron. Astrophys.*, **166**, L8
Cox, P., Mezger, P.G. 1989: Astronomy and Astrophysics Review, **1**, 49
Disney, M., Davies, J., Phillips, S. 1989: *Monthly Notices Roy. Astron. Soc.*, **239**, 939-976
Draine, B. T. 1989: *The Interstellar Medium in Galaxies*, ed. H . A.Thronson and J. M. Shull, Dordrecht, Kluwer
Freeman, K.C. 1970: *Astrophys. J.*, **160**, 811-830
González-Serrano, J.I. and Valentijn E.A. 1990: *Astron. Astrophys. in press*
Holmberg, E. 1975: Stars and Stellar Systems Volume IX, University of Chicago
Issa, M., MacLaren, I., Wolfsdale, A.W. 1990: *Astrophys. J.*, **352**,132
Kruit, P.C. van der 1987: *Astron. Astrophys.*, **173**, 59-80
Lahav, O. 1987: *Monthly Notices Roy. Astron. Soc.*, **225**, 213
Lauberts, A., Valentijn E.A. 1989: *The Surface Photometry Catalogue of the ESO-Uppsala Galaxies* = ESO-LV, ESO, München
Lee, Y., Snell, R.L., Dickman, R.L. 1990: *Astrophys. J.*, **355**, 536
Rowan-Robinson, M. 1986: *Monthly Notices Roy. Astron. Soc.*, **219**, 737
Sandage, A., Tammann, G.A. 1981: *RSA*, Carnegie Institute, Washington
Shane, C.D. and Wirtanen, C.A. 1967: *Publ. Lick Obs.*, Vol. **22**, Part I
Solomon, P.M., Rivolo, A.R., Mooney, T.M., Barret, J.W., Sage, L.J. 1986: in *Star Formation in Galaxies* Editor C.J. Lonsdale Persson
Spitzer, L. 1969: *Astrophys. J.*, **158**, L139
Valentijn, E.A., 1990: *Nature*, **346**, 153
Vaucouleurs, de G., de Vaucouleurs, A., Corwin, H.G. 1976: *RC2*, Univ. Texas Press
Wainscoat, R.J., de Jong, T., Wesselius, P. R. 1987: *Astron. Astrophys.*, **181**, 225
Yahil, A., Walker, D., Rowan-Robinson, M. 1986: *Astrophys. J.*, **301**, L1

RADIO STUDIES OF COSMIC RAYS IN NEARBY GALAXIES

E. HUMMEL
Nuffield Radio Astronomy Laboratories, University of Manchester,
Jodrell Bank, Macclesfield, Cheshire, SK11 9DL, England

ABSTRACT. The constraints on cosmic ray electron sources and propagation as derived from radio continuum observations of galaxies are reviewed. Special attention is paid to the inferences which can be obtained from the radio continuum properties of spiral galaxies seen edge-on.

1. INTRODUCTION

The bulk of the radio continuum emission from spiral galaxies observed below a frequency of 10GHz is non-thermal of origin and as shown by the radio spectra and linearly polarized emission due to the synchrotron emission mechanism. This means that the radio emissivity is proportional to $n_e B^{1-\alpha}$, where n_e is the density of relativistic electrons, B the magnetic field strength and α is the radio spectral index. The latter is directly related to γ, the energy index of the cosmic ray electrons[1]:
$\alpha = (\gamma-1)/2$

This link between the radio continuum emission and cosmic rays, in particular the electron component, features in many studies of the large scale radio continuum structure of spiral galaxies. In short, these studies often focus on: *i)* the sources of relativistic electrons, *ii)* the propagation of these electrons and *iii)* the B-field strength and configuration. Recent reviews of such studies have been given in Hummel (1990), Klein (1990), Krause (1990) and Beck (this volume) and I refer to these papers and the references given there for more details. Here I want to concentrate on some specific *observational* results which are important for studies on the propagation of the relativistic electrons.

---
[1] α defined by $S \propto \nu^{\alpha}$ and γ is defined by $n_e \propto E^{\gamma}$

In Section 2 I discuss briefly the sources of the relativistic electrons. Section 3 deals with the radial distribution of the non-thermal and thermal emission in spiral galaxies. The emphasis is on the results obtained for the nearest face-on galaxy for which this can be studied, IC342. The largest fraction of this paper, Section 4, discusses the results obtained for the radio continuum emission from edge-on spiral galaxies, in particular NGC891 and NGC4631. Such galaxies are of course most suitable to study the propagation of the relativistic electrons perpendicular to the plane of a galaxy and the disk-halo interaction. Section 5 summarizes the more general results which constrain cosmic ray propagation models for galaxies.

2. SOURCES OF RELATIVISTIC ELECTRONS

Radio continuum observations can be used to discuss the sources of cosmic ray electrons. At the moment it appears as if the pendulum is near one of its extremes again: i.e. recent star formation is closely related to the origin of the relativistic electrons. The main reasons for me to adopt this view are *i)* the relation between the dust temperature and the radio *emissivity* in Sbc galaxies (Hummel et al., 1988) and *ii)* the results obtained from a comparison of the far infrared and radio continuum emission from isolated and interacting galaxies (Wunderlich, 1989). The often quoted tight correlation between the non-thermal radio continuum and the thermal dust emission in itself can in this respect only be used as a consistency argument.

Type II supernova remnants (SNRs) are often seen as the sources of the cosmic ray electrons and there is no compelling evidence based on radio continuum results against this view. On the other hand it is certainly not a necessary conclusion. The apparent objections: *i)* the deficiency by a factor 10 if one applies the Σ-D relation to explain the radio continuum emission, *ii)* the spectral index of the integrated non-thermal emission and *iii)* the "smooth" distribution of the non-thermal emission, can be countered by arguments that take into account lifetime and propagation effects. Assuming that SNRs are the sole source of cosmic ray electrons Hummel and van der Hulst (1986) found an efficiency of 1→5% for SNRs in NGC4038/38 to produce cosmic rays (including relativistic electrons).

If the sources of the relativistic electrons are closely related to the more recent star formation then it is conceivable that these sources are distributed like the thermal gas. In principle this distribution can be determined from Hα measurements and multi-frequency radio observations. Another tracer of recent star formation is the thermal dust emission. Bicay et al. (1989) used the distribution of the 60μm emission to define the distribution of the cosmic ray sources. If on the other hand the sources are distributed like the older disk population the source distribution may be derived from optical photometry. As long as it is not known beyond doubt what the sources of cosmic rays are it is probably wise to consider both possibilities.

3. THE RADIAL DISTRIBUTIONS

The main tools provided by radio continuum measurements for the study of the cosmic rays are the brightness distributions at various frequencies (nowadays available at frequencies from 327MHz→10.7GHz) and the resulting spectral index distributions. Usually the emission is averaged in rings in the plane of the galaxy, hence neglecting the fact that the radio emission in general is a function of radius *and* azimuthal angle, to obtain the radial brightness and spectral index distributions. It is of course essential to use measurements which have recorded *all* the emission. Because we do not have an a priori knowledge of the radio continuum structure this is not always so trivial as it seems.

The radio emission we observe is a mixture of thermal and non-thermal emission. This is fortunate but also a nuisance. Fortunate because it allows us to determine the distribution of the relativistic electron sources (provided they are related to recent star formation). A nuisance because in order to study the propagation of cosmic ray electrons the radial brightness and spectral index distributions need to be corrected for the thermal contribution. Note in this respect that the differences in the diffusion models Segalovitz (1977) and van der Kruit (1977) obtained for M51 arose because of the difference in their evaluation of the thermal emission.

The thermal emission distribution and the corrected non-thermal emission distributions are usually obtained from measurements at (only) two frequencies with the assumption of a constant non-thermal spectral index (α_{nth}) out to a certain radius. The choice of α_{nth} is somewhat arbitrary, however it can be constrained by comparing the resulting thermal emission with Hα measurements and with high frequency radio measurements. If α_{nth} is slowly decreasing (spectrum steepening) with radius then the assumption of a constant α_{nth} will lead to an underestimate of the exponential scale length of the thermal emission. However, no attempts have been made yet to incorporate a non-constant α_{nth} in the analysis of the thermal contribution.

As an example I use the analysis of IC342 by Hummel and Gräve (1990). IC342 is the nearest, almost face-on spiral galaxy and also a relatively strong radio continuum source and therefore one of the best candidates to study. A disadvantage is its low galactic latitude which makes it hard to study for instance the Hα emission. For radii <2kpc the emission is dominated by the central radio source. Most of the disk emission is well described by a plateau from 0→5kpc and an exponential distribution beyond r=5kpc. The scale lengths for the latter are 4.2 and 3.5kpc at 327 and 4750MHz, respectively. After taking into account the relatively flat radio spectrum of the central radio source the spectral index between these two frequencies of the disk emission changes gradually from ~-0.65 at r=0 to ~-0.85 at r=15kpc. After correction for the thermal emission α_{nth} is essentially constant with radius (of course partly a result of the assumption made to separate the thermal and non-thermal emission) and is ~-0.8 and the 327MHz scale length can be considered as the scale length of the non-thermal emission (R_{nth}).

Concerning the source distribution we have the two alternatives. Hummel and Gräve (1990) determined the radial distribution of the thermal emission. Beyond r=5kpc its exponential scale length (R_{th}) is ~ 2.1kpc. The radial distribution of the blue light has been measured by Ables (1971) and beyond 5kpc the exponential scale length (R_{opt}) is ~ 3.4kpc. This result $R_{nth} > R_{opt} > R_{th}$ seems to be a more general result (Hummel, 1990).

In terms of propagation models it is not R_{nth} which is of importance but the scale lengths of the cosmic ray electrons and the magnetic field strength (R_e and R_{Bf}, respectively). They are essentially not known, but assuming equipartition we obtain $R_e = [(3-\alpha_{nth})/2]R_{nth}$ and $R_{Bf} = (3-\alpha_{nth})R_{nth}$. This would make the cosmic ray scale length significantly larger than either R_{th} or R_{opt}, requiring for both source distribution hypothesis cosmic ray propagation in the radial direction. A typical radial propagation length in IC342 would be 4→5kpc. An important constraint on the propagation and energy losses is the constancy or modest decrease of the spectral index with radius.

Apart from the uncertainties concerning the energy losses, source distribution, equipartition assumption there is the difficulty of the two-dimensionality of our observations. The radial distributions given in the literature (can) contain emission from both thin- and thick disk or halo components which can be discerned in at least some edge-on galaxies. The results on NGC891 and NGC4631 show that the radial extent of the thick disk or halo component is smaller than the radial extent of the thin disk component.

4. RADIO HALOES

The existence of haloes around spiral galaxies containing cosmic ray particles was first hinted at by cosmic ray physicists. Their evidence is based on measurements of the Be isotopes and the mean path length for cosmic rays of ~ 5g cm^{-2}. A direct proof of the existence of a cosmic ray halo was obtained by Ekers and Sancisi (1977) when they detected a radio continuum halo around the edge-on spiral galaxy NGC4631 at a frequency of 610MHz. A more detailed study of an edge-on galaxy was undertaken by Allen et al. (1978). They found two main components, a thin disk coinciding with the optically visible disk and a thick disk component (halo component) in NGC891.

4.1 The Z-Distribution of the Radio Emission

In Hummel (1990) I give a list of (well) studied edge-on spiral galaxies. This list can be extended by the results obtained from a recent survey of edge-on galaxies (Hummel et al., 1990). For about 20 galaxies we have enough information to determine an half power width (FWHM of a gaussian distribution) of the radio distribution perpendicular to the plane (z-direction). A typical value for the FWHM, which is corrected for beam broadening and inclination effects, is ~1kpc (distances consistent with H=100km/sec/Mpc). However, it appears that the FWHM is not (always) a good parametrization of the z-distribution. There is the complication of two (or more) distinct z-components but only very few galaxies have been observed in enough detail to try a decomposition. In addition, in the few cases for which observations with high enough resolution and sensitivity exist it turned out that the z-distributions are better described by an exponential than by a gaussian distribution.

The two cases that have been studied in most detail are NGC891 and NGC4631 (I assume distances to these galaxies of 7.2 and 7.5Mpc, respectively). Allen and Hu (1985) noted that the z-distribution in NGC891 can be described by a gaussian thin disk and an exponential thick disk component. For the latter they found an exponential scale height (Z_{nth}) of 0.9kpc at 1.4 *and* 4.9GHz out to their detection limit at ~3kpc. This is fully corroborated by recent and more sensitive measurements (see also Allen, this volume). Hummel et al. (1990) find a significant asymmetry in the sense that Z_{nth}=0.8 and 1.0kpc on the E and W side, respectively. These scale heights are valid over a z-range of at least 4→5 scale heights. Similar results are obtained for NGC4631. Hummel and Dettmar (1990) find scale heights of 2.0 and 2.3kpc at 1.5GHz for the N and S side over a range of again 4→5 scale heights.

Apart from these large scale asymmetries NGC891 and NGC4631 show kpc scale structures in their haloes. The E side of NGC891 shows a clear peanut like morphology, and in the SW there is a prominent spur-like feature which extends to a z-height of ~6.3kpc. NGC4631 shows two main spurs, with lengths of ~10kpc, in the north.

4.2 The Spectral Index Distributions

Multi-frequency measurements with good enough resolution to determine the spectral index distribution (α-distribution) perpendicular to the plane of a galaxy have been made for NGC891 and NGC4631. The radio spectra of the total emission

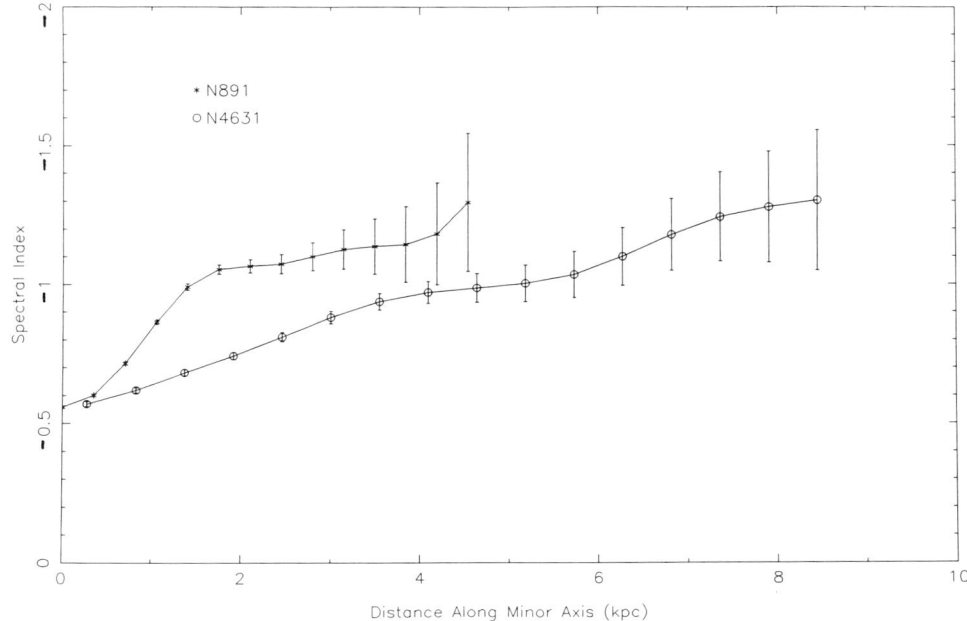

Figure 1: *The spectral index distributions perpendicular to the optical major axis of NGC891 and NGC4631. They were determined from measurements at 610MHz and 1.5GHz with an angular resolution of 40″.*

from NGC891 and NGC4631 show a change in spectral index of ≥ 0.2 and ≥ 0.3, respectively. In both cases the "break" in the spectrum occurs at ~ 0.8GHz. Hence it appears that the relevant α-distributions to construct are the ones using the "low" frequencies 327 and 610MHz and the "high" frequencies 1.5→10.7GHz. However, for various reasons I will use the measurements at 610MHz and 1.5GHz for the determination of the "high" frequency α-distributions.

In Figure 1 I show these "high" frequency α-distributions for NGC891 and NGC4631. In case of NGC891 there is good agreement with the results given by Allen and Hu (1985), which were based on 1.4 and 5.0GHz measurements. The thin disk in NGC891 appears to have a $\alpha \sim -0.4$. The halo component has a rather constant $\alpha \sim -1.05$ from z=0 to 3kpc. Beyond that the spectrum steepens to $\alpha \sim -1.3$ at z$\sim$4.5kpc. The α-distribution of NGC4631 does not show, like the brightness distributions, evidence for a major thin disk component. The spectrum gradually steepens from $\alpha \sim -0.55$ at z=0 to $\alpha \sim -1.3$ at z=8.5kpc.

The "low" frequency α-distributions do not yet have such a good resolution as the "high" frequency distributions have and the work on these is still in progress. However, it is clear that the "high" frequency spectra in the planes of NGC891 and

NGC4631 (i.e. on the major axis) are steeper by $\sim$0.25 when compared to the "low" frequency spectra. This difference is also evident at larger z-distances, in particular on the W side of NGC891 and the N side of NGC4631. On these sides the "low" frequency spectral index is constant with z beyond z=3 and 5.5kpc in NGC891 and NGC4631, respectively. On the E side of NGC891 and the S side of NGC4631 the "low" frequency spectrum steepens gradually from $\alpha \sim -0.4$ to $\alpha \sim -1.2$ at z=4 and 9kpc, respectively.

4.3 The B-field in the Halo

A very important parameter when dealing with cosmic ray propagation and which is accessable by radio continuum measurements is the magnetic field (B-field) strength and structure. This will be dealt with in more detail by Beck (this volume) but I cannot leave it unmentioned in the present context. Hummel *et al.* (1988) found a median B-field strength of $\sim 8\mu$G in Sbc galaxies. After subtraction of the thin disk component, and assuming minimum energy conditions Hummel *et al.* (1990) find an average B-field strength in the haloes of NGC891 and NGC4631 of $\sim 8\mu$G and $\sim 5\mu$G, respectively. Although there are ordered B-fields in these haloes the dominant field component is the random one.

The resulting half life times of the relativistic electrons radiating at 1.5GHz (E$\sim$4GeV) are then $\sim 4\ 10^7$ and $\sim 7\ 10^7$yr, respectively, and assuming a diffusion coefficient $D_o = 10^{29}$cm^2/sec the typical diffusion length scales are $\sim$3.6 and $\sim$4.8kpc. However, because of the significant inverse Compton losses and the stronger B-field strengths in the planes of these galaxies these life times and diffusion length scales are upper limits. More realistic values, provided there is no reacceleration, are $\sim$3 10^7yr and $\sim$3kpc.

5. THE CONSTRAINTS

In this Section I give a brief summary of the more general results obtained from studies of the large-scale radio continuum structure of spiral galaxies and which are of importance for studies on the propagation of cosmic ray electrons. The results given here for IC342 and NGC891 may be considered as representative for spiral galaxies. However, it should be realized that the best studied galaxies are in general those with more than average radio emission. NGC4631, ironically, is most likely an exceptional case among the edge-on galaxies. The relevant findings are:

i) The large scale distribution of the non-thermal radio continuum emission is well described by exponential functions. In the radial direction this appears to be the case beyond a certain radius ($\sim$5kpc in IC342). In that region $R_{nth} > R_{opt} > R_{th}$. In the z-direction there is evidence for at least two components. The thin disk component in NGC891 can be described by a gaussian or an exponential distribution. The thick disk component or halo is best described by an exponential distribution. For NGC891 and NGC4631 the Z_{nth} of the halo component is *larger* than Z_{opt}. If the thin disk coincides with the star forming regions one can conclude that $Z_{nth} > Z_{opt} > Z_{th}$. How severe these constraints are on propagation models depends on the conversion from R_{nth} and Z_{nth} to R_e and Z_e and on the source distribution.

ii) The radio spectra of NGC891 and NGC4631 clearly show a change of slope. The spectra become steeper with frequency and are consistent with a change in slope of $\triangle\alpha=0.5$ (i.e. $\triangle\gamma=1.0$) with the break frequency near 0.8GHz (E~2GeV). The low frequency measurements by Israel and Mahoney (1990) indicate that such a change of slope in the radio spectra could be rather common. The observed spectra do not support a confinement halo. A change in spectral index of $\triangle\alpha=0.5$ is expected for steady state systems from which the cosmic ray electrons can escape by diffusion or convection. Considering the half life time of the relativistic electrons of $\sim 3 \ 10^7$yr the assumption of a steady state is valid for "normal" spiral galaxies. If the breaks in the radio spectra of NGC891 and NGC4631 result from convection by a galactic wind and adiabatic losses as envisaged by Lerche and Schlickeiser (1982) then the adiabatic cooling time is $\sim 4 \ 10^7$yr and the V_∞'s are $\sim$90 and $\sim$150km/sec, respectively.

iii) One of the questions to answer is whether the cosmic ray electrons propagate in a diffusive and/or convective mode. There are some indications that large scale convection is of importance in both radial and z-direction. In case of IC342 and NGC891 it is the constancy of α_{nth} with radius and z-height which suggests that convection is needed while in NGC4631 it is the large z-extent itself. The arguments used involve time scales for the various energy loss processes and assumptions concerning the source distribution and the conversion from R_{nth} and Z_{nth} to R_e and Z_e and are therefore subject to large uncertainties.

In this respect note that with D=10^{29}cm^2/sec and a life time for the cosmic ray electrons of $3 \ 10^7$yr the diffusion length scale and the convection length scale are the same (3.2kpc), when the convection speed is $\sim$100km/sec. This velocity is close to the Alfvén speed in a typical halo medium and the terminal velocity often associated with galactic winds. This means that on the basis of propagation lengths scales and time scales alone it is hard to rule out pure diffusion or pure convection propagation models.

In case of diffusion models with only synchrotron radiation losses the halo extent changes with frequency like $\nu^{-(1-\epsilon)/4}$ where ϵ describes the energy dependence of the diffusion coefficient (e.g. Bulanov and Dogiel, 1974). For NGC891 and NGC4631 it is found that the halo extent changes like $\nu^{-0.3}$, resulting in $\epsilon \sim -0.2$. Cosmic ray studies indicate $\epsilon \sim 0.5$ (Ginzburg and Ptuskin, 1976), suggesting that diffusion models are not compatible with the halo extents as observed.

The presence of spur-like features in the haloes of NGC891 and NGC4631 and the peanut-like morphology of the E side of NGC891 can also be seen as manifestations of large scale convective motions of the cosmic ray electrons.

iv) A common results for NGC891 and NGC4631 is the difference of $\triangle \alpha \sim 0.25$ between the "low" and "high" frequency spectral indices in the planes of these galaxies. For NGC891 this difference is also present at larger z-heights (for NGC4631 this has not been determined yet). Also striking is the asymmetry in the "low" frequency α-distributions. Both galaxies show a more or less constant α on the side which has the largest exponential scale height. On the other side the "low" frequency spectra steepen gradually.

The difference between the "low" and "high" frequency spectra at large z-heights in NGC891 strongly suggests that it is not caused by free-free absorption. Other ways to explain it are the importance of adiabatic losses and/or bremsstrahlung losses in the plane. Because of the low densities bremsstrahlung losses are of less importance at larger z-heights.

6. CONCLUDING REMARKS

From the radio continuum observations of edge-on spiral galaxies it is clear that cosmic ray electrons diffuse and convect away from their source distribution and that a typical propagation length is a few kpc. A large fraction of the cosmic ray electrons leave the source distribution, presumably associated with the thin star forming disk, perpendicular to the plane, hence forming a thick disk or halo component. This halo is *not* a confinement halo and the cosmic ray electrons can escape from the galaxy.

The results obtained for NGC891 and, in particular, NGC4631 on the spectra of the total radio emission, the "low" and "high" frequency spectra and the α-distributions are *consistent* with galactic wind models in which adiabatic losses become important at ν <1GHz (Lerche and Schlickeiser, 1982; Hummel and Dettmar, 1990; Hummel *et al.* 1990). However, this is almost certainly not the only possible consistent model. Detailed modeling, including the various loss mechanisms and using the constraints given by the observations of the edge-on galaxies, is required. Observationally, high resolution, multi-frequency mapping of edge-on galaxies is important.

NGC4631 appears to show the strongest evidence for convection (possibly caused by a galactic wind) perpendicular to the plane. It is also an exceptional case concerning its radio continuum z-extent. Hummel and Dettmar (1990) suggest that this is due to the relatively high star formation rate in the plane of this galaxy and to the gravitational interaction NGC4631 is undergoing. Both circumstances would favour the onset of convection perpendicular to the plane of the galaxy.

An interesting relation that recently emerged in case of NGC891 is the one between the Hα emission and the radio continuum emission in the halo component. The Hα emission shows its largest z-extent in the northern half of NGC891 (Dettmar, 1990, also this volume). This might indicate that convection of cosmic ray electrons is already important at z-distances of $\sim$2kpc and that it depends on the star formation activity in the disk.

REFERENCES

Allen, R.J., Baldwin, J.E., Sancisi, R.: 1978, *Astron. Astrophys.*, **62**, 397
Allen, R.J., Hu, F.X.: 1985, in *New Aspects of Galaxy Photometry* ed. J.L. Nieto, p293
Bicay, M.D., Helou, G., Condon, J.J.: 1989, *Astrophys. J.*, **338**, L53
Bulanov, S.V., Dogiel, V.A.: 1974, *Astrophys. Space Sci.*, **29**, 305
Dettmar, R.-J.: 1990, *Astron. Astrophys.*, **232**, L15
Ekers, R.D., Sancisi, R.: 1977, *Astron. Astrophys.*, **54**, 196
Ginzburg, V.I., Ptuskin, V.S.: 1976, *Reviews of Modern Physics*, **48**, 161
Hummel, E.: 1990, in *Windows on Galaxies*, eds. G.Fabbiano, J.S. Gallagher and A.Renzini, Kluwer, Dordrecht, p141
Hummel, E., Dettmar, R.-J.: 1990, *Astron. Astrophys.*, in press
Hummel, E., Gräve, R.: 1990, *Astron. Astrophys.*, **228**, 315
Hummel, E., van der Hulst, J.M.: 1986, *Astron. Astrophys.*, **155**, 151
Hummel, E., Davies, R.D., Wolstencroft, R.D., van der Hulst, J.M., Pedlar, A.: 1988, *Astron. Astrophys.*, **199**, 91
Hummel. E., Beck, R., Dettmar, R.-J.: 1990, *Astron. Astrophys. Suppl.* in press
Hummel, E., Dahlem, M., van der Hulst, J.M., Sukumar, S.: 1990, *Astron. Astrophys.*, in press
Israel, F.P., Mahoney, M.J.: *Astrophys. J.*, **352**, 30
Klein, U.: 1990, in *Windows on Galaxies*, eds. G.Fabbiano, J.S. Gallagher and A.Renzini, Kluwer, Dordrecht, p157
Krause, M.: 1990, in *Galactic and Intergalactic Magnetic Fields*, eds. R.Beck, P.P.Kronberg and R.Wielebinski, Kluwer, Dordrecht, p187
van der Kruit, P.C.: 1977, *Astron. Astrophys.*, **59**, 359
Lerche, I., Schlickeiser, R.: 1982, *Astron. Astrophys.*, **107**, 148
Segalovitz, A.: 1977, *Astron. Astrophys.*, **61**, 59
Wunderlich, E.: 1989, PhD thesis University of Bonn

MAGNETIC FIELDS IN DISKS AND HALOS OF SPIRAL GALAXIES

RAINER BECK
Max-Planck-Institut für Radioastronomie
Auf dem Hügel 69
D-5300 Bonn 1
Germany

ABSTRACT. Spiral galaxies host interstellar magnetic fields of 4-15 μG total strength. A significant fraction of the field lines shows large-scale structures. At face-on or moderately inclined view, the field lines run generally parallel to the spiral arms, either with uniform direction with respect to azimuthal angle (axisymmetric spiral, ASS), with one reversal along azimuthal angle (bisymmetric spiral, BSS), or with spiral orientation without dominating direction.

At edge-on view, the field is concentrated in a thin disk, often surrounded by a thick radio disk with field lines mostly parallel to the plane, similar to the quadrupole-type dynamo field. Radio polarization data from NGC891 indicate that the thermal gas seen in Hα is responsible for Faraday depolarization. The required scaleheight of the field of ~4 kpc is comparable to the value expected in case of energy equipartition between magnetic fields and cosmic rays. The interacting edge-on galaxy NGC 4631 shows a much larger radio halo with field lines perpendicular to the disk, possibly driven by a strong galactic wind or the result of a dipole-type halo field.

Field lines bending out of the plane are also visible in face-on galaxies as regions with high rotation measures and low star-formation activity. The resemblance to the phenomenon of the solar corona suggests to call them "galactic coronal holes".

1. INTRODUCTION

Most phenomena of the active sun emerge from interaction processes between surface and corona magnetic fields (Parker, 1990). Photospheric fields form magnetic loops. These are footpoints of the "streamers" with a slow, dense solar wind. The fast, dilute solar wind originates in coronal holes with unipolar magnetic fields.

It is tempting to compare these phenomena with disk-halo interactions in spiral galaxies. The main difference is that the magnetic energy density dominates out to several solar radii distance while in the disks of spiral galaxies the energy densities of magnetic field, cosmic rays and turbulent gas motion are similar, and the halo field may dominate only above a certain height (see Sect. 4).

2. MAGNETIC FIELDS IN DISKS OF SPIRAL GALAXIES

Linearly polarized radio synchrotron emission offers the most powerful method to study interstellar magnetic fields. While the intensity of the total synchrotron emission yields an estimate of the *strength* of the *total* field $B_{t,\perp}$ in the plane of the sky, the intensity of the linearly polarized emission can be used to estimate the *strength* of the *uniform* field $B_{u,\perp}$. The degree of polarization is ~75% in case of a completely uniform field; it decreases with increasing contribution of a random field component. The orientation of the E- vector is intrinsically perpendicular to the *orientation* of $B_{u,\perp}$ but is rotated when the radio wave passes through a magnetized plasma cloud (Faraday rotation). The sign of rotation gives the *direction* of the field, and the amount of rotation ("rotation measure") is proportional to the average strength of the uniform field $B_{u,\parallel}$ along the line of sight (see Beck, 1986 and M. Krause, 1990 for details and formulae).

2.1 Large-scale Field Structures

Recognition of large-scale field structures requires the analysis of the rotation measures. A uniform field direction yields a periodic variation of rotation measures along azimuthal angle. The phase of such a variation has to be consistent with the pitch angle of $B_{u,\perp}$ and has to change with radius according to the position angle of the spiral arms. Hence this method is *not* based on a Fourier analysis alone. The assumptions and limitations have been discussed in detail by M. Krause (1990) and Donner and Brandenburg (1990).

For most galaxies the data are still insufficient to determine reliable rotation measures (see Wielebinski, 1990). Only 7 external galaxies have been studied adequately (Table 1). In M31 and IC 342 the dominating field is best described by an axisymmetric spiral (ASS) with constant direction along azimuthal angle and along radius (Beck, 1982; M. Krause et al., 1989a) while M81, M51 (Fig. 1) and possibly M33 show a dominating bisymmetric spiral (BSS) field with one field reversal along azimuthal angle (M. Krause et al., 1989b; Horellou, 1990; Buczilowski and Beck, 1990). In NGC 6946 (Beck et al., 1990, Fig. 2) and probably M83 (Sukumar et al., 1987; Sukumar and Allen, 1989; Sukumar et al., in prep.) the rotation measures are very low in spite of their strong uniform fields: the field in these two galaxies is ordered *without* a constant direction (*neither* ASS *nor* BSS), e.g. forming highly elongated loops or cells.

TABLE 1.
Large-scale structures of *disk* fields

ASS (axisymmetric spiral)	BSS (bisymmetric spiral)	Neither (but spiral)
M31	M51	NGC6946
IC342	M81	M83?
	M33?	Milky Way?

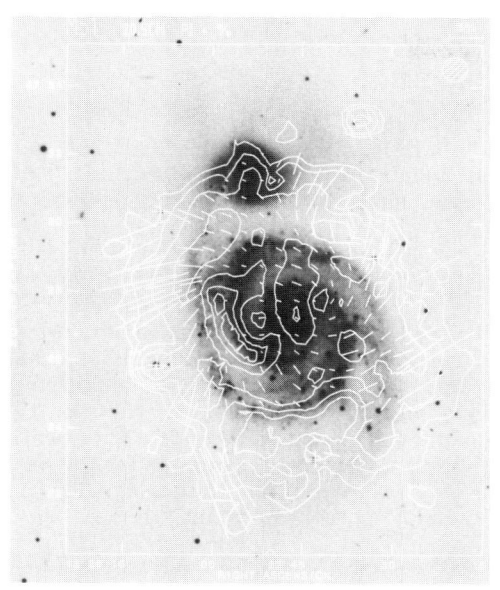

Figure 1. Linearly polarized radio emission from M51 at λ20.5 cm observed with the VLA in its D-configuration. The angular resolution is 43 arcsec, the rms noise is 28 μJy/beam. Contours are 120,200,400,800,1600 μJy/beam. The lengths of the observed E-vectors are proportional to the degree of polarization (from Horellou, 1990).

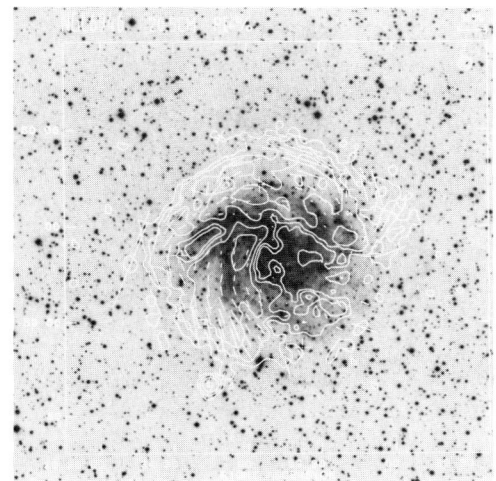

Figure 2. Linearly polarized radio emission from NGC 6946 at λ20.5 cm observed with the VLA in its D-configuration. The angular resolution is 42 arcsec, the rms noise is 20 μJy/beam. Contours are 80, 160,320,640 μJy/beam. The lengths of the observed E-vectors are proportional to the degree of polarization. Faraday rotation at λ20.5 cm is almost constant ~37 rad m^{-2} or 88°, i.e. the vectors approximately show the orientation of the magnetic field component $B_{u,\perp}$ (Beck et al., in prep.).

In our own Galaxy there are at least two field reversals between the solar radius and the center (Vallée, 1988). Two inner reversals are consistent with a BSS field structure only if the pitch angle of the field lines is ≤ 6°. However, the pitch angle of the optical spiral arms is ~17°. If this value is also used for the field lines, the field in our Galaxy could resemble the "neither" cases NGC 6946 and M83 without constant direction. Alternatively, the field may be axisymmetric with reversals at certain radii (Vallée, 1990).

2.2 Detailed Field Structure

The orientation of the interstellar magnetic field is generally parallel to the optical spiral arms. Variations in their pitch angle, e.g. in M51 on the side of its companion galaxy, also appear in the field structure. The unresolved ("turbulent") field B_r (observed as unpolarized radio emission) is strongest in the optical spiral arms but the uniform field B_u is in most galaxies strongest in the *interarm* regions (see Figs. 1 and 2). The degree of field uniformity is anticorrelated with the intensity of CO line emission (Beck et al., 1991). The total field B_t is only slightly enhanced in the spiral arms, leading to the smooth radio disk as observed in total radio intensity. In well-resolved galaxies, B_u is *not* enhanced at the inner edges of spiral arms, i.e. the field is not compressed by a large-scale density-wave shock front. However, an indirect influence of density waves on the dynamo by modification of gas motion (ω-effect) is well possible (see Sect. 3). In M83 blobs of highly polarized radio emission are observed outside the outer optical spiral arm (Allen and Sukumar, 1990). Systematic motions of clouds in a density-wave potential have been invoked to explain this phenomenon.

M31 is an exceptional case: turbulent *and* uniform field are both concentrated in the "10 kpc ring" where most of the star formation takes place. A dust lane in the SW quadrant was studied in detail (Beck et al., 1989). A deviation of $\leq 25°$ from the toroidal field direction occurs near a gas cloud complex.

The general picture of field structure in spiral arms (Beck et al., 1991) is based on the idea that the field lines are anchored in gas clouds. The field structure depends on the motion of the clouds but can also influence the cloud motion. Galaxies with little mass in molecular clouds and hence low star formation rate like M31 show a relatively undisturbed field within the spiral arms and no detectable field outside the arms. A large number of clouds produces a high "turbulence" of the field because the superposition of field loops cannot be resolved in present radio maps. Enhanced turbulent motion of the clouds by collisions or turbulence induced by star formation may further tangle the field lines.

2.3 Field Strengths

The average strength of the total disk field varies between ~4 μG in M31 and M33 and ~12 μG in M51 and NGC 6946. A sample of Sbc galaxies gave $\langle B_t \rangle \simeq 8$ μG (Hummel et al., 1988b). The observed close correlation between the radio continuum and far-infrared luminosities of spiral galaxies (e.g. Wunderlich et al., 1987) could mean that the energy density $B_t^2/8\pi$ of the total field is proportional to the energy density of the stellar radiation field which is dominated by the star formation rate (Hummel et al., 1988b; Völk, 1989). However, the correlation between radio and far-infrared intensities holds within galaxies on scales down to ~1 kpc (Beck and Golla, 1988; Bicay et al., 1989) indicating a cosmic-ray production rate proportional to the star formation rate, plus equipartition between cosmic ray and magnetic field energy densities (Chi and Wolfendale, 1990).

Helou and Bicay (1990) explain the radio-FIR correlation by a tight coupling of the sources of the dust-heating photons and those of cosmic-ray electrons, plus a universal relation between magnetic field strength and gas density of the form $B \propto n^\beta$ ($1/3 \leq \beta \leq 2/3$). Hence, field strength and star formation rate could also be connected via the gas density.

The star formation process may be controlled by the magnetic field via feedback actions (e.g. Chi and Wolfendale, 1990). Magnetic fields are known to be essential for the stability and fragmentation of molecular clouds (e.g. Bash et al., 1981; Mouschovias, 1990) as well as for cloud collisions (Clifford and Elmegreen, 1983).

3. ORIGIN OF DISK AND HALO MAGNETIC FIELDS

3.1 Primordial Origin ("preserve the order")

An ordered, protogalactic field is wound up by differential rotation. Complete wind-up can be avoided if the field slips through the clouds rapidly enough (Kulsrud, 1986). Otherwise, a quasistationary configuration may be achieved by diffusion plus dynamo action (Sawa and Fujimoto, 1986) or by reconnection (Brett and Kahn, 1990). If the rotation axis is perpendicular to the primordial field, a BSS structure is generated; if the rotation axis is along the primordial field, the result is an ASS structure with a field reversal above and below the plane. An ASS field without reversal, however, can be produced only under special conditions, e.g. with a strong gradient of the primordial field (Sofue, 1990).

3.2 Local Origin ("preserve the chaos")

The field is ejected by stars or supernova remnants and wound up by differential rotation (Michel and Yahil, 1973). Field enhancement can be achieved by a local dynamo driven by purely turbulent motion (Molchanov et al., 1985). The resulting field has some alignment, but is subject to frequent reversals within the disk and hence does not show a uniform direction on large scales.

3.3 Dynamo Models ("order out of chaos")

ASS and BSS field structures are interpreted as two different azimuthal modes ($m = 0$ and 1) of the galactic $\alpha\omega$-dynamo (Chiba and Tosa, 1989; F. Krause et al., 1990; Ruzmaikin, 1990; Donner and Brandenburg, 1990). Linear dynamo calculations which assume an axisymmetric gas distribution in galactic disks predict the highest growth rate for the lowest (axisymmetric) mode. A *dominating BSS mode* cannot be explained by linear dynamo models. Deviations from symmetry or from stationarity may preferably lead to BSS modes:
1. Gravitational disturbances (companion, warping of the disk, density waves, central bars) may excite the BSS mode (Chiba and Tosa, 1990).
2. The dynamo field is growing and still reflects the primordial seed

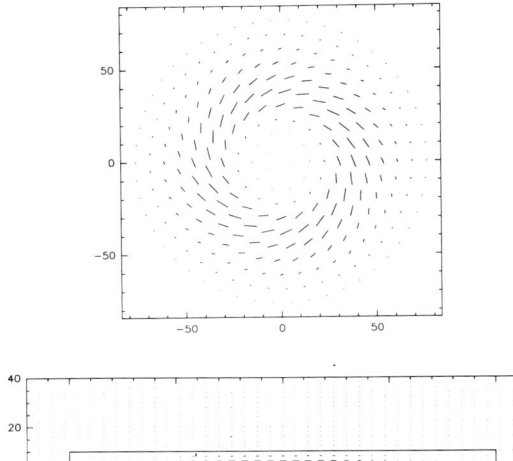

Figure 3. The polarized radio emission from a quadrupole S0 dynamo field in face-on and edge-on view (Elstner et al., in prep.). The density of cosmic-ray electrons was assumed to be constant. The model galaxy is rigidly rotating until 1/3 of the radius r and differentially rotating at larger radii. The α-effect operates only in the disk (up to r/10 height). The conductivity in the halo ($|z| > r/10$) is assumed to be 1/5 of that in the disk.

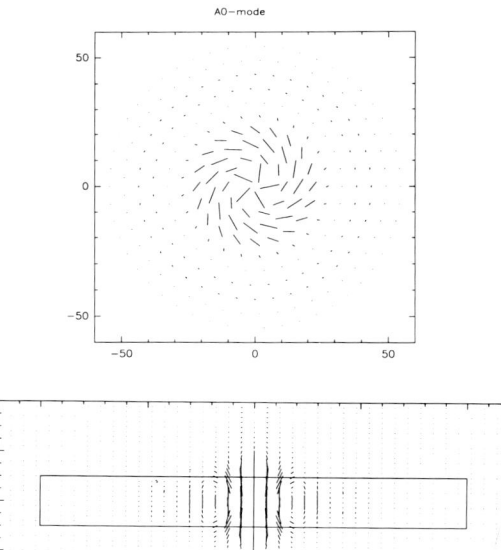

Figure 4. The polarized radio emission from a dipole A0 dynamo field in face-on and edge-on view. Assumptions as in Figure 3.

field which is bisymmetric due to winding-up of the intergalactic or intracluster field (Sawa and Fujimoto, 1986).
3. The BSS mode appears as the result of strong nonlinear effects for fast-growing dynamo fields (Moss and Tuominen, 1990).

The few available data favor the first possibility. The two BSS cases M81 and M51 have massive companions while the ASS and "zero" cases are accompanied only by small galaxies. The presence of density waves or bars seems to be less important because M83 shows no BSS field.

A *superposition* of the ASS and BSS mode was proposed for M31 (Sofue and Beck, 1987; Ruzmaikin et al., 1990) but a systematic variation of field strength, plasma density, or pitch angle of the field lines along azimuthal angle may produce a similar effect (M. Krause et al., 1989b).

Dynamo models distinguish between *even* or *odd* modes (vertical modes), i.e. no reversal or a reversal of the disk field above and below the plane. The *even* ASS field (or symmetric m = 0, S0) can be separated into a toroidal disk field and a quadrupole-type poloidal field in the halo while the poloidal component of the *odd* ASS field (or asymmetric m = 0, A0) is of dipole type. Figures 3 and 4 show the polarized radio emission from S0 and A0 dynamo fields (Elstner et al., in prep.).

In linear theory of the thin-disk $\alpha\omega$-dynamo the ratio of the strengths of the halo and disk field depends on the dynamo numbers R_α and R_ω (Ruzmaikin et al., 1988). Flat rotation curves with $v = v_{max}$ yield $B_h/B_d \simeq (\alpha/v_{max})^{1/2}$. For our Galaxy Ruzmaikin et al. (1988) estimated $R_\omega \simeq 10$, $R_\alpha \simeq 1$ and $B_h/B_d \simeq 0.1$. As synchrotron intensity is proportional to $B_\perp^{1+\alpha_{nth}}$ ($\alpha_{nth} \simeq 1.0$), the halo is normally not expected to contribute much to the radio emission. However, a high star formation rate (high α) and a low rotation velocity v_{max} may lead to higher B_h/B_d and hence to a strong halo field with a large-scale dipole structure (see Sect. 4). In the thick-disk dynamo model by Brandenburg et al. (1990) the dipole mode is indeed the easiest to excite.

Linear dynamo theory does not predict the maximum field strength reached in a galaxy. According to Ko and Parker (1989) the dynamo numbers depend on the turbulent velocity which is thought to increase with star formation rate so that the dynamo grows fast – but this does not necessarily mean that the field is strong. The maximum field strength is probably related to the star formation rate (see Sect. 2.3).

Even and odd dynamo fields can be distinguished with help of radio polarization observations:
I. The rotation measure in odd disk fields is one half of those in even disk fields. Comparison between the strengths of the uniform field as derived from polarized radio emission and from rotation measures indicate that the even ASS mode (S0) is more probable in M31 and IC 342. In M51 the rotation measures (corrected for inclination) are too low compared with the polarized synchrotron emission so that an odd BSS mode (A1) may exist (Horellou et al., in prep.).
II. In face-on or moderately inclined galaxies the dipole-type halo field induces high rotation measures in addition to the periodic rotation measures due to the disk field. This effect can be determined by comparison with the rotation measures of nearby extragalactic radio

sources. However, neither in M51, nor in M81, nor in IC 342 excess rotation measures have been found.

III. The quadrupole-type halo field points perpendicular to the disk only near the center. Near the Galactic center and the centers of active galaxies there is increasing evidence for perpendicular fields (Sofue, 1990) which may originate in central gaseous rings (Lesch et al., 1989). However, the central region of IC 342 shows *no* z-component between 2 and 5 kpc radius at 250 pc linear resolution (Krause et al., in prep.). The perpendicular field component must be restricted to even smaller radii or is too weak to be observable.

IV. In edge-on galaxies the halo field is directly observable (Sect. 4). The quadrupole-type field is expected to have a small scaleheight and runs mostly parallel to the disk (Fig. 3) while the dipole-type field is more extended and runs perpendicular to the disk (Fig. 4).

4. MAGNETIC FIELDS IN HALOS OF SPIRAL GALAXIES

Edge-on galaxies often show vertical dust lanes which may indicate uniform magnetic field lines (Sofue, 1987). Their first detection via polarized radio emission in NGC 4631 by Hummel et al. (1988a) induced a systematic search in several nearby edge-on galaxies. NGC 891 (Hummel et al., 1990b), NGC 3628 (M. Krause et al., in prep.), NGC 4565 (Sukumar and Allen, in prep.) and NGC 5775 (Golla and Beck, 1990; Fig. 5) and several other edge-on galaxies (Hummel, 1990; Hummel, this volume) do not possess extended radio halos but *thick disks* with typically ~1 kpc scaleheights. In most of these galaxies the observed E-vectors are preferably perpendicular to the disk (Table 2). The same result has been obtained for NGC 4945 (Harnett et al., 1989) and NGC 1808 (Dahlem, 1990) but the polarized emission is restricted to two regions on both sides of the plane. Another galaxy with possibly perpendicular E-vectors is NGC 253 (Klein et al., 1983). However, the inclination is too low to exclude a dominant contribution from the disk.

In view of the high frequencies and the regular structure of the polarized emission Faraday rotation is probably small so that most of the field lines run *parallel* to the disk. In the disk itself the polarized emission is negligible due to Faraday depolarization.

NGC 891 is the only edge-on galaxy with polarization data at two

TABLE 2.
Large-scale structures of *thick disk* or *halo* fields

B ∥ disk (quadrupole?)	B ⊥ disk (dipole?)	B ∥ disk (outflow?)	No thick disk detected
NGC 253?	NGC 4631	NGC 1808	M31
NGC 891		NGC 4945	
NGC 3628			
NGC 4565			
NGC 5775			

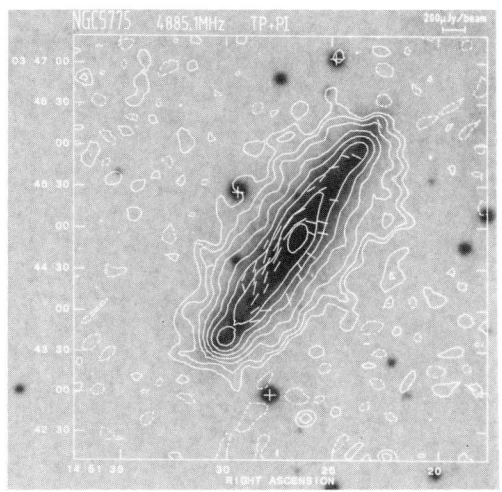

Figure 5. Total radio emission from NGC 5775, observed at λ6.1 cm with the VLA in its D-configuration. The synthesized HPBW is 12".5, the rms noise 25 μJy/beam. Contour levels are 40,80,160,... μJy/beam. The vectors are E-vectors turned by 90° and approximately show the orientations of the magnetic field lines because foreground Faraday rotation is low. The vector lengths are proportional to the polarized intensity (from Golla and Beck, 1990).

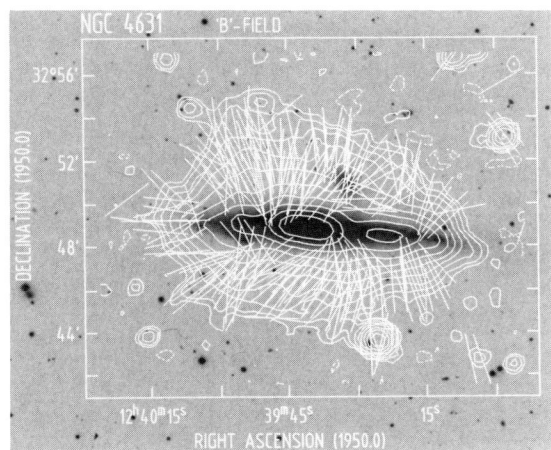

Figure 6. Total radio emission from NGC 4631 at λ20.2 cm, observed with the VLA in its D-configuration. The angular resolution is 40 arcsec, the rms noise is 40 μJy/beam. Contours are 100,200, 400,... μJy/beam. The vectors are E-vectors turned by 90° and approximately show the orientations of the magnetic field lines because foreground Faraday rotation is low (from Hummel et al., 1990a).

frequencies (Hummel et al., 1990a; Allen and Sukumar, 1990). The field lines are again parallel to the plane but flaring out at the edges of the disk (Allen and Sukumar, this volume). A similar behaviour is visible in some cases of a quadrupole dynamo field (see Fig. 3).

The increase of the degree of polarization with height above the disk of NGC 891 has been analyzed by Hummel et al. (1990a). The data can be well modelled by Faraday depolarization in a thermal gas of ~1 kpc scaleheight (assuming a distance of 7.2 Mpc) and a turbulent magnetic field of ~4 kpc scaleheight. The scaleheight of the thermal gas as derived from the radio data agrees well with that observed in Hα light (Rand et al.,

1990; Dettmar, 1990). The fitted thermal electron density in the plane of $\langle N_e \rangle \simeq 0.03$ cm^{-3} indicates a filling factor of $f = \langle N_e \rangle / N_{cloud} \simeq 0.06$. Both $\langle N_e \rangle$ and f are lower than the values derived from Hα observations.

The scaleheight of the turbulent magnetic field in NGC 891 of $z_B \simeq$ 4 kpc is consistent with equipartition between the field and cosmic ray energy densities where $z_B = 2 z_{CR} = (3+\alpha_{nth}) z_{syn}$. With a synchrotron scaleheight of $z_{syn} \simeq 0.9$ kpc and a nonthermal spectral index of $\alpha_{nth} \simeq$ 1.0 (Hummel et al., 1990b) equipartition means $z_B \simeq 3.6$ kpc, in agreement with the polarization data. This is the *first* observational evidence for equipartition in the halo of a galaxy. Hence the magnetic field probably extends much farther out than the thermal gas and the cosmic rays.

NGC 4631 (Fig. 6) is the only case of an extended *radio halo* known until now. The synchrotron scaleheight of ~1.9 kpc at a distance of 7.5 Mpc is about twice as large as for the bulk of edge-on galaxies (Hummel, 1990). The depolarization data indicate that the scaleheight and density of thermal gas is also higher (~1.3 kpc and 0.07 cm^{-3}) than in NGC 891. The distribution of polarized emission and E-vectors observed by Hummel et al. (1988a) is so regular that Faraday rotation is probably small. The magnetic field lines (Fig. 6) are *perpendicular* to the disk. In comparison to all other edge-on galaxies observed, NGC 4631 is an exceptional galaxy. A survey of 181 edge-on galaxies observed with the Effelsberg and VLA radio telescopes (Hummel et al., 1990c) disclosed *no other case* with a halo similar to NGC 4631.

The other extreme is M31: The radio emission from any thick disk is not detectable and has to be at least 200 times weaker than for NGC 891 (Berkhuijsen et al., this volume). The magnetic field in the disk of M31 is almost toroidal with only weak deviations near cloud complexes (Beck et al., 1989). The lack of vertical field lines hampers cosmic ray diffusion to high z and may account for the weak emission there. In terms of dynamo models (Sect. 3.3) a low α-effect and a high rotation velocity lead to a low ratio B_{halo}/B_{disk}. However, this is not true for all regions in M31 (see Sect. 5).

Models of Parker field loops (e.g. Urbanik, 1990) predict the halo field lines to be perpendicular to the disk. Galactic winds would further increase the degree of uniformity of the vertical field. Parallel field lines can be obtained if all field loops reach only a constant height above the disk, e.g. by galactic chimneys (Norman and Ikeuchi, 1989). The scaleheight is sufficient to explain the radio depolarization in NGC 891 but not in NGC 4631 (see above).

The predominance of magnetic fields parallel to the disk (Table 2) can be best interpreted in terms of dynamo theory. Quadrupole-type dynamo fields can explain both the extension and the field structure at high z (Fig. 3). The dipole-type field of NGC 4631 may be the result of high star formation and low rotation velocity (see Sect. 3.3 and Fig. 4).

The high star formation in NGC 4631 may also drive a galactic wind which pushes field lines out of the galactic plane. Extensions of the radio halo in regions of uniform field (Fig. 6) correspond to the "streamers" of the solar corona. If the halo field is in energy equipartition with the cosmic rays, it has a scale height of as large as ~8 kpc. Beyond 8 kpc height above the plane of NGC 4631, where Faraday depolarization becomes negligible, the degree of polarization decreases again. A similar effect is seen in NGC 891. This indicates an increase of field turbulence with increasing height, as predicted by some galactic wind models (Breitschwerdt et al., 1990).

Convection or diffusion of field lines from the disk into the halo has not yet been treated by dynamo models. On the other hand, galactic wind models still have to take the magnetic field into account. As the scale-height of B^2 (at least in NGC 891) is higher than that of the thermal gas, *the magnetic pressure will dominate* above a certain height.

5. GALACTIC CORONAL HOLES

A region in the SE quadrant of M31 with vertical field lines (Berkhuijsen et al., this volume) coincides with low HI emission and with a hole in the CO ring at 10 kpc radius (Dame et al., 1991). Regions with little gas content may allow the field to leave the galactic disk and form open field lines – as in the case of solar coronal holes.

In face-on galaxies coronal holes are observable as regions of high rotation measures with neither enhanced plasma density (Hα emission) nor enhanced field strength (total synchrotron emission). The maps of IC 342 (Krause et al., 1989a) and also of M51 (Horellou, 1990) seem to show such phenomena. The hole in M51 is visible as an extended minimum in polarized emission between the spiral arms traced by Hα and CO (Fig. 1). This region has to be studied with higher resolution.

Of special interest is the SW quadrant of NGC 6946 (Fig. 2): Strong Faraday depolarization and high Faraday rotation measures without indications for enhanced plasma density or uniform field strength allow to conclude that a large fraction of the disk field bends out into the halo. The rotation measures are almost constant over NGC 6946 ("neither" case in Table 1), except in the SW quadrant where both high and low values occur (Beck et al., in prep.). The spiral arms in the SW quadrant of NGC 6946 are more diffuse and the Hα emission is weaker (Bonnarel et al., 1986) compared with the rest of the galaxy: The galactic coronal hole occurs in a region of low star-forming activity.

6. FUTURE OBSERVATIONS

Radio polarization data of both face-on and edge-on galaxies hold the clue to understand the magnetic disk-halo connections. Rotation measures will help to distinguish between the various models (Table 3) while high-resolution polarization observations will show the role of gas complexes and star-forming regions.

Observations in CO lines are required to investigate the relation between cloud distribution and turbulence and the field structure in and above the disk. Optical and X-ray detection of warm and hot halo gas, together with radio polarization data, would allow to determine the halo field strength and its energy density and to compare it with the energy densities of gas and cosmic rays. Finally, the detection of optical polarization caused by the Davis-Greenstein mechanism would directly

TABLE 3.
Observation of magnetic disk-halo connections

	Edge-on galaxies degree of polarization	Edge-on galaxies rotation measure	Face-on galaxies degree of polarization	Face-on galaxies rotation measure
Dynamo fields	high	medium	low	medium
Galactic winds	medium	low	low	low
"Coronal holes"	high	low	low	high

indicate the field orientation with very high resolution. New CCD polarimeters are promising (Scarrott et al., 1990; Neininger et al., 1990).

7. CONCLUSIONS

a. The analysis of radio polarization data shows that interstellar magnetic fields have both a large-scale component with uniform direction as well as small-scale components related to gas clouds and star-formation processes.
b. The *disk* field has an axisymmetric, a bisymmetric or a "neither" structure, in most cases *without* a reversal above/below the plane.
c. The *thick disk* field has a parallel, perpendicular or "neither" orientation with respect to the disk, with a clear dominance of parallel fields.
d. Field lines perpendicular to the disk seem to produce a huge radio *halo* — but this case is very rare.
e. Disk and halo fields can best be interpreted in terms of dynamo models of mostly quadrupole type. The influence of galactic winds or fountains on the (small-scale ?) field structure is not yet understood. Future measurements of Faraday rotation are required.
f. We could learn a lot from solar magnetic fields (Table 4).

TABLE 4.
Magnetic Phenomena

Sun:	Galaxies:
$\alpha\omega$-dynamo	$\alpha\omega$-dynamo, α^2-dynamo
flux tubes	Verschuur's helices? (this volume)
loop prominences	Parker instabilities
reconnection	?
solar wind	galactic wind (NGC 4631)
streamers (slow wind)	extensions into the radio halo (NGC4631)
?	chimneys
coronal transients	outflow? (NGC 1808, NGC 4945)
coronal holes (fast wind)	galactic coronal holes

REFERENCES

Allen, R.J., Sukumar, S. (1990) *Ap. J.* (in press)
Bash, F., Hausman, M., Papaloizou, J. (1981) *Ap. J.* **245**, 92
Beck, R. (1982) *Astr. Ap.* **106**, 121
Beck, R. (1986) *IEEE Trans. on Plasma Science* **PS-14**, 740
Beck, R., Golla, G. (1988) *Astr. Ap.* **191**, L9
Beck, R., Loiseau, N., Hummel, E., Berkhuijsen, E.M., Gräve, R., Wielebinski, R. (1989) *Astr. Ap.* **222**, 58
Beck, R., Buczilowski, U.R., Harnett, J.I. (1990) in *Galactic and Intergalactic Magnetic Fields*, eds. R. Beck, P.P. Kronberg, R. Wielebinski, Kluwer, Dordrecht, p. 213
Beck, R., Berkhuijsen, E.M., Bajaja, E. (1991) in *Dynamics of Galaxies and Molecular Cloud Distribution*, IAU Symp. No. 146, eds. F. Combes, F. Casoli, Kluwer, Dordrecht
Bicay, M.D., Helou, G., Condon, J.J. (1989) *Ap. J.* **338**, L53
Bonnarel, F., Boulesteix, J., Marcelin, M. (1986) *Astr. Ap. Suppl.* **66**, 149
Brandenburg, A., Tuominen, I., Krause, F. (1990) *Geophys. Astrophys. Fluid Dynamics* **50**, 95
Breitschwerdt, R., McKenzie, J.F., Völk, H.J. (1990) *Astr. Ap.* (in press)
Brett, L., Kahn, F. (1990) in *The Interstellar Disk-Halo Connection in Galaxies*, Poster Proc. IAU Symp. No. 144, ed. H. Bloemen, Leiden Observatory, p. 81
Buczilowski, U.R., Beck, R. (1990) *Astr. Ap.* (in press)
Chi, X., Wolfendale, A.W. (1990) *M.N.R.A.S.* **245**, 101
Chiba, M., Tosa, M. (1989) *M.N.R.A.S.* **238**, 621
Chiba, M., Tosa, M. (1990) in *Galactic and Intergalactic Magnetic Fields*, eds. R. Beck, P.P. Kronberg, R. Wielebinski, Kluwer, Dordrecht, p. 131
Clifford, P., Elmegreen, B.G. (1983) *M.N.R.A.S.* **202**, 629
Dahlem, M. (1990) Ph.D. thesis, University of Bonn
Dame, T., Thaddeus, P., Koper, E. (1991), in *Dynamics of Galaxies and Molecular Cloud Distribution*, IAU Symp. No. 146, eds. F. Combes, F. Casoli, Kluwer, Dordrecht
Dettmar, R.-J. (1990) *Astr. Ap.* **232**, L15
Donner, K.J., Brandenburg, A. (1990) *Astr. Ap.* (in press)
Golla, G., Beck, R. (1990) in *The Interstellar Disk-Halo Connection in Galaxies*, Poster Proc. IAU Symp. No. 144, ed. H. Bloemen, Leiden Observatory, p. 47
Harnett, J.I., Haynes, R.F., Klein, U., Wielebinski, R. (1989) *Astr. Ap.* **216**, 39
Helou, G., Bicay, M.D. (1990), in *Galactic and Intergalactic Magnetic Fields*, eds. R. Beck, P.P. Kronberg, R. Wielebinski, Kluwer, Dordrecht, p. 239
Horellou, C. (1990) Diplomarbeit, University of Bonn
Hummel, E. (1990) in *Windows on Galaxies*, eds. G. Fabbiano et al., Kluwer, Dordrecht, p. 141
Hummel, E., Lesch, H., Wielebinski, R., Schlickeiser, R. (1988a) *Astr. Ap.* **197**, L29
Hummel, E., Davies, R.D., Wolstencroft, R.D., van der Hulst, J.M., Pedlar, A. (1988b) *Astr. Ap.* **199**, 91
Hummel, E., Beck, R., Dahlem, M. (1990a) *Astr. Ap.* (in press)

Hummel, E., Dahlem, M., van der Hulst, J.M., Sukumar, S. (1990b) *Astr. Ap.* (in press)
Hummel, E., Beck, R., Dettmar, R.-J. (1990c) *Astr. Ap.* (in press)
Klein, U., Urbanik, M., Beck, R., Wielebinski, R. (1983) *Astr. Ap.* **127**, 177
Ko, C.M., Parker, E.N. (1989) *Ap. J.* **341**, 828
Krause, F., Meinel, R., Elstner, D., Rüdiger, G. (1990) in *Galactic and Intergalactic Magnetic Fields,* eds. R. Beck, P.P. Kronberg, R. Wielebinski, Kluwer, Dordrecht, p. 97
Krause, M. (1990) in *Galactic and Intergalactic Magnetic Fields,* eds. R. Beck, P.P. Kronberg, R. Wielebinski, Kluwer, Dordrecht, p. 187
Krause, M., Hummel, E., Beck, R. (1989a) *Astr. Ap.* **217**, 4
Krause, M., Beck, R., Hummel, E. (1989b) *Astr. Ap.* **217**, 17
Kulsrud, R. (1986) in *Plasma Astrophysics,* ESA SP-251, p. 531
Lesch, H., Crusius, A., Schlickeiser, R., Wielebinski, R. (1989) *Astr. Ap.* **217**, 99
Michel, F.C., Yahil, A. (1973) *Ap. J.* **179**, 771
Molchanov, S., Ruzmaikin, A., Sokolov, D. (1985) *Sov. Phys. Usp.* **28**, 307
Moss, D., Tuominen, I. (1990) *Geophys. Astrophys. Fluid Dynamics* **50**, 113
Mouschovias, T.Ch. (1990) in *Galactic and Intergalactic Magnetic Fields,* eds. R. Beck, P.P. Kronberg, R. Wielebinski, Kluwer, Dordrecht, p. 269
Neininger, N., Beck, R., Backes, F. (1990) in *Galactic and Intergalactic Magnetic Fields,* eds. R. Beck, P.P. Kronberg, R. Wielebinski, Kluwer, Dordrecht, p. 253
Norman, C.A., Ikeuchi, S. (1989) *Ap. J.* **345**, 372
Parker, E.N. (1990) in *Galactic and Intergalactic Magnetic Fields,* eds. R. Beck, P.P. Kronberg, R. Wielebinski, Kluwer, Dordrecht, p. 1
Rand, R.J., Kulkarni, S.R., Hester, J.J. (1990) *Ap. J.* **352**, L1
Ruzmaikin, A. (1990) in *Galactic and Intergalactic Magnetic Fields,* eds. R. Beck, P.P. Kronberg, R. Wielebinski, Kluwer, Dordrecht, p. 83
Ruzmaikin, A., Sokoloff, D., Shukurov, A. (1988) *Nature* **336**, 341
Ruzmaikin, A., Sokoloff, D., Shukurov, A., Beck, R. (1990) *Astr. Ap.* **230**, 284
Sawa, T., Fujimoto, M. (1986) *P.A.S. Japan* **38**, 133
Scarrott, S.M., Rolph, C.D., Semple, D.P. (1990) in *Galactic and Intergalactic Magnetic Fields,* eds. R. Beck, P.P. Kronberg, R. Wielebinski, Kluwer, Dordrecht, p. 245
Sofue, Y. (1987) *P.A.S. Japan* **39**, 547
Sofue, Y. (1990) in *Galactic and Intergalactic Magnetic Fields,* eds. R. Beck, P.P. Kronberg, R. Wielebinski, Kluwer, Dordrecht, p. 227
Sofue, Y., Beck, R. (1987) *P.A.S. Japan* **39**, 541
Sukumar, S., Allen, R.J. (1989) *Nature* **340**, 537
Sukumar, S., Klein, U., Gräve, R. (1987) *Astr. Ap.* **184**, 71
Urbanik, M. (1990) in *The Interstellar Disk-Halo Connection in Galaxies,* Poster Proc. IAU Symp. No. 144, ed. H. Bloemen, Leiden Observatory, p. 111
Vallée, J.P. (1988) *A. J.* **95**, 750
Vallée, J.P. (1990) *Ap. J.* (in press)
Völk, H.J. (1989) *Astr. Ap.* **218**, 67
Wielebinski, R. (1990) in *The Interstellar Medium in Galaxies,* eds. H.A. Thronson, J.M. Shull, Kluwer, Dordrecht, p. 349
Wunderlich, E., Klein, U., Wielebinski, R. (1987) *Astr. Ap. Suppl.* **69**, 487

SYNCHROTRON EMISSION AS A TRACER OF THE OUTFLOW IN M82.

E.R. SEAQUIST
Department of Astronomy
University of Toronto
Toronto, Ontario M5S 1A1, Canada

NILS ODEGARD
General Sciences Corporation
Goddard Space Flight Center
Code 685.3, Greenbelt, MD 20771, U.S.A.

ABSTRACT. We report the discovery of radio synchrotron emission from the outflow in M82. The brightness morphology and radio spectral index distribution add new insights into the physical processes and origin of the wind, which are briefly discussed in this paper.

1. INTRODUCTION

It is now well established that M82 is undergoing mass outflow from its nuclear region, in accordance with evidence from optical spectroscopic data, molecular line data, and the morphology of its X-ray emission (eg. McCarthy, Heckman, and van Breugel 1987, Nakai et al. 1987, and Fabbiano 1988). The wind is possibly in the form of a bipolar outflow at an optically determined speed of 600 km s^{-1} (Bland and Tully 1988).

We report in this paper the discovery of radio synchrotron emission at several wavelengths associated with this outflow detected with the VLA and Westerbork radio telescopes*. The emission takes the form of a substantial nonthermal radio halo surrounding M82. The relativistic electrons which emit this radiation must necessarily be convected outward by the wind, and therefore must trace its morphology.

The synchrotron halo is extremely faint compared to the bright nuclear emission. The key factor in the success of our observation is the high dynamic range achieved, which exceeds 10,000 in all maps. In this paper we confine our attention primarily to the VLA 6 cm and 20 cm maps.

*The VLA is part of the National Radio Astronomy Observatory which is operated by Associated Universities Inc., under a cooperative agreement with the National Science Foundation. The WSRT is operated by the Foundation for Research in Astronomy which is financially supported by the Netherlands Organization for Scientific Research.

2. RADIO MAPS AND THEIR INTERPRETATION

Figure 1 shows the radio data in various forms. The top panel shows the 20cm map made from data taken with the C and D configurations combined and superposed on a continuum photograph. The radio halo shown is about 8 kpc in diameter which is far larger than any radio emitting feature of M82 so far known. Of particular interest is the asymmetry with respect to the disk, which we return to at the end of the paper. The middle panel shows the same radio map (cropped) superposed on an H alpha photograph. The radio supernova remnants studied by Kronberg et al. (1985) are confined to a region roughly bounded by the inner contour of this map. The bottom panel shows a 6-20 cm spectral index map overlaid on the same photograph and scale as the middle panel. This map was made from 6 cm D-Array and 20 cm C-Array maps, comprising essentially identical uv plane coverage. Note that the spectral index distribution is plotted only on the inner region brightness distribution since the 6 cm emission becomes too weak in the outer region for a reliable measurement to be obtained.

The spectral index varies from about -0.5 in the nuclear region to about -1.0 in a plateau within a distance of 1 kpc from the nucleus. The steepest gradient is

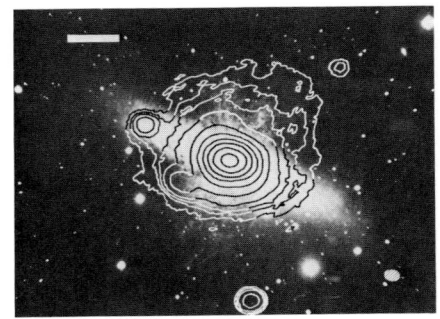

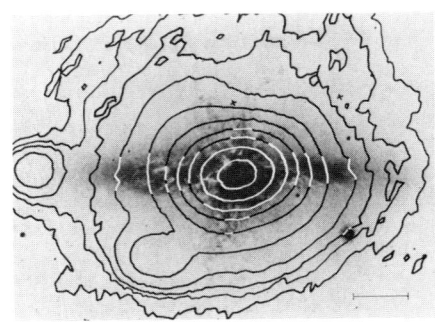

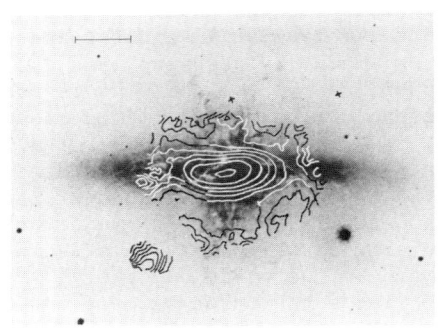

Figure 1. Top: 20 cm C+D configuration VLA map superposed on an optical continuum photograph of M82 from Sandage and Miller (1964). The bar at the upper left is 2 arcmin. Contours are: 1.0 mJy x (1.3,1.9,2.5,4,10, 30,100,300,1000,3000). Middle: The same map superposed on an H alpha photograph from Lynds and Sandage (1963). The bar at the lower right is 1 arcmin. Bottom: 6-20 cm spectral index map prepared from maps with identical uv plane coverage superposed on the Lynds and Sandage photograph with same scale as above (middle). Contours are: -1.3, -1.2, -1.1, -1.0, -0.9, -0.8, -0.7, -0.6, -0.5. The maximum contour at the nucleus is -0.5.

roughly parallel to the minor axis of M82. Similar variations in spectral index were found between 49 and 21 cm (Westerbork data) and 90 and 20 cm (VLA data). These maps are not shown. An important point here is that this spectral index gradient cannot be caused primarily by different admixtures of nonthermal and thermal emission. The amount of thermal emission in the disk is too small (cf Seaquist, Bell, and Bignell 1985; Ho, Beck, and Turner 1990). Therefore the variation must be predominantly caused by a spatial variation in the nonthermal spectral index, which in turn implies a variation in the energy spectral index of the relativistic electrons. Such a gradient would be anticipated in the presence of the M82 outflow because of Inverse Compton (IC) scattering by these electrons off the IR photons from the starburst region in the nucleus. In fact the losses by this mechanism would be expected to exceed those due to synchrotron losses which would produce a similar steepening (eg Rieke et al. 1980).

3. AN ILLUSTRATIVE MODEL FOR THE SPECTRAL INDEX VARIATION

We consider a simple model to account for the variation in spectral index with radius along the minor axis. The following assumptions are made:

(1) Relativistic electrons are transported by convection alone.
(2) The electrons lose energy by IC losses and adiabatic expansion.
(3) The radiation field and wind density follow an inverse square law. The IR luminosity is 1.82×10^{44} erg s^{-1}.
(4) The magnetic field follows an inverse linear law (similar to the behaviour of the equipartition field).
(5) No particle re-energization occurs in the wind.

Figures 2(a) and 2(b) show the results of the model calculations for two forms of the inverse square law. The form in Figure 2(b) is a softened version of the inverse square law (quantities $\propto (r+a)^{-2}$ with a = 200 pc. This form was used as a more realistic representation near the origin (the nucleus) where the sources are extended. Comparison with observation yields two noteworthy points.

The first is that the model does not predict the plateau in the spectral index at a value of -1.0. Possible implications of this are that either particle re-energization in the wind is occuring or that particle diffusion plays an important role in the transport mechanism. The spectral index behaviour is not unlike that expected for the 1-D models of Lerche and Schlickeiser (1980) incorporating both convection and diffusion. The second is that a comparison between the initial decline in spectral index with the model curves suggests outflow speeds in the range of 1000-3000 km s^{-1} are plausible. Note that such speeds are significantly higher than that inferred from optical data.

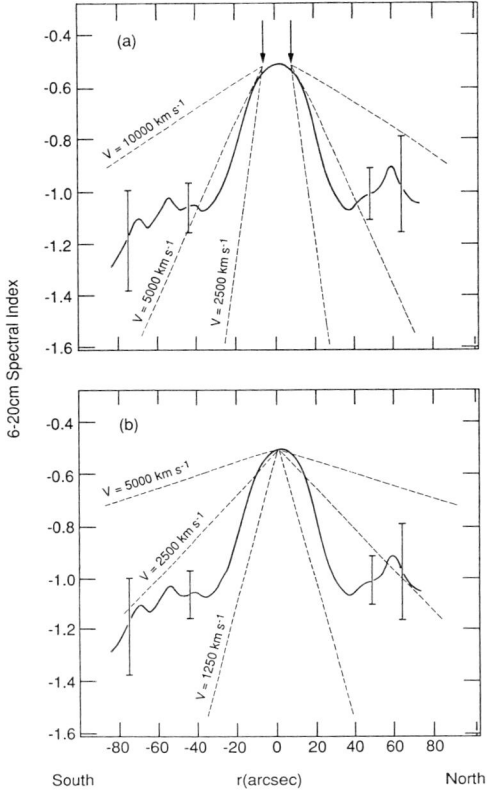

Figure 2. Profile of the 6-20 cm spectral index along the minor axis (solid curve) compared with that predicted for the emissivity from the simple outflow models described in the text (dashed curves). The error bars shown are +/- 0.1 and +/- 0.2 and are due to noise in the maps. Errors at lower radii are essentially negligible. The top panel (a) is for inverse square laws for the densities of the wind and radiation field energy, and an inverse linear law for the magnetic field. Arrows at $r_0 = 100$ pc mark the assumed particle injection site. The bottom panel (b) is identical except for the addition of a "softening parameter" a = 200 pc to the radius r in the above laws. This gives a more realistic representation of the dependences at small r, and the particle injection site is r = 0.

4. MODEL INDEPENDENT CONCLUSIONS

There are some conclusions that we have made that are not based on the above simple model. By assuming equipartition conditions in the wind, we can estimate the number of relativistic electrons available to produce IC scattering. We conclude that only a few percent of the X-ray emission in the halo of M82 could be produced by IC scattering. Therefore, most of the X-ray emission must be produced by hot gas. Secondly, the ratio of the energy in the form of relativistic particles to that in the form of thermal and kinetic energy in the wind is about 2 percent. This is consistent with the corresponding ratio in supernova remnants and is therefore consistent with a supernova origin for the wind and associated relativistic particles.

Finally, we return to the question of the asymmetry in the brightness distribution. Possible explanations include an asymmetry in the distribution of supernovae with respect to the disk of M82, or a termination shock in the outflow produced by a southward movement of the galaxy with respect to the intergalactic medium (IGM). The latter explanation is particularly appealing because of the sharp gradient in brightness at the southern edge. Gottesman and Weliachew (1977) mapped the region in HI and showed indeed that there is an HI envelope surrounding both M81 and M82. The ambient density in this region is plausibly 2×10^{-3} cm^{-3} obtained by inspecting their column density map. This density would be sufficient to produce a termination shock in the observed region if the IGM at this density were flowing northward relative to M82 by about 100 km s^{-1}.

5. CONCLUSION

Our conclusions may be summarized as follows:

(a) The outflow in M82 is visible in synchrotron radiation.
(b) IC losses are evident in the steepening of the radio spectrum, though this process is not a major contributor to the observed halo X-ray emission.
(c) Particle acceleration and/or diffusion in the wind may be important.
(d) Outflow speeds of 1000-3000 km s^{-1} are inferred from the spectral index profiles, which significantly exceeds the optically determined values.
(e) The fraction of the wind energy in the form of relativistic particles is consistent with a supernova origin.
(f) A termination shock produced by IGM may be shaping the observed halo of M82.

ACKNOWLEDGMENTS

This research was supported by an operating grant from the Natural Sciences and Engineering Research Council of Canada.

REFERENCES

Bland, J. and Tully, R.B. (1988) Nature 334, 43
Fabbiano, G. (1988) Ap. J. 330, 672
Gottesman, S.T. and Weliachew, L. (1977) Ap. J. 211, 47
Ho, Paul T.P., Beck, Sara C., and Turner, Jean L. (1990) Ap. J. 349, 57
Kronberg, P.P., Biermann, P., and Schwab, F.R. (1985) Ap. J. 291, 693
Lerche, I. and Schlickeiser, R. (1980) Ap. J. 239, 1089
Lynds, C.R. and Sandage, A.R. (1963) Ap. J. 137, 1005
McCarthy, P.J., Heckman, T., and van Breugel, W. (1987) A.J. 93, 264
Nakai, N., Hayashi, M., Handa, T., Sofue, Y., Hasegawa, T., and Sasaki, M. (1987) Pub. Ast. Soc. Japan 39, 685
Rieke, G.H., Lebofsky, M.J., Thompson, R.I., Low, F.J., and Tokunaga, A.T. (1980) Ap. J. 238, 24
Sandage, A.R. and Miller, W.C. (1964) Science 144, 405
Seaquist, E.R., Bell, M.B., and Bignell, R.C. (1985) Ap. J. 294, 546

NGC 891: A SUMMARY OF OBSERVATIONS

R. J. ALLEN
Space Telescope Science Institute
3700 San Martin Drive, Baltimore, MD 21218

S. SUKUMAR
Astronomy Department and
National Center for Supercomputing Applications
University of Illinois, Urbana IL 61801

ABSTRACT. Three questions are posed, the answers to which are relevant to our understanding of the physical processes which shape the radio continuum morphology of normal spiral galaxies. Observations of the edge-on galaxy NGC 891 have been made for many years by several groups with the intention of contributing at least partial answers to these questions. We review here the work which we have recently done on this subject.

1. INTRODUCTION

What can we learn about the subject of this symposium, "The Interstellar Disk-Halo Connection in Galaxies", from detailed observations of the radio continuum emission from a galaxy seen edge-on? We may break this (too) general question down into three specific questions:

1. Is there any relationship between the radio continuum surface brightness and the optical emission at high Z?

2. How does the spectrum of the radio continuum emission vary with Z?

3. What is the orientation of the magnetic field at high Z?

Before we can understand why the answers to these three questions may be interesting, we have to review some facts about the nature of the nonthermal radio continuum emission from the interstellar medium (cf. for example Ruzmaikin, Shukurov, and Sokoloff 1988):

1. The synchrotron volume emissivity η at radio frequency ν for a power-law distribution of relativistic elecrons $N(E) = N_0 E^{-\gamma}$ in a magnetic field H is $\eta \sim N_0 H_\perp^{1-\alpha} \nu^\alpha$ where $\alpha = (1-\gamma)/2$ is typically in the range -0.6 to -1.3 or so. The

electron energy loss rate is greater at higher energies, leading in general to a progressive steepening of the spectrum with age unless an efficient re-acceleration mechanism exists.

2. Synchrotron radio emission is highly (linearly) polarized, the degree of polarization $p_0 = (3\gamma + 3)/(3\gamma + 7) \approx 0.75$ for $\alpha \sim -1$ (i.e. $\gamma \sim 3$), and the direction of maximum intensity is perpendicular to the magnetic field.

3. The polarization observed at a linear size scale S where L > S > d is reduced if only part of the total field $\mathbf{H} = \mathbf{B} + \mathbf{b}$ is uniform (B) on scales of "L" and the rest random (b) on scales of "d", according to $p = p_0(B_\perp/H_\perp)^2$.

4. An unresolved, clumpy foreground "Faraday Screen" (or depolarizing material mixed in with the source) leads to a strong dependence of the degree of polarization p on wavelength. Typical model geometries lead to $p \sim \exp(-2R^2\lambda^4 L/d)$, with the rotation measure $R = 0.8 \times \int_l n_e H_\parallel dl$, for n_e in cm^{-3}, H in μG, l in pc, and λ in m. There is very little change in the direction of maximum polarization with λ at longer wavelengths in this situation (see e.g. Laing 1984, Cioffi and Jones 1980).

2. THE RADIO AND OPTICAL CONTINUUM EMISSION FROM NGC 891

The first high-resolution radio continuum observations of NGC 891 (Allen et al. 1978) made with the Westerbork Synthesis Telescope (WSRT) clearly showed the existence of two main components to the Z-distribution of the emission: a thin disk, which was barely resolved in the 7.2″ beam at 6 cm, and a thick disk with a FWHM of 76″ or 2.7 kpc (1′ = 2.1 kpc at an assumed distance of 7.2 Mpc). Figure 1 shows these features. Of the two components, the thin disk is the more extended along the major axis (see also Figure 1 in Sancisi and Allen 1979), and corresponds roughly with the main dust lane there.

The thick disk is quite "fat", and at first glance did not seem to correspond well with the optical light at high Z. However, prints of galaxy images made from telescope plates usually accentuate the discrete features and do not accurately reproduce the faint extended emission, so that a quantitative comparison is necessary to settle the matter. This was first carried out by Hu et al. (1987), using calibrated surface photometry of NGC 891 made at optical wavelegths by Van der Kruit and Searle (1981) and 20 cm continuum data obtained from unpublished WSRT HI synthesis observations by Sancisi. Owing to the limited radio sensitivity, Hu et al. had to average the data in strips parallel to the major axis. A correlation was found of the form $I^{U'}_{opt} \sim (I^{21}_{rad})^{1.05}$ over the range of from 7 to 35 mJy/arcmin^2. The optical U′ band covers the wavelength range 340-425 nanometers. This relation between the radio and optical emission is reminiscent of the result found for the total continuum power from normal galaxies of $P^B_{opt} \sim (P^{21}_{rad})^{1.0\pm0.2}$ in the survey by Hummel (1981).

More recently, a further improvement in the radio sensitivity has permitted a test of this correlation over a wider range of surface brightness (Allen et al. 1990).

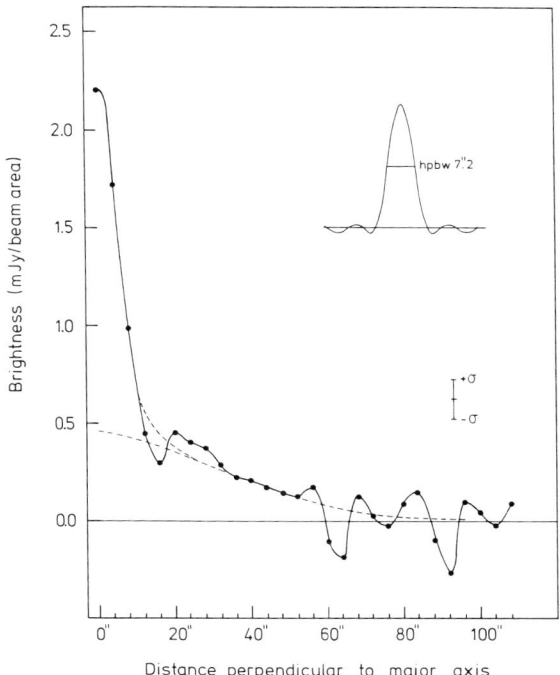

Figure 1. Distribution of radio surface brightness perpendicular to the major axis of NGC 891 at 6 cm obtained by averaging in strips 4' long parallel to the major axis and 7.2" wide. From Allen et al. 1978.

Star images were first cleared from a strip in the optical image 200" × 24' centered on the nucleus and stretching along the minor axis, and the zero level adjusted to the level at the extremities of this strip. The radio data required removal of a few discrete sources and a similar base-level re-adjustment. The data were smoothed to a common resolution (16" × 12" at PA = 0°) and registered to a common grid. The radio image could then be compared pixel-for-pixel with the optical data, at least down to 1 mJy/beam (excluding the optically-obscured central strip), and by averaging in 200" strips parallel to the major axis the brightness limit could be lowered further to 20 microJansky/beam. The final result is shown in Figure 2. The data show a correlation which now extends over nearly three orders of magnitude in surface brightness and is of the form $I_{opt}^F \sim (I_{rad}^{21})^{1.2}$, covering a range in Z from 12" to 140" (420 pc to 4.9 kpc). The optical photometric F band used for this comparison covers the wavelength range 580-690 nanometers.

We can now answer the first question from the Introduction in the affirmative. In the thick disk at high Z above the optical obscuration of the plane of NGC 891, the radio surface brightness can be predicted from the optical surface brightness to within about a factor of 2. The result is surprising, given the complicated

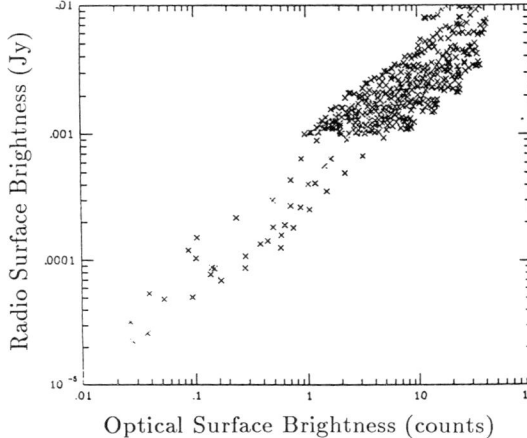

Figure 2. Correlation between the optical F band and the 21-cm radio continuum surface brightnesses in NGC 891. From Allen et al. 1990.

dependence of η on N(E) and H as recalled in the Introduction above. At the present time, the "standard theory" for the generation of the radio synchrotron emission does not couple η with the local density of starlight. Here is a challenge to the theoreticians.

Note that several other face-on galaxies (e.g. M51, Tilanus et al. 1988; NGC 6946, Van der Kruit et al. 1977) also show some degree of correlation between the nonthermal radio and the optical brightnesses, although the ratio I_{opt}/I_{rad} appears to vary a lot from one galaxy to another. The data for these other galaxies have also not been analyzed on a detailed pixel-by-pixel basis as has been done for NGC 891. Here is a challenge to the observers.

3. THE VARIATION OF SPECTRAL INDEX WITH Z

Using newer WSRT data at 6 cm and the 21 cm continuum data from Sancisi's HI synthesis, Allen and Hu (1985) showed that the two components identified earlier had distinctly different spectral indexes, and that there was no convincing case for a change of those spectral indexes with Z. Figure 3 shows the results of fitting the two-component model (right panel) to the data (left panel) averaged in strips parallel to the major axis. The model has the thin disk extending from 0 <Z< 15" (0 <Z< 500 pc) with $\alpha \approx -0.5$ and the thick disk dominating in the range 15" <Z< 2' (0.5 <Z< 4.2 kpc) with $\alpha \approx -1$. There is some indication in the data that the spectral index of the thick disk steepens at still higher Z, but it is very difficult to be sure of this since the infamous "short spacing problem" in radio synthesis data can easily produce an artificial steepening.

Quite recently, Hummel et al. (1990) have presented new VLA observations of NGC 891 which support the simple two-component model of Allen and Hu. Figure 4 shows their VLA results and, although their resolution (40") is not as good as that obtained by Allen and Hu (20"), the separation into two components is clearly

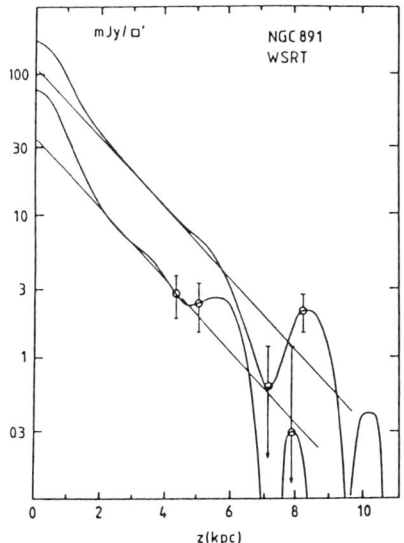 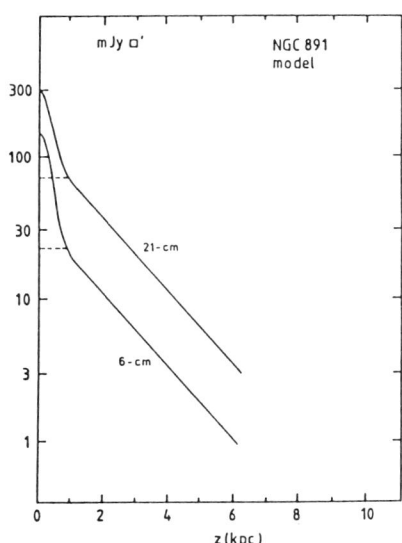

Figure 3. Z-distribution of surface brightness in NGC 891 at 6 and 21 cm observed with a 20″ beam (left panel), and the two-component model (right panel) with a thin and a thick disk which fits the data. From Allen and Hu 1985.

a very good approximation. The possible steepening at high Z is also seen, but the same caveats about the "missing short spacings" apply here as well.

Parenthetically, we would like to draw attention to two practical limitations which have adversely affected some observations, including our own. For the first example the earlier results of Allen et al. (1978) on the Z-variation of spectral index (cf. their Figure 6) gave an erroneous impression that the spectral index was steepening smoothly with increasing Z; this is a result of insufficient angular resolution. Secondly, even if the resolution is sufficient, the "short spacing problem" can also produce a spurious steepening of the spectral index above Z of about 45″ (cf. their Figure 5).

The second question from the Introduction is, therefore, to be answered in the negative. There is no convincing evidence that the spectral index of the thick disk steepens with increasing Z, at least up to 2′ (4.2 kpc) above the plane. There is a hint that the spectrum of the thick disk may steepen at still higher Z, but every time we look harder for this feature it recedes to greater Z distances. The flattening of the spectral index below Z of 15″ is a consequence of the increasing importance of the thin disk.

Note that some face-on galaxies also show little variation of their radio spectral indexes with radius (M51, Van der Kruit 1977; NGC 6946, Van der Kruit et al. 1977). The edge-on case is, however, especially tricky to explain in the "classical"

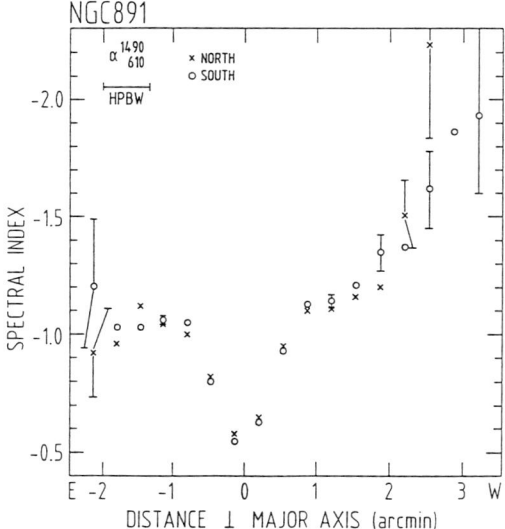

Figure 4. Radio continuum spectral index along the minor axis of NGC 891, from Hummel et al. 1990.

model, where the cosmic ray electrons are produced and accelerated in the plane and diffuse out of the plane losing energy along the way. One alternative is to postulate some mechanism for re-accelerating the electrons, but this mechanism would require some fine tuning in order to keep the spectral index substantially constant as the electrons diffuse over many kiloparsecs in Z.

4. THE ORIENTATION OF THE MAGNETIC FIELD AT HIGH Z

Hummel and Dahlem (1990) have recently reviewed the available results concerning the direction of the magnetic field in NGC 891 and NGC 4631 at high Z. In the case of NGC 4631, there is a modest amount of polarization (about 10%) observed at 20 cm wavelength in the thick radio disk at $Z \sim 3'$ (4.5 kpc at a distance of 5.2 Mpc) and a clear indication that the magnetic field is oriented perpendicular to the plane, which the authors take as supporting the galactic wind model for the transport of cosmic rays out of the plane. In NGC 891 the situation as reported by Hummel and Dahlem is less clear, with less large-scale order apparent and no great degree of uniformity in the field direction over the thick disk.

At 6 cm any effects of Faraday rotation and depolarization will be considerably reduced (cf. point 4 in the Introduction), so we may expect that the observed degree of polarization would be greater and the position angle of maximum polarization would be a more accurate indicator of the true direction of the magnetic field. Figure 5 shows the distribution of total intensity in NGC 891 which we have recently observed with the VLA at 6 cm. The brightest peak in the total intensity contours

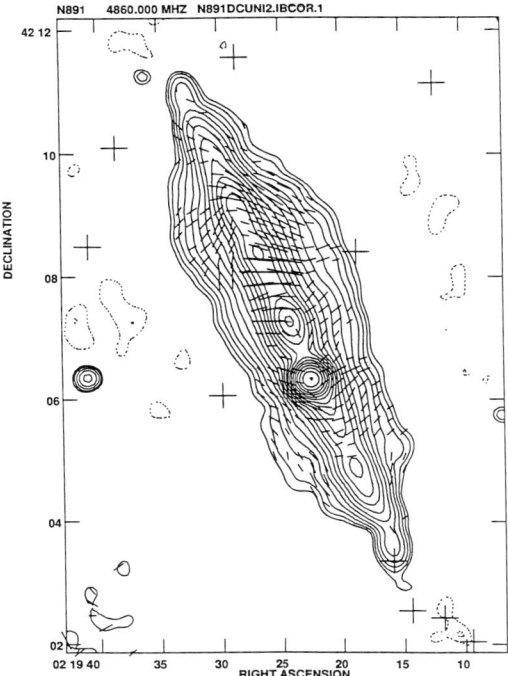

Figure 5. Contours of the total radio continuum emission at 6 cm from NGC 891, starting at 150 microJanskys beam^{-1} and increasing by $2^{N/2}$. The line segments have lengths proportional to the linearly polarized surface brightness at their centers (1″ on the plot is 8.33 microJanskys beam^{-1}), and directions indicating the position angle of maximum polarization. The restoring beam width is 20″ (FWHM, circular), and the r.m.s. noise is about 25 microJanskys beam^{-1}.

is SN 1986J, located about 1′ southwest of the nucleus. The polarization, indicated by the short line segments, reaches more than 30% in the thick disk. The direction of the magnetic field, which is perpendicular to the line segments, shows a large-scale symmetry not visible on the 20-cm map of Hummel and Dahlem, with the field largely parallel to the plane in the central regions of the galaxy and an indication that the lines of force "flare" upwards out of the disk at larger radial distances.

A comparison of our 6 cm data with the 20 cm data shown by Hummel and Dahlem reveals surprisingly little rotation of the direction of maximum polarization with wavelength. From this, and the larger degree of polarization at 6 cm, we conclude that the thick disk of NGC 891 is substantially "Faraday thick" at 20 cm, becoming progressively thin as we move to 6 cm and shorter wavelengths. By way of illustration, we can do the computation for a Z height of 30″ (1 kpc). At this height, the electron density from the observations of Rand et al. (1990) and Dettmar et al. (1990) is about 0.01 - 0.03 cm^{-3}. For L = 10 kpc, d = 100 pc, H

= 2 μG and B = b = $\sqrt{2}\mu$G we find from the equations in the Introduction that R(regular) = 200 rad m^{-2} and R(random) = 2 rad m^{-2}. The intrinsic degree of polarization p $\approx$ 40%, is hardly affected at 6 cm, but is reduced to 10% at 20 cm.

So we have the answer to the third question posed in the Introduction: The magnetic field in the thick disk of NGC 891 is oriented largely parallel to the plane in the central regions of the galaxy, possibly curling upwards out of the plane at larger radii. The morphology of the polarization is strongly affected by Faraday depolarization, which can explain not only the increase of p with Z but also the great differences in the appearance of the polarized emission at 6 and 20 cm. A more detailed description of these results and a discussion of the implications is in preparation.

ACKNOWLEDGMENTS

We are grateful to Rainer Beck for sharing his unpublished results with us, and to Ko Hummel for helpful discussions.

REFERENCES

Allen, R. J., Baldwin, J. E., and Sancisi, R. (1978), *Astron. Astrophys.*, **62**, 397-409.
Allen, R. J., Sukumar, S., Hu, F. X., and Van der Kruit, P. C. (1990), in *Galactic and Intergalactic Magnetic Fields*, ed. R. Beck, P. P. Kronberg, and R. Wielebinski (Kluwer, Dordrecht), 223-224.
Allen, R. J., and Hu, F. X. (1985), in *New Aspects of Galaxy Photometry*, ed. J.-L. Nieto (Springer, New York), 293-296.
Cioffi, D. F., and Jones, T. W. (1980), *Astron. J.*, **85**, 368-375.
Dettmar, R.-J. (1990), *Astron. Astrophys.*, **232**, L15-L18.
Hu, F. X., Allen, R. J., Van der Kruit, P. C., and You, J. H. (1987), *Astrophys. Space Science*, **135**, 389-392.
Hummel, E. (1981), *Astron. Astrophys.*, **93**, 93-105.
Hummel, E., Dahlem, M. (1990), in *Galactic and Intergalactic Magnetic Fields*, ed. R. Beck, P. P. Kronberg, and R. Wielebinski (Kluwer, Dordrecht), 219-222.
Hummel, E., Dahlem, M., Van der Hulst, J. M., and Sukumar, S. (1990), *Astron. Astrophys.* (submitted).
Laing, R. A. (1984), in *Physics of Energy Transport in Extragalactic Radio Sources*, ed. A. H. Bridle and J. A. Eilek (NRAO, Green Bank) 90-98.
Rand, R. J., Kulkarni, S. R., and Hester, J. J. (1990), *Astrophys. J.*, **352**, L1-L4.
Ruzmaikin, A. A., Shukurov, A. M., and Sokoloff, D. D. (1988), *Magnetic Fields of Galaxies* (Kluwer, Dordrecht).
Sancisi, R., and Allen, R. J. (1979), *Astron. Astrophys.* **74**, 73-84.
Tilanus, R. P. J., Allen, R. J., Van der Hulst, J. M., Crane, P. C., and Kennicutt, R. C. (1988), *Astrophys. J.*, **330**, 667-671.
Van der Kruit, P. C. (1977), *Astron. Astrophys.*, **59**, 359-366.
Van der Kruit, P. C., Allen, R. J., and Rots, A. H. (1977), *Astron. Astrophys.*, **55**, 421-433.
Van der Kruit, P. C., and Searle, L. (1981), *Astron. Astrophys.*, **95**, 116-126.

THE DIFFUSE IONIZED GAS PERPENDICULAR TO THE PLANE OF NGC 891

R.-J. DETTMAR[1], J.W. KEPPEL[2], M.S. ROBERTS[3], J.S. GALLAGHER[4]
(1) Radioastr. Inst. Uni. Bonn, Auf dem Hügel 71, D-5300 Bonn, FRG
(2) 603 Pauley Dr., Prescott, AZ 86301, USA
(3) NRAO, Edgemont Rd., Charlottesville, VA 22901, USA
(4) AURA Inc., 1625 Massachusetts Ave., Washington, DC 20036, USA

ABSTRACT. Hα images and long–slit spectra of NGC 891 show the presence of a thick layer of diffuse ionized gas (DIG) in this edge–on galaxy. Hα emission originating in the DIG of NGC 891 is detected out to several kpc above the midplane with a typical scale height for the electron density $n_e(z)$ of $h_e \sim 600$ pc. The distribution of the gas is very inhomogeneous with a large scale asymmetry and filamentary structure. Individual Hα emitting structures can be traced out to 4–5 kpc above the midplane. The emission line ratios of [NII] and [SII] to Hα vary with height above the disk. We compare these properties with the analogous gas phase in our Galaxy.

The distribution of the DIG is related to star formation in the underlying galactic disk and we conclude that in NGC 891 the interstellar medium is heavily effected by processes related to star formation.

1. INTRODUCTION

Under excellent seeing conditions the edge–on galaxy NGC 891 exhibits a very complex network of filamentary structures in the dust lane. Some of these dust filaments are almost perpendicular to the stellar disk and can be traced out to at least 1.5 kpc above the plane. NGC 891 also possesses a "thick" radio continuum emitting disk and it therefore is a very promising candidate for studies of the so-called "disk–halo–connection".

In an study of the structure and the properties of the diffuse ionized interstellar medium (DIG) perpendicular to the plane of disk galaxies we have obtained Hα images and long slit spectra of NGC 891.

2. OBSERVATIONS

Hα images were obtained with a 2:1 focal reducer at the 42–inch telescope at Lowell Observatory (Dettmar, 1990). Long slit spectra at intermediate resolution were

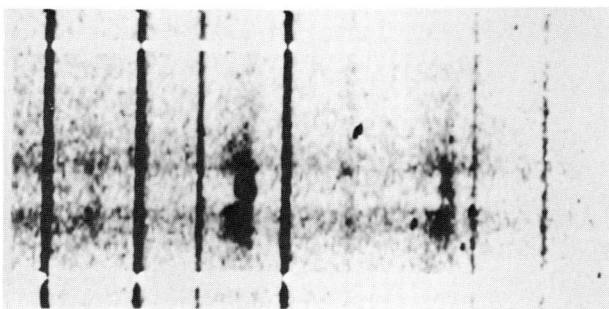

Figure 1. Hα and [NII] emission lines in the CCD–echelle–spectrum taken perpendicular to the plane of NGC 891 (PA 112°) 65″ NE of the nucleus. Night sky lines cover the slit length of 2.25′. The wavelength coverage is λ6540Å-λ6610Å and the binned pixel size 0.24Å×1.51″ resulting in a velocity resolution of ∼23 km sec^{-1}.

collected with the Cassegrain spectrographs at the KPNO 4–m Mayall telescope and the 72–inch Perkins telescope at Lowell Observatory. In addition spectra were obtained with the Echelle-spectrograph at the 4–m (see Fig. 1).

3. RESULTS

The most remarkable property of the Hα emission line in NGC 891 is its extent out of the plane of the galaxy: we are able to measure the Hα line out to more than 30″ (1.4 kpc, we use a distance of D=9.5 Mpc throughout this paper) from the midplane. Individual features can be traced out to ∼4.5 kpc above the plane.

The large z-extent of the ionized gas is confined to the inner half of the optical disk. In this inner region the Hα distribution also shows a filamentary structure of the diffuse ionized medium. These filaments, sticking out of the plane, originate in HII regions in the plane. Some of the filaments are correlated with dust features perpendicular to the plane. Typical linear dimensions for these structures are diameters of 300 pc and they reach out to ≥1 kpc.

The large scale distribution and the filamentary structure of the Hα emission is confirmed by similar imaging observations presented by Rand et al. (1990) but we definitely failed to detect the very extended component with a scale height of several kpc reported by them. As the dust obscures the emission near the midplane we have applied a correction for dust absorption before estimating the scale height of the DIG. A typical exponential scale height for the electron density $n_e(z)$ is $h_e \sim$ 600 pc. But in one outstanding region ∼2 arcmin NE of the nucleus this scale height is $h_e \sim 1$ kpc.

Our long slit spectra confirm the large scale distribution of the Hα–emitting ionized gas but they are not sensitive enough to probe the very extended Hα emitting component found by Rand et al. (1990). The [NII] (Fig. 1) and [SII]

($\lambda\lambda 6816/31$) to Hα emission line ratios vary with height above the plane in the sense that they become larger outside the HII regions of the central plane.

4. DISCUSSION

The scale height and the large scale distribution of the Hα emitting gas in NGC 891 is similar to the DIG observed by Reynolds (1990) in the Galaxy. The filamentary structures seen in the dust and Hα-emission may represent the "chimneys" in supernova–dominated models of the interstellar medium (e.g. Norman and Ikeuchi, 1990). Dust structures perpendicular to plane of the Galaxy, which would be comparable to those observed in NGC 891, have been identified from the *IRAS*–survey by Koo (1990). For the extended component in NGC 891 the Hα flux from the high z–component is $\sim$10% of the Hα flux emitted from the plane, a again ratio very similar to the Galactic value.

The observed changing emission line ratios exclude the possibility that the large scale height of the emission is due to scattering of disk emission by dust high above the plane. The line ratios rather vary in the same sense as they do in the "Reynolds–layer" of the Galaxy, indicating a low-excitation ionization by photons that can escape from the HII regions in the midplane (Mathis, 1986). From a detailed comparison of the line profiles in the high resolution spectra with the observed HI velocity field (Sancisi and Allen,

1979) we can rule out that the high z-extent of the Hα emitting gas is just due to geometrical effects, like warping of the disk. The measured velocities in the Cassegrain spectra and the line profiles of the high resolution spectra also show that the DIG at high z is corotating.

All the observed properties of the DIG in NGC 891 mentioned above are comparable with the analogous gas phase in the Galaxy. The main difference is that the surface density of the DIG in NGC 891 is about twice that of the solar neighbourhood.

The large extent of the DIG out of the plane in the region $\sim 2'$ NE of the nucleus is most probably related to star formation activity in the underlying disk as can be shown by a comparison of star formation tracers with the Hα distribution (Dettmar and Dahlem, 1990). Also the polarized radio continuum intensity as deduced from new VLA observations at 1.49 GHz is surprisingly well correlated with the observed diffuse Hα–emission. Though Faraday depolarization effects are visible towards the midplane this indicates that the magnetic field structure in this region is very regular on scales of a few kiloparsec.

Additional long–slit spectra were taken perpendicular to the planes of several other edge–on galaxies with varying radio continuum properties (NGC 3628, NGC 4244, NGC 4631). This data allow to determine the scale heights of the diffuse ionized gas and show a possible relation of the extent of the DIG with the presence of a radio continuum emitting "thick" disk. The transport of relativistic electrons into the halo and the sources of ionization of the DIG both are related to actively star–forming regions in the underlying galactic disks. Most probably these independent phenomena are coupled by the presence, orientation, and order of magnetic fields.

5. CONCLUSIONS

The inhomogeneous distribution and the filamentary structure of the diffuse ionized gas in NGC 891 are very different from the stratified layers in hydrostatic equilibrium models. This observational evidence rather supports a more dynamical view of the interstellar medium where the energy input from star forming processes strongly affects the gas phase and connects the interstellar medium in the disk to the galactic halo.

ACKNOWLEDGEMENTS

In the course of this work RJD was supported by the Deutsche Forschungsgemeinschaft (De 385/2–1), the Lowell Observatory Research Fund, and the MPIfR.

REFERENCES

Dettmar, R.-J. (1990) *Astron. Astrophys. Lett.* **232**, L15
Dettmar, R.-J., Dahlem, M. (1990) *this conference, Poster Proceedings*, ed. H. Bloemen, Sterrewacht Leiden, p. 41
Koo, B.-C., Heiles, C., Reach, W. T. (1990) *this conference*
Mathis, J. S. (1986) *Astrophys. J.* **301**, 423
Norman, C. A., Ikeuchi, S. (1990) *Astrophys. J* **345**, 372
Rand, R. J., Kulkarni, S. R., Hester, J. J. (1990) *Astrophys. J. Lett.* **352**, L1
Reynolds, R. (1990) in *IAU Symposium No. 139, Galactic and Extragalactic Background Radiation*, eds. S. Bowyer and C. Leinert.
Sancisi, R., Allen, R. J. (1979) *Astron. Astrophys.* **74**, 73

A CO SURVEY OF THE HALO OF NGC891

S.GARCÍA-BURILLO[1,2], M. DAHLEM[3] AND M.GUÉLIN[1]
[1] IRAM, 300 rue de la piscine, 38406 St-Martin-D'Heres
[2] Centro Astronómico de Yebes, Apdo.148 E-19080 Guadalajara,Spain
[3] Max Planck Institut für Radioastronomie,Auf dem Hügel,69 D-6300 Bonn 1

ABSTRACT. We report ^{12}CO J=2-1 line observations of the edge-on galaxy NGC 891, made with the IRAM 30 m telescope. These observations show that the molecular gas probably extends to large distances from the galactic plane.

The distribution of the molecular gas perpendicularly to the plane of spiral galaxies is poorly known. Distance uncertainties plague most Milky Way studies (see however Grabelsky et al. 1987), whereas small angular sizes and low intensities limit those of extragalactic systems.

We have used the IRAM 30 m telescope to study the z-distribution of CO in the edge-on galaxy NGC 891. At 230 GHz, the IRAM 30 m telescope allies a high angular resolution with a high sensitivity to extended sources. We selected NGC 891, because it looks similar to the Milky Way, it is seen perfectly edge-on (incl$\geq$ 89°) and it is not too distant from us (between 8 and 14 Mpc, e.g. Sancisi and Allen 1979, Rand et al. 1990). This galaxy is known to give rise to strong CO emission (Solomon 1983, Sofue et al. 1987) and to possess an ionized halo, visible in radio continuum (Allen et al. 1978) and Hα emission (Rand et al. 1990, Dettmar et al. 1990). The halo of NGC 891 has not been detected in HI (Sancisi and Allen, 1979) and, so far, there was no indication of any out-of-the-plane extension of the neutral gas. Optically, NGC 891 appears as a bright, elongated ellipse, 10' long, 1' wide, divided along its major axis by a narrow dust lane, and oriented almost north-south (P.A.= 23°).

Twenty cuts, perpendicular to the galactic plane (i.e. in the z direction), were observed. They consisted of strip maps of typically 8–10 points spaced by 6" in z. Six of these cuts were made at intervals of 6" along the major axis (x direction), providing a fully sampled map of the central bulge. The other cuts, made every 60" or 30" in x, were aimed at observing further out the z-extent of the gas. The cuts at x=$\pm$108" and $\pm$102" from the centre, were slightly shifted from the regular grid, in order to cover a conspicious dust spur, or "chimney", raising at right angle from the major axis.

Each cut was observed 3–6 times. The individual strip maps were followed by a pointing/focussing session on the nearby radio-galaxy 3C 84, one of the brightest continuum sources of the sky at 230 GHz, then repeated. In this way, integration times of 8–15 min

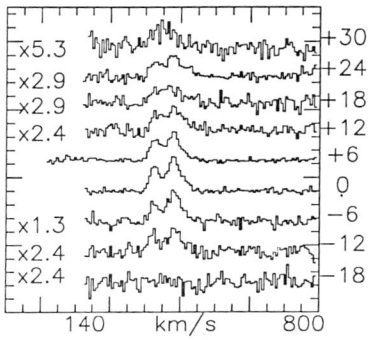 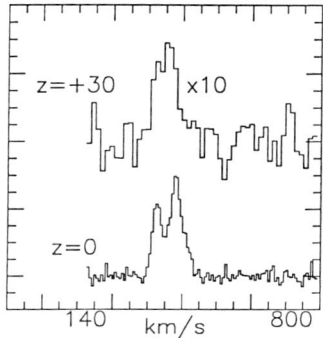

Figure 1.a: Cut perpendicular to the plane of NGC 891, made at +108" to the north of the centre. The z offsets, in arc sec., are relative to the mid-molecular disk. The intensities of the spectra have been multiplied by the factors indicated on the left and are not at the same scale. The velocity resolution is 6.5 kms^{-1}. **b:** The upper (z=+30") spectrum of Fig. 1a, smoothed to a resolution of 13 kms^{-1}, compared to the spectrum at z=0.

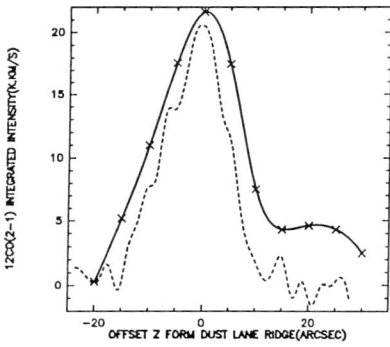

Figure 2. / *full line:* The velocity-integrated intensity distribution perpendicular to the plane of NGC 891 (offsets in z, in arc second, are relative to the mid molecular disk ; they are positive to the southeast, negative to the northwest) / *dashed line:* The intensity distribution across 3C 84 in the same direction (see text).

on-source per point could be achieved, while keeping a good pointing; moreover the shape of the telescope beam could be monitored almost in real time. This proved to be important as the observations were carried out during day time: although the weather was clear and dry and the source elevation high (40° most of the time) the beam pattern was found dirtier than observed routinely during the night.

The cuts at x= +108" and −108" were observed respectively when the source, still high in the sky, was setting down and rising up. Fig. 2 show the CO integrated intensities, observed along the direction perpendicular to the galactic disk, compared to the average of 18 cuts across 3C 84, made in the same direction (elevation) after the individual strips. The beam in this direction shows an asymmetric coma-like extension, or "tail", possibly due to a distortion of the dish caused by sunshine; it is somewhat broadened (FWHP=14") by atmospheric turbulence (Altenhoff et al. 1987) and/or by changes in focal length caused by thermal gradients. In the orthogonal direction (azimuth), the beam is symmetrical and cleaner (FWHP= 13"). The broadening of the telescope beam, observed on 3C 84, has been taken into account in the data analysis. In the case of the cuts of Fig. 2, the elevation "tail", located below the main beam, tends to enhance the signals to the east of the galactic disk. It could explain the broadening of the CO disk observed at negative zs, but not the extended emission observed to the west at positive zs.

The relatively high integrated intensity observed at z= 30" argue against an error-beam contamination. The 30 m telescope error-beam response to an uniformly bright, 6" thick, CO ridge (such as that suggested by our in-plane observations and the recent interferometric measurements of Handa et al. 1990), should be much wider in frequency and a factor of 5 weaker than observed on Fig. 2. Hence, although additional (e.g. nightime) observations would be needed to determine the exact distribution of CO in the z direction, we can probably conclude that CO extends far outside the bright CO ridge.

The results of the other cuts supports the picture emerging from the discussion above. A bright and narrow ridge of CO emission (6–7" wide, after beam deconvolution), follows closely the dust lane which marks the near edge of NGC 891's disk. It is relayed to the west by a weak CO extension, or "plateau".

The coincidence of the CO ridge with the dust lane is remarkable and supports the image of a thin gaseous disk, seen almost perfectly edge-on. Both CO and the dust exhibit a slight warp, "raising" westwards, 3' to the south of the centre, and "falling back" further out (i.e. at x$\simeq$ $-4'$). At places, the ridge emission shows a complex velocity structure (double or triple-peaked spectra), partly arising from molecular arms or rings.

Table 1. Out of the plane gas detections

Strip positions along major axis(arcsec)	Out of the plane detections (arcsec)	Indicative distance from the plane(kpc)
+102"	+24"	0.8 kpc
−102"	+27"	1 kpc
+108"	+30"	1.2 kpc
−108"	≥24"	≥0.8 kpc
+150"	+27"	1 kpc
−150"	+24"	0.8 kpc
+210"	+18"	0.5 kpc
−210"	+12"	−
centre	+27"	1 kpc

Table 1. limits in **z** of the "plateau" component (detections at $> 5\sigma$ in the CO J=2-1 line)

The data relative to the extended CO "plateau" is summarized in Table 1, which gives the limits in the z-direction of the detections at $> 5\sigma$. CO is detected at z$\geq$ +24" (to the west of the galactic plane) on all inner ($|x|\leq 150''$) high sensitivity cuts. On several cuts, CO is observed 30" "above" the mid-ridge. In the case of Fig. 2, this corresponds, after beam deconvolution, to a projected distance in the plane of the sky of $\geq$ 1 kpc. Taking into account the shape of the telescope beam, CO is marginally or not at all detected at negative zs.

The question raises whether the CO "plateau" component comes from a halo, or from the edges of the molecular disk, which could be warped, flaring, or simply less inclined than indicated by the optical image. As can be seen on Fig. 1b, the profile at (108,30) has a mean velocity (first moment) slightly lower than that of the (108,0) disk profile (350 vs 370 kms^{-1}). It should have had a larger velocity (i.e. much closer to the systemic velocity of 510 kms^{-1}) if it were arising near the edges of the molecular disk. In fact, assuming axial symmetry and using the rotation curve and the CO distribution observed along the major axis, we calculate that the gas at the edges of the disk has a velocity of 470 kms^{-1}. The plateau component appears thus to follow the motion of the gas on the major axis and should lie some 1 kpc above the molecular disk.

The average integrated intensity in the (2–1) line, at z= +30", is 3 K.kms^{-1}, a factor 7–8 times lower than the intensity at z= 0. The corresponding ratio of the H$_2$ column densities could be somewhat larger, judging from the low value of the CO to H$_2$ conversion factors derived for the galactic high latitude clouds.

Besides the z-distribution of CO, our observations allow to study the kinematics and in-plane distribution of the molecular gas, as well as, from (2–1)/(1–0) and ^{12}CO/^{13}CO line intensity ratios, the gas excitation at selected positions (Garcia-Burillo et al. in preparation). The ^{12}CO/^{13}CO ratio is found to vary within the line profile, a behaviour which can be explained by a higher value of this ratio in the interarm regions. A similar effect is observed in M51 (Garcia-Burillo,S. and Guélin,M. 1990).

ACKNOWLEDGEMENTS:S.G.B. thanks the Spanish CICYT (project PB88-0453) for the financial support during part of this work

REFERENCES:
Allen, R.J., Baldwin, J.E. and Sancisi, R. (1978) *AaA* **62**, 397
Altenhoff,Baars,Downes,Wink (1987) *AaA* **184**, 381
Dettmar R.-J. (1990) *AaA* **232** ,L15-L18
Garcia-Burillo,S.,Dahlem,M.,Guélin,M. and Cernicharo,J. (1990) *in prep*
Garcia-Burillo,S. and Guélin,M. (1990) *proceedings IAU symposium No. 146*
Grabelsky, D.A., Cohen, R.S., Bronfman, L., Thaddeus, P. and May, J. (1987) *Ap.J.* **315** ,122.
Handa,T.,Kawabe,R.,Ishizuki,S.,Ikeuchi,S. and Sofue,Y. (1990) *poster proceedings IAU symposium N 144*
Rand,R.J. and Kulkarni S.R. (1990) *Ap.J.* **352** ,L1-L4
Sancisi,R. and Allen,R.J. (1979) *AaA* **74** ,73-84
Sofue,Y.,Nakai,N., and Handa,T. (1987) *Pub. Astr. Soc. Japan* **39**, 47
Solomon,P.M. (1983) *proceedings IAU Symp No. 100* , p.35

COOL IONIZED GAS IN GALAXY THICK DISKS

F.P. Israel
Sterrewacht, Postbus 9513
2300 RA, Leiden, The Netherlands

ABSTRACT. Spiral galaxies whose radio continuum emission is dominated by their disks show a flattening of the radio continuum spectrum at frequencies well below 1 GHz. The effect appears to be stronger for edge-on than for face-on galaxies The most feasible explanation appears to be that the flattening reflects free-free absorption of nonthermal emission by a very low temperature ionized gas. This gas is probably highly clumped, and must be well-mixed with the nonthermally emitting plasma.

1. Radio Continuum Spectra.

For a long time, reliable radio continuum data at frequencies below 400 MHz, or indeed below 1 GHz have been lacking for most normal galaxies. This spectral region is now becoming accessible. With the presently defunct Clark Lake Radio Observatory 68 galaxies were detected at 57.5 MHz (Israel and Maloney, 1990), while the Cambridge array has been used to observe galaxies at 151 (6C) and 38 MHz. A priori, one might expect radio continuum spectra to steepen at lower frequencies: in composite spectra such as those of galaxies, the steepest spectral component dominates at the lowest frequencies. Thus, thermal emission becomes negligible below 1 GHz, whereas steep-spectrum haloes should become dominant. In fact, the opposite is observed: below a few hundred MHz, the continuum spectra of many, but not all galaxies flatten noticeably. This was already noticed by others, see for instance Slee (1972) and Lerche and Schlickeiser (1982). Few, if any, galaxies exhibit a steepening.

An accurate determination of galaxy radio continuum spectra is not easy. In principle, single dish observations as well as interferometer (aperture synthesis) observations should yield reliable flux-densities. However, at high frequencies (> 5 GHz) galaxy angular extent and low surface brightness tend to conspire to yield integrated flux-densities that underestimate the true values. At low frequencies, especially single dishes have large beamsizes that may lead to flux-densities higher than actual by the inadvertent inclusion of unrelated strong background sources. These problems are not insurmountable, but in using data from the published literature, great care must be taken to avoid unreliable or erroneous determinations that unfortunately abound.

2. The CLRO Survey.

In 42 fields, 133 galaxies were surveyed, yielding 68 detections. The majority of nondetections refers to galaxies weak at any frequency. Slightly over half the detected galaxies have also reasonable flux-determinations in the GHz range. Eighteen of these are spiral galaxies without significant central emission at least at 1.4 GHz. For all galaxies, the high-frequency spectrum was extrapolated to the observing frequency of 57.5 MHz and compared with the actually measured value. As usually 1.4 and 5 GHz measurements are the most reliable, these measurements were used to define the high-frequency spectrum. On average, only about two-thirds of the extrapolated flux was detected, with little difference between disk-dominated galaxies and those with significant central emission. This is an upper limit, because the 1 - 5 GHz spectrum was extrapolated without taking into account possible, indeed probable, steepening below 1 GHz due to the decreasing influence of flat-spectrum thermal components and the increasing influence of steep-spectrum nonthermal components.

A plot of the ratio of observed over extrapolated 57.5 MHz flux-densities versus optical axial ratio (a measure for galaxy inclination) revealed for the disk-dominated galaxies a correlation between flux discrepancy and inclination not shared by galaxies with significant (1.4 GHz) central emission. Although determinations for individual galaxies have large possible errors, the relation for the disk-dominated sample appears well-established, in the sense that edge-on galaxies appear to emit a smaller fraction of the expected flux density than face-on galaxies.

For the Local Group galaxy M33 we determined not only the 57.5 MHz flux density, but also fluxes at four other frequencies between 20 and 75 MHz, as well as new flux densities at 151, 327 and 610 MHz (Israel, Maloney & Howarth, 1990, submitted to A&A). M33 is an extended galaxy of rather low surface brightness, so that a reliable spectrum is hard to construct (cf. section 1). The Bonn telescope single dish maps by Buczilowski (1988) in the range 840 to 4750 MHz define a steep high-frequency spectrum with a spectral index α = -0.9 and a spectral break to α = -0.1 around 800 MHz. A less extreme case may be argued in which the high-frequency spectral index is of order $\alpha = -0.6$ and a spectral break to $\alpha = -0.1$ is introduced at 300 MHz. The presence of a spectral break is unmistakeable, and the magnitude of the break (range $\alpha = 0.8$ to 0.5) as well as its location (range 300 - 800 MHz) can be determined by verifying published integrated flux-densities at specifically 600 MHz, or 5 GHz and higher (cf. Israel, Mahoney and Howarth, 1990).

3. Galactic Winds versus Free-free Absorption.

In principle, breaks in galactic radio continuum spectra can be caused by a variety of mechanisms. In practice, limited magnetic field strengths and radio surface brightness of galactic disks reduce the number of possible explanations. If galaxies produce winds perpendicular to the plane, a spectral break of magnitude $\alpha = 0.5$ occurs at a critical frequency below 1 GHz if the convection-deceleration 'halo' model applies; no such break occurs in static 'halo' models (Lerche and Schlickeiser, 1982). Thus, if M33 indeed has a relatively moderate spectral index of about $\alpha = -0.6$, the low frequency data are consistent with a galactic wind and a convection-decelaration halo. However, if the Bonn data are correct, the change in spectral index ($\alpha = 0.8$) appears to large for such an explanation.

Also, the apparent relation between inclination and the ratio of observed to expected flux density of disk-dominated galaxies is not a feature predicted by the galactic wind model. An intriguing alternative explantion is that at low frequencies, the nonthermal emission is absorbed by a thermal plasma (Israel and Maloney, 1990). The observations, in conjunction with known properties of spiral galaxies then place limits on the physical condition of both the thermal and nonthermal components. Efficient free-free absorption at the observed frequencies can only take place if the absorbing gas has a low electron temperature. Absorption by normal HII regions can be ruled out completely. If the thermal absorber is an extended, diffuse gas, constraints on the available ionizing power and on the Hα brightness of the gas lead to electron temperatures of less than a few hundred K, and probably 100 K or less.

If the absorbing thermal gas is very clumpy, electron temperatures may be as high as 500 - 1500 K without exceeding the constraints. Clump volume filling factors should be of order 10^{-3}, and clump densities should be in the range 0.1 - 2.0 cm^{-3}. Emission from such a cold ionized gas at higher frequencies is negligible. Both the small observed to expected flux density ratio and the shape of the M33 low frequency spectrum imply that the absorbing thermal gas must be well-mixed with the nonthermal emitting plasma. This condition is naturally met if the interstellar medium is characterized by a range of clump sizes and densities maintained for instance by compressing shock waves.

It is important to note that in the free-free absorption interpretation the absorbing **must** fill a large fraction (> 60%) of nonthermally emitting plasma, and the absorbing gas **must** have a low electron temperature, independent of the actual detailed geometry.

4. Comparison with the Galaxy.

At present, there is no clearcut Galactic evidence for a similar low electron temperature component in the Galaxy. The lowest electron temperatures inferred for HII regions in the inner Galaxy are of order 3000 K. In a few directions, the presence of a diffuse, local ionized gas with 500 K < T_e < 3000 K has been inferred from H166α recombination lines (Lockman, 1980), but generally it appears that a cold ionized component is not present in the Solar Neighbourhood. An extended diffuse component appears to be absent in the plane of the Galaxy, but a highly clumped, cool ionized component would not contradict the low frequency SNR absorption measurements by Dulk and Slee (1975). It would be interesting to determine the spectral index distribution of the Galaxy in the 50 - 400 MHz range, in particular in the anticentre, centre and polar directions. It should be noted that even if the cool ionized gas is absent in the thin disk (Galactic plane), it still can be a major component of the thick disk, which at 408 MHz contributes 90% of the total flux density of the Galaxy (see Beuermann et al, 1985).

The inclination-dependence of the observed to expected flux ratio, and the magnitude of the spectral break in the M33 spectrum favour the free-free absorption interpretation, and argue against the convection-deceleration halo interpretation. If these arguments were to disappear, both explanations would be a priori equally probable, and independent evidence for or against either galactic winds or cool ionized gas would have to decide the issue.

References.

Beuermann, K., Kanbach, G., Berkhuysen, E.M.: 1985 Astr. Ap. **153**, 17
Dulk, G.A., Slee, O.B.: 1975 Ap.J. **199**, 61
Israel, F.P., Maloney, M.J.: 1990 Ap.J. **352**, 30
Lerche, I., Schlickeiser, R.: 1982 Astr. Ap. **107**, 148
Lockman, F.J., 1980 in 'Radio Recombination Lines' Ed. P. Shaver (Dordrecht: Reidel), p. 185
Slee, O.B.: 1972 Proc. Astr. Soc. Australia **2**, 159

SLOW ROTATION OF GAS IN THE HALOS OF EDGE-ON GALAXIES M82 AND NGC 4631

Y. SOFUE[1], N. NAKAI[2], T. HANDA[1],

G. GOLLA[3], H.-P. REUTER[3], R. WIELEBINSKI[3]
1. Institute of Astronomy, Univ. of Tokyo, Tokyo 181, Japan
2. Nobeyama Radio Observatory, Nagano 384-13, Japan
3. MPIfR, Auf dem Hügel 69, Bonn 1,FRG

ABSTRACT: The rotation velocity of molecular gas in the halos of M82 and NGC4631 decreases with the height from the galactic plane. The slower rotation of halo gas can be explained if the gas is supplied from the central region of the galaxies due to some ejection.

1. INTRODUCTION

Molecular gas observations in the CO line emission of the edge-on galaxies M82 and NGC 4631 have been extensively made with the IRAM 30-m telescope and the Nobeyama 45-m telescope (Loiseau et al 1990; Sofue et al 1990). The galaxies are rich in molecular gas, and the high angular resolution maps in the CO line emission have given opportunity to study the kinematics of the halo gas. Observations of the $^{12}CO(J = 2 - 1)$ line of M82 and NGC 4631 were made from 1987 through 1990 using the 30-m telescope of IRAM. The antenna had a HPBW of 13″ at 230 GHz.

2. MOLECULAR GAS OUT OF THE PLANE

(a) **M82**: Obtained $^{12}CO(J = 2 - 1)$ and $^{12}CO(J = 1 - 0)$ line spectra shows that there exists a shift of V_{LSR} at high latitudes: V_{LSR} varies with the distance from the galactic plane. Position-velocity diagrams at different latitudes for M82 show clearly that the rotation is slower in the halo than in the disk. The rotation velocity of the disk is about 100 km s^{-1}, while the rotation at $Y = 300$ pc is about a half, ~ 50 km s^{-1}. Fig. 1 shows the rotation curves at different heights in M82.
(b) **NGC 4631**: Fig. 2 shows $^{12}CO(J = 2 - 1)$ spectra for NGC 4631. Note that V_{LSR} at $X = -42″$ varies with the distance from the plane. Data are not good enough to make position-velocity diagrams. However, from the shift of V_{LSR} with latitude, we can see that the rotation velocity is slower in the halo region than in the disk:.The maximum rotation in the disk is about 150 km s^{-1}, while it decreases to 70–80 km s^{-1} at $Y = -24″$ (-600 pc).

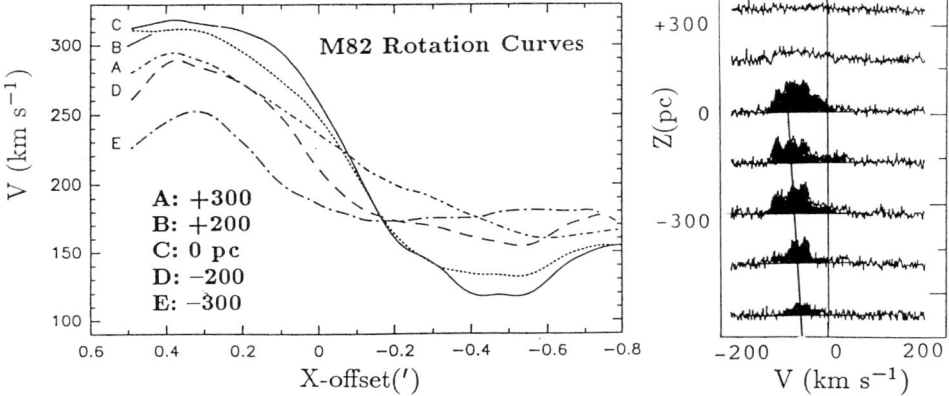

Fig. 1 Rotation curves of M82. Fig. 2 NGC 4631, $X = -1$ kpc.

3. DISCUSSION

In the two edge-on galaxies, M82 and NGC 4631, the rotation velocity of molecular gas in the halo at the height of a few hundred pc is slower than that in the disk. The following two possibilities can be considered to explain the slower rotation in the halo:

(1) Outflow model: The halo molecular gas has been supplied from the central region of the galaxy. Because of the transfer of smaller angular momentum from the central region, the gas attains a slower rotation compared to the disk when the gas reached at larger radius. In this case the gas flow should be not only perpendicular to the disk but also radial.

(2) Primordial halo molel: The molecular gas observed in the halo is supplied from the galactic plane driven by activities like, SN explosions, magnetic inflations, etc.. If there exists a gaseous halo which is the remnant of a primordially slowly rotating halo, the gas supplied from the disk will suffer from braking by the existing halo gas. It is unlikely, however, that the observed molecular gas itself is the primordial halo gas because of its heavy-element content.

In the present case of M82 and NGC 4631, the former model seems more plausible, becasue both the galaxies show activity in the central molecular disk with active star formation. Namely, M82 is the well-known starburst galaxy with an intense outflow, and NGC 4631 is known by its high-luminosity radio continuum disk and a large nonthermal halo with magnetic field.

references

Loiseau, N., Nakai, N., Sofue, Y., Wielebinski, R., H.-P. Reuter, and Klein, U. 1990, *Astron. Astrophys.*, **228**, 331

Sofue, Y., Handa, T., Golla, G., and Wielebinski, R. 1990, *Publ. Astron. Soc. Japan*, in press.

BOILING-STEAMING GALACTIC DISK: VERTICAL DUST JETS IN THE DISK-HALO INTERFACE

Y.SOFUE[1], K.WAKAMATSU[2], and D.F.MALIN[3]

1. *Institute of Astronomy, University of Tokyo, Japan*
2. *Physics Department, Gifu University, Japan*
3. *Anglo-Australian Observatory, Australia*

ABSTRACT: Optical photographs of highly-tilted, dust-rich nearby spiral galaxies like NGC253 have revealed numerous vertical dark filaments which we call vertical dust jets (VDJ). The VDJ exdend more than a few kpc from the disk in an almost coherent manner, while they are as thin as a few tens of pc. They are most likely due to boiling-steaming galactic disk, which ejects gas into the halo. The coherency suggests that VDJ trace large-scale poloidal magnetic lines of force.

1. INTRODUCTION

Vertically extending filamentary structures are often seen in optical photographs of tilted galaxies. The vertical filaments may provide information about the large-scale circulation of gas from the disk into halo and vice versa. They may also trace vertical magnetic fields (Sofue 1987). We present result of a systematic study of vertical filaments based on un-sharp masked optical photographs of nearby, dust-rich galaxies.

2. PHOTOGRAPHS

We have undertaken optical imaging of the nearby galaxis, NGC 253, NGC 4945, NGC 4594, and NGC 1808. The photographs have been taken with the AAT 4-m telescope and the Las Campanas 2-m telescope. The plates have been unsharp masked, so that faint, filamentary structures are enhanced.

NGC 253 - Central Region: Vertical dust lanes most extensively emerge from the central 1-2 kpc region. The jets are as long as a few minutes of arc, or 2 - 3 kpc above the galactic plane, while their widths are as narrow as a few arc sec (a few tens of pc). Fig. 1 shows a sketch of the vertical dust jets in NGC 253.

NGC 253 - Outer Disk: Vertical dust lanes are also well found in the outer disk at galacto-centric distances up to 6–8 kpc. Their sizes are comparable to those found in the central region. However, the outer dust lanes are more complicated, more disturbed, and look something like steam from a hot dish.

NGC 4945: This is also a dust-rich galaxy, showing complicated dusty filaments emerging from the disk. They are not well aligned as in NGC 253. The vertical extents of the filaments are estimated to be an order of magnitude larger than the

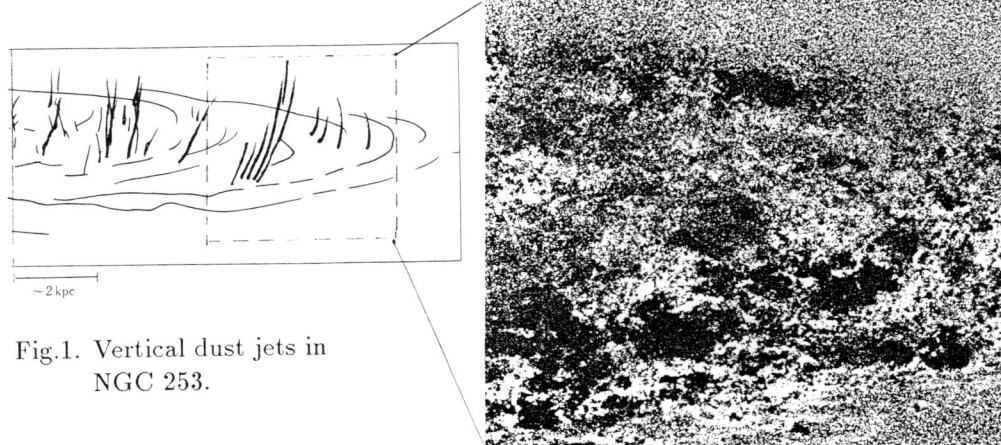

Fig.1. Vertical dust jets in NGC 253.

main disk thickness, i.e. 1–2 kpc.

NGC 4594: This is an almost edge-on Sc galaxy. Dust is well confined in a thin galactic layer. However, here we can also see many dust filaments emerging almost perpendicular to the disk plane for about 1 kpc.

NGC 1808: The central region shows jet-like dust lanes, well collimated, opening, and extend for a few kpc. [For more other galaxies, see Sofue (1987)]

3. BOILING-STEAMING GALACTIC DISK

The VDJ phenomenon seems common in any disk galaxies, in so far as we have examined published optical photographs (Sofue 1987). The VDJ phenomenon is most probably manifestation of ejection of dusty interstellar matter from the galactic plane into halo. We propose the following models to explain VDJ:

Boiling-Steaming Disk: Galactic gas disks are "boiling". The energy sources are supernova explosions, stellar winds, and inflating magnetic bubbles with cosmic rays. The boiling disk should then give "steam" (vapor) toward the halo, similarly to a hot plate. In the halo the steam will cool down and become molecular gas. Radiation pressure by star light on dust grains will also act as a driving force (Ferrini et al 1990).

Vertical Magnetic Fields: The very coherent alignment of the dust jets suggests that they trace some guid lines, which are likely poloidal magnetic lines of force (Sofue and Fujimoto 1987).

references

Ferrini, et al 1990, in this issue.
Sofue, Y. 1987, *Publ. Astron. Soc. Japan*, **39**, 547
Sofue, Y., and Fujimoto, M. 1987, *Publ. Astron. Soc. Japan*, **39**, 843

III

Theory and Modelling

THE GALACTIC GRAVITATIONAL POTENTIAL

M. CREZE
Observatoire de Strasbourg, CNRS URA 1280
11, rue de l'Université
67000 Strasbourg, France

ABSTRACT. Observational constraints on the galactic potential and modelling aspects are reviewed. The following conclusions come out. The mass distribution generating the galactic gravitational potential should include a bulge to account for the innermost features of the rotation curve, an extensive massive corona responsible for a flat rotation at large radius, and a disc able to produce a rotation curve as given by Rohlfs and Kreitschmann (1988) and a vertical force law K_z as from Bienaymé et al. (1987). We still know very little about the potential more than 1500 pc above the galactic plane.

1. WHAT ARE POTENTIALS GOOD FOR, HOW CAN WE GET THEM

This talk is not meant to address in great detail the complex mathematical problem of modelling the galactic potential. I would like to focus on more practical questions: How much do we actually know about the galactic potential? What are the most stringent observational constraints so far available? Within which limits is the galactic potential bound?

Gravitational potentials are of practical importance in two respects. First, they follow the trend of the total mass distribution, given by the Poisson equation

$$\nabla^2 \Phi = 4\pi G \rho. \qquad (1)$$

Second, they govern the motions of particles and gas on large scales. The relevant equation is then the Boltzmann equation, which we write here in its collisionless form

$$v_x \cdot \frac{\partial f}{\partial x} + \frac{\partial \Phi}{\partial x} \cdot \frac{\partial f}{\partial v_x} + ...(y,z)... = 0, \qquad (2)$$

where $f(v_x, x, ...)$ is the phase space distribution of the particles.

One possible approach would be to observe the mass distribution ρ, then to integrate (1) to get the force law $\nabla \Phi$ and then study the behaviour in this force field of any specific population, for instance the interstellar matter away from the galactic plane, which is the subject of this symposium. Unfortunately ρ is usually not observable since some of the mass happens to be dark. So investigators would rather search for tracer populations which have a distribution function in the phase space f that can be observed in situations simple enough to recover $\nabla\Phi$ through equation (2). Then equation (1) would provide an estimate of the total dynamical mass. One major difficulty in this process follows from equation (2). The Boltzmann equation sets a local dependence between the components of the force and the derivatives of the tracer distribution function with respect to space and velocity. This means that wherever the tracer sample is too small or too noisy to estimate the derivatives of the phase space distribution, we get no information about the force field at this place, neither do we get any information about the local density from (1).

The general trend of ideas initiated by J. Oort (1932) has been to collect pieces of evidence that might help choosing a realistic mathematical formula describing either f or Φ with a small number of free parameters, then to use tracer samples to constrain the free parameters.

2. ROTATION CURVES AND OVERALL MASS DISTRIBUTION

The first piece of evidence is provided by the galactic rotation. There is a number of indications that in our galaxy, as well as in others of similar type, the bulk of the mass is roughly axisymmetric. Then, whatever tracer which follows the circular velocity provides one direct evidence for the radial force law through

$$V_c^2 = R\partial\Phi/\partial R. \tag{3}$$

In the spherical approximation, the total mass within radius R follows directly from

$$M(R) = G^{-1}RV_c^2(R). \tag{4}$$

Even in this oversimplified approximation, what we get is just the total mass inside a sphere, with very little information about the distribution inside.

In order to determine what happens at distance R, one should face equation (5), which again requires the estimation of a local derivative, $\partial V/\partial R$:

$$\partial M/\partial R = (4\pi G)^{-1}V^2/R^3(1 + 2(R/V)\partial V/\partial R). \tag{5}$$

Looking into the details of rotation velocity data (for instance in Rohlfs et al., 1986), it is clear that the large scale features of the rotation curve are reasonably well constrained, but existing data are far from tracing accurately velocity gradients at any radius. Therefore, most investigations in this field start with constructing mass models. An extensive review of this topic can be found in Binney and Tremaine (1988).

There are two separate questions involved in this dynamical modelling: one is the mathematical aspect, which we will adress briefly in the chapter dealing with Stäckel potentials, the other is the link between model parameter estimation and the constraints given by observations. Regarding the second question, recent models, such as the ones by Rohlfs and Kreitschmann (1988) or Haud and Einasto (1989), do improve the situation because they are based on improved knowledge of the rotation curve and improved rotation parameters, but there is no guarantee for the validity of the mathematical representation for other purposes than fitting the rotation curve. A striking illustration of this fact has been given by Bahcall, Soneira and Schmidt (1982), who show that a simplified analysis, based on single spherical component models, may overestimate the dynamical mass by a factor 2.6 within 10 kpc and still 1.2 within 20 kpc, irrespective of the quality of the fit of the rotation curve.

The link between model parameters and observational constraints has been investigated in great detail by Caldwell and Ostriker (1981). These authors include as observational constraints not only the rotation curve and the local rotation gradient, as defined by the Oort's constants A and B, but also the surface density of our galaxy at $R_\odot$. They also introduce remote constraints, such as the velocity distribution of distant globular clusters which requires an extended halo of very low luminosity matter.

In Bienaymé et al. (1987), we include both the rotation curve constraints and others derived from the vertical distribution of disc stars in a single iterative solution. Details of this solution are discussed in chapter 3.

From this very raw and partial survey of the rotation constraints I conclude that

a) remote constraints require an extended corona.
b) the innermost parts of the rotation curve require a massive central bulge.
c) the mass between bulge and corona should be distributed such that it produces a flat rotation curve at about 200 km s^{-1}, from 5 kpc outwards.
d) more details about how the mass involved in producing this rotation curve splits into different components should be derived from studies of the vertical structure of the galaxy.
e) a drastic improvement of the determination of the Oort's constant would be very important.

3. THE VERTICAL STRUCTURE

Here, again, a quick look at formulae that are valid in a simplified situation does illustrate the main problems. As long as the mass distribution can be represented by a series of infinite parallel layers, the potential can be separated in radial and vertical components

$$\Phi = \Phi_R + \Phi_z. \tag{6}$$

This is a good approximation at moderate distances from the galactic plane in the neighbourhood of the sun. Moderate means $z \ll R_\odot$, which we will consider as satisfied below 1 kpc. The collisionless Boltzmann equation (2) can then be written separately for the z component of the phase space distribution (the index i refers to the chosen tracer population)

$$v_z \cdot \partial f_i(v_z, z)/\partial z + \partial \Phi_z/\partial z \cdot \partial f_i(v_z, z)/\partial v_z = 0. \tag{7}$$

This means that observing the phase space distribution f_i of a tracer population in the direction of the galactic poles will give sufficient information on the vertical force law.

In case isothermal tracers can be identified (iosthermal means in this case $\partial \sigma_{ivz}/\partial z = 0$, where σ_{ivz} is the rms velocity), then there is a straightforward solution of equation (7) for the vertical component of the force law, known as K_z

$$K_z = \partial \Phi_z/\partial z = -\sigma_{ivz}^2/\rho_i \partial \rho_i/\partial z. \tag{8}$$

The phase space distribution is specified by the velocity dispersion σ_{ivz} and the trend of the volume density $\rho_i(z)$.

Assuming that we do have a good (presumably spectral) criterion to identify the members of such an isothermal population, observations of the nearby members will give the velocity dispersion and the luminosity function. A spectral survey together with photometric observations towards the polar directions will give the apparent magnitude distribution $a_i(m)$. The observational effort required to produce the necessary spectral survey is so heavy that the work of Upgren (1962) has been the basis of most investigations over more than twenty years. From $a_i(m)$ and the luminosity function, the inversion of an integral equation will lead to $\rho_i(z)$. Here, again, the resulting K_z is just as reliable as the density derivative is. An important by-product of the K_z estimate is the dynamical estimate of the local volume density ρ_o, which follows from the Poisson equation in its simplified form:

$$\partial K_z/\partial z = 4\pi G \rho_o. \tag{9}$$

Clearly, ρ_o, which depends on the local derivative of K_z, is even more sensitive to the scarcity of most density tracers near the galactic plane.

The different aspects of this quest have been reviewed in a colloquium held in Danbury one year ago (Davis Philip and Lu, 1989). A sample of the many K_z curves obtained over nearly fifty years is printed on the cover of the colloquium proceedings. I do not like at all this picture, which conveys the impression that any new result in this field is definitely condemned to add to the general confusion. The reason for this confusion is that very little effort has been made to put meaningful error bars on the results. Most investigations based on the direct numerical estimation of the tracer density from star counts produce K_z curves that steeply rise in the first 100 pc and then abruptly fall beyond. A closer look at the tracer sample shows that the density law responsible for the steep rise is derived from one or two dozen stars, which means that there is indeed no information on the slope at low z. This is well illustrated by the negative slopes of the K_z curves beyond 100 pc, which are physically impossible (they would imply layers of negative mass).

A synthetic approach that imposes the density laws of the tracers to be realistic (Oort 1932, Oort 1960, Hill 1960) does avoid this difficulty, although it should not be forgotten that imposing realistic mathematical shapes does not create information where there is no data to constrain them. At larger distances, the difficulties are to safely identify complete samples of tracer population members, and also to measure radial velocities of faint stars. As a result, it is uncertain whether the density tracers at large z belong indeed to the isothermal component they are supposed to belong to. In addition, halo stars sometimes are not recognized. Beyond 1.5 kpc, the plane parallel approximation is dramatically wrong. Also, both synthetic and direct numerical inversions are sensitive to errors in the luminosity function (absolute magnitude calibration). Three recent papers have substantially improved the situation.

3.1 Bahcall (1984)

used a mass model including a double exponential disc, a *de Vaucouleurs* spheroid, and a corona. He tentatively introduced different hypothetical unseen mass discs to reconcile the dynamical mass required to fit the tracer distributions with the observed mass.

His analysis was based on a self-consistent solution of the combined Poisson-Boltzmann system for isothermal components in the presence of a spheroid. The tracer samples used were the Upgren's F-dwarfs and K-giants towards the North galactic pole (Upgren 1962). Furthermore, Bahcall rediscussed both the luminosity functions of the tracer stars and also the validity of the isothermal decomposition.

The K_z curve resulting from Bahcall's best fit solution is presented in Figure 1. A major consequence of this result is that it implies an unseen mass disc as heavy as the visible one: the total local mass density is found to be about 0.2 solar mass per cubic parsec, which is about twice what we actually see, and requires ten times more local dark matter density than the corona imposed by the rotation curve. Two investigations by Crézé et al. (1989) and by Gould (1989) lead to the same conclusions: the very large amount of missing mass in the solar neighbourhood found by Bahcall is a result of the poor constraints imposed by the scarce tracer data at low z. In other words, a model with no unseen mass disc would have fitted the raw data as well as the one chosen.

3.2 Bienaymé, Robin, and Crézé (1987)

followed a slightly different approach. Admitting that it is not easy to identify members of a predefined isothermal component of the galaxy, they avoided the difficulty by using a complete decomposition of all not-too-young disc stars in coeval isothermal components.

The basic mass model is not very different from Bahcall's, only the stellar disc is made of a series of Einasto ellipsoids, each associated with an age range. Then a scenario of galaxy evolution (including star formation, stellar evolution and progressive heating of the disc) is used to predict the absolute magnitude and colour

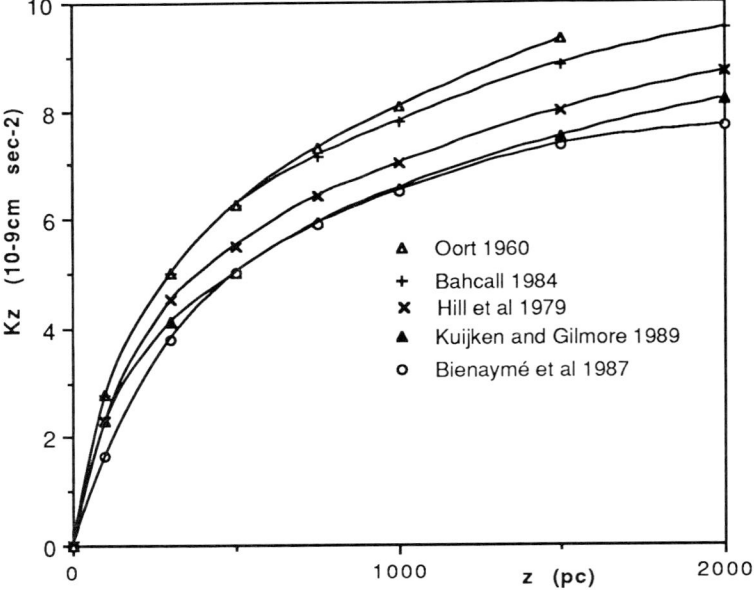

Figure 1. The galactic force law perpendicular to the galactic plane.

distribution in each coeval population (Robin and Crézé 1986). Then the whole disk population can be split into a series of isothermal components, each component being associated to an age. The reasonably well known age/velocity dispersion relation in the solar neighbourhood (Mayor 1974) is used to assign each component an isothermal velocity distribution. The dynamical closure ties the characteristic scales of the density ellipsoids to the velocity dispersions through the potential.

This approach is based on
- the idea that stars born at the same epoch should have experienced the same dynamical history
- the observation that in the solar neighbourhood, whenever a good age criterion enables selection of a good coeval sample, the velocity distribution turns out to be gaussian.

The main consequence of this approach is that it makes general star counts a tracer of the potential, because in this global modelling we do not need to know which among the observed stars belong to a certain component. Only the overall magnitude distribution of the model predictions should fit the observed one. Also, the modelling being three-dimensional, samples at intermediate galactic latitudes can be used to set additional constraints. The samples are far larger than the usual tracer samples and the statistical errors are negligable. Systematic errors in the luminosity function for stars taken from the whole HR diagram or errors in assigning velocity dispersions to subpopulations, can hardly be suspected to affect the resulting K_z. No selection error can be made in the observation of the tracers (as long as the disc dominates).

The resulting K_z is given in Figure 1. It requires much less mass than the K_z distributions derived by Bahcall and by Oort. Near the galactic plane, the local volume density is 0.1 ± 0.01 solar mass per cubic parsec, which is in good agreement with the observed mass.

This result is not very well established beyond say 1500 pc, because the dynamical modelling is not fully three-dimensional, even though the mass model is. Furthermore, beyond this distance from the midplane the contribution of non-disc stars becomes important and the prediction of this contribution in the model is not very well constrained. The result has been questioned, mainly because it involves a lot of modelling.

3.3 Kuijken and Gilmore (1989)

produced a new survey of K-dwarfs towards the south galactic cap. New photometric and radial velocity data have been obtained for several hundreds of stars over an extended range in z. Their approach involves nearly no modelling hypothesis, although substantial progress has been made in the theoretical analysis. They adopted a simple parameterization of the potential (instead of modelling the densities). Then they directly derived the velocity distribution of the tracer stars from their space distribution in each hypothetical potential according to equation (7). Hence they avoided the problem of isothermal decomposition. The likelihood of the observed velocity distribution for a certain model is used to choose the best potential.

The resulting K_z is presented in Figure 1. It is not significantly different from the distribution obtained by Bienaymé et al. The KG sample is again too scarce at low z to provide a good constraint on the local volume density. Above 1500 pc, both approaches suffer from not being fully three-dimensional. The local surface densities obtained are quite similar and both results do not require any hidden mass in the solar neighbourhood.

Hence, two independent results, obtained by Bienaymé et al. (1987) and Kuijken and Gilmore (1989), based on different data and different theoretical approaches, are in such a good agreement that one can hardly escape the conclusion that substantial progress has been made in the knowledge of the vertical component of the galactic potential, at least below 1500 pc.

4. STACKEL POTENTIALS

The final step to fully understand the galactic potential requires full three-dimensional modelling that is valid at any z. The solution may be found in the Stäckel potentials. One important property of these potentials is that they are separable in spheroidal coordinates, which means that one-dimensional solutions of the Boltzmann equation can be worked out without any loss of generality for spheroidal systems. They are already being extensively used in the study of triaxial ellipticals. Their axisymmetric oblate version, which would be suitable for discs, has been studied by Eddington (1915), Kuzmin (1953), Hori (1962), Van de Hulst (1962), Dejonghe and de Zeeuw (1988). They have been shown to fit (at least locally) any general potential with good precision, and to provide integrals of motion (de Zeeuw and Lynden Bell 1985). They are suitable to address the K_z problem (Statler 1989), because they release the limitations imposed by the plane-parallel approximation and they give a good formalism to analyse distributions in the phase space.

There are, however, two limitations which make the final solution still steps ahead. An important constraint for choosing the appropriate potential out of the extensive Stäckel family is still missing. We should know what the velocity ellipsoid looks like at high z. Another limitation comes from the unfriendly mathematics involved.

REFERENCES

Bahcall, J., Soneira R., Schmidt, M. (1982) *Ap. J.* **258**, L23.
Bahcall, J. (1984a) *Ap. J.* **276**, 156.
Bahcall, J. (1984b) *Ap. J.* **276**, 169.
Bienaymé, O., Robin, A.C., Créze, M. (1987) *Astron. Astrophys.* **180**, 94.
Binney J., Tremaine, S. (1987) " *Galactic Dynamics* " , Princeton Univ. Press, Princeton.
Caldwell, J.A.R. , Ostriker, J.P. (1981) *Ap. J.* **251**, 61.
Créze, M. ,Bienaymé, O.,Robin, A.C. (1989) *Astron. Astrophys.* **211**, 1.
Davis Philip, A.G., Lu, P., (Eds) (1989) "*The gravitational force perpendicular to the galactic plane* ", L.Davis Press Schenectady, New York.
Dejonghe, H., de Zeeuw, T. (1988) *Ap. J.* **329**, 720
de Zeeuw, T.,, Lynden Bell, D. (1985) *M.N.R.A.S.* **215**, 599
Eddington, A.S. (1915) *M.N.R.A.S.*, **76**, 37
Einasto, J., Haud U. (1989) *Astron. Astrophys.* **223**, 89.
Gould A., (1989) in "*The gravitational force perpendicular to the galactic plane* ", L.Davis Press Schenectady, New York, p 19
Haud U., Einasto, J. (1989) *Astron. Astrophys.* **223**, 95.
Hill, G. Hilditch, R., Barnes, J.V. (1979) *M.N.R.A.S.* **186**, 813.
Hori, G. (1962) *P.A.S.J.* **14**, 353
Kuijken, K. and Gilmore, G. (1989) *M.N.R.A.S.* **239**, 571
Kuijken, K. and Gilmore, G. (1989) *M.N.R.A.S.* **239**, 605
Kuijken, K. and Gilmore, G. (1989) *M.N.R.A.S.* **239**, 651
Kuzmin, G. (1953) *Tartu Astr. Obs. Teated*, **1**,1
Mayor, M.,(1974) *Astron. Astrophys.* **32**, 321.
Oort , J. (1932) *Bull. Astron. Inst. Netherlands* **6**, 249.
Oort , J. (1960) *Bull. Astron. Inst. Netherlands* **15**, 45.
Robin, A., Créze M. (1986) *Astron. Astrophys.* **157**, 71.
Robin, A.C., Créze, M. ,Bienaymé, O. (1988) in XXIIIth Rencontre de Moriond, "*Dark Matter*", Eds. J. Audouze and J. Tran Thanh Van, p 239.

Rohlfs, K., Chini, R., Wink, J.E., Böhme, R.(1986) *Astron. Astrophys.*, **158**, 181
Rohlfs, K., Kreitschmann, J. (1988) *Astron. Astrophys.*, **201**, 51
Statler , T. (1989) *Ap. J.* **344**, nnn.
Upgren, A.R. (1962) *Astron. J.* **67**, 37.
Upgren, A.R. (1978) *Astron. J.* **83**, 626.
Van de Hulst, H.C. (1962) *Bull. Astron. Inst. Netherlands* **16**, 235

HYDROSTATIC EQUILIBRIUM OF GAS-FIELD SYSTEM IN THE GALAXY AND ITS STABILITY

S.A. STEPHENS
Tata Institute of fundamental Research
Homi Bhabha Road, Bombay 400005, India

ABSTRACT. The stability of the gas-field system in the Galaxy is examined to show that it is difficult to maintain a stable configuration due to numerous energy releasing phenomenon occurring in the Galaxy. However, it is argued that, as in the case of our atmosphere, hydrostatic equilibrium is expected in the Galaxy on a global scale. A summary of the advances made in the study of the global equilibrium is presented. The major postulates and predictions from some of these studies are compared with the observations. The usefulness of detailed observation of nearby edge-on galaxies in radio and X-ray regime is brought out.

1. INTRODUCTION

Interstellar gas, magnetic field and cosmic rays are coupled to each other in the Galaxy. They interact among them exchanging energies. Cosmic rays being charged particles, are tied to the magnetic field, which is confined to the gas due to the ionized component. The gas is attracted towards the plane by the gravitational potential, which results from the distribution of stars and matter, both seen and unseen. In the plane of the Galaxy effects due to galactic rotation, spiral density wave etc. would further complicate this coupling. While these effects are not very important away from the galactic plane, phenomenon such as infalling of gas, galactic wind may influence the dynamics of the system.

The important role played by the magnetic field and cosmic rays in the dynamics of the gas-field system was emphasized by Parker (1966, 1969) two decades ago. Progress in this field is rather slow over these years mainly due to the lack of observational tests to many of the theoretical expectations. In this review I briefly introduce the general problem with a simple approach and examine the stability of the gas-field system. It appears that there can be no stable system. The consequences of these instabilities are many including the formation of cloud complexes and small scale turbulences. However, it is expected that globally the system should be in a state of hydrostatic equilibrium. The advances made in this field are briefly discussed and the results from such studies are summarized. Though observational tests for these models are difficult to realize, some of the predictions are compared with the available observations. They provide the description of the present state of art in this field of research and help us to see what can be done in the future to have a greater understanding of the gas-field system.

2. STABILITY OF THE GAS-FIELD SYSTEM

Hydrostatic equilibrium of the gas-field system perpendicular to the galactic plane (z-direction) can be described by the well known relation (Parker, 1969) between the internal pressure of the system and the gravitational force, as given by

$$d[P_G(z) + P_{MF}(z) + P_{CR}(z)]/dz = - \rho(z)g(z) \qquad 1$$

Here, P's are the pressures due to gas (G), magnetic field (MF) and cosmic rays (CR), ρ is the density of gas and g is the acceleration due to gravity. The total internal pressure $P_T = P_G + P_{MF} + P_{CR}$ is given by the integral

$$P_T(z) = \int_z^\infty \rho(z)g(z)dz \qquad 2$$

Parker considered a simple case in which z distributions of all three components of the internal pressure are similar, such that $P_{MF} = \alpha P_G$ and $P_{CR} = \beta P_G$ and the gas is isothermal. For a constant value of $g(z)$, the distribution of gas, MF and CR is exponential with a scale height $H = (1. + \alpha + \beta)u^2/<g>$, where u is the rms velocity of the gas.

This simple hydrostatic equilibrium state was tested for stability by Parker(1966) using two dimensional perturbation of the type $\exp(ik_y y + ik_z z)$, where k is the wave number and the direction of y is parallel to the initial MF in the plane. He had shown that the above perturbation can lead to instability if the adiabatic index γ of the composite fluid satisfies the relation

$$\gamma < 1. + \beta + \alpha[0.5 - 8.0(k_y^2 + k_z^2)H] \qquad 3$$

For small values of k, the above relation becomes $\gamma < (1. + \beta + \alpha/2.)$. It can be seen that for large values of k_y, the system is stable against perturbation due to the tension in the field lines. On the other hand, instability arises at long wavelengths due the slackness of the field lines. As a result thermal gas slides down along the field into the depressions, which in tern makes the field lines rise further due to CR pressure. The fastest growing instabilities are triggered by three dimensional perturbations with $(1. / k_x) << H$. The reason for this growth of instability is as $k_x > k_y$, the raised portion of the magnetic field expand into the room available by the depression of lines on either side. This kind of instability would generate small scale turbulences.

Shu(1974) examined Parker instability in some detail by including differential rotation with an axis perpendicular to the magnetic field lines. He concluded from his analysis that no finite amount of shear and rotation can stabilize the system. On the other hand, Mouschovias(1974, 1975) argued that, even for small k_y perturbations, the inflation can not continue for ever. The flux tube will be highly deformed and the inflation will be arrested when the tension in the field lines becomes the major confining force of CR. Cosmic rays can only hasten the initial inflation, but at the latter stages the CR pressure gradient becomes small as the CR density decreases with the expansion of the MF. He had shown that the ratio of the MF to CR pressures becomes proportional to $H^{.333}$. The greater the inflation, the larger the value of H and thus CR will not overwhelm the field. However, the main assumption in the above argument is the conservation of CR particle number in a flux tube. It was pointed out by Cesarsky (1980) that the above assumption need not be correct because of the possible presence of CR sources which continues to inject CR into the flux tube during the period of inflation.

The instability criteria derived by Parker can not be used in a situation, in which α and β are functions of z. Using hydrodynamic energy principle, a more general criteria for testing the instability was derived by Lachieze-Rey et al (1980). They showed that the critical adiabatic index γ_c for stability should be

$$\gamma_c = -\Gamma P_G(z)/P_{CR}(z) + \rho^2 g(z)/P_G(z)(d\rho(z)/dz) \qquad 4$$

where Γ is the polytropic index for CR. When $\gamma < \gamma_c$, the system becomes unstable. If CR re-distributes rapidly along the field lines, $\Gamma = 0.$, otherwise it could be as large as 4/3. From the above equation one notices that the polytropic behaviour of CR tends to stabilize the system. Therefore, the maximum value of the critical adiabatic index would be without the first term in Eqn.4.

The growth rate of these instabilities are of particular interest in the understanding of these dynamical effects. In the simple model of Parker, the instability grows in time scale comparable to the free fall time of the gas over one scale height. This is about $H/c_s \sim 10^7$ yrs, where c_s is the the sound speed. Zeweibel and Kulsrud (1975) examined the effect of including microturbulent magnetic field produced by cloud motion and showed that the entangled magnetic field acts as a viscous fluid and tends to stabilize the system. As a result, the growth time becomes longer by an order of magnitude. A generalization of the above derivation was carried out by Lachieze-Rey et al (1980), who derived the growth time for the horizontal equilibria of Badhwar and Stephens (1977). The values obtained for those states, which were found to be unstable, are in the range of a few times 10^7 yrs.

It is interesting to note that these growth times are of the order of a few tens of million years and appear to be rather very large. If one consider supernova explosion of once in 20 yrs in the Galaxy, a total of 1.5 million explosions are to take place over a surface area of 700 kpc^2 during 3×10^7 yrs. Taking an approximate size of $(100pc)^2$ for the surface area of an instability, about 20 supernovae are exploded during the growth period. Numerously more small scale energy releasing phenomena such as, novae, star formation and stellar winds from OB associations are taking place throughout the Galaxy. These are to be considered in examining the growth pattern of instabilities, and to see how these activities trigger instabilities and drive those already initiated either by small perturbations or by violent events. It is clear to me that instabilities can not be avoided and 'there is generally no complete stable equilibrium state for a gas-field system confined by gravity'- Parker (1969).

3. STUDY OF HYDROSTATIC EQUILIBRIUM

Our experience of living inside the Earth's atmosphere tells us that the atmosphere is always in a unstable state locally. There are disturbances from small to large scale, both in extent and in intensity. However, the atmosphere is in a state of global hydrostatic equilibrium. The locally induced non-equilibrium conditions, though appear to be violent at times, are not strong enough to disturb the gross equilibrium state. Therefore, it is natural to expect that the gas-field system in the Galaxy to be in a global equilibrium state. One may also make a note that the time scale for the growth of disturbances in the atmosphere is in general larger than for their decay. This is important while examining the stability of the gas-field system in the Galaxy.

It is very useful to study the global equilibrium states in order to compare with the observation and to understand the physical state of the gas-field system in the Galaxy. In this review I do not consider hydrostatic equilibrium studies carried out on self gravitating cloud of gas, but confine to the study of the tenuous interstellar medium which extends to the halo. Table 1 summarizes the important studies made in this area and in the following I discuss a few of them briefly. In the pioneering work, Parker(1966) first showed that the observed scale height of gas, the CR density and MF strength in the solar neighbourhood is roughly consistent with the expectation from the hydrostatic equilibrium. He made three assumptions, which are (a) the distributions of gas, MF and CR perpendicular to the galactic plane are similar, (b) the gravitational force acting on the gas is constant over one scale height, and (c) the gas is isothermal: ie the rms velocity u of the gas is independent of z. As a result, the derived gas distribution deviates from that observed beyond about 100 pc. Improvements have been made during the last two decades over this work. Instead of the assumption (b), Kellman (1972) made use of $g(z)$ determined by Oort(1960), who made use of the observed velocity and density distribution of K giants. However, the results of Kellman were only marginally different from those of Parker and he was able show that the half thickness of gas in the outer parts of the Galaxy should increase.

The deviation of the calculated gas distribution from the observed was attributed (eg. Daniel and Stephens,1975; Thielheim,1975; Stephens,1979) to the result of assumption (a). The next major step was undertaken by Badhwar and Stephens (1977), who dropped all assumptions (a) to (c). They made use of the observed gas distribution, the galactic model (Schmidt, 1965) for g(z) and the two component model of the interstellar gas, to derive horizontal equilibrium states. The form of $\{P_T(z) - P_G(z)\}$ derived from Eqn.2 was found to be very different from $P_G(z)$. They calculated the radio emission in the halo by assuming that $P_{CR}(z) = P_{MF}(z)$ and the CR electron spectral shape at all z is the same as that near the solar vicinity. A comparison of the calculated the integral radio emission towards the galactic pole with the observed spectrum suggested the need for increasing $P_T(z)$ and hence a halo gas component was introduced to account for the radio observation. This halo component has a scale height H_h of a few kpc with a density of about 10^{-2} atom/cc at z = 0. A typical model is shown in Figure 1 for a condition that the clouds are not dynamically coupled to the system and a halo gas with H_h = 2 kpc. In this figure the behaviour of P_T, P_G and the radio emissivity at 400 MHz are shown as a function of z. The flattening noticed beyond 10 kpc is due to the inclusion of intergalactic gas with a constant density of 10^{-5} atom/cc.

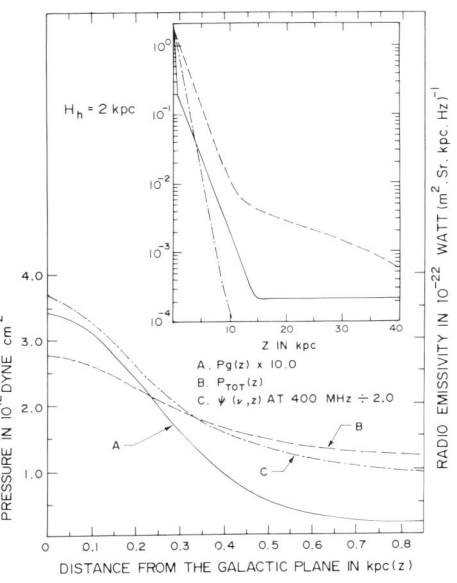

Fig.1 The distribution of P_G, P_T, and radio emissivity at 400 Mhz as a function of z for a horizontal equilibrium model with H_h = 2 kpc.

The above horizontal equilibrium model was able to satisfy all the available observations in a consistent manner at that time. Two important aspects came out of this study. (i) The need for the halo gas with a column density of about 10^{20} atom/cc. This value is found to be in agreement with the free electron density of $(0.8 - 1.4) \times 10^{20}$ cm^{-2}, determined from dispersion measures from pulsars located in clusters 47 Tuc and M15 (Reynolds, 1989). (ii) It can be seen from Figure 1 that the distribution of radio emission is approximately exponential with a scale height about half that of the gas. It was clear then that a radio halo in the conventional sense of a volume with uniform radio emissivity cannot exist, but the halo can be identified up to a few kpc by a high resolution telescope with a dynamic range of at least 50. Indeed, the later observations of edge on galaxies showed that the intensity falls of rapidly with distance from the plane (eg. Sukumar and Velusamy, 1985). However, this model of Badhwar and Stephens was found to be stable only for > 1.0 (Lachieze-Rey et al, 1980).

An improvement was introduced by Ghosh and Ptuskin (1983) in this study by determining P_{CR} using diffusion model. They assumed that hot coronal gas and random MF to dominate in the halo and equipartition of energy densities to exist between them. However, they did not take into account the MF tension arising from random field orientation. Their major conclusion was that though P_{CR} and P_{MF} are not the same at all values of z, they are of the same magnitude. Because of the assumption that $P_G = P_{MF}$ in the halo, P_{CR} becomes smaller for larger values of halo gas temperature. The calculated radio emission does not show an exponential decrease with z.

TABLE 1.
Summary of the study carried out on the hydrostatic equilibrium of gas-field system in the Galaxy.

Author/s	Outcome of the study
Parker(1966)	Showed the consistency of derived P_G, P_{CR}, P_{MF} in the solar neighbourhood with the observation.
Kellman(1972)	Explained the observed increase of the thickness of gas disk in the outer parts of the Galaxy.
Badhwar and Stephens(1977)	Postulated halo gas with $N_G \sim 10^{20}$ / cm^2. Predicted extended distribution of CR & MF in the halo and derived ($P_{MF} + P_{CR}$) & radio emissivity, E_R, as a function of z. Predicted E_R to decrease exponentially with z.
Fuchs and Thielheim(1979)	Showed that gas clouds may not be coupled to the interstellar MF.
Ghosh and Ptuskin(1983)	Postulated the random MF pressure to be in equilibrium with the hot halo gas. Showed that $P_{MF} \sim P_{CR}$ and calculated P_{MF}, P_{CR} & E_R as a function of z.
Chevalier and Fransson(1984)	Demonstrated the possible support of coronal hot gas in the halo by CR.
Bloemen(1987)	Postulated the existence of hot halo gas & predicted its temperature profile. Derived the distribution of ($P_{CR} + P_{MF}$) in the halo and a condition relating the $H_{h,G}$ and $n_{h,min}(0)$.
Boulares and Cox(1990)	Invoked magnetic tension to support the system. Derived distributions of P_{MF}, P_{CR}, orientation of MF. Calculated the velocity dispersion profile of warm halo gas.

A detailed study was undertaken by Bloemen (1987), who introduced more recent information available on g(z) and on the gas distribution of atomic and molecular hydrogen. His analysis was motivated by the desire to obtain stable equilibrium states. He calculated $P_T(z)$ from Eqn.2 for various values of H_h and $n_h(0)$. Figure 2 shows the derived distribution of P_T for H_h = 6kpc. He also calculated the minimum gas pressure $P_{G,min}$ required to have stability by setting in Eqn.4, =0. and $_c$ = 1. These are also shown in the same figure as a set of curves starting from the same point in the plane. It is clear from this figure that, while $P_{G,min}$ is very much smaller than P_T at small values of z, it becomes equal to P_T beyond a few kpc. As a result , $\{P_T - P_{G,min}\}$, which is the pressure due to MF and CR decreases to 0 beyond a few kpc. Consequently, it is expected that the radio emission does not follow an exponential distribution in the halo.

He set constraints from the integrated radio emission towards the pole at 30 MHz, by taking $\{P_{MF}(0) + P_{CR}(0)\}$ = 1.5 x 10^{-12} dyne / cm^2. From this he derived a relation $n_{h,min}(0)$ $(H_h / 1kpc)^2$ = 0.17 cm^{-3} between the minimum halo gas density at z = 0 and the H_h. From this we find that H_h > 5kpc, in order to match the observed column density of free electrons. He also derived the temperature of halo gas as a function of z using the value of P_G for various halo gas distribution. These are shown in Figure 3. An interesting feature that one notices from this figure is that the gas temperature has a minimum value of about (2 - 3)x10^5K between 1 and 2 kpc, which is in agreement with the UV absorption line measurement from the halo. The temperature of the halo gas for H_h = 5 kpc is about 8x10^5K. It is clear that unless the radiative cooling of the halo gas is compensated by energy input from sources, equilibrium can not be maintained over a sufficiently long period of time. It may also be noted that the apparent behaviour of $\{P_{CR}+P_{MF}\}$ is forced upon by demanding minimum stability condition and setting $\{P_{CR}(0) + P_{MF}(0)\}$ = 1.5x10^{-12} dyne.cm^{-2}, which is uncertain by a factor 2. Cox (1990)

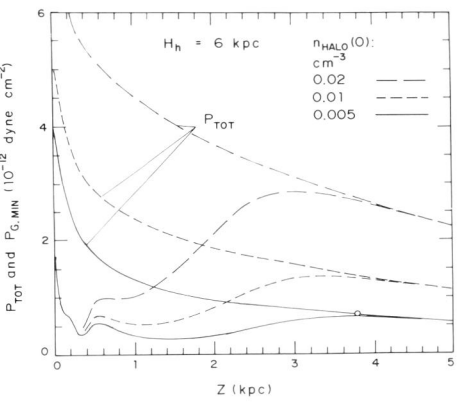

Fig.2 The distribution of P_T and $P_{G,min}$ as a function of z for various values of the halo gas density.

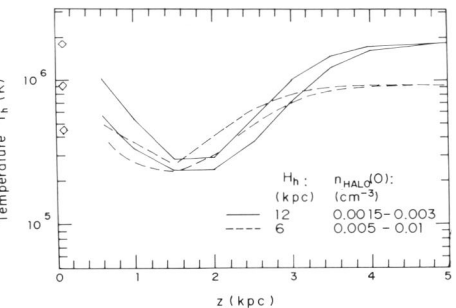

Fig.3 The temperature profile of the halo gas for different values of H_h and $n_h(0)$.

pointed out that the existence of hot coronal gas in the halo is not very well supported by the observations. Therefore, Boulares and Cox(1990) examined the equilibrium condition assuming a warm gas to exist in the halo, as in the case of Badhwar & Stephens. They invoked the MF tension resulting from the orientation of MF to provide support to the gas. They derived the distribution of variety of parameters in the halo as shown in Table 1.

It may be pointed out that in all the above discussions, the hydrostatic equilibrium is examined only in the solar neighbourhood. It has been brought to notice by de Boer (1990) that the situation is very different in the inner parts of the Galaxy, where it is difficult to provide support to the gas-field system in the halo. In the inner parts of the Galaxy, the gravitational field is much stronger and the gradient is also large. Solution to this problem lies in the study of nearby edge-on galaxies, where one can directly compare the expectations with the observations.

4. OBSERVATIONAL TESTS AND DISCUSSION

While some of the above deductions are only consistency checks for the validity of hydrostatic equilibrium, others are predictions which can be tested by observations. A detailed analysis of the radio profile of the edge-on galaxy NGC 4631 was carried out by Stephens and Velusamy (1990) at 1465 MHz. They examined the radio brightness as a function of z at different positions along the major axis. They found that, except near the plane, the distribution is exponential up to about 5 kpc. The radio scale height H_R varies from 1.25 to 2.5 kpc; the inner galaxy having smaller H_R. The true variation may be much larger as a function of R since the observations relate to the integrated emission along the line of sight. This appears to be consistent with the idea that variation of $g(z)$ in the inner parts of the galaxy would result in a steeper gradient of P_T. Exponential distribution of E_R is also seen in NGC 891 (Hummel,1990). Thus, the radio emission in the halo seems to follow an exponential distribution. Stephens and Velusamy (1990) noticed a flattening of the radio profile from the exponential distribution in some regions of NGC 4631. This behaviour can be expected from the interaction of infalling gas with the halo gas, leading to possible turbulence and acceleration of CR.

Free electron distribution in the halo provides evidence to the existence of halo gas, because it is expected that the electrons constitutes the electron component of the ionized gas. Pulsar dispersion measures directly give the column density of free electrons. An analysis of the pulsar data shows (Reynolds,1990) that the column density of free electrons is ~ 7×10^{19} cm^{-2} and the observations can be fitted with a scale height $H_{h,e}$ of ~ 900 pc. However, the UV absorption studies show that the distribution of high ions in the halo has a larger scale height of ~ 3 kpc (Salvage, 1990). It is essential that a large sample of pulsar at high z is required to estimate the value of $H_{h,e}$ and to determine its dependence on R. It appears from these studies that the gas in the halo is predominately ionized.

At this stage one may compare the calculations with the observations. As an example, I have shown in Table 2 some of the predictions or postulates along with the observations. Though this table reflects the sad state of the available observational tests, it is clear that one can test the model predictions by making proper observations both in our Galaxy and in external galaxies. It may be noted that the value of $H_{h,G}$ is very important input parameter in the study of the hydrostatic equilibrium. The difference in the derived values from UV absorption lines and from pulsar dispersion measures is rather large. This difference may be considerably narrowed in the near future. It ia also clear that the external galaxies show that the radio distribution is exponential in the halo and thus sets very strong constraint on the hydrostatic equilibrium models.

TABLE 2.
Summary and observational test

Parameter	Badhwar & Stephens	Bloemen	Boulares & Cox	Observation
$P_{MF} + P_{CR}$ dyne/cm^2	~2.5x10^{-12} (predicted)	~1.5x10^{-12} (assumed)	~(2-4)x10^{-12} (assumed)	good within a factor of 2
$H_{h,G}$	~2 kpc (postulated)	~5 kpc (expected)	1.5 kpc (assumed)	0.9 kpc# ~3 kpc*
$H_{h,radio}$	$H_{h,G}/2$.	not exponential	not predicted	exponential
$T_{h,Gas}$	warm (assumed)	hot (variability predicted)	warm (dispersion predicted)	?

\# electrons; * high ions

It is clear that observation of edge-on galaxies in radio wave lengths with a high angular resolution and flux contrast at a level of about a 500 will provide in the future valuable information on the dynamics of the gas-field system. Continuum measurements furnish information on the halo structure and the spectral shape, while 21 cm observation provides valuable details on the neutral gas distribution. Using the next generation of high resolution X - ray telescopes, information can be obtained on the temperature and column density of halo gas as a function of distance from the plane. Future measurements of the gamma rays in the Galaxy at high latitudes would also furnish information on the gas distribution in the halo, which is mostly ionized. These are the optimistic expectations of the future. Theoretical breakthrough can be also made by including dynamical effects, such as the CR sources, radiative cooling and heating by sources and from hydrodynamic interactions, and the effect of small and large scale explosive phenomena in the study of the stability of the system.

REFERENCES

Badhwar, G.D., Stephens, S.A. (1977) Ap. J., 212, 494.
Bloemen, J.B.G.M. (1987) Ap. J., 322, 694.
Boulares, A., Cox, D.P. (1990) to appear in Ap. J.
Chevelier, R.A., Fransson, C. (1984) Ap. J., 279, L43.
Cox, D.P.(1990) this Symposium.
de Boer, H. (1990) this Symposium.
Daniel, R.R., Stephens, S.A. (1975) Sp. Sci. Rev., 17, 45.
Fuchs, B., Thielheim, K.O. (1979) Ap. J., 227, 801.
Ghosh, A., Ptuskin, V.S. (1983) Ap. Sp. Sci., 92, 37.
Hummel, E.(1990) this Symposium.
Lachieze-Rey, M., Asseo, E., Cesarsky, C., Pellat, R. (1980) Ap. J., 238, 175.
Kellman, S.A. (1972) Ap. J., 175, 353.
Mouschovias, T.Ch. (1974) Ap. J., 192, 37.
Mouschovias, T.Ch. (1975) Astr. Ap., 40, 191.
Oort, J.H. (1960) Bull. Astr. Inst. Neth., 15, 45.
Parker, E.N. (1966) Ap. J., 145, 811.
Parker, E.N. (1969) Sp. Sci. Rev., 9, 651.
Reynolds, R.J. (1989) Ap. J., 339, L29.
Reynolds, R.J. (1990) this Symposium.
Savage, B.D. (1990) this Symposium.
Shu, F.H. (1974) Astr. Ap., 33, 55.
Stephens, S.A. (1979) in 'Non-Solar Gamma Rays' eds. Cowsik, R., Wills, R.D., Pergamon Press, p.191.
Stephens, S.A., Velusamy, T. (1990) Proc. 21st Int. Cosmic Ray Conf., Adelaide, 3, 221.
Sukumar, S., Velusamy, T. (1985) MNRAS, 231, 367.
Thielheim, K.O.(1975) in 'Origin of Cosmic Rays' eds. Osborne, J.L, Wolfendale, A.W., Reidel, Dordrecht, p.165.
Zeweibel, E.G., Kulsrud, R.M. (1975) Ap. J., 201, 63.

SOME PROBLEMS FOR GALACTIC HYDROSTATIC EQUILIBRIA

H. DE BOER
Space Research Leiden, P.O. Box 9504
2333 AL Leiden, The Netherlands

ABSTRACT. Although locally a static, stable equilibrium seems consistent with observations of the interstellar medium (ISM), the extrapolation towards other galactic regions is not straightforward. Basically, the variation with galactic radius of the gravitational acceleration cannot be reconciled with the approximate constancy of the HI scaleheight. Moreover, halo gas located in the outer galaxy is prone to thermal instability and subsequent collapse towards the galactic plane.

1. INTRODUCTION

Ever since the classical work of Parker (1966,1969) on stratified equilibria for the galactic gas, magnetic fields (MF) and cosmic ray (CR) particles, the local hydrostatic equilibrium has been the subject of numerous stability analyses, often involving unrealistic simplifications. Bloemen (1987, hereafter Paper I) applied a general stability criterion for stratified equilibria (Lachièze-Rey et al. 1980) to a model based on an observational description of the ISM, and concluded that a stable hydrostatic equilibrium is not precluded by observations.

We have investigated similar models for $R = 5$ and $R = 15$ kpc ($R_\odot = 10$ kpc), which are representative radii for the "inner" and "outer" galaxy. Detailed results will be presented elsewhere; here only the (im-)possibility of the radial extrapolation is discussed.

2. METHOD AND INGREDIENTS

From the equilibrium equation

$$\frac{d}{dz}[P_{GAS}(z) + P_{MF}(z) + P_{CR}(z)] = -\rho(z)g(z), \tag{1}$$

the total hydrostatic pressure

$$P_{TOT}(z) = \int_z^\infty \rho(x)g(x)\,dx \tag{2}$$

is calculated. Here $P_{MF} = B^2/8\pi$ is ascribed to a systematic MF ($\perp z$, the distance to the galactic plane); P_{CR} is mainly due to CR protons. (Generally, the net tension due to a random MF can be neglected (e.g. Parker 1969).) The total density, $\rho(z)$, is the sum of the observed gas density distributions. For $R = 5$ kpc and $R = 10$ kpc we adopt the distributions listed in Table I. Note that, in contrast to Paper I, an ionized medium is now included (Reynolds 1989), with the same distribution for all considered galactocentric radii.

At $R = 15$ kpc, the average HI density profile with respect to the warped midplane was derived from the density atlas of Burton and te Lintel Hekkert (1986), and molecular hydrogen was neglected.

For the acceleration perpendicular to the galactic plane, $g(z)$, we adopt Oort's (1960) determination; this is a good average of recently proposed curves and is consistent with the massive halo distribution ($\epsilon(R)$ below) derived by Bahcall et al. (1982). The radial extrapolation of g is based on a constant vertical (e.g. van der Kruit and Searle 1982; van der Kruit and Freeman 1984; Lewis and Freeman 1989) and exponential radial (e.g. de Vaucouleurs and Pence 1978) stellar density distribution for spiral galaxies. The result is summarized by

$$g(R,z) \approx [\sigma^2(R)/z_0]\{\tanh(z/z_0) + \epsilon(R)(z/z_0)\} \qquad (3)$$

with $z_0 = 250$ pc, $\sigma^2(R) = (15.4 \,\mathrm{km\,s^{-1}})^2 \exp(-(R-R_\odot)/0.44 R_\odot)$ and $\epsilon = 0.04$, 0.07 and 0.14 for $R = 5$, 10 and 15 kpc respectively.

The equilibrium is stable if $P_{GAS} \geq -g\rho^2[\gamma(d\rho/dz)]^{-1}$, assuming a very simple large scale equation of state for the gas: $P \propto \rho^\gamma$ (we take $\gamma = 1$). Model dependent stabilizing effects (CR diffusion, turbulent MF) are not taken into account, so that the above criterion defines a conservative lower limit on the gas pressure.

TABLE 1.
Parameters of the vertical gas distribution, for $R = 5$ and $R = 10$ kpc.
Gauss: $n(z) = n(0)\exp\{-\frac{1}{2}(z/h)^2\}$; Exponent: $n(z) = n(0)\exp\{-z/h\}$.

Component	Distribution	$n(0)$ (cm^{-3})	h (pc)	$\sqrt{<V_z^2>}$ (km s^{-1})
Cold HI	Gauss	0.3	135	6–7
Warm HI	Gauss	0.07	135	9–11
	Exponent	0.1	400	9–15
H$_2$	Gauss	1.6*–0.6**	60*–70**	5–6
HII	Exponent	0.025	1500	~ 20

*($R = 5$ kpc) **($R = 10$ kpc)

3. OBSERVATIONAL CONSTRAINTS

3.1 Mid Plane Pressures

The various pressure contributions of the ISM near the galactic plane are listed in Table II.

The gas pressure is the sum of the turbulent and thermal contributions of all gas phases. Note that the velocity dispersion probably increases with latitude (*e.g.* Kulkarni and Fich 1985); therefore the last column of Table I gives a *range* of dispersion values encountered in literature.

The radial unfolding of the *COS-B* and *SAS-2* γ-ray data, indicates that the scalelength of both CR electrons and protons amounts to $\simeq 15$ kpc (Bloemen 1989). Together with the 4 kpc radial scalelength of the synchrotron emissivity (Beuermann *et al.* 1985), this implies a scalelength of $\simeq 5$ kpc for B^2. The radial gradient of P_{MF+CR} thus derived is based on energy independent scalelengths for the CR particles. For energy dependent CR distributions the gradient will generally be smaller.

3.2 Radio Continuum Data

The radio continuum intensity towards the galactic poles can be reconciled with a stable equilibrium if a gaseous halo is included. For an exponential halo distribution the constraint $\left(n_{halo}(z=0)/0.01\,\mathrm{cm}^{-3}\right)\left(H_{halo}/1\,\mathrm{kpc}\right)^2 > \Gamma$ was derived in Paper I. When an ionized medium is included, we find $\Gamma \simeq 8$, ($\sim$ half the value found by Bloemen). In short, the models of Paper I required the halo thermal pressure for support of the colder gas components, thus allowing a long tail in the MF and CR distribution (albeit constrained by the stability criterion).

4. RESULTS AND DISCUSSION

We find that, within the uncertainty of our knowlowdge of the ISM parameters and their radial variation, a locally consistent hydrostatic equilibrium cannot be extrapolated to other R. Table II learns that, near $z = 0$, a discrepancy of $\sim$ a factor 2 exists at $R = 5$ kpc between hydrostatic and seemingly available pressure, even when the CR and MF pressures are maximized. More conservative values of P_{MF+CR} imply a discrepancy larger than a factor 3 ! The problem cannot be solved by altering the local parameters under the constraint of local equilibrium. If the "missing" pressure at small R is contributed by a halo gas, the minimum halo temperature must be 4.5×10^6 K, implying an overpressure $>$ a factor 20 with respect to the other ISM gas components. Moreover, the halo cooling rate versus possible supernovae energy input (*e.g.* Heiles 1987) cannot be rendered consistent if the halo mid-plane density and scaleheight are weak functions of galactocentric radius. Particularly, halo gas located at large R will cool significantly in $\sim 10^6$ yr,

TABLE 2.
Pressures at $z = 0$, in $10^{-12}\,\mathrm{dyne\,cm^{-2}}$ (corrected for He).

$R\,(\mathrm{kpc})$	5	10	15
P_{TOT}*	15.5	4.8	2.4
P_{MF}	2.7–5.2	1.0–1.9	0.4–0.7
P_{CR}	0.7–1.4	0.5–1.0	0.4–0.7
P_{GAS}	1.8–2.7	1.3–1.9	0.9–1.9

* (from equation (2), excluding halo gas)

even if all available SNR power is dissipated in the halo; the estimated ISM pressure at $R = 15\,\mathrm{kpc}$ is however consistent with the hydrostatic pressure.

The hydrostatic equilibrium condition thus seems to be violated on a galactic scale, if the gravitating material has a constant scaleheight. Only if the gas velocity dispersion is strongly dependent on R, simple hydrostatics can be maintained. This result does not depend on stability considerations. The stability criterion mainly prescribes the halo density and temperature profile (which depends also strongly on the gas velocity dispersions). We want to stress, that the found discrepancies may be partly rooted in the naive application of equation (1), which ignores the details of gas-phase interactions in the ISM.

ACKNOWLEDGMENTS

I am very thankful to Hans Bloemen, for both his encouragement and his active participation in many instructive discussions.

REFERENCES

Beuermann, K., Kanbach, G., and Berkhuyzen, E.M. 1985, *Astr. Ap.* **153**, 17
Bloemen, J.B.G.M. 1987, *Ap. J.* **322**, 694 (Paper I)
Bloemen, J.B.G.M. 1989, *Ann. Rev. Astr. Ap.* **27**, 469
Burton, W.B., and te Lintel Hekkert, P. 1986, *Astr. Ap. Suppl.* **65**, 427
Heiles, C. 1987, *Ap. J.* **315**, 555
van der Kruit, P.C., and Freeman, K.C. 1984, *Ap. J.* **278**, 81
van der Kruit, P.C., and Searle, L. 1982, *Astr. Ap.* **110**, 61
Kulkarni, S.R., and Fich, M. 1985, *Ap. J.* **289**, 792
Lachièze-Rey, M., Asséo, E., Cesarsky, C.J., and Pellat, R. 1980, *Ap. J.* **238**, 175
Lewis, J.R., and Freeman, K.C. 1989, *Astron. J.* **97**, 139
Oort, J.H. 1960, *Bull. Astr. Inst. Netherlands* **15**, 45
Parker, E.N. 1966, *Ap. J.* **145**, 811
Parker, E.N. 1969, *Space Sci. Rev.* **9**, 651
Reynolds, R.J. 1989, *Ap. J. (Letters)* **339**, L29
de Vaucouleurs, G. and Pence, W.D. 1978, *Astron. J.* **83**, 1163

THE GLOBAL MASS, ENERGY AND PHOTOIONIZATION BALANCE OF THE DISK-HALO INTERACTION

C. A. NORMAN
Johns Hopkins University, Department of Physics and Astronomy, and
Space Telescope Science Institute

ABSTRACT. The observational evidence for chimney models of the interstellar medium is reviewed. The variation of the state of the interstellar medium of an external galaxy between the three-phase mode, the chimney mode and the two-phase mode as a function of: galaxy type, galactocentric radius within a given galaxy, and the time varying star formation rate is discussed.

In the context of the chimney models it is shown how the photoionization of the halo can be achieved by utilizing both direct ionizing rays from the central OB association and diffuse reradiation from the chimney walls that will create an extended low ionization halo around chimneys with SII/H$\alpha \sim 0.5$.

A number of considerations for future studies that are a direct result of the work discussed at this conference are outlined.

1. INTRODUCTION: THE BASIC MODEL

Observational studies of the interstellar medium in nearby galaxies are now revealing a wealth of details and structures. Early work on M31 by Brinks and Bajaja (1986) indicated that large holes were present in its interstellar medium and the natural explanation is that correlated type II supernovae are the primary cause. If correlated explosions occur in gas disks with exponential scale heights, they will accelerate rapidly outwards normal to the plane and break out of the gas layer forming chimney like structures (c.f. Norman and Ikeuchi 1989). The walls of these chimneys are seen in our own Galaxy where they are identified with Heiles' supershells (c.f. Heiles 1987, 1989). Large structures consistent with the chimney morphology in HI and Hα are seen in the LMC (Meaburn et al. 1987, Dopita et al. 1985) , M31 (c.f. Brinks 1990 , Braun 1990, Walterbos 1990), M33 (Deul and den Hartog 1990), NGC 891 (Rand, Kulkarni and Hester 1990, Dettmar 1990) and M101 (Kamphuis 1990).

The chimneys models explicitly indicate that the flow of mass, energy, momentum and magnetic flux from the disk to the halo is funneled through the chimneys that have been created by $\sim 10^2$ correlated type II supernovae per superbubble.

The flow back down to the disk is the result of cooling and infall and occurs over the whole disk.

When considering the multiphase structure of the interstellar medium of external galaxies a number of parameters are significant, namely, the level of star formation, the amount of clustering of the energy input from type II supernovae, the mean ambient density, the galaxy type, and the variation of the physical quantities with Hubble type.

It turns out that depending on the state of these quantities a galaxy can be in either the three-phase mode (McKee and Ostriker 1977), the chimney mode (Norman and Ikeuchi 1989) or the two-phase mode (Field, Goldsmith and Habing 1969). These states can change as a function of galactocentric distance, Hubble type and time as the level of star formation changes. In extreme cases such as starbursts huge central starburst driven winds can be blown out of galaxies with outflow rates up to $100 M_\odot$ yr^{-1}(Heckman 1990). More typical numbers are an energy flow into the halo of $\sim 10^{40} - 10^{41}$ erg s^{-1}, and a mass circulation rate of $\sim 1 M_\odot$ yr^{-1}. Detailed modelling of the observational signatures of the interstellar medium has been undertaken by Li and Ikeuchi (1990).

The ionization requirement of the extended ionized layer above the disk of our Galaxy indicates that a significant fraction of the total ionizing photons of all the OB stars in the disk (Reynolds 1990) must be utilized and therefore must be able to escape from their original location in the thin star forming part of the disk. The vertical extent of the ionized layer is ~ 1500 pc and, if this warm ionized component of the interstellar medium is at a temperature $T \sim 10^4$K then the power required to ionize it is 4×10^{-5} erg s^{-1} cm^{-2}. Similar extended ionized disks have been seen in other galaxies and the effect is most apparent in NGC 891 where the layer extends to ~ 4.5 kpc above the plane and has an estimated mass of $4 \times 10^8 M_\odot$ (Dettmar 1990, Rand, Kulkarni and Hester 1990). The radial extent of the extended warm ionized medium is apparently $\sim 30 kpc$ and there are clear indications of chimneys blown out by multiple supernovae explosions occurring in OB associations. The estimate of the superbubble formation rate necessary to power this extended ionized gas component in NGC 891 is an order of magnitude greater than that of our Galactic disk (Heiles 1987, Norman and Ikeuchi 1989, 1990, Rand et al. 1990). Correlations of OB associations with HI in the gas distribution of the Magellanic Clouds, M31, M33, IC 10 and 1C 1613 have been summarized by Brinks (1990). It is worth emphasizing that in the LMC where it is possible to study such details the observed Hα shells lie just inside the HI shells.

The general problem of the ionization of the halo has been studied by Chevalier and Fransson (1984) using background ionizing radiation from quasars and Bregman and Harrington (1986) using photons from OB stars that have leaked out into the halo. Other sources of radiation that have been suggested are planetary nebulae, white dwarfs (Panagia and Terzian 1984), and the soft x-ray background. Thuan (1975) presented indications of how OB stars with a population scale height of ≤ 200 pc could have burnt a hole in the HI layer and has estimated that about 20% of the ionizing photons can escape. Mathis (1986) has estimated that about 10% of the diffuse ionizing photon field associated with O stars in the disk can achieve significant distances above the galactic plane. Thus the observations of the widespread SII/Hα ratio of ~ 0.5 corresponding to diffuse ionized gas can be satisfactorily explained. Finally, Sciama (1990) has recently suggested that the decay of dark matter in the halo accompanied by UV photon emission is the natural explanation of the ionization power budget problem.

Here we shall show that OB associations embedded in chimneys can directly ionize the halo gas above the top of the chimney. The chimney walls are also ionized and the diffuse ionizing radiation re-emitted by the walls can also ionize halo gas over a much wider range of subtended angle than that available to the directly ionizing rays. Details are given in Norman and Panagia (1990). The characteristic ionization structure above a chimney is that of a directly ionized hard HII region with species such as HeI, OIII, SIII and a softer more extended halo due to the diffuse radiation from the chimney walls associated with OII, SII, NII, CIII. Using canonical chimney model parameters the filling factor of the softer spectrum is found to be of order unity. The spatial variation between hard and soft spectra associated with chimney should be readily observed in our own galaxy and external galaxies.

The most important aspect of this meeting was the large number of excellent observations that now lead to a number of very interesting studies that require immediate further study. These include the nature and significance of the hot gas component in the interstellar medium, the effect of magnetic fields, the question of realistic dynamo models in realistic interstellar media, the general ionization state of our Galaxy and external galaxies, the morphology of the cool component of the interstellar medium, the physical significance of the morphological characterization of chimneys, walls, worms etc. (i.e. are they the same or different) the general characterization of the interstellar medium with Hubble type and the propagation of cosmic rays in the various types of interstellar media.

A significant question was asked at the meeting concerning whether or not chimneys break out of the extended diffuse ionized layer. The answer is only rarely for the most powerful superbubbles even though all respectable superbubbles can break out of the thin neutral layer and undergo acceleration to form chimneys, holes etc. As I argue in section III, the answer is that the superbubbles rising up above the thin disk probably feed mass into the extended Reynold's layer and therefore sustain it.

2. IONIZATION STRUCTURE ABOVE CHIMNEYS

The concept of an OB association embedded in a superbubble or chimney that is powered by multiple supernovae has been much discussed (Norman and Ikeuchi 1989, MacLow et al. 1989). We analyze here the escape of ionizing photons into the halo in such a chimney structure. Since we wish to avoid very specific geometries we have studied three particular cases of an ionizing source where the OB association is placed at the center of a cube, a hemisphere and a cylinder. We solve the integral equation for the ionization of the walls and the subsequent re-emission of the diffuse ionizing radiation from them. The direct ionizing radiation is penetrating a density profile that is exponential given by $n_e = n_{e0}e^{-z/H}$ where $n_{e0} \sim 0.025$ cm^{-3} and H is ~ 1500 pc (Reynolds 1990).

To simplify the analysis of the emission and reabsorption of ionizing photon at a chimney wall we use a plane parallel approximation at the chimney wall interface. In terms of the standard recombination formula given in, say, Osterbrock (1989) and Spitzer (1978) we define ξ as the ratio of the recombination coefficient to the ground state over the total recombination coefficient. Such recombinations to the ground state will re-emit ionizing photons at 13.6 eV $+\delta$, where δ is of order ~ 1 eV, which originates from the kinetic energy of the recombining electrons. Estimates of ξ depend on temperature but are in the range of 0.4. The efficiency of conversion of directly absorbed ionizing radiation into re-radiated diffuse ionizing radiation is given by, ϵ, the ratio of outgoing to incoming ionizing flux is found to be

$$\epsilon = \frac{2}{\xi}\left((1-\frac{\xi}{2})-(1-\xi)^{\frac{1}{2}}\right) \qquad (1)$$

For numerical estimates we adopt a typical value here of 13%.

Let us describe the solution of the hemi-spherical case in spherical coordinates (R,Θ,Φ). Details are given in Norman and Panagia (1990). The equation for the intensity radiated by the walls $I_w(z)$ is

$$I_w(z) = \frac{\epsilon S_o}{2R^2(1-z)} + 2\pi\epsilon \int_{-1}^{z_0} dz \frac{I_w(z)}{(z-z')^2} \qquad (2)$$

where $z = cos\Theta$, S_0 is the luminosity of the central source and R is the radius of the hemisphere. The first interaction is obtained by using the source term as the zeroth approximation for $I_w(z)$ in the integral. This gives

$$I_w(z) = \frac{\epsilon S_0}{2R^2(1-z)}\left[1+\frac{2\pi\epsilon}{(1-z)}\left[\ln\left(\frac{2(z_0-z)}{(z_0-1)(z+1)}\right)-\frac{(1+z_0)(1-z)}{(z_0-z)(1+z)}\right]\right] \qquad (3)$$

The total intensity at a point outside the hemispherical surface $I(r,\theta,\phi)$ can be well approximated for $\theta > 45^0$ by

$$I(r,\theta,\phi) = \frac{\epsilon So}{r^2}\left(\frac{\pi}{2}\right)\ln(1+\sin 2\theta). \qquad (4)$$

Similar results hold for the open cylinder where the zeroth order solution is

$$I(r,\theta,z) = \frac{\epsilon So}{r^2}\pi\left[\arctan\left(\frac{L}{R}\right)-\arctan\left(\frac{L}{R}-\frac{2z}{r}\right)\right] \qquad (5)$$

The ionization sphere for the directly ionized material above the chimney is given by (Spitzer 1978 §5-1),

$$r_{s1} = \left(\frac{S_0}{\frac{4}{3}\pi n_e^2 \alpha}\right)^{\frac{1}{3}}. \tag{6}$$

For the diffuse ionizing radiation from the walls, the ionizing sphere in the non-directly (i.e. diffusely) ionized region ($\theta \geq 45^0$) has a shape given by

$$r_{s2} = \left(\frac{\epsilon S_0}{\frac{4}{3}\pi n_e^2 \alpha}\right)^{\frac{1}{3}} \left(\ln(1+\sin 2\theta)\right)^{\frac{1}{3}} \approx \epsilon^{1/3} r_{s1}. \tag{7}$$

Results for cylinders and cubes give similar answers. For canonical halo parameters we choose (Reynolds 1990) ($\alpha \sim 2 \times 10^{-13}$ cm^3 s^{-1}, $n_e \sim 0.025$ cm^{-3} up to a scale height of 1500 pc and $S_0 \sim 10^{50} - 10^{51}$ s^{-1}) we find

$$r_{s1} = 2.4 \text{kpc} \left(\frac{S_0}{10^{50}\text{s}^{-1}}\right)^{\frac{1}{3}} \left(\frac{0.025\text{cm}^{-3}}{n_e}\right)^{\frac{2}{3}} \left(\frac{2 \times 10^{-13}\text{cm}^3\text{s}^{-1}}{\alpha}\right) \tag{8}$$

which shows that for canonical numbers the free electron layer directly above the chimney can be ionized readily. In fact, because the density distribution is exponential the hard ionization region above the chimney is density bounded.

The characteristic size of the ionization region created by the diffuse radiation from the walls is

$$r_{s2} \sim \epsilon^{\frac{1}{3}} r_{s1}$$

$$\sim 1.2 \text{kpc} \left(\frac{\epsilon}{0.16}\right)^{\frac{1}{3}} \left(\frac{S_0}{10^{50}\text{s}^{-1}}\right)^{\frac{2}{3}} \left(\frac{0.025\text{cm}^{-3}}{n_e}\right)^{2/3} \left(\frac{2 \times 10^{-13}\text{cm}^3\text{s}^{-1}}{\alpha}\right)^{\frac{1}{3}}. \tag{9}$$

Therefore the effective cross section of the softer diffuse radiation associated with a chimney has a cylindrical radius of ~ 1.2 kpc.

The filling factor of the ionizing radiation from OB associations embedded in chimneys is calculated assuming that the diffuse radiation above a chimney is cylindrical with scale height $1.5 kpc$ and radius 1.2 kpc. For N such chimneys in a galaxy the filling factor is given by

$$Q = \frac{\pi N r_{s2}^2}{\pi R_g^2} \sim \left(\frac{N^{\frac{1}{2}} r_{s2}}{R_g}\right)^2$$

$$\approx 1 \left(\frac{N}{100}\right) \left(\frac{\epsilon}{0.16}\right)^{\frac{2}{3}} \left(\frac{S_0}{10^{50}\text{s}^{-1}}\right)^{\frac{2}{3}} \left(\frac{0.025\text{cm}^{-3}}{n_e}\right)^{\phi} \left(\frac{15\text{kpc}}{R_g}\right)^2 \tag{10}$$

Since the superbubble and chimney formation process is stochastic some regions will find themselves without an ionization source at any given time. The recombination time is short so that significant cooling and possibly infall may occur. (It is conceivable that this could be a self regulating process with the cooling and infall triggering a further massive burst of star formation).

For edge-on systems there should be a characteristic change from hard to soft ionization species as one scans across the top of a chimney. For face-on systems a characteristic hard core with an extended soft halo should be seen across the holes blown by the chimney.

3. FURTHER STUDIES

The most important aspect of the meeting for me was the large number of further studies that need to be undertaken as a result of the new observational studies presented at the meeting many of which are based on outstanding new observations of external galaxies. I will try to summarize some of the most important of these problems needing further and urgent study.

The hot gas component content of our Galaxy seems to have a very uncertain role at least if one believes the discussions at this meeting where its filling factor and contribution to the energy balance of the interstellar medium ranged from very important to irrelevant. Clearly ROSAT and AXAF observations of the observed holes, bubbles, worms and chimneys in both face-on and edge-on external galaxies will be crucial here in helping us understand the significance and extent of the hot gas component in the energy balance of the interstellar media of galaxies.

The effect of magnetic fields on the propagation of supernovae and superbubbles needs to be closely studied. It is still uncertain whether the effect is of minor importance or dominant. The general dynamo theory for disks with realistic interstellar media is still awaiting a full development. In simple models it is the Parker instability that allows transport of field normal to the disk, with Coriolis force twisting the flux tubes followed by reconnection and subsequent infall back to the disk thus closing the dynamo action loop. Clearly, with chimneys the flux would be forced up the chimneys and the dynamo action with the standard twisting and reconnection would be less clear. Also important here is a better understanding of the structure of magnetized clouds and the relation of the field structure to the cloud structure and cloud distribution.

Important observational studies to be performed are those that study the ionization structure of the disk and halo. It is important to pin down observationally the contribution to the ionization of the halo of the various candidates including white dwarfs, planetary nebulae, OB associations and the metagalactic radiation field. Detailed observations in HI 21 cm, SII and $H\alpha$, etc. of a large sample of nearby galaxies will eventually tell us the source of the ionization of the halo. It will be a fairly complicated data set and comparison with simplified models such as that presented here in section 2 could prove useful.

The general morphology of the components of the interstellar medium seems most intriguing. Fascinating data presented at this meeting indicated how filamentary and sheetlike was the HI component with a covering factor near unity and a filling factor that is very small. It is not yet clear whether this could naturally arise from the cooled intersecting shells of supernovae and superbubbles or whether it is necessary to invoke a general cooling process in a magnetized medium where the filaments would naturally occur along the filaments.

The many observations of holes, worms, chimneys in various wavebands their association with regions of massive star formation require some systematic classification scheme with regard to their velocity fields, morphology, general physical conditions as revealed by line ratios, radio fluxes, polarization, dust content etc. The interplay between models and observations both current and planned are potentially very fruitful here. The physical distinctions and similarities between holes, worms, and chimneys need to be clarified and given a sound physical basis.

In this context it is important to undertake large surveys of a wide range of galaxy types and relative orientations in all possible wavebands including ROSAT imaging, HST absorption line studies, 21cm line studies, radio continuum, infrared imaging, narrow band imaging in the optical and UV from the ground and in space, and molecular line studies in the millimeter and submillimeter bands. The range of galaxies should extend from large edge-on systems such as NGC891 through more face on systems such as M33 to more active galaxies and starbursts such as NGC3079 through to blue compact dwarf galaxies such as 1Zw18.

The propagation of cosmic rays in the different types of interstellar media described here will differ in a number of respects. The damping of Alfven waves in a predominantly neutral medium such as is envisaged in the two- phase and chimney models can allow relatively free streaming of cosmic rays out of the galactic plane. In contrast in the three phase models the cosmic rays remain locked to the hot fluid. The chimney models could have large scale shock acceleration in the halo due to interaction with the large scale shocks propagating above the chimneys. The electron component is still somewhat of a mystery and the source of the cosmic ray electrons and their subsequent reacceleration in both the disk and the halo needs considerable further study.

An important issue at the meeting became whether or not the superbubbles could break out of the extended ionized layer discovered by Reynolds. There is no question that superbubbles do break out of the smaller scale height neutral layer. As discussed at this meeting it is only the most rare and energetic explosions that can break out of the Reynolds layer. I see no real problem here as the natural explanation is that the superbubbles break out of the disk forming holes, worms and chimneys as they propagate upwards. Their combined action is to *produce* the Reynolds layer. This would explain why the superbubbles don't normally propagate through the layer since they *form* it and also can explain the origin of the layer itself.

A related issue concerns the source of ionization for the chimney model of the ionization of the Reynolds layer. It is obvious that the massive stars that created the chimney through which the direct and diffuse components are emanating are not the source since they would have died earlier. The source is likely to be the massive stars triggered later by the propagation of the superbubble. This mode of star formation is given considerable credence by the observations of the inner shells of $H\alpha$ lining the walls of the supershells seen in the Magellanic clouds.

It is a pleasure to thank Robert Braun, Eli Brinks, Don Cox, Laura Danly, Don Garnett, Tim Heckman, Jeff Hester, Jim Pringle, Ron Reynolds Nick Scoville and Rene Walterbos for interesting discussions and help in understanding some of these difficult problems. I especially thank both Nino Panagia who shared his wisdom on HII regions with me in our collaboration on the ionization of the halo and Satoru Ikeuchi for many profound insights on the physics of the interstellar medium. It was a pleasure to be back in Leiden where some of my thoughts on this subject were developed in the early eighties. It was also good to hear such excellent talks here from a number of those who were graduate students during that time and who are now major contributors to the field.

REFERENCES

Braun, R. (1990) *Ap. J. Suppl.* **72**, 755
Bregman, J.N., Harington, J.P. (1986) *Ap. J.* **309**, 833
Brinks, E. (1990) in *The Interstellar Medium in External Galaxies*, ed. H.A. Thronson and J. M. Shull
Brinks, E. (1990) in *The Interstellar Medium in Galaxies*, ed. H.A. Thronson and J. M. Shull, p. 39 (Kluwer)
Brinks, E. and Bajaja, E. (1986) *Astr. Astrophys.* **169**, 14
Chevalier, R.A., Fransson, C. (1984) *Ap. J. (Letters)* **275**, L71
Dettmar, R.J. (1990) *Astron. Astrophys.* **232**, L15
Deul, E.R. and den Hartog, R.H. (1990) *Astron. Astrophys.* **229**, 362
Dopita, M.A., Mathewson, D.S. and Ford, V.L. (1985) *Ap. J.* **297**, 599
Field, G.B., Goldsmith, D.W. and Habing, H.J. (1969) *Ap. J. (Letters)* **155**, L149
Heiles, C. (1990) *Ap. J.* **354**, 483
Heiles, C. (1987) *Ap. J.* **315**, 555
Kamphuis, J. (1990) poster presented at this meeting
Li, F. and Ikeuchi, S. (1990) *Ap. J. Suppl.* **73**, 401
Mathis, J.S. (1986) *Ap.J.* **301**, 423
McKee, C.F. and Ostriker, J.P. (1977) *Ap. J.* **218**, 148
Meaburn, J., Marston, A.P., McGee, R.X. and Newton, L.M. (1987) *M.N.R.A.S.* **225**, 591.
Norman, C. A. and Ikeuchi, S. (1989) *Ap. J.* **345**, 372
Norman, C.A. and Panagia, N. (1990) *Ap. J.*, in preparation
Osterbrock, D.E. (1989) *Astrophysics of Gaseous Nebulae and Active Galactic Nuclei* (University Science Books)
Rand, R.J., Kulkarni, S., and Hester, J.J. (1990) *Ap. J. (Letters)* **352**, L1
Reynolds, R.J. (1990) *Ap. J.* **349**, L17
Sciama, D.M. (1990) *M.N.R.A.S.* **244**, 1p
Spitzer, L. (1978) *Physical Processes in the Interstellar Medium* (Wiley-Interscience)
Thuan, T.X. (1975) *Ap. J.* **198**, 307
Walterbos, R.A. (1990) these proceedings

COSMIC RAY POWERED FOUNTAINS AND WINDS

H.J. VÖLK

Max-Planck-Institut für Kernphysik, P.O. Box 103980
D-6900 Heidelberg, West Germany

ABSTRACT. The dynamics of galactic halos is discussed with special emphasis on the role of Cosmic Rays. Whereas Cosmic Rays should at best play a minor role for galactic fountain flows they appear essential in driving a wind from galaxies like ours. Active galaxies with a very hot interstellar medium (T $\geq 10^7$K) have winds due to the thermal gas alone.

1. INTRODUCTION

Let me start by changing the title of this talk and cross out fountains from the title. This is because I do not believe that Cosmic Rays (CRs) play an important role in powering Galactic fountains. And I will give the corresponding physical arguments in section 2.3 below.

The question of the dynamics of the Galactic halo, especially that of winds and fountains has been discussed mostly without regard to CRs (e.g. Mathews and Baker, 1971; Habe and Ikeuchi, 1980; Kahn, 1981; Cox, 1981). These considerations included as an essential ingredient the radiative cooling of the thermal gas in a gravitational field that is determined by the stars of total mass M_* and, sometimes, in addition, by a distributed dark matter halo (with a mass $M_{Halo} \approx 3M_* \approx 8 \times 10^{11} M_\odot$). Assuming a spatially uniform disk, thermally driven winds exist only if the gas temperature in the disk is high enough, roughly, if T_{disk} exceeds 4.10^6K, and 8.10^6K at a galactocentric distance of 12 kpc, and 1 kpc, respectively. This holds for the case of a massive dark halo. Without a dark matter halo the minimum disk temperatures are smaller by about a factor of 2. For lower temperatures of gas, injected at disk midplane, the gas cools radiatively and falls back to the disk. This is the so-called Galactic fountain as described in the original paper by Shapiro and Field (1976).

If the disk gas is allowed to be nonuniform, both spatially and temporally (e.g. Norman and Ikeuchi, 1989, and references therein), then the energy input can be taken as correlated, as is for instance the case for Type II-SN explosions in OB associations (Heiles, 1984, 1989; Kennicutt, Edgar, and Hodge, 1989). The corresponding bubbles of hot gas can break through the extended neutral disk gas (Lockman, Hobbs, and Shull, 1986) into the halo which I define here as the region of fully ionised gas, commencing on average at a height $|z| \approx 1$ kpc above (and below) the disk midplane. A given energy input and mean disk gas density can either lead to winds, i.e. gas secularly escaping from the galaxy, or to fountains or to both, at different positions. For our Galaxy at the present epoch, Norman and Ikeuchi prefer the fountain alternative.

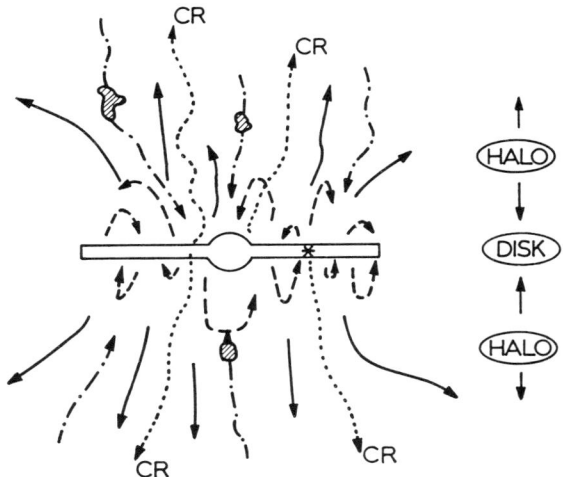

Figure 1: The various halo flows above the disk of a galaxy are shown schematically: Gas moving upwards and falling back to the disk, the fountains (dashed curves); hot gas moving secularly away from the disk, galactic wind (solid lines); infall of circum- or extragalactic matter (dashed-dot lines) in the form of "clouds" (hatched); CRs escaping the galaxy (dotted curves).

These are not the only possible flows in the halo of a galaxy (Figure 1): material might, in addition, fall in from large distances "outside" the galaxy (e.g. Mirabel, 1989), and finally the nonthermal gas component, the CRs, escapes the Galaxy. The latter fact and its consequences will be the main subject of this talk.

2. "STATIC" HALO INCLUDING COSMIC RAYS

2.1 THE NONTHERMAL CR COMPONENT OF THE INTERSTELLAR MEDIUM. The CRs constitute the nonthermal, to a large extent even relativistic component of the Interstellar Medium (ISM). The dynamically relevant CRs with (individual particle) energies $\leq 10^{15}$ eV probably originate at the shocks of Supernova Remnants (SNRs). These can convert some 10 to 30 percent of the total hydrodynamic energy released in a Supernova explosion into relativistic nucleons (Drury, Markiewicz, and Völk, 1989) essentially independent of the conditions of the ambient ISM (Markiewicz, Drury, and Völk, 1990). Within a factor of a few Supernovae thus appear to be standard candles for CRs wherever they explode! For our Galaxy these particles then populate the ISM for timescales (at energies E of a few GeV/ nucleon) $t_{esc} \leq 10^8$ yr, decreasing with increasing E, before they escape. For the CR pressure p_c we have $p_c \approx 3.10^{-13}$

dyn/cm^2 $\leq p_g \approx 6.10^{-13}$ dyn/cm^2, where p_g is the pressure of the thermal gas in the disk.(This value for p_c is about a factor of 3 smaller than the one used by de Boer (1990) in the preceding talk). The average CR energy flow leaving the disk of the Galaxy is $F_{co} \approx 10^{41}$ erg/sec, roughly equal to a tenth of the total energy input from Galactic Supernovae. Since the mean random velocities of the CR particles approach the speed of light, their scale height h_c as a gas tends to exceed h_g, the scale height of the thermal gas. In sharp contrast to the gas, CR nucleons suffer essentially no radiative cooling. Compared to typical spatial scales of the Galactic disk, their gyroradii are minute of order 10^{-6} pc at E = 5 GeV in a 5μG magnetic field; they can escape only along magnetic field lines. Thus the conclusion is clear: the field lines must be partially open to extragalactic space. The question that remains is whether this escape is "bloodless", or whether the CRs drag some of the gas with them! The answer to this question depends on the transport properties of CRs. Their observed near-isotropy in momentum space shows that the particles are well scattered on fluctuations (waves) of the magnetic field $\underline{B}$ leading to spatial diffusion in the wave frame. This implies strong momentum coupling to the gas. Therefore, we can now rephrase the above question in an alternative form: is the gas dragged away by the CRs, at least at great heights in the halo, or do the CRs merely diffuse and drift (with the waves) away through the static gas (static halo)? Let me first discuss static halos.

2.2 STATIC COSMIC RAY HALOS (LOW DENSITY GAS) Dynamically selfconsistent static CR halos with selfexcited magnetic fluctuations and adiabatic thermal gas have been considered by Dougherty, McKenzie, and Westergaard (1985), and by Ko, Dougherty, and McKenzie (1990). These authors confined themselves to a plane halo geometry with straight field lines $\underline{B}$ = const. perpendicular to the plane of the Galactic disk. Ko et al. find that CR diffusion is negligible compared to Alfvénic drift as long as the (Rosseland) mean diffusion coefficient $\bar{\kappa} \leq 10^{29}$ cm^{2}sec^{-1}, a number which has traditionally been considered as an upper bound for kinematic, purely diffusive and static CR halos (see e.g. the review by Ginzburg and Ptuskin, 1985). Thus diffusion in the wave frame can be neglected and then, as already shown by Dougherty et al., the dynamical equations for p_g, p_c, ϱ, and $<(\delta\underline{B})^2>$ can be integrated analytically; here ϱ and $<(\delta\underline{B})^2>$ denote the mass density and mean square magnetic fluctuation amplitude, respectively. It is interesting that this strong coupling solution can actually be generalized to any open magnetic field configuration with arbitrary flux tube crossection A(z), in particular to a halo that becomes spherically symmetric at distances large compared to the disk radius. I will not pause here for the proof but simply give a typical result (independent of A(z) in Table 1), where the suffices ∞ and o denote quantities for $|z| = \infty$, and $|z| = $ 1kpc at galactocentric distances R_o. The quantitiy p_w denotes the wave pressure $<(\delta\underline{B})^2>/8\pi$. The parameters at $|z| = 1$ kpc were chosen corresponding to a hot interstellar medium: $n_o = 3.10^{-3}$ H-atoms/cm^3, $p_{go}/K = 4000$ K/cm^3, $p_{co} = 2.7 \times 10^{-13}$ dyn/cm^2, $B_o = 1\mu$G. The escape speeds to infinity at the galactocentric distances $R_o = 1$, and 10 kpc were taken as $u_{esc}(z=\infty) = 420$, and 600 km/sec, respectively.

Thus, even if we were to assume that the dominant wave pressure p_w could ultimately be dissipated away by some nonlinear damping or mode coupling process, the ratio $p_{c\infty}/p_{co}$ is quite large and probably too large to be in equilibrium with the average intergalactic pressure. However $p_{c\infty}$ is presumably quite a bit lower than the total pressure in gas-rich clusters of galaxies. Since the gas should be fully ionized at $|z| > 1$kpc by OB-stars, halo and globular stars, and quasars (therefore assuring strong momentum coupling of gas and CRs) we expect such static halos to normally give way to winds except in clusters with high pressure hot intracluster gas.

TABLE 1

Gas pressure $p_{g\infty}$, CR pressure $p_{c\infty}$, and wave pressure $p_{w\infty}$ at infinity for a static CR halo, normalised to the values $p_{go} = 5.5 \times 10^{-13}$ dyn/cm^2 and $p_{co} = 2.7 \times 10^{-13}$ dyn/cm^2 at the reference level $|z| = 1$kpc, for two galactocentric distances $R_o = 1$ and 10 kpc in the disk.

	$p_{g\infty}/p_{go}$	$p_{c\infty}/p_{co}$	$p_{w\infty}/p_{co}$
$R_o = 1$kpc	10^{-4}	2.8×10^{-2}	8×10^{-2}
$R_o = 10$kpc	3×10^{-3}	10^{-1}	1.5×10^{-1}

2.3 "HIGH DENSITY" GAS (FOUNTAINS). The case of "high density" gas is much more complicated because radiative cooling and recombination play an important role. The borderline in density to an essentially non-cooling, fully ionized "low density" halo gas is of course somewhat uncertain, since this depends also on the history of an element of gas and the spatial distribution of heating and ionizing agents. However, the cooling shells of SNRs or, perhaps, expanding wind bubbles, certainly belong to the "high density" gas. The recombination of the gas then leads to wave damping due to ion-neutral friction (Kulsrud and Pearce, 1969) and increased diffusion of CRs through the gas. One wonders if this is not the average situation near the disk for $|z| < 1$kpc (see, e.g. the review by Cesarsky, 1980). Yet, galactic fountains are overwhelmingly driven by the gas internal and dynamic pressures until radiative cooling starts to dominate the internal energy loss. Two mechanisms appear to me most likely for fountains: either the fountain is a priori ballistic, e.g. cooling SNR shell fragments continue an inertial motion upwards at already high densities and fall down subsequently. Or, hot gas wells up at somewhat lower but still "high densities" to cool, recombine, and fall down subsequently. Cosmic rays should be unimportant under both circumstances. In the first case simply due to the high inertia of such shell fragments, above everything else, and in the second still because of diffusive decoupling of CRs from gas. Thus I believe that for fountain flows the CRs are at best a nuisance, rather than a dynamical element.

Low density, ionized gas on top of high density gas should, however, be blown away by the CRs. This should then constitute a galactic wind.

3. GALACTIC WINDS

3.1 MODEL ASSUMPTIONS. The first explicit theory of CR driven winds is due to Ipavich (1975). He restricted his attention to a spherically symmetric, stellar type geometry, and assumed the wave pressure to be negligible, the waves being dissipated locally and completely. The case of a plane geometry near the disk, changing over to a spherical geometry at distances exceeding the disk radius R_{gal} = 15 kpc in a dark matter halo, and explicitly including the wave pressure, was considered by Breitschwerdt, McKenzie, and Völk (1987,1990), and is briefly summarized here. The model assumptions are as follows: Only the halo dynamics is considered, with fully ionized gas, above a reference level $|z_0|$ of order 1 kpc. The gas is considered adiabatic - not a very critical assumption at the low reference level densities of order 10^{-3} H-atoms/cm^3 - in a 3-component flow where the gas, the CRs, and the wave field interact nonlinearly. The scattering waves are assumed to propagate outwards, amplified by the outwards streaming CRs. CR diffusion is assumed negligible in the wave frame that moves with Alfvén velocity $\underline{v}_A = \underline{B}/(4\pi \rho)^{1/2}$ through the gas. Some magnetic flux tubes above the disk are open towards intergalactic space with flux tube area $A(z) = A_o(1+(z/z_1)^2)$, where z_1 = 15 kpc, and the dynamics is considered only for those flux tubes. These flux tubes will "touch" for $|z| \gg 1$ kpc since the combined CR and wave pressures within the flux tubes exceeds the sum of gas pressure between and average field tension on flux tubes beyond such distances (Figure 2).

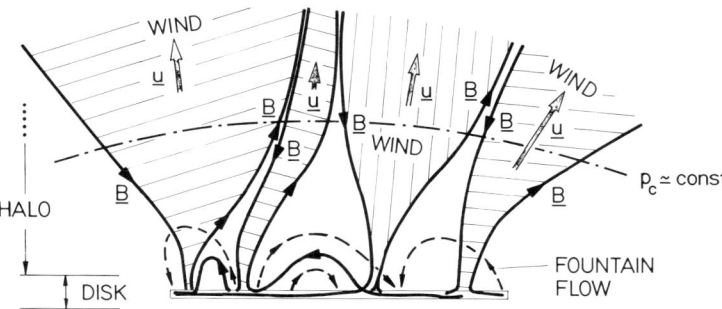

Figure 2: Schematic picture of the Galactic halo open flux tubes with CRs (hatched) which carry the galactic wind. They touch each other due to the dominance of the CR pressure p_c at greater heights. Wind speeds $\underline{u}$ can vary from one flux tube to the other. At low heights above

the disk fountain flows and closed magnetic field lines $\underline{B}$ dominate the 3-dimensionally and magnetically structured halo.

At low $|z|$ closed field lines as well as fountain flows are abundant, but disjoint from the open flux tubes; the magnetic structure is correlated over times of order 10^6 to 10^7yrs, the lifetime of SNRs and Superbubbles (e.g. Heiles, 1989), respectively. In such a structure the closed loops could for example be compared with those discussed by Boulares and Cox (1989). The fundamental difference is that in the model of Breitschwerdt et al. a large part of the hot gas and all the CRs exist on the open flux tubes. This relieves the closed structures of excessive particle pressure forces and, in the extreme, allows it to relax to a force free configuration. Closer in spirit is the characterisation of the halo in terms of superbubbles breaking out of the disk for which Norman and Ikeuchi (1989) invented the term "chimneys". However, I believe that also Supernovae exploding at heights $|z| > 100$ pc, i.e. outside the dense gas disk, should often be able to push field lines open and to initiate an outflow.

Finally, we neglect galactic rotation - a conservative assumption - and assume a steady state.

3.2 WIND SOLUTIONS WITH COSMIC RAYS. The dynamics is described by the overall momentum balance

$$\rho u \frac{du}{dz} = - \text{grad}\, (p_g + p_c + p_w) - \rho \cdot \text{grad}\, \phi$$

plus appropriate energy balances for the gas, the CRs and the waves; the quantities u and ϕ denote the mass velocity and the gravitational potential, respectively. Assuming the intergalactic pressure to be negligible the equations admit outflow solutions with a subsonic - supersonic transition, i.e. supersonic winds. The square of the appropriate sound speed

$$c_*^2 = \gamma_g \cdot \frac{p_g}{\rho} + \frac{(M_A + 1/2)^2}{(M_A + 1)^2} \cdot \gamma_c \, \frac{p_c}{\rho} + \frac{(3M_A + 1)}{2(M_A + 1)} \cdot \frac{p_w}{\rho}$$

contains a sum of terms proportional to the pressures of the 3 dynamical components, where $M_A = u/v_A$ is the Alfvén Mach number, $\gamma_c \approx 4/3$ the adiabatic index of the CRs, and $\gamma_g = 5/3$ that of the thermal gas.

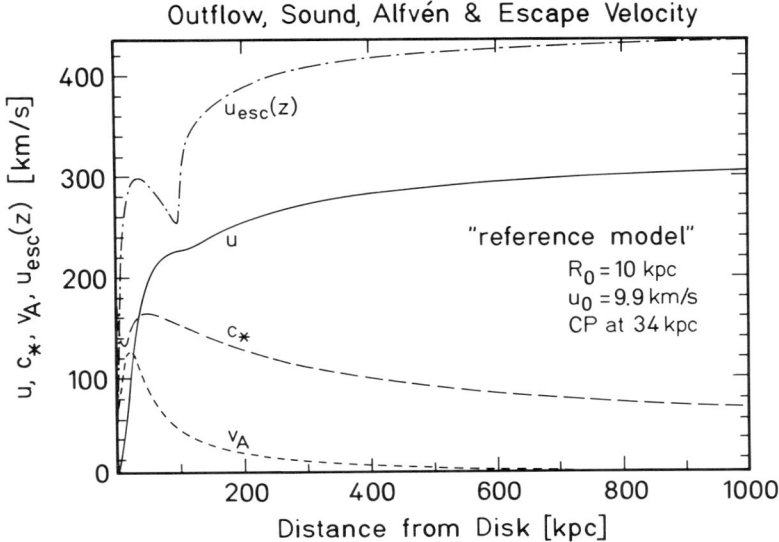

Figure 3: A plot of flow speed u, effective sound speed c_*, Alfvén velocity v_A, and escape speed $u_{esc}(z)$ to height z, as a function of the distance z from the disk. The "reference model" describes a flux tube orginating at R_o at 10 kpc and $|z_o|$ =1 kpc with $p_{go}=2.8\times10^{-13}$dyn/cm^2, $p_{co}=1\times10^{-13}$dyn/cm^2, ϱ_o = 1.67 x 10^{-27} g/cm^3, B_o = 1 μG, and a negligible wave pressure level p_{co} = 4 x 10^{-16}dyn/cm^2. The critical point (CP), where u= c_A, lies at 34 kpc; the resulting flow speed at $|z_o|$ equals u_o= 9.9 km/sec. The kink in $u_{esc}(z)$ occurs because the dark matter halo was assumed to terminate at 100 kpc distance (from Breitschwerdt et al., 1990).

Figure 3 contains a representative solution for a "reference model" (Breitschwerdt et al., 1990), characterised by a flux tube foot point at R_o = 10 kpc, $|z_o|$ = 1 kpc. The critical point (CP) is at $|z|$ = 34 kpc, far from the disk. For the same parameters, but at R_o = 1 kpc, the asymptotic outflow velocity is 1500 km/sec.

3.3 GALACTIC MASS LOSS. The total energy loss, in the case of the Galaxy, is of the order of the CR energy flux from the sources of CRs. Taking the asymptotic outflow velocity as being of the order of the escape velocity u_{esc}, the mass loss rate from a given flux tube is $\dot{M}_{\text{flux tube}} \approx F_{co} \cdot A_o / u_{esc}^2$, where the suffix o indicates the reference level value. The total mass loss rate from the Galaxy at the present epoch is then $\dot{M}_{gal} = \Sigma \dot{M}_{\text{flux tubes}} \approx 1 M_\odot$/yr. This rate would correspond to a mass loss over the age t_{gal} of the Galaxy $\dot{M}_{gal} \cdot t_{gal} \approx$

$10^{10} M_\odot \cdot [t_{gal}/10^{10} yr] \approx M_{gas}^{present}$, of the order of the entire present ISM mass. Higher star formation rates in the past would have resulted in correspondingly higher mass loss rates.

3.4 CHEMICAL EVOLUTION OF GALAXIES. The galactic winds contain Supernova-processed hot gas. If single SNRs contribute to the mass loss, then the corresponding CRs, coming from the SNR's outer shock, have interstellar abundances. The CRs from superbubbles, however, where successive SN explosions accelerate thermal particles from the debris of the last generation SNRs, would contain an excess of heavy elements. For the gas we should, for both cases, though to different degrees, expect preferential removal of heavy elements. Thus the chemical evolution of the Galaxy should show a "slowdown" relative to naive expectations from star formation history. This appears to be indeed the case. The observed lack of heavy element enrichment in the CRs should give a limit on the Superbubble contribution to CR production, at least at the present epoch in the solar neighbourhood.

3.5 EXTERNAL GALAXIES From what has been said before, I believe that for field galaxies winds should be common, even for galaxies as relatively quiescent as ours. For starburst galaxies, like M 82, where the gas is hot ($T_g \geq 10^7 k$), the gas alone is able to drive a wind, as also the present theory predicts (for much earlier calculations, see Chevalier and Clegg (1985)). Then the CRs make a 10 to 30 percent contribution which is not essential. In gas-rich clusters of galaxies, on the other hand, the large intracluster pressure should inhibit winds from individual cluster members. However, generalizing our earlier arguments, we then expect a wind from the cluster as a whole.

Observations of galactic winds are severely hampered by low surface brightness. This is perhaps not too surprising for normal galaxies if we remember that the Solar Wind cannot be observed by its photon radiation even from the earth. For spectacular starburst galaxies like M 82 on the other hand, a combination of X-ray and radio continuum observations shows unequivocally the existence of massive and fast winds. This has been discussed by other speakers at this symposium and therefore I will not go into it here any further except to say that M 82's mass loss is consistent with a SNR origin of the CRs (Völk, Klein, Wielebinski, 1988, 1990). Also the radio observations of closeby edge - on galaxies have been discussed in dedicated talks at the meeting (Hummel, 1990; Beck, 1990). These observations show not only good evidence for radial (= open?) field lines in the case of NGC 4631 but also interesting $|z|$ -dependencies of the polarization of the synchroton emission. Due to the spatially increasing amplitudes of the wave field, we expect a drop of the polarization at large $|z|$ from its rise from lower $|z|$. It would be interesting to verify such an expectation for a sample of nearby edge-on galaxies.

It seems that we will not have to wait too long to get these observations. Then galactic winds, even from normal galaxies, will be more than a compelling theoretical prediction. The consequences for the mass and energy balance of the extragalactic medium need to be investigated in the future.

ACKNOWLEDGEMENTS. I would first of all like to thank D. Breitschwerdt and J.F. McKenzie for the continued collaboration on galactic winds we enjoy together. My thanks are also due to R. Beck, C.J. Cesarsky, M. Dougherty, E. Hummel, and C.M. Ko for discussions on various aspects of this paper.

REFERENCES

Beck, R. (1990) *these proceedings*
Boulares, A., Cox, D.P. (1989) *Ap. J.* submitted
Bregman, J.N. (1980) *Ap. J.* **236**, 577
Breitschwerdt, D., McKenzie, J.F., Völk, H.J. (1987) in *"Interstellar Magnetic Fields"*, eds. R. Beck, R. Gräve, Springer Heidelberg, p. 131
Breitschwerdt, D., McKenzie, J.F., Völk, H.J. (1990) submitted to *Astron. Astrophys.*
Cesarsky, C.J. (1980) *Ann. Rev. Astron. Astrophys.* **18**, 289
Chevalier, R.A., Clegg, A.W. (1985) *Nature* **317**, 44
Cox, D.P. (1981) *Ap. J.* **245**, 543
de Boer, H. (1990) *these proceedings*
Dougherty, M.K., McKenzie, J.F., Westergaard, N.J. (1985) *Proc. 19th. Int. Cosmic Ray Conf., La Jolla* **3**, 83
Drury, L. O'C., Markiewicz, W.J., Völk, H.J. (1989) *Astron. Astrophys.* **225**, 179
Ginzburg, V.L., Ptuskin, V.S. (1985) *Sov. Sci. Rev. E. Astrophys. Space Phys.* 4, 161
Habe, A., Ikeuchi, S. (1980) *Prog. Theor. Phys* **64**, 1995
Heiles, C. (1984) *Ap. J. Suppl.* **55**, 585
Heiles, C. (1989) in *Proc IAU Coll. No. 120.* eds. G. Tenorio–Tagle, M. Moles, J. Melnick, Springer, Heidelberg, p. 484
Hummel, E. (1990) *these proceedings*
Ipavich, F. (1975) *Ap. J.* **196**, 107
Kahn, F.D. (1981) in *"Investigating the Universe"*, eds. F.D. Kahn, Reidel, Dordrecht, p. 1
Kennicutt, R.C. Jr., Edgar, B.F., Hodge, P.W. (1989) *Ap. J.* **337**, 761
Ko, C.M., Dougherty, M.K., McKenzie, J.F. (1990) *Astron. Astrophys.* in press
Kulsrud, R.M., Pearce, W.D. (1969) *Ap. J.* **156**, 445
Lockman, F.J., Hobbs, L.M., Shull, J.M. (1986) *Ap.J.* **301**, 380
Markiewicz, W.J., Drury, L. O'C., Völk, H.J. (1990) *Astron. Astrophys.* in press
Mathews, W.G., Baker, J.C. (1971) *Ap. J.* **170**, 241
Mirabel, I.F. (1989) in *Proc. IAU Coll. No. 120*, eds. G. Tenorio– Tagle, M. Moles, J. Melnick, Springer, Heidelberg, p. 396
Norman, C.A., Ikeuchi, S. (1989) *Ap. J.* **345**, 372
Shapiro, P.R., Field, B.G. (1976) *Ap. J.* **205**, 762
Völk, H.J., Klein, U., Wielebinski, R. (1988) *Astron. Astrophys.* **213**, L2
Völk, H.J., Klein, U., Wielebinski, R. (1990) to appear in *Astron. Astrophys.*

THE X-RAY APPEARANCE OF SUPERNOVA REMNANTS IN TENUOUS MEDIA

DENIS F. CIOFFI
Physics Department
North Carolina State University
Raleigh, NC 27695-8202 USA

ABSTRACT. By estimating the X-ray intensities of old supernova remnants in the disk-halo interface, deductions can be made about the ability of isolated supernovae to have created a coronal halo.

1. INTRODUCTION

The relative lack of obscuration to the halo permits some confidence that no high-z supernovae (SNe) have escaped detection in recent human history (500 - 1000 yr), but some halo massive stars (e.g., Tobin, this volume), or Type I's in binary systems, presumably have exploded in the past as SNe. The remnant of SN1006, with galactic latitude $b = 14.°6$, and the Lupus Loop ($b = 15.°0$) are the known supernova remnants (SNRs) with the greatest heights above the plane (Green 1988). Do the more ancient remnants still emit X-rays, and how important are these SNRs to the large-scale structure of the halo?

First consider how the environmental conditions peculiar to the halo affect, in general, the evolution of SNRs. Most importantly, the low density will result in a larger expansion than disk SNRs, with lower interior pressure and lower luminosity. The lower metallicity of the swept material also means a less-luminous remnant. A smaller amount of dust means that infrared radiation will have even less of an effect on the dynamics than it might have for disk SNRs (Dwek 1981). Finally, when compared to the two-dimensionality of the disk (Heiles 1987), the full three-dimensionality of the halo allows a straightforward porosity calculation (§4).

2. THE CALCULATION

In extremely tenuous media SNRs will never evolve beyond the Sedov-Taylor stage (Sedov 1959; Taylor 1950; Cioffi, McKee, Bertschinger 1988, hereafter CMB), but in the disk-halo interface (DHI) the investigation of dynamically old remnants generally refers to SNRs that have made the transition to the radiative stage and formed a cold shell (CMB; see Cioffi [1990] for a simple description of SNR evolution throughout all stages). Here I sketch the basic physics behind the luminosity calculations, with details to be found in Cioffi and McKee (1988, 1991).

If one can approximate the cooling of a hot gas with a power law in temperature T such that the cooling rate (ergs cm^3 s^{-1}) is proportional to $T^{-1/2}$, then, at time t, the entropy of a parcel of gas that was shocked at time t_s depends only on its initial entropy (i.e., on the speed of the shock at t_s) and on the time interval $t - t_s$ (Kahn 1976). For a gas of cosmic abundances we can make this cooling approximation for $5 \lesssim \log T \lesssim 7.5$. The time t_s thus "flags" all gas elements that were originally at a distance $R_s(t_s)$ from the center of the spherical explosion.

If n_o is the hydrogen density of the ambient medium, and R_s and v_s are the radius and the velocity of the SNR shock, then the volume occupied by those gas elements that have been shocked prior to any chosen time $t_s \leq t$ is given by

$$\mathcal{V}(t, t_s) = \int_0^{t_s} \frac{n_o}{n(t, t'_s)} 4\pi R_s^2(t'_s) v_s(t'_s) dt'_s;$$

when $t_s = t$, the volume $\mathcal{V}$ equals the total volume of the SNR, $V = 4/3\pi R_s^3$. With a uniform pressure throughout the SNR (CMB), one can find the density $n(t, t_s)$ and then transform from the volume $\mathcal{V}$ to a radius r to produce the profiles $T(r)$ and $n(r)$. The product $n^2 T^{-1/2}$ is proportional to the emissivity, and we introduce a rough spectral dependence by multiplying by $\exp(-\epsilon/k_b T)$, where k_B is Boltzmann's constant and ϵ has been set to 0.1 keV in this work .

Non-equilibrium ionization will not be important at late times, but including magnetic fields and thermal conduction (see Cox, this volume) will result in a cooler interior, so the luminosities that I calculate here are upper limits.

2.1 Relevant Parameters

The scale heights of the thick disk of Population II stars (Gilmore, this volume), the neutral hydrogen layer (Lockman, this volume), and the ionized hydrogen layer (Reynolds, this volume) are all $\sim 1 - 2$ kpc, so in the DHI[1] a SN probably evolves into densities of the order $0.1 - 0.01$ cm^{-3}. The mean metallicity (from K giants) is ~ 0.25 solar (Gilmore *et al.* 1989). Burton (1991, this volume) has pointed out that conditions in the outer Galaxy duplicate those in the DHI.

3. X-RAYS FROM DHI SNRS

At densities $\log n_o = 0, -1, -2, -3$, Figure 1 shows the radius $r_{0.1}$ within which the temperature is above 1.16×10^6 K (i.e., 0.1 keV), as a function of time from the shell formation time, t_{sf} (CMB), to $10 t_{sf}$. Although the radius of the outer shell doubles in this time, $r_{0.1}$ increases by only $\approx 50\%$: the volume increase of the hot gas is less than half the total increase of volume enclosed by the SNR. Figure

---

[1] OB stars may exist at much greater heights, $z \sim 6$ kpc (e.g., Keenan *et al.* 1986).

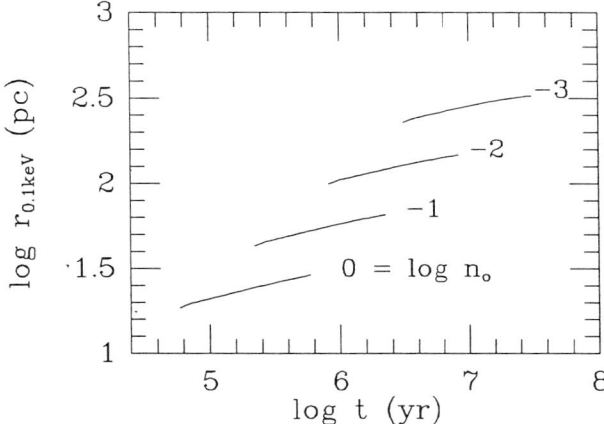

Figure 1. Log radius (pc) within which gas temperature is above 1.16×10^6 K. Merger with ISM probable before large radii shown at $n_o = -3$ dex (CMB).

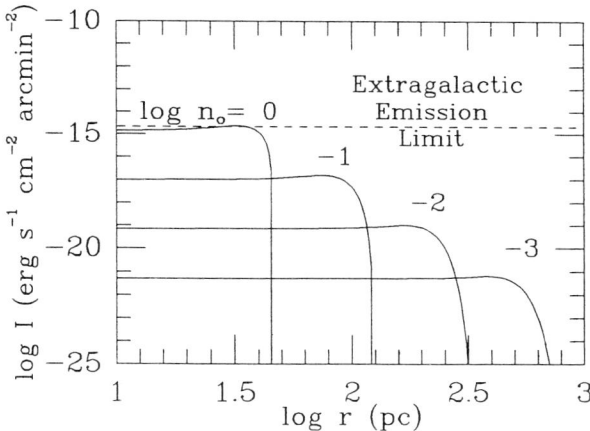

Figure 2. Intensity of old ($t = 10 t_{sf}$) SNRs as a function of size. Most are below background. (At a distance of 3.44 kpc, 1 pc subtends 1 arcmin on the sky.)

2 shows the intensity (ergs s^{-1} cm^{-2} arcmin^{-2}) across the spherical surface. From Wu et al. (1991) the number for the extragalactic X-ray background is also noted.

4. CONSEQUENCES

X-ray luminosity peaks at $t \approx t_{sf}$. Figure 2 indicates that we will not detect older halo SNRs; even if X-ray emission is observed, the lack of substantial limb brightening may prevent recognizing the SNR as a distinct object.

A three-dimensional porosity calculation (e.g., Heiles 1987) should use the volume of the hot gas, not the entire SNR. The porosity parameter Q gives the volume fraction of hot gas through $Q/(Q+1)$. Obtaining $Q = 1$ in the DHI would require the volumetric SN rate to be ~ 0.25 times the disk rate (van den Bergh et al. 1987), an unlikely number: individual SNRs in the halo will leave behind isolated hot volumes that will not create a global coronal medium.

ACKNOWLEDGEMENTS

"At-home," I thank Eric Sharpe for intensity integrations and Steve Reynolds for many discussions. "On the road," I thank Hans & the L.O.C. for a most pleasant, well-organized, and stimulating meeting, where discussing this particular topic with Mike Shull, Jeff Hester, and Paul Shapiro was both enlightening and enjoyable.

REFERENCES

Burton, W. B. (1991) this volume
Cioffi, D. F. (1990) in *Physical Processes in Hot Cosmic Plasmas*, eds. W. Brinkmann, A. C. Fabian, F. Giovannelli, Kluwer, Dordrecht, p. 1
Cioffi, D. F., McKee, C. F. (1988) in *Proc of IAU Coll. 101, Supernova Remnants and the Interstellar Medium*, eds. R. S. Roger, T. L. Landecker, C.U.P., Cambridge, p. 435
—————————————— (1991), in preparation
Cioffi, D. F., McKee, C. F., Bertschinger, E. (1988) *Ap. J.* **334**, 252 (CMB)
Cox, D. P. (1991) this volume
Dwek, E. (1981) *Ap. J.* **247**, 614
Gilmore, G. (1991) this volume
Gilmore, G., Wyse, R. F. G., Kuijken, K. (1989) *Ann. Rev. Astr. Ap.* **27**, 555
Green, D. A. (1988) *Ap. & Sp. Sci.* **148**, 3
Heiles, C. (1987) *Ap. J.* **315**, 555
Kahn, F. D. (1976) *Astr. Ap.* **145**, 50
Keenan, F. P., Lennon, D. J., Brown, P. J. F., Dufton, P. L. (1986) *Ap.J.* **307**, 694
Lockman, F. J. (1991) this volume
Sedov, L. I. (1959) *Similarity & Dimensional Methods in Mechanics*, Academic Press, NY
Taylor, G. I. (1950) *Proc. Roy. Soc. London* **201A**, 159
Tobin, W. (1991) this volume
Wu, X., Hamilton, T., Helfand, D. J., Wang, Q. (1991), preprint
van den Bergh, S., McClure, R. D., Evans, R. (1987) *Ap. J.* **323**, 44

STATIC VERSUS DYNAMICAL COSMIC-RAY HALOS

FRANK C. JONES
Laboratory for High Energy Astrophysics, Code 665
NASA/Goddard Space Flight Center
Greenbelt, MD 20771, USA

ABSTRACT. The dynamical halo of the Galaxy offers a natural explanation for the form of the variation of cosmic-ray path length with energy. The variation above 1 GeV per nucleon can be understood as due to the variation of the diffusion coefficient, and hence the resident time in the galaxy, with energy. The flattening of the curve below 1 GeV per nucleon is seen to mark a transition to a convection dominated regime where the variation of the diffusion coefficient is no longer a determining factor. It is possible that the random motion of the cosmic rays about the galaxy that prevents us from seeing their sources in a clear manner may enable us to extract information about the galaxy at large and learn something about its large scale motions.

1. INTRODUCTION

It was pointed out by Shklovsky (1952) that the observed high latitude distribution of non-thermal radio emission could be explained by postulating an extensive halo of radio emissivity surrounding the disk of the Galaxy. Pikel'ner (1953) suggested that a gaseous halo with a magnetic field of $\approx 3 \times 10^{-6}$ gauss would be needed to retain the particles responsible for the emission. Soon ideas of this sort of galactic halo had permeated the thinking of those concerned with the problem of the propagation of cosmic rays in the Galaxy (Bierman and Davis, 1958.)

The notion of storage of cosmic rays in a volume larger than the galactic disk was attractive because it gave a natural answer to the question of why the radiation was so isotropic. Furthermore, it offered an explanation for the fact that the combination of mean age of the cosmic rays obtained from the study of unstable isotopes and the mean path length from the primary to secondary ratios yielded a matter density over the particles lifetime of 0.06 particles cm^{-3} while the density in the galactic disk was believed to be about 1 particle cm^{-3}.

The notion of a large, static, low density region surrounding the galaxy that was able to trap a significant number of energetic particles proved, however, to be untenable. Several authors showed that, if nothing else, the energetic cosmic rays themselves had more than enough pressure to blow away this galactic halo or corona. If such a halo existed it would have to be a dynamical one, with a constant galactic wind constantly replenished from the higher density disk region of the galaxy.

In the following we shall examine the ways in which the dynamical nature of the halo will manifest itself in the properties of the cosmic-ray electrons and nucleons and try to determine what measurements of these particles can tell us about the nature of the halo of our galaxy.

2. COSMIC RAY ELECTRON HALO

The primary manner in which a galactic halo that traps or holds energetic electrons manifests itself is through the region of radio emission surrounding the galaxy. As one might imagine, the difference between a dynamic halo with outflow and a static one could be rather subtle in such a circumstance, manifesting itself only through the detailed manner in which the electron spectrum, and hence that of the radio emission, changes with position in the halo.

A very detailed development of the theory of electrons in an outflowing halo has been presented in a series of papers by Lerche and Schlickeiser (1980, 1981a-c)

Werner (1988) has presented results of observations of the edge on galaxy NGC4631 at several different frequencies and concluded that the predictions of Lerche and Schlickeiser were confirmed. The confirmation was based on the observation of a break in the radio spectrum (reflecting a corresponding break in the electron spectrum.)

2.1 A Simple Theory

Since the expressions of Lerche and Schlickeiser are quite complicated it would be useful to derive expressions which, although greatly simplified, can show more clearly the basic difference between the electron spectra to be expected in the dynamical and static halos. The difference arises from the different time behavior of diffusive and convective transport which are working against the continuous change of the electron spectrum produced by the synchrotron radiation process which is responsible for the radio emission itself.

First we should note that electrons that are emitting radiation via the synchrotron process (or inverse Compton scattering as well) lose energy at a rate proportional to the square of their energy, *ie.*

$$\frac{dE}{dt} = -bE^2. \tag{1}$$

If we integrate equation (1) from an initial energy E_1 to a final energy E_2 we obtain

$$bt = \frac{1}{E_2} - \frac{1}{E_1}.$$

and setting E_1 equal to ∞ we obtain

$$E_c = \frac{1}{bt}. \tag{2}$$

Equation (2) shows that after a time t no particles can remain above the critical energy E_c. In other words any injection spectrum whatsoever will exhibit a cutoff at E_c after a time t.

A power law spectrum of the form

$$N(E, t = 0) = AE^{-q}$$

will have the form at time t

$$N(E, t) = AE^{-q}(1 - bEt)^{q-2}. \tag{3}$$

From this we can see that a reasonable approximation to the correct spectrum after a time t has passed is to multiply an injection spectrum of any form by a step function $U(E, bt)$ that has a value of unity for $E < 1/bt$ and is zero for $E > 1/bt$.

With this approximation it is easily seen that if we have a distribution of particle ages rather than a discrete time after injection our step function $U(E, bt)$ should be replaced by

$$U(E) = \int_0^\infty f(t)U(E, bt)dt = \int_0^{1/bE} f(t)dt \tag{4}$$

where $f(t)$ is the age distribution function.

2.2 A Halo Model

Before proceeding to look at the age distributions that are produced by static and dynamical halo models we should first see what a theoretician's galaxy looks like.

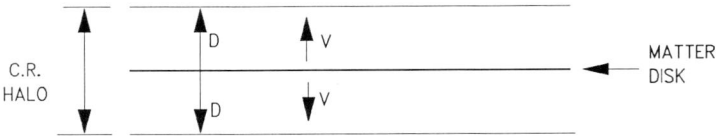

Figure 1. Cosmic-Ray Theoretician's View of a Galaxy

Figure 1 shows that the essential components of the galaxy are the disk, usually considered to contain all or almost all of the matter, and the halo which is flowing away from the disk on both sides with a (possibly variable) velocity V. The matter in the halo is considered to be significant only in that it carries the irregular magnetic fields which convect the energetic particles and through which they diffuse. In most cases the extent of the disk and halo in directions parallel to the disk is considered to be infinite thus reducing the problem to one dimension.

If particles are released at the position of the disk † they are convected a distance x out into the halo in a time t with $x = Vt$ where V is the halo wind speed. While their mean position is being convected the particles are also diffusing about this center so that their density as a function of x and t is given by

$$N(x,t) = \frac{\exp\left(\frac{-(x-Vt)^2}{2\kappa t}\right)}{\sqrt{2\pi\kappa t}} \tag{5}$$

where κ is the diffusion coefficient of the particles in the flowing halo magnetic fields.

Characteristic Length and Time At this point it would be advisable to note that there are two parameters in the problem (κ and V) that have different dimensions. This means that certain characteristic dimensions can be formed from these parameters. A more physical way of looking at this is to realize that in diffusion a particle travels a distance x in a time x^2/κ whereas in convection it travels the same distance in a time x/V. There is clearly a characteristic distance given by $x_c = \kappa/V$ for which these times are equal and the corresponding characteristic time is $t_c = \kappa/V^2$. For distances shorter than x_c diffusion is the quickest way to travel and we can say that we are in a diffusion dominated regime. However if the lengths of interest are greater than x_c convection is faster than diffusion and we are in a convection dominated regime.

---

† All models that I am familiar with assume that the energetic cosmic rays are produced in the disk portion of the galaxy, presumably by supernova related phenomena

Obviously if the size of the halo, D is smaller than x_c the entire halo will be diffusion dominated and it will be very difficult to talk in a meaningful way about a dynamical halo as far as the cosmic ray particles are concerned. We, therefore, require for a dynamical halo in the energetic particle sense that $x_c < D$ (actually we really need $x_c \ll D$ for convective effects to be pronounced enough for unambiguous observation.)

Employing the characteristic length and time that we have defined we may express equation (5) in dimensionless units as

$$N(x', t') = \frac{1}{x_c \sqrt{\pi}} exp\left(\frac{-(x' - t')^2}{2t'}\right) \qquad (6)$$

where $x' = x/x_c$ and $t' = t/t_c$.

Time Distribution for $x' \gg 1$ If we look out in the halo a distance that is large compared with the characteristic distance x_c we find that convection dominates at all times and a pulse of width $\approx x'^{-1/2}$ is swept by us at a time $t' = x'$ as shown in Figure 2.

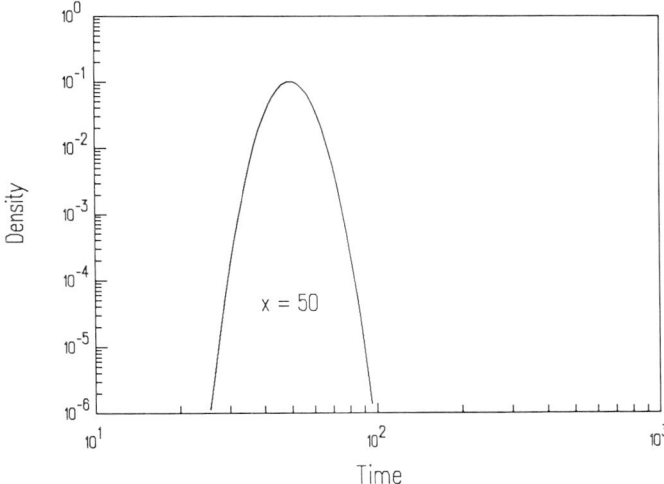

Figure 2. Time Dependance of Density for $x' \gg 1$

Time Distribution for $x' \ll 1$

In Figure 3 we see plotted the density of particles as a function of time (or age) at a position well inside the characteristic length for the outflowing halo we see that the time profile looks exactly like a purely diffusive one until we reach

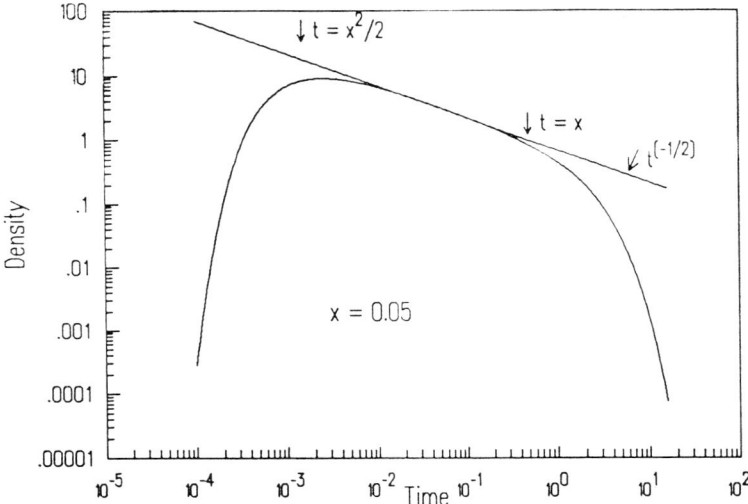

Figure 3. Time Dependance of Density for $x' \ll 1$

a time when convection effects can reach the observation point ie. when $t' = x'$. After this time the particle density no longer decays with a $t'^{-1/2}$ time dependance that is characteristic of diffusion but instead is swept away with a $\exp(-Vt)$ time dependance typical of convection. This is strictly true only in the one dimensional, infinite extent halo; a higher dimensionality or an absorbing boundary will change this as we will see later.

We note that in this case there are two times of importance $t' = x'^2/2$ at which time the density rises sharply from zero and $t' = x'$ at which time the density drops rapidly to zero again with a $t'^{-1/2}$ dependance in between.

We can now see that for the convection dominated case there is essentially one time for each position in the dynamical halo and the electron spectrum at that point is unchanged below the characteristic energy E_c and is rather sharply cut off above that energy. The time is given by the relation $t = x/V$ and therefore $E_c = V/bx$, the characteristic energy is lower the further out in the halo one looks.

On the other hand, whenever one looks in a region where diffusion dominates the two characteristic times produce a somewhat different spectrum. It is straightforward to see that once again the spectrum is unchanged from its injection form for energy below $E_c = V/bx$. However, above that energy the spectrum will be multiplied by a "filter" function proportional to $E^{-1/2}$ or, if it is a power law spectrum, steepened by a one half power. The cutoff occurs at an energy $E_{c2} = 2\kappa/bx^2$. So we see that the signature of diffusion is the extension of the cutoff with a half power steepening to an energy characteristic of the diffusion time scale.

If there is no outflow at all (a static halo) the half power steepening will continue down to the lowest energy in the injection spectrum. Such a condition would be hard to distinguish from the convection dominated case were it not for the fact that the cutoff in the spectrum varies with distance out into the halo by a different power of distance in the two cases. In the pure diffusion case the cutoff energy is inversely proportional to the square of the distance while in the convection dominated case it is inversely proportional to the distance.

It should also be noted that the bend in the spectrum is a signature of convection *only* in the case of a one dimensional, infinite extent halo. Any deviation from this situation such as an absorbing boundary or a transition to three dimensional diffusion will produce a characteristic time, t_c such that for $t > t_c$ the time profile drops off faster than $t^{-1/2}$ and produces such a bend in the spectrum. In this case the variation of the energy of the bend with distance is the *only* signature that can distinguish convection from diffusion. In this type of diffusion the characteristic time for the non $t^{-1/2}$ behavior to set in would be the time required for the particles to diffuse to the boundary of the halo plus the time for the effects to propagate back to the point of observation, *ie.* $t_c = (2D - x)^2/2\kappa$ and the spectral bend comes at $E_c = 2\kappa/b(2D - x)^2$.

3. COSMIC RAY NUCLEON HALO

We have seen that the appearance of an electron halo as seen through its radio emission depends on how the actual extent of the halo D compares with the characteristic distance of a diffusing/convecting medium $x_c \equiv \kappa/V$. Unfortunately nothing that we have discussed so far gives us an idea of the magnitude of the quantities that we need to know. It so happens that considering the effect that a dynamical halo can have on the nuclear component of the galactic cosmic rays can give us a handle on this question.

3.1 Parker Instability

Although the idea that a galactic halo could be important for the propagation of cosmic rays was around almost from the beginning (Bierman and Davis 1958), it wasn't until Parker's (1965, 1966, 1968, 1969) "bubble gum" picture that the notion was introduced that the cosmic rays could effect the creation of such a halo. In these papers Parker showed that the gas, magnetic field, cosmic ray system in the galactic disk was unstable. The instability occurred whenever a particular portion of a magnetic field line became slightly elevated. The cooler gas would drain away to a lower region leaving the hot, high pressure cosmic ray gas to inflate the field into an ever increasing bubble and hence escape the galactic disk. This process could be responsible for dragging the magnetic field and some of the gas out of the disk and forming a halo for the galaxy. It should be noted that this process of halo formation by its very dynamical nature necessarily produces a dynamical halo.

This idea that the cosmic rays are responsible for driving a dynamical halo has been carried further by others the first of whom was Ipavitch (1975) who discussed a cosmic ray driven, spherical galactic wind. Others have considered this problem with more complicated geometries.

3.2 Effect of Outflow on Disk Cosmic Rays

Although studying the effect of the cosmic rays on the dynamics of the galactic halo is important in its own right we are most interested in examining the effect that a dynamical halo would have on the cosmic rays that we observe in the disk. It is only by studying these effects that we may hope to gain information about the dynamical properties of the halo of our galaxy.

Storage Place A galactic halo has been proposed as a storage place for cosmic rays first to produce the observed isotropy (Bierman and Davis, 1958) and later by Prischep and Ptuskin (1975) and by Ginzburg et al. (1980) to reconcile the cosmic ray age determination via measurements of unstable nuclei such as ^{10}Be with the mean grammage derived from observed secondary to primary ratios.

Aid to Lateral Diffusion Another effect that halo storage can have on the disk population of cosmic rays is that of facilitating lateral diffusion in the disk. In the case of isotropic diffusion it is intuitive that cosmic rays could not diffuse laterally to the disk much further than they can diffuse perpendicular to the disk before they escape the galaxy. In other words, a thick halo will allow cosmic-ray particles to diffuse much further from their sources during their residence time in the galaxy than will a thin one.

This idea was analyzed quantitatively by Stecker and Jones (1977) employing SAS-2 gamma ray data and CO line emission data to estimate the cosmic ray density in various regions of the disk. Assuming a source distribution similar to the supernova remnant distribution, they concluded that the halo could not be much thicker than 3 kpc. A reanalysis of this problem using COS-B data has recently been reported by Bloemen (1989) who finds a thicker halo, $\approx$ 20 kpc. This discrepancy appears to come from the difference between the SAS-2 and COS-B data rather from any difference in the method of analysis.

The above analysis was called into question by the work of Jokipii (1976) and Owens and Jokipii (1977a,b) who showed that an outflowing halo could have an effect on parameters of the disk cosmic rays such as their age distribution. However, it was shown by Jones (1978) that the effect of a dynamical halo on the particles observed in the disk was to replace the actual extent of the halo with the characteristic length $x_c = \kappa/V$ discussed in section 2.2.

Energy Variation of κ is The Key It was the realization that the variation of the diffusion coefficient κ with energy could cause a transition from diffusion dominated transport to convection dominated transport that was the key to relating cosmic ray data to parameters of the halo. The data seemed to indicate (Ormes and Freir,

1978) that while the mean path length was a decreasing function of energy, varying as $\approx E^{-1/2}$ for energies above a few GeV, below this energy the path length seemed to be constant $\approx$ 5-6 g cm^{-2}. This fact would seem to indicate that ≈ 1 GeV marks the transition between the two modes of transport.

It has been stated (Blandford and Ostriker, 1980) that if convection dominates the removal of cosmic rays from the galaxy then the escape time is independent of the energy of the particles. This is only partly true. When convection dominates the mean time spent in a fixed region of the galaxy is independent of energy, however, mean age of a particle is given by $t_c \approx \kappa/V^2$ and is therefore, dependent upon energy through the diffusion coefficient, κ. During this mean lifetime the particle has been confined to a distance $x_c = \kappa/V$ on either side of the disk and thus the amount of time spent in a disk of thickness a is given by $t_c a/x_c = a/V$ which is independent of κ and hence independent of energy.

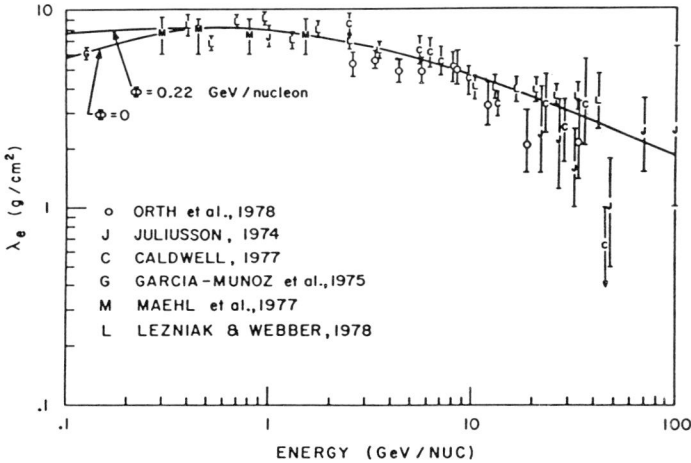

Figure 4. Mean Path Length vs. Energy for Cosmic Rays at Earth

Dynamical Halo Model of Our Galaxy Analysis of this situation by Jones (1979), Freedman *et Al.* (1980), and Kota and Owens (1980) indicated that the data could be fit by a dynamical halo model with a wind speed of 8 - 20 km s^{-1} the particular value depending on the value of the disk thickness in g cm^2 assumed by the authors. Although the particular values of κ and the size of the halo were much harder to obtain due to the uncertainty of the actual mean age of the cosmic rays. The main point of the analysis can be seen from Figure 4 and that is that for energies above 1 GeV per nucleon or rigidities above 3.4 GV the galactic halo of our galaxy is diffusion dominated. This is true if the low energy flattening of the path length vs. energy curve is due to a dynamical halo *regardless* of the numerical value of any quantity derived from the theory.

4. CONCLUSION

This, of course, implies that the transport electrons above 3 or 4 GeV is also diffusion dominated and that, assuming a magnetic field of $1 \cdot 10^{-6}$g in the halo, no obvious convection effects can be seen at frequencies above about 100 MHz. One must look for the spectral cut off that represents the first arrival time of the electrons at the point of observation and try to determine whether this scales with the inverse square of the distance from the disk or as the simple inverse. In addition one may identify a bend in the spectrum but realizing that this is *not* an unambiguous sign of convection one must try to determine its behavior as a function of position in the halo.

I believe that convection effects can be detected in radio halos but it will not be done by identifying some single feature in the spectrum. Rather one will have to map out the halo region and find how spectral features move from one place to another to determine whethte the move adout by convection or diffusion.

5. REFERENCES

Bierman, J. L., Davis, L. (1958) *Zs. Ap.* **51**, 19
Blandford, R. D., Ostriker, J. P. (1980) *Ap. J.* **237**, 793
Bloemen, J. B. G. M. (1989) *Adv. Space Res.* **10**, (2)199
Freedman, I., Giler, M., Kearsey, S., Osborne, J. L. (1980) *Astron. Ap.* **82**, 110
Ginzburg, V. L.,Khazan, Ya. M., Ptuskin, V. S. (1980) *Ap. Space Sci.* **68**, 295
Ipavitch, F. M. (1975) *Ap. J.* **196**, 107
Jokipii, J. R. (1976) *Ap. J.* **208**, 900
Jones, F. C. (1978) *Ap. J.* **222**, 1097
Jones, F. C. (1979) *Ap. J.* **229**, 747
Kota, J., Owens, A. J. (1980) *Ap. J.* **237**, 814
Lerche, I., Schlickeiser, R. (1980) *Ap. J.* **239**, 1089
Lerche, I., Schlickeiser, R. (1981a) *Ap. J. Suppl.* **47**, 33
Lerche, I., Schlickeiser, R. (1981b) *Astron. Ap.* **107**, 148
Lerche, I., Schlickeiser, R. (1981c) *Ap. Lett.* **22**, 31
Owens, A. J., Jokipii, J. R. (1977a) *Ap. J.* **215**, 677
Owens, A. J., Jokipii, J. R. (1977b) *Ap. J.* **215**, 685
Parker, E. N. (1965) *Ap. J.* **142**, 584
Parker, E. N. (1966) *Ap. J.* **145**, 811
Parker, E. N. (1968) in *Stars and Stellar Systems, Vol. VII: Nebulae and Interstellar Matter*, eds. B. M. Middlehurst, L. H. Allen , University of Chicago Press, Chicago, Chapt. 14
Parker, E. N. (1969) *Space Sci. Rev.* **9**, 651
Pikel'ner, S. B. (1953) *Doklady Akad. Nauk SSSR* **88**, 229
Prischep, V. L., Ptuskin, V. S. (1975) *Ap. Space Sci.* **32**, 265
Shklovsky, I. S. (1952) *Astr. Zh.* **29**, 418
Stecker, F. W., Jones, F. C. (1977) *Ap. J.* **217**, 843

THE INFLUENCE OF EXTENDED SOURCE DISTRIBUTIONS ON COSMIC RAY SPECTRAL INDEX VARIATIONS IN THE GALACTIC WIND MODEL

M.POHL
Max-Planck-Institut für Radioastronomie
Auf dem Hügel 69, 5300 Bonn, FRG

ABSTRACT. The solution of the steady-state transport equation describing the propagation of relativistic electrons perpendicular to the galactic plane in a galactic wind is discussed for extended source distributions. The wind velocity is assumed to be zero in the plane and to increase with galactic height. The electrons undergo simultaneously diffusion, convection, adiabatic deceleration, radiative losses and injection. We contrast the resulting spectra with those derived for an infinitely thin source distribution and discuss the dependence of the variation of the spectral index inside the source distribution on the form of this distribution in any transport model. Apart from a new region at very small energies the asymptotic spectra agree with those of the line-source model (LS). Inside the source distribution the spectral index break is doubled from $\Delta\gamma_{lin}=0.5$ to $\Delta\gamma_{ext}=1.0$.

1. INTRODUCTION AND BASIC EQUATIONS

The interpretation of observations of nonthermal radio emission depends strongly on our understanding of the dynamics of relativistic electrons in their environment. This led to the introduction of electron transport models for static halos, in which the electrons undergo radiative losses and diffusion (Webster, 1970; Bulanov and Dogiel, 1974). In the early Eighties the picture of the ISM changed radically (Savage and DeBoer, 1981; York, 1982; McCammon et al, 1983). It exists a hot (10^5 K), tenuous ($10^{-3}cm^{-3}$) coronal phase, which occupies a large fraction of the ISM. This coronal phase is maintained by high-velocity shocks ($V > 20km/sec$) from supernova explosions or from powerful winds of early-type stars. Thus, transport models have to include the effects of convection and adiabatic deceleration.

Following Lerche and Schlickeiser (1982,LS) we consider the 1-dim. stationary continuity equation for the differential number density N(E,z) of relativistic electrons including terms for diffusion, convection, adiabatic deceleration and radiative losses

$$\frac{\partial}{\partial z}\left(D(E,z)\frac{\partial N}{\partial z} - V(z)N\right) + \frac{\partial}{\partial E}\left(\left[\frac{1}{3}\frac{\partial V}{\partial z}E - \dot{E}(E,z)\right]N\right) + q(r,\phi,z,E) = 0$$

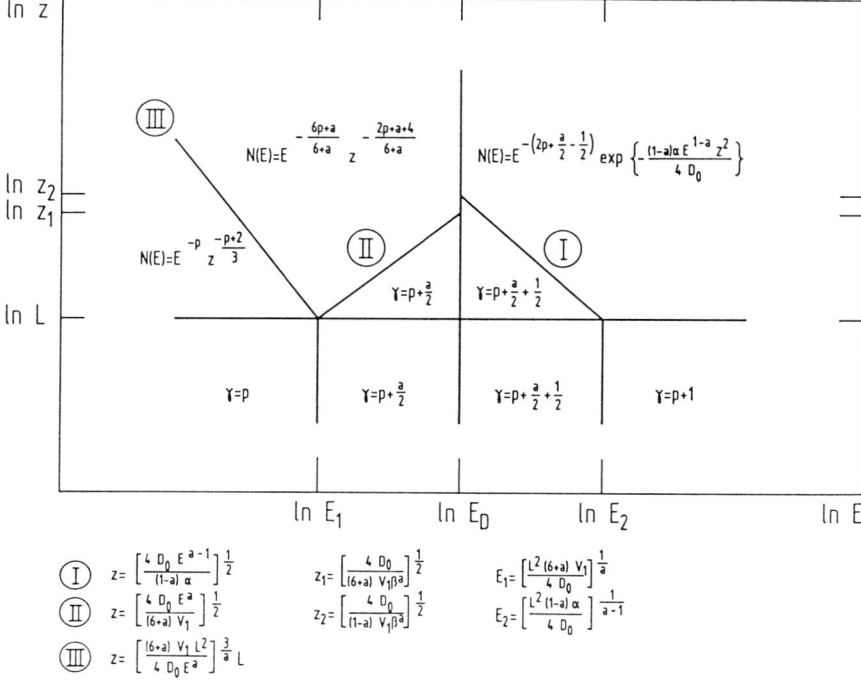

Figure 1. Schematic illustration of the nine regions in z-E-space considered for the assumption of a uniform diffusion d(z)=1, and thus $\mu = z$, for which asymptotic spectra have been derived. One may also replace z by μ yielding the original μ-E-diagram. The asymptotic number density and the spectral index, respectively, are given.

which is analytically solvable under the assumptions discussed in LS.

Relaxing on their line-source assumption we present asymptotic solutions for extended source distributions, e.g. the step function

$$q(r,\phi,\mu,E) = q_0(r,\phi)S(\mu)E^{-p} \quad \text{with} \quad S(\mu) = \begin{cases} 1 & \text{for } |\mu| < L \\ 0 & \text{otherwise} \end{cases}$$

2. RESULTS AND CONCLUSIONS

The solution of the continuity equation yields (for details see Pohl and Schlickeiser 1990,PS)

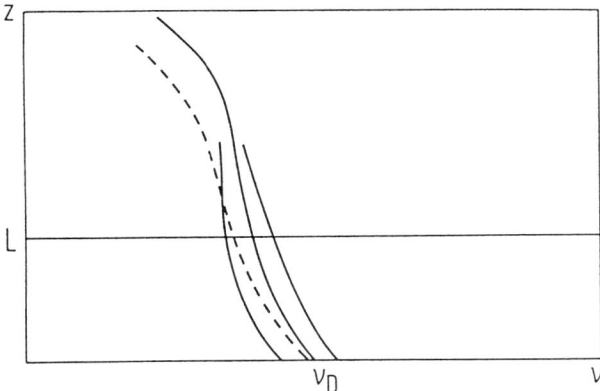

Figure 2. Variation of the break frequency ν_D in the case of a scale height of the magnetic field much smaller than L. The dashed line is valid for very small scale heights, whereas the solid line marks the case that ν_D does not vary with galactic height over a wide range of z resulting in a step in the z-variation of the spectral index. The accompanying solid lines mark the breaks around E_1 and E_2.

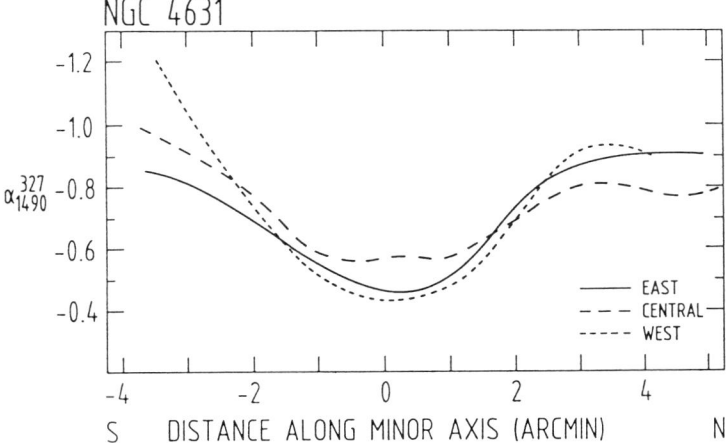

Figure 3. Variation of the spectral index between 327 and 1490 MHz for NGC 4631 as obtained by Hummel and Dettmar (1990). On the northern side a stepwise steepening is visible.

$$N(\mu, E) = \frac{q_0\, E^{-(p+a/2)}}{2(1+\beta E)^4 \sqrt{\pi\, V_1\, D_0}} \int_{-\infty}^{\infty} d\mu'\, S(\mu') \int_1^{\infty} \frac{d\rho\, \rho^{-p}}{\left(\int_1^{\rho} \frac{dx\, x^{5+a}}{(1+\beta E x)^7}\right)^{1/2}}$$

$$\times \exp\left[-\frac{V_1\left(\mu'\left(\frac{1+\beta E}{1+\beta E \rho}\right)^3 \rho^3 - \mu\right)^2}{4 D_0\, E^a (1+\beta E)^6 \int_1^{\rho} \frac{dx\, x^{5+a}}{(1+\beta E)^7}}\right]$$

with $\beta^{-1} = E_D = V_1/\alpha$

We derive asymptotic solutions for regions in which one of the following timescales is smallest (see figure 1)

$$\tau_r = (\alpha E)^{-1} \quad \text{radiative losses}$$

$$\tau_D = \frac{\mu_0^2}{D_0\, E^a} \quad \text{diffusion}$$

$$\tau_A = V_1^{-1} \quad \text{adiabatic deceleration}$$

Inside the source distribution the spectra can be determined with the concept of the catchment sphere introduced by Webster (1970).

For a comparison of our results with the observations one needs a translation of the μ-E-diagram into a z-ν-diagram. The best way to do this is a hydrodynamical calculation of realistic galactic wind patterns and therefore the spatial variations of parameters like the magnetic field strength and the wind speed gradient. For large values of L (e.g. secondary electrons, reacceleration) and an exponentially decreasing magnetic field strength the synchrotron break frequency ν_D related to the break energy E_D does not vary with the galactic height over a wide range of z (see figure 2). Then in a specific frequency range around ν_D one observes a more or less stepwise steepening of the spectra with z. This behaviour is observed on the northern side of NGC 4631 (figure 3).

REFERENCES

Bulanov, S.V., Dogiel, V.A. (1974) *Astroph. Spa. Sci* **29**, 305
Hummel, E., Dettmar, R.J. (1990) *Radio observation and optical photometry of the edge-on galaxy NGC 4631*, submitted
Lerche, I., Schlickeiser, R. (1982) *Astron. Astroph.* **107**, 148, **LS**
McCammon, D., Burrows, D.N., Sanders, W.T., Kraushaar, W.L. (1983) *Astroph. Jour.* **269**, 107
Pohl, M., Schlickeiser, R. (1990) *Astron. Astroph.*, in press, **PS**
Savage, B.D., De Boer, K.S. (1981) *Astroph. Jour.* **243**, 460
Webster, A.S. (1970) *Astroph. Lett.* **5**, 189
York, D.G. (1982) *Ann. Rev. Astron. Astroph.* **20**, 221

DYNAMICAL IMPLICATIONS OF DIFFUSIVE AND CONVECTIVE COSMIC RAY PROPAGATION IN GALACTIC HALOS

D. BREITSCHWERDT[1], J.F. MCKENZIE[2] AND H.J. VÖLK[1]
[1]MPI für Kernphysik, D-6900 Heidelberg, Postfach 103980, FRG
[2]MPI für Aeronomie, D-3411 Katlenburg-Lindau, Postfach 20, FRG

ABSTRACT. On the basis of our present knowledge about Cosmic Ray (CR) propagation, it is argued that galactic halos should mainly consist of two parts, namely a *lower* region, extending from the disk-halo interface to a few kpc, in which CR diffusion prevails and an *upper* region, where convection dominates. The upper part is the possible site of galactic wind formation due to the strong coupling of the CRs to the thermal plasma via (mainly outwards propagating) Alfvén waves as a mediator. In the lower halo and also in the disk, the gas will be effectively static, the wave field will be almost random in its direction (due to stochastic gas motions), and the CRs must diffuse through the gas to escape. We present a model that describes both the upper and the lower halo and briefly discuss simple analytic solutions for the diffusion and numerical results for the convection region.

1. INTRODUCTION

It has been argued that there should be a constant efflux of mass and energy into the halo, presumably powered by supernovae (SNe) and stellar winds (SWs). Qualitatively, for the model presented here, it is not essential whether this occurs in a correlated (Norman, 1990) or a distributed (Kahn, 1990) fashion. The important facts are the existence of hot, ionized gas and CRs, as shown by the observations of diffuse X-rays (Nousek et al., 1982) and "thick" radio disks (Hummel et al., 1984), respectively. Presumably at $|z| >$ few kpc the gas will expand laterally and try to attain pressure equilibrium. Magnetic lines of force will be pushed open by the combined pressure of gas and CRs (cf. Breitschwerdt et al., 1990a; Völk, 1990). The activity of star formation in the disk below and the break-out of shells and/or supershells will provide a sufficient level of turbulence, so that the wave field in the lower halo will be isotropic to lowest order and therefore the ensemble averaged Alfvén speed $\langle v_A \rangle \approx 0$. This leaves only *diffusion* as the mechanism for CRs to propagate into the upper halo. Another observational fact is that the CRs do eventually have to escape from the Galaxy, probably along magnetic field lines pointing away from it. There will then be a resonant growth of waves, caused by the CRs streaming along these field lines (self-excited waves). The now dominantly outwards propagating waves will cause a considerable momentum transfer of the CRs to the plasma in z during scattering and thus lead to *convective flow* in the upper halo, combined with an Alfvénic *drift* of the CRs.

2. SIMPLE MODEL FOR THE DIFFUSIVE HALO REGION

This region extends from the Galactic disk up to the reference level ($0 \leq z \leq 1$ kpc), z being the coordinate perpendicular to the disk. Since the CR disk has at least a half thickness of 300 pc, as inferred from the CR propagation properties, the above value for the reference level should be considered as a lower limit. The hot gas flowing out into the halo is essentially the one swept up by supernova remnants (SNRs) and has a temperature of $T \leq 10^6$ K. The majority of SNRs are the result of Type II SN explosions and are decreasing exponentially by number with a scale height of $H_{SN} \sim 55$ pc (Bregman 1980). Therefore there will be a mass loading of the lower halo with hot gas at a rate $Q(z)$. The simplest representation, reflecting such an exponential decrease of sources is one which is just proportional to the local gas density ρ, i.e. $Q(z) = \alpha \rho(z)$, where α^{-1} is a characteristic time scale for the mass loading. The expanding gas will cool adiabatically, but there is also an input of energy by shock heating of Type I SNe, as well as from Type II from runaway O stars. Therefore for $|z| \leq 1$ kpc, an isothermal equation of state seems convenient and more appropriate. We note that in the momentum equation, there has to appear also a term for the drag force per unit mass Qu/ρ due to the mass loading. Because of $\langle v_A \rangle \approx 0$ and small wave pressure, diffusion dominates convection as long as the mass velocity u satisfies $|u| \ll \bar{\kappa}/h_c = 65$ km/s, where typical values have been adopted for the averaged diffusion coefficient $\bar{\kappa} = 10^{29}$ cm²/s and the CR pressure scale height $h_c \sim 5$ kpc. Applying the above assumptions and taking z as the only independent variable, the equations in their steady-state form read:

$$\frac{d}{dz}(\rho u) = \alpha \rho, \tag{1}$$

$$\frac{d}{dz}(\rho u^2 + p_g + p_c) = -\rho g_{\text{eff}}, \tag{2}$$

$$-\frac{\bar{\kappa}}{\gamma_c - 1}\frac{dp_c}{dz} = \text{const.}, \tag{3}$$

$$T(z) = \text{const.}. \tag{4}$$

Here p_g, p_c, T, g_{eff} and γ_c denote the gas pressure, the CR pressure, the gas temperature, the effective gravitational acceleration, and the CR adiabatic index (4/3 for an ultra-relativistic gas), respectively. We have shown (Breitschwerdt et al., 1990b) that the specific form of the mass loading term allows for the separation of the variables ρ and u, to give

$$\frac{d^2\rho}{dz^2} + \frac{2\alpha^2}{c^2}\rho + \frac{1}{c^2}\frac{d}{dz}(\rho g_{\text{eff}}) = 0, \tag{5}$$

where the isothermal speed of sound is $c^2 = k_B T/\mu$, with k_B being Boltzmann's constant and μ the mean mass per particle. To solve equation (5) we have to impose the following boundary conditions (BCs) at $z = 0$: $\rho = \rho_d$, $(d\rho/dz)_{z=0} = -\rho_d g_{\text{eff}}/c^2 + p_{cd}/(h_c c^2)$, whereby subscript 'd' refers to quantities evaluated in or close to the galactic disk. We note that according to equation (5) and the BCs, there are three different scale heights involved in the problem, viz. $l = c^2/g_{\text{eff}}$ (thermal gas), c/α (mass loading) and h_c (CRs). Therefore the structure of the lower halo can be quite different from a simple hydrostatic model. To discriminate between the various solutions, it is convenient to define a characteristic dimensionless parameter $\Delta = 8l^2\alpha^2/c^2$. There are three different classes of solutions, according to whether $\Delta < 1, = 1, > 1$, respectively, i.e.:

$$\frac{\rho}{\rho_d} = \frac{\exp(m_+ z)}{(m_- - m_+)}\left[m_- + \frac{1}{l} - k\right] + \frac{\exp(m_- z)}{(m_+ - m_-)}\left[m_+ + \frac{1}{l} - k\right], \qquad (6a)$$

$$\frac{\rho}{\rho_d} = \exp(-z/2l)\left[1 + \left(k - \frac{1}{2l}\right)z\right], \qquad (6b)$$

$$\frac{\rho}{\rho_d} = \exp(-z/2l)\left[\cos(\beta z) + \beta^{-1}\left(k - \frac{1}{2l}\right)\sin(\beta z)\right], \qquad (6c)$$

where $k = p_{cd}/(h_c \rho_d c^2)$, $m_{\mp} = -(1/2l)(1 \mp \sqrt{1-\Delta})$ and $\beta = \sqrt{(2\alpha^2/c^2) - (1/4l^2)}$. The flow speed u then follows from integration of the continuity equation (1): $u(z) = (\alpha/\eta)\int_0^z \eta dz'$, $\eta \equiv \exp\left(-\int H^{-1} dz'\right)$, $H^{-1} \equiv -(1/\rho)(d\rho/dz)$. To illustrate these solutions, we use some typical values for the parameters: $g_{\text{eff}} = 10^{-8}$ g/cm^2 (at a Galactocentric radius $R_0 = 10$ kpc), $T = 10^6$ K, and α we estimate to be the ratio of the total mass loss rate into the halo to the total halo mass, i.e. $\dot{\mathcal{M}}_H/\mathcal{M}_H$. Both of these numbers are highly uncertain and $\dot{\mathcal{M}}_H$ varies from $2.1 M_\odot$/yr to $22 M_\odot$/yr (cf. Heiles, 1987), while $\mathcal{M}_H$ is supposed to lie between $7 \times 10^7 M_\odot$ (Bregman, 1980) and $9.8 \times 10^7 M_\odot$ (Heiles, 1987). Taking the lower values we get: $\alpha = 9.5 \times 10^{-16}$ s^{-1}, $l = 1.65 \times 10^{22}$ cm and $\Delta = 11.6$. Since $\beta z < 1$ we can approximate equation (6c) to obtain $\rho/\rho_d = \exp(-z/2l)\left[1 - (z/2l) + p_{cd} z/(\rho_d c^2 h_c)\right]$. For $p_{cd} < \rho_d c^2$ we get $\rho_0 = \rho(z = 1 \text{ kpc}) = 0.8\rho_d$. Because of the uncertainties in the parameters and since Δ is not too different from unity, each of the solutions (6a) - (6c) may be relevant in practice.

3. TRANSITION TO THE CONVECTIVE HALO REGION

At distances $|z| \geq 1$ kpc from the disk, the dominant source of waves will be self-excitation, and we then expect Alfvénic drift to be so efficient, that convection will dominate diffusion in the CR transport; at least this is our assumption for the sequel. The change in geometry, from plane parrallel to spherically symmetric flow, as $|z|$ increases has been modelled by a simple flux tube geometry. All this has been described in detail by Breitschwerdt et al. (1987, 1990a); we will therefore present only the most important results. The model calculations have been carried out for a flux tube located at $R_0 = 10$ kpc and a set of paramters given at reference level $|z| = 1$ kpc: $\rho_0 = 1.67 \times 10^{-27}$ g/cm^3, $p_g = 2.8 \times 10^{-13}$ dyne/cm^2, $p_c = 1.0 \times 10^{-13}$ dyne/cm^2, $B_0 = 1 \mu$G for the magnetic field strength in $|z|$-direction and a mean fluctuating field component $\langle \delta B \rangle = 0.1 B_0$. It turns out that the initial mass velocity is quite low, $u_0 = 10$ km/s, and that the flow accelerates with increasing $|z|$, attaining an asymptotic value of $u_f = 310$ km/s. We emphasize that the low base velocity u_0 (subalfvénic) is consistent with the flow properties derived for the lower halo and therefore guarantees a continuity of the two types of solutions.

4. CONCLUSIONS

The presence of CRs in galactic halos is important, both for its density structure and its dynamics. In particular, we have shown that the propagation properties of the CRs, in a simple minded picture, give rise to *two* essentially different, i.e. *diffusive* and *convective*, regions which can be connected in a smooth fashion. Since there is much more observational information about the lower halo, it is certainly worth while incorporating more physics into our simple model, such as momentum addition, heating and cooling and linear wave damping. So far, we can only assess that if the CRs mainly diffuse through the halo, their support of the plasma against gravity due to their large scale height is rather weak. However, there is also mass loading from the sources, which modifies the thermal scale height. Once Alfvénic drift becomes important, the plasma is accelerated considerably against the drag of the gravitational field and the "overburden" will be lifted to infinity (*galactic wind*).

REFERENCES

Bregman, J.N. (1980) *Astrophys. J.* **236**, 577
Breitschwerdt, D., McKenzie, J.F., Völk, H.J. (1987) *in "Interstellar Magnetic Fields"*, Beck, R., Gräve, R. (eds.), Springer-Verlag, Heidelberg, p. 131
Breitschwerdt, D., McKenzie, J.F., Völk, H.J. (1990a) *Astron. Astrophys. (submitted)*
Breitschwerdt, D., McKenzie, J.F., Völk, H.J. (1990b) *Proc. 21st Int. Cosmic Ray Conf. (Adelaide)* **3**, 315
Heiles, C. (1987) *Astrophys. J.* **315**, 555
Hummel, E., Sancisi, R., Ekers, R.D. (1984) *Astron. Astrophys.* **133**, 1
Kahn, F.D. (1990) *these proceedings*
Norman, C.A. (1990) *these proceedings*
Nousek, J.A., Fried, P.M., Sanders, W.T., Kraushaar, W.L. (1982) *Astrophys. J.* **258**, 83
Völk, H.J. (1990) *these proceedings*

PARTICLE ACCELERATION IN THE DISK-HALO SYSTEM

REINHARD SCHLICKEISER
Max-Planck-Institut für Radioastronomie
Auf dem Hügel 69
D-5300 Bonn 1, F.R.G.

ABSTRACT. The recent observations of the nonthermal properties of the halo of our Galaxy at radio and γ-ray wavelengths are summarized. Radio and γ-ray data show a similar spectral flattening with Galactic height towards the anticenter direction, which is interpreted as a cosmic-ray effect. Several theoretical explanations for the flattening of the energy spectra of the radiating cosmic-ray electrons (in the radio) and nucleons (in γ-rays) are reviewed including propagation of cosmic rays in an accelerating Galactic wind and the presence of cosmic-ray sources with flat energy spectra in the halo.

1. INTRODUCTION

The formation of cosmic-ray halos in galaxies is a still unsolved problem of high-energy astrophysics: do the cosmic-ray particles diffuse or convect away from their sites of origin which for relativistic electrons definitely and for cosmic-ray nucleons probably are located inside galaxies? Some insight may be gained from the recent observations of significant cosmic-ray spectral differences in various regions of the disk-halo system reported both for cosmic-ray electrons from studies of the radio continuum background (Reich and Reich, 1988b) and for cosmic-ray nucleons from studies of the diffuse Galactic gamma-ray emission (Bloemen, 1987; Bloemen et al., 1988). Radio studies exist for the edge-on galaxies NGC 4631 (Hummel and Dettmar, 1990) and NGC 891 (Hummel, 1990). After summarizing the relevant Galactic observations I investigate several theoretical interpretations of these measurements.

2. SUMMARY OF RELEVANT OBSERVATIONS

2.1 Radio continuum background at 408 and 1420 MHz

Based on radio continuum surveys at 408 MHz (Haslam et al., 1982) and 1420 MHz (Reich, 1982; Reich and Reich, 1986) Reich and Reich (1988a) have presented a map of spectral indices of the northern sky with an angular resolution of 2° and an absolute spectral index error of ~0.1. At these frequencies the radio continuum is the sum of thermal free-free emission in the ionized interstellar medium (intensity spectrum $I \propto \nu^{-\alpha}$, $\alpha = 0.0-0.1$) and synchrotron emission of relativistic electrons of Lorentz factor $\gamma \simeq 1.5 \cdot 10^4 \ [\nu(\mathrm{GHz})/B(\mu G)]^{1/2}$ in Galactic magnetic fields of strength B whose intensity spectrum is of power-law type $I \propto \nu^{-\alpha}$, $\alpha = (s-1)/2$, if the relativistic electron energy spectrum is of power-law type, $N(\gamma) \propto \gamma^{-s}$. The radio study shows significant variations of the spectral index of the brightness temperature $T_b \propto \nu^{-\beta}$, $\beta = 2+\alpha$, along the Galactic plane and from the plane toward higher Galactic latitudes. Most noteworthy is the reported *flattening* of spectral indices with increasing latitude both in the inner and outer Galaxy (see Figure 1). Towards the outer Galaxy $\beta(|b| = 0°) = 2.7-2.8$ near the plane and β reduces to $2.5-2.6$ at $|b| \simeq 30°$. If correcting for the thermal emission in the plane Reich and Reich (1988) have found that the nonthermal spectral index β_{nth} flattens from a value 2.85 in the plane $|b| \simeq 0°$ by $\Delta\beta = 0.35 \pm 0.2$ with increasing latitude. Towards the inner Galaxy the nonthermal spectral index varies from $\beta_{nth} \simeq 3.1$ in the plane to values of $\beta_{nth} \simeq 2.7$ at high latitudes.

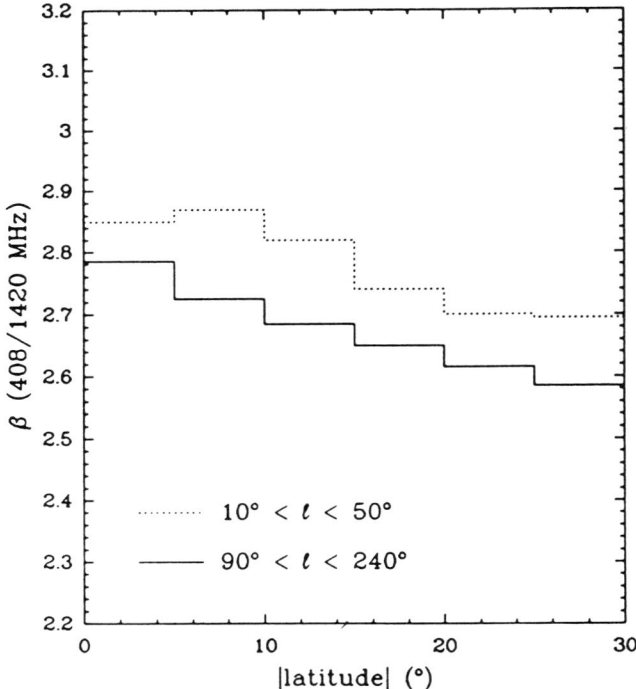

Figure 1. Latitude distributions of the spectral index β of the Galactic radio emission at 408 and 1420 MHz ($T_b \propto \nu^{-\beta}$) where regions above and below the Galactic plane are combined (from Bloemen et al., 1988).

2.2 Diffuse galactic gamma-ray emission above 300 MeV

Cosmic gamma rays with energies above 300 MeV originate predominantly (≥ 80 percent) from the decay of neutral pions produced in inelastic p-p, p-He, α-p collisions of cosmic-ray nucleons of energy greater than 3 GeV (Stecker, 1973; Dermer, 1986) with atoms and molecules of the interstellar gas, with a minor (≤ 20 percent) contribution from nonthermal bremsstrahlung of relativistic electrons (0.6-10 GeV) where most of these electrons may be of secondary origin from the decay of charged pions $\pi^\pm \to \mu^\pm \to e^\pm$ produced in the same inelastic collisions as the neutral pions (Schlickeiser, 1981, 1982). Inverse Compton scattering of low-frequency microwave, infrared and starlight photons by cosmic-ray electrons plays a negligible role (contribution less than a few percent). The energy spectrum of the generated γ-rays at high photon energies (> 0.5 GeV) reflect the energy spectrum of the cosmic-ray nucleons: if the latter is of power-law type $\propto E_{CR}^{-s}$ the γ-ray number spectrum is also of power-law type $\propto E_\gamma^{-\Gamma}$ with $\Gamma = (s-2b)/(1-b)$ where b denotes the energy dependence of the pion multiplicity $\xi \propto E_{CR}^b$ (Stecker, 1971). Measured variations in Γ directly indicate variations in the nucleon spectral index s.

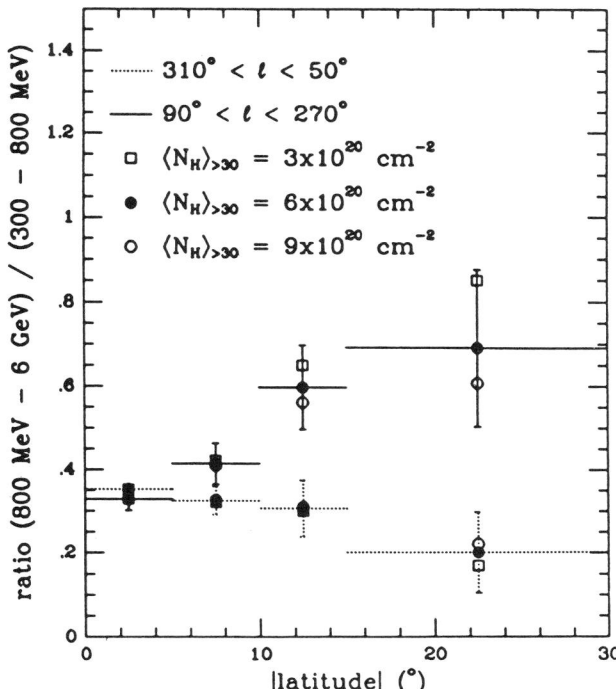

Figure 2. Latitude distribution of the γ-ray colour towards the inner and outer Galaxy after subtraction of the background levels (from Bloemen et al., 1988)

Using the COS-B data above 300 MeV, excluding regions with radius 8° around the well-known γ-ray point sources Crab pulsar, Vela pulsar and CG195+4 and subtracting the isotropic (extragalactic and instrumental) background, Bloemen et al. (1988) determined spectral variations in Γ from the γ-ray colour C = I(0.8-6 GeV)/I(0.3-0.8 GeV) calculated from the γ-ray intensity in two energy intervals. The variation of the γ-ray colour with Galactic latitude in the inner and outer Galaxy is shown in Figure 2. The rising values of C in the outer Galaxy from 0.33±0.03 at $|b| \simeq 0°$ to 0.68±0.18 at $|b| \simeq 15°-30°$ suggest a strong flattening of the nucleon spectrum with latitude. Towards the inner Galaxy the data are consistent with a constant value of C ≃ 0.30.

2.3 Conclusions from observations

Since radio and γ-ray data show a similar gradual flattening in their spectral behaviour we interpret this as a cosmic-ray effect that the energy spectra of the radiating cosmic-ray electrons and nucleons flatten with increasing latitude. Figure 3 shows the derived spectral flattening in the respective energy distributions $n(E) \propto E^{-\Gamma}$, indicating in both cases a flattening by $\Delta\Gamma \simeq 0.4-0.6$ from $|b| \simeq 0°$ and 30° in the outer

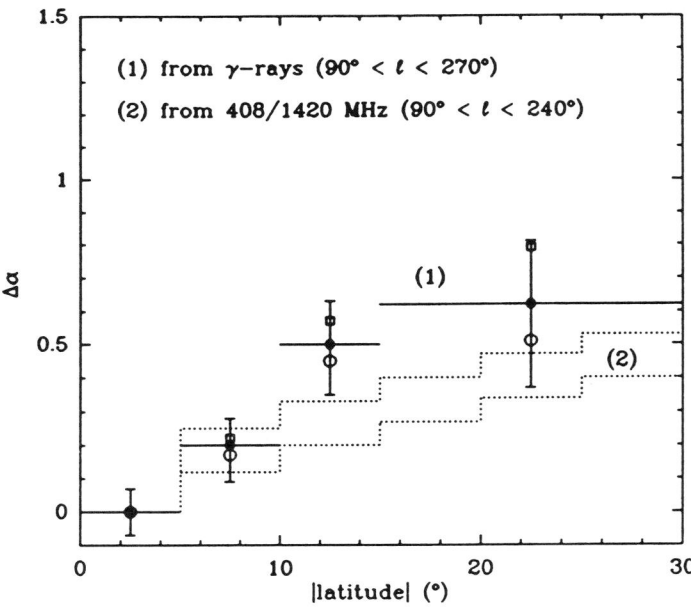

Figure 3. Variation of the cosmic-ray spectral index α ($n(E) \propto E^{-\alpha}$) as a function of Galactic latitude. $\Delta\alpha(b) \equiv \alpha(0)-\alpha(b)$ is estimated from the γ-ray colour for cosmic-ray nucleons and from the radio continuum emission at 408 and 1420 MHz for cosmic-ray electrons (from Bloemen et al., 1988)

Galaxy. First, this result implies that the measured cosmic ray spectra at the position of the solar system are *not* representative for the whole Galaxy which affects total cosmic-ray power estimates of our Galaxy. Secondly, this finding is an important constraint on cosmic-ray acceleration and transport models, which needs to be explained by any viable model of cosmic-ray origin. Bloemen et al. (1988) have pointed out that their results can be accounted for in remarkable detail by the Galactic wind model for cosmic-ray propagation where spectral flattenings have been predicted (Lerche and Schlickeiser, 1981, 1982a,b) due to the combined action of adiabatic deceleration in the accelerating Galactic wind and energy-dependent spatial diffusion. It is the purpose of the remainder of this paper to investigate additional interpretations of these observations.

3. PARTICLE ACCELERATION IN THE DISK-HALO SYSTEM

The radio and γ-ray observations summarized above refer to the steady-state equilibrium spectrum of cosmic rays $N(\gamma,\vec{r})$ that results from the balance of sources (injection by point sources as supernova remnants, pulsars), sinks (leakage from the Galaxy, catastrophic losses as fragmentation and spallation) and interaction processes of cosmic rays with cosmic matter, radiation and electromagnetic fields on their way from their sites of origin to us. This balance is described by the convection-diffusion cosmic-ray transport equation

$$-\frac{dN}{dt} = \text{div }(\kappa(\gamma,\vec{r})\text{ grad }N - \vec{V}(\vec{r})N) + \frac{\partial}{\partial \gamma}\left[\left\{\frac{1}{3}\text{ div }\vec{V}(\vec{r})\gamma - \dot{\gamma}(\gamma,r)\right\}N\right]$$

$$+ \frac{\partial}{\partial \gamma}\left[\gamma^2 \, a_2(\gamma,\vec{r}) \frac{\partial}{\partial \gamma}\left\{N\gamma^{-2}\right\}\right] - \frac{N}{T_F(\gamma,\vec{r})} = -Q(\gamma,\vec{r}) \quad (1),$$

that contains terms representing spatial diffusion with diffusion coefficient $\kappa(\gamma,\vec{r})$, convection and adiabatic deceleration determined by the cosmic-ray bulk speed $\vec{V}(\vec{r})$, energy diffusion with diffusion coefficient $a_2(\gamma,\vec{r})$, spontaneous energy loss processes $\dot{\gamma}(\gamma,\vec{r})$ and catastrophic losses with loss time $T_F(\gamma,\vec{r})$. The source term $Q(\gamma,\vec{r})$ represents injection from point sources. A recent derivation of equation (1) can be found in Kirk et al. (1988) and Schlickeiser (1989). Spatial and energy diffusion result from the interaction of cosmic rays with interstellar magnetohydrodynamic turbulence. For Alfvén waves propagating parallel and antiparallel to the ordered magnetic field the two diffusion coefficients are related as (Dung and Schlickeiser, 1990a,b)

$$\kappa(\gamma,\vec{r}) \, a_2(\gamma,\vec{r}) = V_A^2 \, \gamma^2 \, G(h_c,\sigma^{\pm}) \, F(h_c,\sigma^{\pm}) \quad (2),$$

where the two dimensionless functions G and F depend on the magnetic helicity of the parallel (σ^+) and antiparallel (σ^-) waves and the cross helicity (h_c) of the Alfvénic turbulence, that is related to the fractional abundance of the parallel to total waves r as $h_c = 2r+1$. $V_A = 2.18.10^{11}$ $B(G) n_e^{-1/2} (cm^{-3})$ is the Alfvén speed.

The cosmic-ray bulk velocity is the sum of the interstellar gas velocity $\vec{U}(\vec{r})$ and some fraction of the Alfvén velocity $\vec{V}_A(\vec{r})$,

$$\vec{V}(\vec{r}) = \vec{U}(\vec{r}) + V_A(\vec{r}) \, H(h_c, \sigma^{\pm}) \qquad (3),$$

where the dimensionless function $H(h_c, \sigma^{\pm})$ depends again on the helicities of the interstellar turbulence, and is restricted to values between -1 and $+1$ (Dung and Schlickeiser, 1990b). For vanishing interstellar gas velocity, $\vec{U} = \vec{0}$, there still are convective terms in equation (1) if $H \neq 0$.

The functions $\gamma(\gamma, \vec{r})$ and $T_F(\gamma, \vec{r})$ describe continuous and catastrophic energy loss processes and differ for cosmic-ray electrons and nucleons (see the discussion in Schlickeiser, 1986). For relativistic electrons $T_F^e = 0$ and

$$-\dot{\gamma}_e = 6.10^{-14} n_e \left\{ 18.56 \left[1 + 1.35.10^{-2} \ln \frac{\gamma}{n_e} \right] + 2.32.10^{-3} \gamma (\ln \gamma + 0.36) + \frac{4}{9} \gamma^2 \frac{W_{ph} + \frac{B^2}{8\pi}}{m_e c^2 n_e} \right\} \qquad (4)$$

in a fully ionized interstellar medium of density n_e in cm^{-3}, target photon energy density W_{ph} and magnetic field strength B.

An additional constraint for any model of cosmic-ray origin is the requirement to reproduce the measured cosmic-ray nucleon and electron energy spectra at the position of the solar system, which at energies above 1 GeV are straight power laws over more than 4 decades in energy for nucleons (Burnett et al., 1983; Grunsfeld et al., 1988) and over more than 2 decades in energy for electrons (Nishimura et al., 1990).

There are several theoretical alternatives on the basis of equation (1) to explain the given set of observations but only few of them have been worked out thoroughly. They can be divided into two classes (a) propagation effect, (b) source effect. Let us consider each in turn.

3.1 Propagation effect

Lerche and Schlickeiser (1981, 1982a,b) have pointed out that spectral flattenings are signatures of the presence of convective terms in the cosmic-ray transport equation coupled with energy-dependent spatial diffusion. And it is indeed remarkable how well their original theoretical predictions match the now available observations (see the discussion in Bloemen et al., 1988). Recently their calculations have been extended analytically to thick disk source distributions by Pohl and Schlickeiser (1990), see also Pohl (1990), and numerically by Van der Walt (1990).

3.2 Source effect

3.2.1 Anomalous Halo Cosmic-Ray Component.
Another interpretation of the spectral flattenings is the existence of a new component of cosmic rays in the halo which has a flatter energy spectrum than cosmic rays at the solar system. This Anomalous Halo Cosmic Ray Component (AHCRC) may well be associated with particle acceleration at the Galactic wind termination shock (Jokipii and Morfill, 1985, 1987). As an aside note that in the interplanetary medium there exists an anomalous cosmic-ray component associated with the solar wind termination shock (for review see Webber, 1989) and it is fair to speculate that a similar phenomenon occurs on Galactic scales. The AHCRC cannot make a strong contribution to the locally measured cosmic-ray flux, which can be accounted for by the Galactic modulation of the AHCRC in the outwardly accelerating Galactic wind (Ahlen et al., 1982), if the location of the Galactic wind termination shock is far enough away from the Galactic plane.

I want to make two critical remarks to this model:
(1) the explanation by particle acceleration at the termination shock relies on the existence of a Galactic wind and a dense enough intergalactic medium to generate a termination shock. At the moment it is not clear that these conditions indeed exist in the interstellar and intergalactic medium. Moreover, the concept of a Galactic wind already offers an explanation of spectral flattenings as a propagation effect, as noted in Section 3.1, so there is no necessity to postulate this new cosmic-ray component;
(2) in the halo region radiation losses dominate ionization, Coulomb and bremsstrahlung losses for Lorentz factors larger than

$$\gamma > 290 \, (n_e/10^{-3} \, cm^{-3})^{1/2} \tag{5},$$

since the halo gas densities are small ($n_e \simeq 10^{-3}$ cm^{-3}) and $[W_{ph}+(B^2/8\pi)] \geq W(2.7K) = 0.25$ eV cm^{-3}. In order to obtain a flat steady-state equilibrium electron spectrum $N(\gamma) \propto \gamma^{-2.3}$, the electron source spectrum has to be even flatter as $Q(\gamma) \propto \gamma^{-1.3}$ to account for the spectral index steepening by $\Delta\alpha = 1.0$ by the radiation losses. Such flat power-law spectra for the electrons of the AHCRC are difficult to reconcile with the standard theory of diffusive shock-wave acceleration (for review see Drury, 1983; Blandford and Eichler, 1987), which yields $Q(\gamma) \propto \gamma^{-s}$ and $s = (r+2)/(r-1)$ being determined by the shock's compression ratio $r \leq 4$. The smallest s can get is $s = 2$ for $r = 4$. Either this explanation does not work or the standard theory of shock-wave acceleration has to be modified. An important modification has been proposed by Dröge et al. (1987) and Schlickeiser and Fürst (1989), which is based on the inclusion of energy diffusion of particles in the Alfvénic turbulence near the shock wave. They have shown that in the case of low values of the upstream plasma beta $\beta_p = c_{s1}^2/V_{A1}^2 = 8\pi nkT/B_0^2$ this modification is important and produces very flat power-law spectra ($s \to 1$ for $\beta_p \to 0$) for the accelerated particles. This brings us to a second explanation of spectral flattenings which relies on flat-spectrum sources in the halo.

3.2.2 Flat-Spectrum Sources in the Halo. Observations of the radio spectral indices of shell-type supernova remnants in the Galaxy and the Magellanic Clouds show dispersion around $\langle \alpha \rangle = 0.5$ with a width $\sigma_\alpha \simeq 0.3$, which points to a dispersion in the energy spectra of the radiating electrons around $\langle s \rangle = 2.0$ with a width $\sigma_s = 0.6$. While this dispersion is difficult to understand by the standard model of shock-wave acceleration the inclusion of energy diffusion in the turbulence near the shock provides a straightforward explanation (see discussion in Dröge et al. (1987) and Schlickeiser and Fürst (1989)).

Brecher and Burbidge (1972) have emphasized an important consequence of dispersion in the spectral indices of the source spectra $Q(\gamma) \propto \gamma^{-p}$. If the probability of a certain value of p is determined by a Gaussian distribution

$$n(p) = \frac{n_0}{\sqrt{2\pi\mu}} \exp\left[-\frac{(p - \langle p \rangle)^2}{2\mu}\right] \qquad (6),$$

the equilibrium spectrum $N(\gamma)$ for the very simple version $N(\gamma)/T(\gamma) = Q(\gamma)$ of equation (1) is

$$N(\gamma) = T(\gamma) \int_0^\infty dp \, (\gamma/\gamma_1)^{-p} n(p) \propto T(\gamma) \, \gamma^{-\Gamma(\gamma)} \qquad (7)$$

with $\Gamma(\gamma) = \langle p \rangle - (\mu/2) \ln(\gamma/\gamma_1)$. With increasing energy the effective spectral index $\Gamma(\gamma)$ becomes flatter. This is a consequence of the fact that sources with the flattest spectra dominate in the superposition at large energies. So this simple argument can explain the spectral flattenings in the halo.

However, there is a contradiction with the measured energy spectra at the solar system, which apparently do not show this effect. And in fact Brecher and Burbidge (1972) have used this contradiction to argue against a Galactic origin of cosmic rays. But I think we can still use this effect of flattening if we can argue that sources with flat spectra occur only in the halo region, i.e. that low values of the plasma beta of the interstellar medium preferentially occur in the halo region than in the disk region. It would be also interesting to investigate whether flat-spectrum shell-type supernova remnants are preferentially located at large Galactic heights. This theoretical alternative can turn out to be very useful but more detailed theoretical and observational work is needed to assess its importance.

4. CONCLUSIONS

We have summarized the recent observations of the nonthermal properties of the halo in our Galaxy. Most noteworthy are the spectral flattenings of both the radio continuum and diffuse γ-ray background emission with Galactic height towards the anticenter direction. Since the radio and γ-ray data simultaneously show this flattening this is certainly a cosmic-ray effect that the energy spectra of the radiating cosmic-ray electrons (in the radio) and nucleons (in γ-rays) flatten with increasing latitude. We investigate several theoretical interpretations of these measurements including the well-known spectral flattening resulting from the presence of convective terms (convection and adiabatic deceleration in an accelerating Galactic outflow) in the cosmic-ray transport equation coupled with energy-dependent spatial diffusion, the possible existence of a new anomalous halo cosmic-ray component with a flat energy spectrum, and the flattening resulting from the dispersion in the power-law spectral indices of the cosmic-ray sources. While some interpretations still have to be worked out more thoroughly before a final assessment, we think that the interpretation as a propagation effect in an accelerating Galactic outflow is at present the best approach, since this model can remarkably well account in a quantitative fashion for the observed properties of the nonthermal halo.

ACKNOWLEDGEMENTS. I thank Ms. G. Breuer for the careful typing of the manuscript. Work on Galactic winds in Bonn is supported partially by the Deutsche Forschungsgemeinschaft (Fa 97/8-2) which is gratefully acknowledged.

5. REFERENCES

Ahlen, S.P., Price, P.B., Salamon, M.H., Tarle, G. (1982) *Ap. J.* **260**, 20
Blandford, R.D., Eichler, D. (1987) *Phys. Rep.* **154**, 1
Bloemen, J.B.G.M. (1987) *Ap. J.* **317**, L15
Bloemen, J.B.G.M., Reich, P., Reich, W., Schlickeiser, R. (1988) *Astr. Ap.* **204**, 88
Brecher, K., Burbidge, G.R. (1972) *Ap. J.* **174**, 253
Burnett, T.M. et al. (1983) *Phys. Rev. Lett.* **51**, 1010
Dermer, C.D. (1986) *Astr. Ap.* **157**, 223
Dröge, W., Lerche, I., Schlickeiser, R. (1987) *Astr. Ap.* **178**, 252
Drury, L.O.C. (1983) *Rept. Progr. Phys.* **46**, 973
Dung, R., Schlickeiser, R. (1990a) *Astr. Ap.* (in press)
Dung, R., Schlickeiser, R. (1990b) *Astr. Ap.* (in press)
Grunsfeld, J.M., L'Heureux, J., Meyer, P., Müller, D., Swordy, S.P. (1988) *Ap. J.* **327**, L31
Haslam, C.G.T., Salter, C.J., Stoffel, H., Wilson, W.E. (1982) *Astr. Ap. Suppl.* **47**, 1
Hummel, E. (1990) these proceedings
Hummel, E., Dettmar, R.-J. (1990) *Astr. Ap.* (in press)
Jokipii, J.R., Morfill, G.E. (1985) *Ap. J.* **290**, L1
Jokipii, J.R., Morfill, G.E. (1987) *Ap. J.* **312**, 170
Kirk, J.G., Schneider, P., Schlickeiser, R. (1988) *Ap. J.* **328**, 269
Lerche, I., Schlickeiser, R. (1981) *Ap. Lett.* **22**, 161
Lerche, I., Schlickeiser, R. (1982a) *Astr. Ap.* **107**, 148
Lerche, I., Schlickeiser, R. (1982b) *M.N.R.A.S.* **201**, 1041
Nishimura, J. et al. (1990) *Proc. 21st Intern. Cosmic Ray Conf. (Adelaide)*, Vol. **3**, p. 213
Pohl, M. (1990) these proceedings

Pohl, M., Schlickeiser, R. (1990) *Astr. Ap.* (in press)
Reich, W. (1982) *Astr. Ap. Suppl.* **48**, 219
Reich, P., Reich, W. (1986) *Astr. Ap. Suppl.* **63**, 205
Reich, P., Reich, W. (1988a) *Astr. Ap. Suppl.* **74**, 7
Reich, P., Reich, W. (1988b) *Astr. Ap.* **196**, 211
Schlickeiser, R. (1981) *Fortschr. d. Phys.* **29**, 95
Schlickeiser, R. (1982) *Astr. Ap.* **106**, L5
Schlickeiser, R. (1986) in *Cosmic Radiation in Contemporary Astrophysics*, ed. M.M. Shapiro, Reidel, Dordrecht, p. 27
Schlickeiser, R. (1989) *Ap. J.* **336**, 243
Schlickeiser, R., Fürst, E. (1989) *Astr. Ap.* **219**, 192
Stecker, F.W. (1971) *Cosmic Gamma Rays*, Mono Book Corp., Baltimore
Stecker, F.W. (1973) *Ap. J.* **185**, 499
Van der Walt, D.J. (1990) *Ap. Space Sci.* **168**, 23
Webber, W.R. (1989) in *Cosmic Abundances of Matter*, ed. C.J. Waddington, AIP Conf. 183, p. 100

THE STRUCTURE OF THE INTERSTELLAR MEDIUM

JOEL N. BREGMAN and GREGORY A. ASHE
Astronomy Department, University of Michigan
Ann Arbor, MI 48109-1090, USA

ABSTRACT

We have compared theoretical HI models to a large area (18°x12°) HI channel map centered at l = 205°, b = 0°. One set of models were calculated in which uncorrelated clouds of HI populate space. Of the cloud models considered, which include "Spitzer"-type clouds and the McKee-Ostriker formulation, none were able to reproduce the data successfully. In another class of models, holes and holes with shells permeate an otherwise continuous HI medium. Although not entirely successful, these models contain characteristics that are similar to the data. In order to reproduce the amplitude of the HI variations on angular scales of >1°, the volume occupied by ionized gas must be less than 30%, in sharp conflict with the models that suggest that the volume of the ISM is dominated by hot gas.

1. INTRODUCTION

The structure of the interstellar medium in the disk of the galaxy has a profound effect on the evolution of energetic events in the disk, such as supernovae. Supernovae produce hot gas and cosmic rays which, depending upon the structure of the ISM, can escape into the halo. The evolution of supernovae and other energetically important phenomena, such as the closely related and very powerful stellar winds (van Buren 1986), is determined by the properties of the interstellar gas that occupies most of the volume. At this time, we lack a consensus as to the volume occupied by the most common component and its geometry.

1.1 Theoretical Models

In theoretical models of the ISM that were developed in the mid-1970s and early 1980s, the principal energetic component was randomly placed supernova remnants (Cox and Smith 1974; McKee and Ostriker 1977; Cox 1981; Cowie, McKee, and Ostriker 1981). These models applied the results of spherically symmetric supernova remnant calculations to the disk in a statistical manner in order to determine the filling factor of the hot supernova remnants. Given the estimates for the supernova rate and the energy deposition per supernova, these efforts suggested that most of the volume was occupied by a hot, dilute gas that was responsible for the soft X-ray emission seen in the Galaxy. In this picture, supernovae shocks propagate mainly through the dilute hot medium, and this hot gas escapes freely into the halo.

A change in this theoretical perspective began when it was argued that supernovae occur primarily in associations of young stars, which give rise to large reheated supernova remnants, known as superbubbles (Weaver et al. 1977; Bruhweiler et al. 1980; Tomisaka and Ikeuchi 1986; McCray and Kafatos 1987; Tenorio-Tagle, Bodenheimer, and Rozyczka 1987; Mac Low and McCray 1988; Tenorio-Tagle and Bodenheimer 1988; Mac Low, McCray, and Norman 1989). Typically, 30 supernovae lead to a single superbubble, which grows in size to several hundred parsecs (diameter) and eventually breaks out of the disk. If most of the supernovae participate in the growth of superbubbles, then the fractional volume occupied by the hot medium is likely to be less than 50%, possibly in the 10-20% range (Norman and Ikeuchi 1989).

1.2 Observational Constraints

From observations, it was difficult to estimate the filling factor of the various phases of the hot medium. For external galaxies, X-ray observations of disk galaxies failed to detect a diffuse emission component from a hot dilute gas (McCammon and Sanders 1984), although this may simply indicate that the gas is cooler than 6×10^5K. A search for HI holes in the nearby galaxies M31 and M33 (Brinks and Shane 1984; Brinks and Bajaja 1986; Deul 1988) was more successful, demonstrating that hundreds of large bubbles could exist in a single galaxy. These surveys, which were sensitive to holes larger than about 100 pc (about the size of a small superbubble, but larger than a single supernova remnant), showed that about 5% of the surface area of a galaxy was covered by such objects.

In the Milky Way galaxy, several prominent ISM components either have significant mass or volume fractions. For warm ionized gas (10^3-10^4 K), a comparison between the Hα emission measure and the pulsar dispersion measure suggests a volume filling factor of about 20% (e.g., Cox and Reynolds 1987; Reynolds 1989). For the hot material (10^6K), there is an anticorrelation between the soft X-ray emission and the HI. This effect, which can be explained as the displacement of one phase by another, would not have been noticed if either medium occupied an insignificant volume of the ISM. This suggests that both the hot and cold components occupy >10% of the volume, although it is difficult to obtain more precise constraints because the phases are not uniform and we detect mainly local hot gas (Cox and Reynolds 1987). The HI bubbles seen by Heiles (1984) is analogous to the HI holes seen in external galaxies and supports the concept of hot gas displacing the HI.

The pervasiveness of the 21 cm HI emission line has prompted most observers to conclude that this component must occupy a significant amount of the volume (e.g., Kulkarni and Heiles 1988, Dickey and Lockman 1990, and references therein). However, most of these arguments have been qualitative in nature. We have been developing a model that will permit a more quantitative analysis of the volume occupied by the HI and hotter components. This model also permits us to discuss the geometric structure of the HI.

2. THE MODEL

Our approach is to calculate HI maps of a part of the disk of the Galaxy that can be compared directly to observations (HI channel maps). A phenomenological model interstellar medium can be constructed by specifying different "forms" that may be present in the HI layer. Possible forms of HI are clouds (with or without a dense core), a uniform medium (with a vertical scale height and radial scale length), holes in the uniform medium, and expanding shells. The properties of each member of each form within the Galaxy are specified (location, random velocity, size, density, etc.) and then the emission image as viewed from the Sun can be calculated and compared to the observed surface brightness distribution. We regard this as a phenomenological model because the gas properties are assumed (or taken from observations) rather than calculated from first principles using detailed hydrodynamics.

In constructing such a model, we require that the mean observed density distribution be reproduced in latitude, longitude, and velocity. In doing so, we adopt the mean HI vertical distribution given by Dickey and Lockman (1990) in which there are two Gaussian distributions

(FWHM of $h_1 = 212$ pc and $h_2 = 530$ pc) and an exponential distribution (scale height $h_3 = 403$ pc). In addition, the flaring of the HI disk is included by having the scale heights be proportional to the distance:

$$h_i(R) = h_i(R_\circ) \qquad \text{for } R < R_\circ$$
$$h_i(R) = \frac{R}{R_\circ} h_i(R_\circ) \qquad \text{for } R \geq R_\circ$$

This leads to a final mean distribution for the HI of

$$n(z) = (0.40\, e^{-\frac{z^2}{2h_1^2}} + 0.11\, e^{-\frac{z^2}{2h_2^2}} + 0.06\, e^{-\frac{z^2}{h_3}})\, e^{\frac{-(R-R_\circ)}{h_R}}$$

which implies a total column perpendicular to the disk of 6.2×10^{20} cm^{-2}. This gas is assumed to be in circular rotation about the Galaxy with a velocity of 220 km s^{-1} (flat rotation curve). The warp of the HI layer is not taken into account because the longitude region that will be the point of comparison is nearly along the line of nodes.

We have produced models in the longitude and latitude range 196-214° and ±6°, which is the region where Lockman and Ganzel (1983) have obtained spectra with 20' resolution. Within this volume, we distribute the interstellar medium in particular forms and then calculate the appearance of the HI emission, producing a three dimensional data cube, with ordinates of longitude, latitude, and velocity. An example of this modeling is an interstellar medium composed of an ensemble of uncorrelated HI clouds. These clouds are defined by their size, density, internal velocity dispersion, and random velocity. When determining whether a cloud exists at a particular location, a random number generator is first used to select a location. Then, the probability that a cloud will be created is proportional to the mean density n(R,z) given above. Clouds are added to the volume until the total emissivity, integrated over longitude, latitude, and velocity, matches the observed quantity.

From this ensemble of clouds, in order to convert the column density into a brightness temperature (T_B), opacity corrections are introduced. Largely for convenience, we have assumed that the HI is at a constant temperature of $T_{spin} = 100$K. The simple radiative transfer equations become

$$T_B(v) = T_{spin}(1 - e^{-\tau(v)})\, \Delta v$$

$$\tau(v) = \frac{N_H(v)}{1.83 \times 10^{18}\, T_{spin}}$$

where Δv is the channel width and v is the central velocity of the channel. In practice, we use $\Delta v = 2$ km s^{-1}, and this calculation is carried out for angular bins that are 5' on a side. The image at a particular velocity, $T_B(l,b,v)$, is convolved with a Gaussian to a resolution of 20', which is the resolution of the data set of Lockman and Ganzel.

The resulting models or data cubes are meant to be representative of an ISM with a particular form (e.g., clouds) rather than a detailed reproduction of the observations. Consequently, it is necessary to characterize the models and the data in statistical terms. We have calculated surface area as a function of brightness temperature as well as structure functions and Fourier power spectra of strips across the channel maps. These statistics place on a quantitative footing the information evident to the eye in Figures 1, 2.

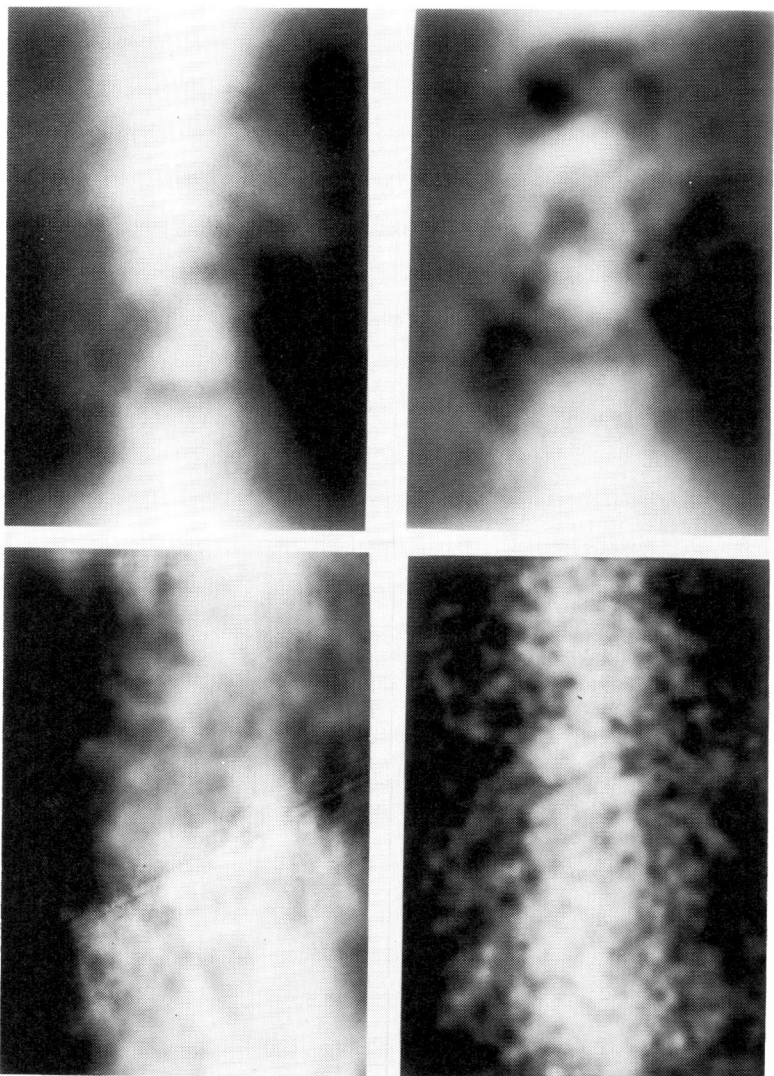

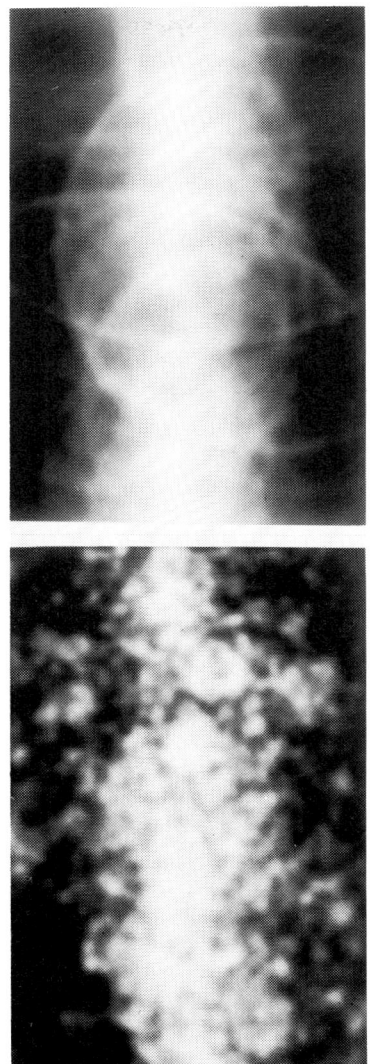

Figure 1. These gray scale representations of the total brightness temperature, integrated in velocity, for the region of the Galaxy 196-214°. In a clockwise fashion beginning with the upper left are the observations, the 20% hole model, the 33% hole model, the 20% holes plus shells model, the two-component uniform cloud model, and the non-uniform cloud model (exponential clouds).

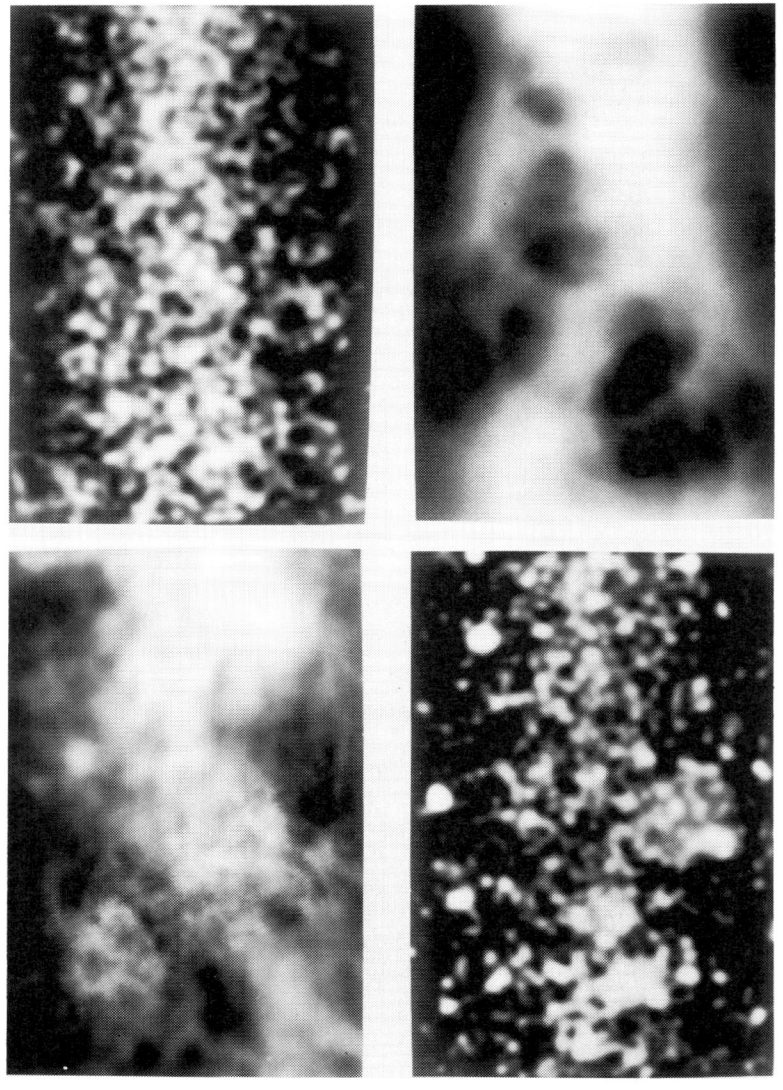

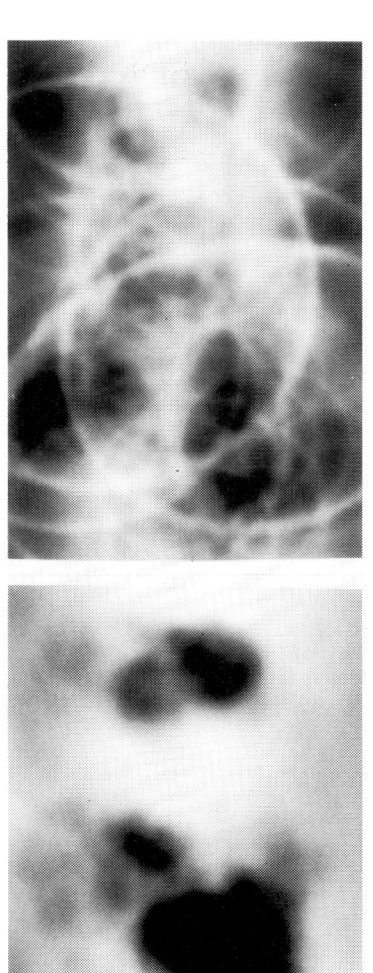

Figure 2. These gray scale representations of the brightness temperature in channel maps centered at 20 km s^{-1}. The order is slightly different than in Figure 1, where, beginning at the top left and going clockwise, we have the observations, the non-uniform cloud model, the 20% hole model, the 20% hole plus shell model, the 33% hole model, and the two-component hole model.

2.1 Cloud Models

Six different neutral hydrogen models were considered, although they can be classified either as cloud models or pervasive models. For the cloud models, calculations were made of (1) uniform spherical clouds of constant size, (2) nonuniform spherical clouds of constant size, and (3) two component clouds of varying sizes, which is representative of the McKee-Ostriker model. In all cases, the velocity dispersion of the clouds are taken to be 6 km s^{-1} and the same value is used for the internal velocity dispersion of the gas within a cloud. For the uniform spherical cloud model, the radius of the clouds is taken to be 5 pc, the density is 20 cm^{-3}, and the temperature is 100 K unless otherwise stated. This is similar to the standard "Spitzer" cloud (Spitzer 1978). In the second model, the density decreases radially outward in a cloud with a density law given by

$$n_{cloud}(r) = n_{cloud}(r=0)\, e^{-\frac{r}{r_{cloud}}}$$

where $r_{cloud} = 3$ pc and $n_{cloud}(r=0) = 20$ cm^{-3} (the exponential cloud model); the masses of these clouds and of the uniform clouds are the same.

The two-component cloud case, model 3, has a cold uniform cloud (T = 80 K, n = 42 cm^{-3}) surrounded by a larger warm low-density uniform cloud (T = 8000 K, n = 0.37 cm^{-3}). This model is suggested by McKee and Ostriker (1977), and following their prescription, the radius of the warm component is 2.5 times that of the cold component (r_{cold}). There is a range of cloud sizes where the number of clouds of a particular size is proportional to r^{-4}, and the minimum and maximum values for r_{cold} are 1.5 pc and 10 pc respectively. This minimum radius is a bit larger than the value suggested by McKee and Ostriker (0.5 pc), but this difference should not lead to a qualitative change in the results. In calculating the opacity for this model, the contribution from the warm component is neglected, which is a good approximation since its optical depth is small along all lines of sight due to its high temperature.

2.2 Continuous Models

In contrast to the cloud models, we examined models in which there is a continuous HI distribution given by the above equation for the mean gas density plus either holes or holes and expanding shells in that distribution. The motivation for this model is that supernovae can create bubbles or superbubbles in which HI is converted into hot gas or displaced. Holes with a range of sizes exist in a single model, with the range of radii being given by

$$0\, pc < r_{hole}(R) < 100\, \frac{R}{R_\odot}\, pc$$

On average, there are equal numbers of holes of all sizes. The hole radii are scaled to the galactic radius because the larger superbubbles are seen at greater radii (McCray and Kafatos 1987). When a hole is placed in the HI distribution, the remaining gas density is 2% of the original density.

Models were calculated in which 0%, 20%, 33%, and 50% of the volume of the interstellar medium is occupied by holes. As the percentage of the volume in holes increases, so does the HI density in the remaining HI substrate (after all, we still reproduce the total observed emissivity).

The final model considered is where the material in the holes is placed in an expanding shell at the boundary of the shell. The expansion velocity is taken to be 30 km s^{-1} for all shells. Calculations were made when 20% and 33% of the volume of ISM is in the form of holes, which are surrounded by shells.

3. RESULTS

There is a qualitative similarity between the cloud models, all of which provide a poor representation of the data (Fig. 1,2). This is especially evident in the channel maps of the cloud models in Figure 2, where much of the flux is concentrated in a small fractional area. The striking difference between the observations and the cloud models that the power is dominant on small scales in the cloud models while in the observations, it occurs both on large and small scales; power spectrum analysis confirms this inference. This difference in the angular size distribution of the emission structure is greatest between the data and the two-component cloud models.

The reason for the apparent failure of the cloud models to reproduce the data is that there are too few clouds that contribute to each resolution element in a map (especially the channel maps). This problem is greatest in the low velocity channel maps (gas within 2-3 kpc of the Sun) and at latitudes greater than 2°. For this set of conditions, there are only a few clouds per resolution element (telescope beam), so the beam to beam variations can be large. This effect is greatest in the two-component cloud model, where the few large clouds can dominate the emission locally. Although it may be possible to reproduce some of the large-scale structure in the cloud models by introducing an agent that groups the clouds (e.g., supernovae), the problem with the small scale structure will remain. These calculations suggest that a theoretical understanding of the interstellar medium should not be based on cloud structures.

The continuous models of the ISM reproduce some of the characteristics of the data with greater success. In Figure 1, a comparison between the observations and the 20% and 30% hole models shows that the large-scale structure is naturally reproduced. The small-scale structure so dominant in the cloud models is much less pronounced in these continuous models (actually, there is too little power on small scales). Because the location of the holes are uncorrelated, they can occasionally line up over a velocity range comparable to the velocity dispersion of the gas (0.5 kpc). When this occurs, there is a local minimum in the channel map (dark spot). Local minima of similar depth occur in the data and in the 20% hole model, but the variations seen in the 33% hole model are already becoming so large as to be inconsistent with the observations.

The continuous model composed of 20% holes plus shells shows very pronounced emission from shells, in contrast to the observations. In these calculations, the thickness of the shells is small so that when seen edge-on, they are always unresolved. At a typical distance of the shells in Figure 2, any shell thinner than 15 pc would be unresolved and appear like the shells in this calculation. Shells thinner than this are expected in both supernova and supershell calculations, although with less mass than we have employed. The infrequence of such features in the data indicate either that shells are much thicker than assumed, less frequent (not associated with every hole), or much less massive (most likely, a combination of all three effects).

To summarize the results of our models, the geometry of the interstellar medium is not described by an ensemble of clouds, either uniform, non-uniform (exponential), or the two-component clouds (the McKee-Ostriker ISM). The neutral ISM is better described by a continuous HI distribution with holes with sizes characteristic of both supernova remnants and superbubbles (diameters up to 200 pc). The amplitude of the large-scale variations (>1°) places limits on the volume occupied by these holes, suggesting that no more than about 30% of the ISM is in ionized material. This result is consistent with the observations of other ISM phases if about 10-20% of the volume is occupied by hot X-ray emitting gas and about 20% of the volume is occupied by the warm ionized component seen by Reynolds (some of this could be cospatial with the HI, so it might not require a displacement of the neutral phase).

ACKNOWLEDGEMENTS

This project is deeply indebted to Jay Lockman for the use of his data and for his advice. JNB gratefully acknowledges support for this project from NASA LTSA-89-033.

REFERENCES

Brinks, E., and Bajaja, E. 1986, Astr. Ap., **169**, 14
Brinks, E., and Shane, W.W. 1984, Astr. Ap. Suppl., **55**, 179.
Cowie, L.L, McKee, C.F., and Ostriker, J.P. 1981, Ap. J., **247**, 908.
Cox, D.P. 1981, Ap. J., **245**, 534.
Cox, D.P., and Reynolds, R.J. 1987, Ann. Rev. Astr. Ap., **25**, 303.
Cox, D.P., and Smith, B.W. 1974, Ap. J. (Letters), **189**, L105.
Dickey, J.M., and Lockman, F.J. 1990, Annual Reviews of Astronomy and Astrophysics, 28, in press.
Deul, E.R. 1988, Ph.D. Thesis, Sterrewacht Leiden.
Heiles, C. 1984, Ap. J. Suppl., **55**, 585.
Kulkarni, S.R., and Heiles, C. 1988, in "Galactic and Extragalactic Radio Astronomy", ed. G.L. Verschuur and K.I. Kellermann (Springer: Heidelberg), p. 95.
Lockman, F.J., and Ganzel, B.L. 1983, Ap. J., **268**, 117.
Mac Low, M.-M., McCray, R. 1988, Ap. J., **324**, 776.
Mac Low, M.-M., McCray, R., and Norman, M.L. 1989, Ap. J., **337**, 141.
McCammon, D., and Sanders, W.T. 1984, Ap. J., **287**, 167.
McCray, R., and Kafatos, M. 1987, Ap. J., **317**, 190.
McKee, C.F., and Ostriker, J.P. 1977, Ap. J., **218**, 148.
Norman, C.A., and Ikeuchi, S. 1989, Ap. J., **345**, 372.
Reynolds, R.J. 1989, Ap. J. (Letters), **339**, L29.
Spitzer, L. 1978, "Physical Processes in the Interstellar Medium", (Wiley: New York) p. 44.
Tenorio-Tagle, G., and Bodenheimer, P. 1988, Ann. Rev. Astr. Astrophys., **26**, 145.
Tenorio-Tagle, G., Bodenheimer, P., and Rozyczka, M. 1987, Astron. Ap., **179**, 219.
Tomisaka, K., and Ikeuchi, S. 1986, Pub. Astr. Soc. Japan, **38**, 697.
Van Buren, D. 1986, Ap. J., **306**, 538.
Weaver, R., Castor, J., McCray, R., Shapiro, P., and Moore, R. 1977, Ap. J., **218**, 377.

GRAIN EVOLUTION IN THE FRAMEWORK OF DISK–HALO INTERACTIONS

F. FERRINI
Dipartimento di Fisica
Università di Pisa
56100 Pisa, Italy

ABSTRACT.
The presence of dust grains at high galactic latitude as well as in the Halo of external galaxies has received substantial observative support in the last few years. Besides intense hydrodynamics stirring phenomena, like Supernovae expanding shells, stellar winds and Galactic fountains, the removal of dust from the Disk can be ascribed to the global galactic radiation field. The continuous sputtering in the hot Halo gas may explain the large scale height found for refractory elements (Edgar and Savage, 1989).

1. GRAINS AT HIGH GALACTIC LATITUDES.

Dust is now recognized to be a fundamental component of the galaxy on all length scales; from local dense clouds to molecular cloud concentrations, accompanying OB associations to fully global distributions. Indeed the IRAS observations at 100 μm revealed that, in the Milky Way, dust is distributed similarly to the HI component (Burton, 1990). The presence of grains is not limited to the Disk but dust is pervasive also at high galactic latitudes, as it results evident from the extended search by Désert et al. (1988) for clouds with enhanced infrared emission with respect to the normal population of diffuse clouds. Isolated dust clouds have been studied in detail by Rohlfs et al. (1989) and Herter et al. (1990).

Clear views of the large scale distribution of gas and dust out of the Disk have been obtained for external galaxies. Véron–Cetty and Véron (1985) dividing a blue image by an infrared image of NGC 1808, noticed that the bright central nucleus is embedded in a filamentary reddish structure, obscuring the arm. Sofue (1987), from an analysis of the photographs of the Hubble Atlas of Galaxies reported the presence of dark filaments vertically emerging toward the Halo from the 3–4 kpc molecular ring in NGC 253 and NGC 7331, up to 1–2 kpc on the Disk plane. Similarly, Rand et al. (1990) have revealed numerous faint vertical filaments extending up to ~ 2 kpc off the plane of the edge–on spiral NGC 891 (see also Allen and Dettmar, this Conference).

Furthermore, intergalactic extinction, due to both a diffuse medium and individual clouds (Rudnicki et al. 1989) is a subject of renewed interest, because of its connection with cosmological observations and theories.

The frequence of dusty clouds out of the Disk suggests that they are rather resistant to the violent phenomena which are responsible for a removal of material from the Disk, or that a soft fountain is connected with their presence. Furthermore, the fate of naked grains in the Halo ambient deserves investigation.

2. FORCES ON GRAINS.

The dynamical evolution of small solid particles in interstellar space has been studied, beginning with some pioneering works in the 30's (see Spitzer 1941 and references therein), under a variety of different conditions. Aside from the intensity of the radiation field, a key parameter for the evolution is the coupling between gas and dust. This coupling depends on the grain properties and the physical state of the gas (e.g. Spitzer 1978; Draine and Salpeter 1979): the time scales for grain–gas interactions are sensitive functions of the gas density and grain charge. If gas drag is neglegible, the pressure from strong radiation sources can accelerate dust grains to large velocities (e.g. Wolfe et al. 1950). Similarly, bare grains reaching the scale height of the gaseous Disk can be ejected out of their host galaxy by starlight (e.g. Chiao and Wickramasinghe 1972; Barsella et al. 1989; Ferrara et al. 1989). When the gas–dust coupling is important, a net momentum transfer from the radiation field to the gas takes place and the gas is accelerated (radiation exerts a neglegible force directly on the gas; Pecker 1972). This process seems to be the driving agent of massive winds from cool giant stars (e.g. Salpeter 1974; Kwok 1975), internal holes in HII regions (e.g. Cochran and Ostriker 1977), and large scale interstellar structures such as the Barnard Loop (e.g. O'Dell et al. 1967).

2.1 Early Studies on Dust Motion.

Only two papers by Pecker (1972, 1974) and one paper by Chiao and Wickramasinghe (1972) have taken up large scale dust motion. In all cases particle motions are considered far enough from stars to disregard any local effect (winds, etc.). Pecker (1972) considered, at first, the effect of individual stars of various spectral types on grains, ranging from 10^{-3} to 10 μm, varying the star spectral type (and consequently their effective temperature). The optical properties of the grains, that is their radiation pressure coefficient Q_{pr}, were considered in a rather qualitative way.

He concluded that grain size plays a larger role than its composition, resulting in a particularly efficient expulsion for grains with radii $\simeq 0.1$ μm; the early type stars (O5) can efficiently expel any grain. Similar results have been obtained by Divari and Reznova (1970).

The subsequent step in Pecker's calculation was the evaluation of the forces exercised by a 'synthetic' galaxy on dust particles. His model takes into account, schematically at least, the presence of "invisible matter" in the Disk as well as a distribution of visible matter and light in the Disk.

The result is a generalized attraction of dust grains, but this is true for very high temperatures of the stars in the Disk. Pecker thus concluded that the time variation of the ratio M/L is a fundamental factor in determining the behaviour of grains, especially as regards the early phases of evolution of our galaxy.

Chiao and Wickramasinghe (1972) studied the problem in a similar way, first in the interior of the Disk, by calculating the contributions of single stars, with gravitation and radiation forces being calculated in a simple way. In the discussion of forces inside the Disk, these authors correctly take into account the drag force due to gas clouds and the magnetic drift exerted on charged grains. They concluded that the expulsion of dust grains is very efficient for spiral galaxies.

The fact that the conclusions of these authors are opposite to each other is a clear indication that, although the physics of this problem can be formulated very simply, the end result depends critically on the adopted description both of the dust and of the galaxy properties.

2.2 Recent Studies on Dust Motion.

Our group considered recently in some more detail the problem of the motion of dust particles in the galactic radiation field (Greenberg *et al.* 1987; Ferrini *et al.* 1988; Barsella *et al.* 1989; Ferrara *et al.* 1989 and 1990). For the gravitational force the problem is extremely easy: the force on a grain of mass m_g may be written:

$$\vec{F}_G(\vec{r}) = m_g \vec{G}(\vec{r}) \qquad (1)$$

where $\vec{G}(\vec{r})$ is the gravitational field intensity at the point $\vec{r}$.

The formula for the radiation pressure force is more complicated, being an integral over the radiation frequency dependent Q_{pr}:

$$\vec{F}_R(\vec{r}) = \pi a^2 \int d\vec{\rho} \int d\nu Q_{pr}(a,\nu) \vec{\Psi}(\vec{r},\vec{\rho},\nu) \qquad (2)$$

$\vec{\Psi}$ is the radiation field due to the galactic element at $\vec{\rho}$ on the grain at position $\vec{r}$ at frequency ν.

The wavelength dependence of the luminosity is assumed independent of the position. Therefore, the radiation field function may be written:

$$\vec{\Psi}(\vec{r},\vec{\rho},\nu) = \vec{\Xi}(\vec{r},\vec{\rho})\Omega(\nu)$$

and the integration over ν in (2) yields the formula:

$$Q_{pr}^*(a) = \int d\nu \, Q_{pr}(a,\nu)\Omega(\nu) \qquad (3a)$$

$$\vec{F}_R(\vec{r}) = \pi a^2 Q_{pr}^*(a) \int d\vec{\rho}\, \vec{\Xi}(\vec{r},\vec{\rho}) = \pi a^2 Q_{pr}^*(a)\vec{\Gamma}(\vec{r}) \qquad (3b)$$

The expression for the radiation pressure force is reduced, therefore, to a form similar to that for the gravitational force.

The principal populations of interstellar grains may be divided optically into two categories: dielectric and metallic. The dielectric grains are represented by either silicate core–organic refractory mantles (Greenberg and Chlewicki, 1983) or by pure silicate grains (Mathis et al. 1977). We find convenient to use the available Draine and Lee (1984) "astronomical silicate" properties as representative of the dielectrics. Metallic grains are represented by graphite, whose optical constant are taken from Tosatti and Bassani (1970) and Phillip (1977).

A detailed knowledge of the luminosity and matter distribution of the underneath galaxy is required to compute the grain evolution. A spiral galaxy may be characterized by three components: Bulge, Disk and Halo. The Disk is modelled as an infinitely thin axially symmetric exponential distribution of luminosity and matter. The Bulge is considered as a massive and luminous addition to the center of the galactic Disk, and the Halo is considered completely dark with a spherically symmetric mass structure. We considered the detailed frequency dependence of the galactic radiative flux, $\Omega(\nu)$ different for the various Hubble types of galaxies, following Pence (1976) and Yoshii and Takahara (1988), which give the spectrum for λ between 1400 to 8000 Å.

We used as a test galaxy, NGC 3198, which is a fairly well studied Sc galaxy (Wevers 1984, Burstein and Rubin 1985, van Albada et al. 1985), to analize the spatial distribution of radiative and gravitational forces, considering then only the static aspect of the problem. A typical example of the fate of a grain located at the border of the Disk is shown in Fig. 1.

The main results of our analysis, extended to other 15 galaxies, for which we have good models for the luminous and matter distributions, and to an extended range of grain radii, may be briefly summarized as follows:

i) Graphite grains with radii in the 0.02–0.2 μm range are expelled from the studied galaxies.

ii) Silicate grains of intermediate radii range (0.05–0.2 μm) have, in general, equilibrium positions high on the galactic Disk, inside the mass distribution of the Halo. The equilibrium positions range from 2/10 to 8/10 of the Halo radius.

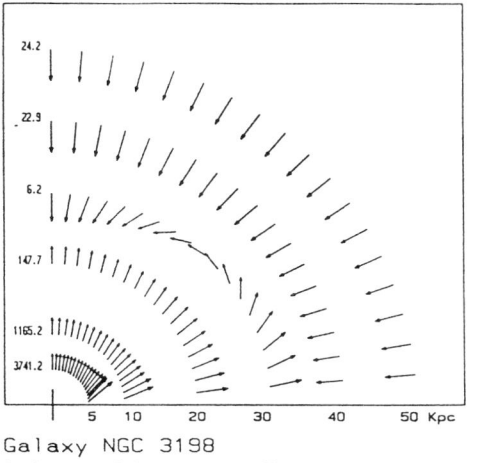 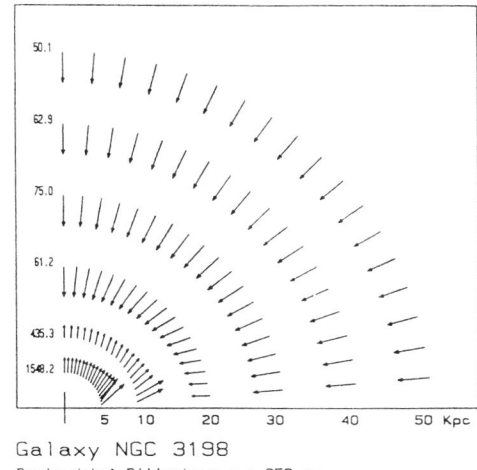

Figure 1. The distribution of total resulting force on silicate grains (radii 0.1 and 0.25 μm) in the plane perpendicular to the galactic Disk is shown for the galaxy NGC 3198. It is clear that the grain can be removed from the Halo–Disk interface; a static equilibrium position is present at an height of about 25 and 15 kpc respectively on the galactic polar axis.

3. PHOTOLEVITATION OF DIFFUSE CLOUDS.

The overall effect of radiation pressure on the existing interstellar phases depends on the gas–grain coupling. For regions in which the hotter and rarefied phases dominate, grains can drift through the gas along the B-field lines and will tend to diffuse out of the region (*e.g.* Chiao and Wickramasinghe 1972). In the denser and cooler phases where dust particles can be stopped by drag, the final results depend on the gas column density and radiation field. Massive molecular clouds are not affected by the momentum transfer from the general radiation field, but small diffuse clouds can receive a net acceleration. If one considers that the interstellar gas is distributed in clouds with a given velocity dispersion in the z-direction, a fraction of these clouds can be raised above the main gaseous Disk by radiation pressure of starlight. This "photolevitation" effect has been introduced by Franco *et al.* (1990).

The radiation pressure on dust grains located at an optical depth τ from the edge of a cloud is

$$P(\tau) = \frac{Q_p F \, e^{-\tau}}{c}$$

where $Q_p F = \int Q_{pr}(\lambda) F_\lambda d\lambda$ is the "effective" flux for radiation pressure. The total confining pressure is

$$\int_0^{\tau_c} P(\tau) d\tau \simeq \frac{F Q_p}{c}(1 - e^{-\tau_c})$$

where τ_c is the total optical depth of the cloud.

Grains located at $\tau \geq 1$ do not feel any substantial pressure and, thus, the effectiveness of the momentum transfer from the photon field is restricted to regions with moderate column density values, say, of about a few times 10^{20} cm^{-2} (Franco and Cox, 1986).

The extent to which this radiation pressure can be transmitted to the gas depends, of course, on the dust–gas coupling. For charged dust particles, the mean free path for stopping a grain via electric and viscous interactions with gas particles is (*e.g.* Draine and Salpeter 1979)

$$\lambda_c \simeq \frac{1}{n\mu} \left(\frac{m_d}{A}\right) f(s)$$

where $s = v/v_{th}$. The least favorable case for the coupling results when $f(s) = 1$ (*i.e.* fast grains) and the corresponding gas column density (assuming cosmic abundances) is

$$N = n\lambda \simeq 10^{19} \left(\frac{a}{10^{-5} \text{ cm}}\right) \left(\frac{\rho_d}{2 \text{ g cm}^{-3}}\right) \text{ cm}^{-2}$$

which fixes the minimum column density for an efficient transfer of momentum from dust to gas.

Assuming the optical constant for graphite and "astronomical silicate" given by Draine (1987) and the interstellar radiation field at the solar circle given by Mathis *et al.* (1983), the effective flux results about $(F Q_p)_\odot \simeq 5 \times 10^{-3}$ erg cm^{-2} s^{-1}, which translates into a confining pressure of about $P_\odot \sim 1.7 \times 10^{-13}$ dyn cm^{-2}. This value is about a half of the one derived from the diffuse cloud data (*e.g.* Spitzer 1978) and indicates that radiation could play a significant role in the structure of these clouds.

When the photon field is anisotropic the cloud receives a net acceleration. Clouds located in the neighborhood of star clusters are accelerated away from the cluster (*e.g.* Mathews 1967; Krishna Swamy and O'Dell 1967; Elmegreen and Chiang 1982), and clouds located above midplane can be pushed by the radiation field from the Disk to even larger heights. In this latter case, which is the one explored by Franco *et al.* , the clouds are flattened during the acceleration process and can even be ejected from the Disk. The sputtering time scales for grains drifting with $v \leq 5 v_{th}$ under diffuse cloud conditions is well in excess of 6×10^9 yr (*e.g.* Draine and Salpeter 1979), indicating that grains can survive a wide range of radiative forces and are not destroyed during cloud acceleration.

It is easy to estimate the minimum height above which clouds can be levitated, evaluating the ratio between F_{down} the average energy flux passing through a point located at a height z from the plane and directed towards midplane, and F_{up} the corresponding one directed outwards the Disk. Adopting for the average optical scale height a value of 150 pc for a self–gravitating or gaussian Disk, diffuse semi–opaque clouds can levitate when located above $z \simeq 25$ pc.

The principal contributions to the equation of motion for the cloud are the gravitational acceleration in the z–direction and the total outward radiation pressure, due to the average field and to nearby stellar cluster: $P_{rad} = P_{av} + P_{cl}$, with

$$P_{av} = \frac{Q_p(F_{up} - F_{down})}{c}(1 - e^{-\tau_c}), \qquad P_{cl} = \frac{LQ_p(1 - e^{-\tau_c})}{4\pi z^2 c}$$

where $F_t = F_{up} + F_{down}$ is the total output per unit surface from one face of the Disk and L is the luminosity of the nearby cluster.

Franco et al. do not introduce the magnetic field, considering that it has random fluctuations and presents numerous open channels, and solve the equation of motion for moderate values of the drag from the ambient medium.

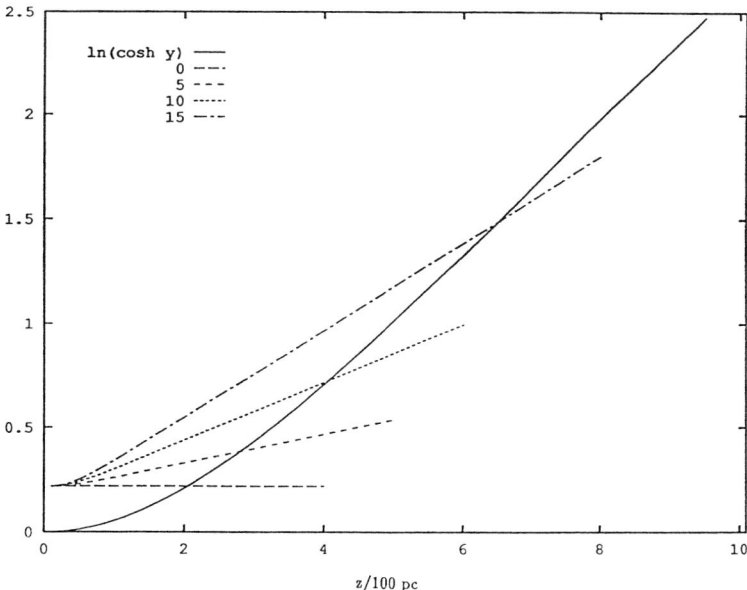

Figure 2. Graphical solution of equation of motion. The solid line is the gravitational term; the dashed lines are the radiative terms for the gaussian light distribution. For R equal to 0, 5, 10 and 15, the maximum heights are 200, 280, 410, and 650 pc, respectively.

A typical graphical solution is shown in Fig.2, where the maximum height reached by the levitated cloud is given by the intersection of the gravitational term with the radiative term for an initial cloud velocity of 15 km s^{-1} (the effective velocity near a stellar cluster with 10^6 L$_\odot$). The curves differ for the ratio of effective flux for radiation pressure to the solar circle value ($R = (F_{up} + F_{down})Q_p/5 \times 10^{-3}$ erg cm^{-2} s^{-1}). R can reach high values in sites with intense star formation and in spiral arms.

4. CONCLUSIONS.

Radiation pressure on dust grains may play an important role in determining some features of the interstellar medium. In particular, small dusty clouds with $N < 5 \times 10^{20}$ cm^{-2} can be raised to considerable heights above the galactic plane and would be observed as small local features emerging out of the Disk. Grain drift inside the accelerated clouds may be expected. If this is the case, some grains can leave the cloud and will continue their evolution in a rarefied hot medium with a much lower drag. These bare grains are easily accelerated to very high latitudes and can even be expelled out of the galaxy (*e.g.* Chiao and Wickramasinghe 1972; Barsella *et al.* 1988; Ferrara *et al.* 1990).

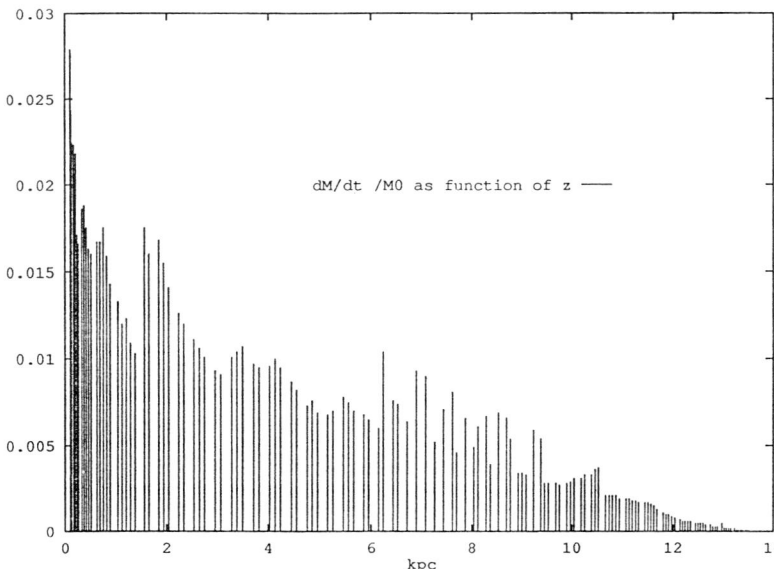

Figure 3. The fractional release of matter from a dust grain (silicate, 0.1 μm radius) moving in a hot hydrostatic isothermal (5×10^5 K) gasous Halo vs. the height on the galactic plane.

If sputtering is important, the grains will be destroyed somewhere in their evolution through the Halo (Ferrara et al. 1989). Similar results are expected if the original clouds evaporate as they evolve through the coronal gas. In both cases, however, the grains will act as chemical pollutants of the Halo or the intergalactic medium.

In Fig. 3, I show the fraction of mass released by the grain every 10^5 yrs, as a function of the height on the galactic plane. These processes could explain the large scale heights found for highly refractory elements at low ionization stages (e.g. Edgar and Savage 1989 and Savage, this Conference). Moreover, such a possible connection between Disk and Halo may be relevant in the chemical evolution of the interstellar and intergalactic medium.

ACKNOWLEDGEMENTS

I thanks warmly my collegues and friends with whom the arguments treated in this paper have been studied: S. Aiello, B. Barsella, A. Ferrara, J. Franco and J.M. Greenberg. It was a pleasure to give this talk in Leiden, where this work began in 1986, during a two months visit to the Huygens Laboratorium. Thanks to D. Boschiglio for her assistance in writing this paper.

REFERENCES

Barsella, B., Ferrini, F., Greenberg, J. M., and Aiello, S. 1989 *Astr. Ap.*, **209**, 349.
Burstein, D., Rubin, V.C.: 1985, *Ap. J.*, **297**, 423
Burton, W.B.: 1990, in *Chemical and Dynamical Evolution of Galaxies*, eds. F. Ferrini, J. Franco and F. Matteucci, Giardini, Pisa, p.657
Chiao, R.Y., and Wickramasinge, N.C. 1972, *M.N.R.A.S.*, **159**, 361.
Cochran, W. D., and Ostriker, J. P. 1977, *Ap. J.*, **211**, 392.
Désert, F. X., Bazell, D., and Boulanger, F. 1988, *Ap. J.*, **334**, 815.
Divari, and Reznova, 1970 *Sov. Astron.*, , .
Draine, B. T. 1987, Princeton Obs. POP-213.
Draine, B. T., and Lee, H. M. 1984, *Ap. J.*, **285**, 89.
Draine, B. T., and Salpeter, E. E. 1979, *Ap. J.*, **231**, 77.
Edgar, R. J., and Savage, B. D. 1989, *Ap. J.*, **340**, 762.
Elmegreen, B. G., and Chiang, W. H. 1982, *Ap. J.*, **253**, 666.
Ferrara, A., Ferrini, F., Barsella, B. and Aiello, S. 1990, *Astr. Ap.*, in press
Ferrara, A., Franco, J., Ferrini, F., and Barsella, B. 1989, *IAU Colloquium 120, Structure and Dynamics of the Interstellar Medium*, eds. M. Moles and G. Tenorio–Tagle, Springer Verlag, Berlin, p.454.
Ferrini, F., Barsella, B., Greenberg, J.M. 1988, *Dust in the Universe*, eds. M.E. Bailey and D.A. Williams, Cambridge Univ. Press, p.513
Franco, J., and Cox, D. P. 1986, *P. A. S. P.*, **98**, 1076.
Franco, J., Ferrini, F., Ferrara, A., and Barsella, B. 1990, *Ap. J.*, in press
Greenberg, J.M., Chlewicki G., 1983, *Ap. J.*, **272**, 563
Greenberg, J.M., Ferrini, F., Barsella, B., Aiello, S., 1987, *Nature*, **327**, 214
Herter, T., Shupe, D.L., and Chernoff, D.F. 1990, *Ap. J.*, **325**, 149.

Krishna Swamy, K. S., and O'Dell, C. R. 1967, *Ap. J.*, **147**, 529.
Kwok, S. 1975, *Ap. J.*, **198**, 583.
Mathews, W. G. 1967, *Ap. J.*, **147**, 965.
Mathis, J. S., Rumpl, W., and Nordsiek, K. H. 1977, *Ap. J.*, **217**, 425
Mathis, J. S., Mezger, P. G., and Panagia, N. 1983, *Astr. Ap.*, **128**, 212.
O'Dell, C. R., York, D. G., and Henize, K. G. 1967, *Ap. J.*, **150**, 835.
Pecker, J.-C.: 1972, *Astr. Ap.*, **18**, 253
Pecker, J.-C. 1974, *Astr. Ap.*, **35**, 7.
Pence, W. 1976, *Ap.J.*, **203**, 39
Phillip, H.L.: 1977, *Phys. Rev.*, **B16**, 2896
Rand, R.J., Kulkarni, S.R., and Hester, J.J. 1990, *Ap. J.*, **325**, L1.
Rohlfs, R., Herbstmeier, U., Mebold, U., and Winneberg, A. 1989, *Astr. Ap.* **211**, 402.
Rudnicki, K., Wszolek, B., Masi, S., De Bernardis, P., and Salvi, A. 1989, *Comm. in Ap.*, **13**, 171
Salpeter, E. E. 1974, *Ap. J.*, **193**, 585.
Savage, B.D. 1989, in *Evolution of the Interstellar Medium*, *P.A.S.P*, in press
Sofue, Y. 1987, *Publ.Astr.Soc.Jap.*, **39**, 547.
Spitzer, L. 1941, *Ap. J.*, **94**, 232.
Spitzer, L. 1978, *Physical Processes in the Interstellar Medium*, (New York: Wiley and Sons).
Tosatti, E., Bassani, F.: 1970, *Nuovo Cimento*, **65B**, 161
van Albada, T.S., Bahcall, J.N., Begeman, K., Sancisi, R.: 1985, *Ap. J.*, **295**, 305
Véron–Cetty, M.P. and Véron, P. 1985, *Astr. Ap.*, **145**, 425.
Wevers, B.M.H.R.: 1984, Ph. D. Thesis, Groningen University
Wolfe, B., Routly, P., Wightman, A., and Spitzer, L. 1950, *Phys. Rev.*, **79**, 1020.
Yoshii, Y., and Takahara, F. 1988, *Ap.J.*, **326**, 1

NUMERICAL SIMULATIONS OF GALACTIC OUTFLOW AND INFLOW PHENOMENA

KOHJI TOMISAKA
Faculty of Education, Niigata University,
Niigata 950-21, JAPAN

ABSTRACT. The recent progress of numerical studies on outflow phenomena from the galactic disk to the halo is summarized. Firstly, a galactic-scale outflow is considered. If the high-velocity cloud is formed from the radiatively cooled gas, which was originally ejected from the disk as a hot gas, the temperature and density at the base of the halo should be $\sim 10^6$ K and 10^{-3} cm^{-3}. Next, recent results of numerical simulations of the evolution of superbubbles, through which hot gas flows out to the halo, are reviewed. In the case of a thin disk whose density scale height is $H \simeq 100$ pc, the shell begins to be accelerated upwardly after several dynamical time scales. After that, the polar cap of the shell is broken and the hot gas flows away into the halo. In the case of a thick disk ($H \simeq 500$ pc) or a magnetized disk with a magnetic field parallel to the disk ($B \simeq 5\mu$G), the shell is not accelerated and never shows blow–out.

1. INTRODUCTION

In recent years, there has been increasing interest in the study of the disk-halo connection, in relation to the structure of the interstellar medium. The volume filling factor of hot tenuous matter, which is essentially heated by supernova (SN) explosions, was estimated to be as high as $f \simeq 75\%$ from the assumption of global pressure and mass equilibrium (McKee and Ostriker 1977). When their model is applied to the disk and the halo, hot gas spreads over several kpc above the disk, and the mass exchange rate between the disk and the halo is expected to be $\sim 1~M_\odot$ yr^{-1}. Even if the volume filling factor of the hot gas is much smaller than unity, clustered or bunched type II SNe occurring in OB association form superbubbles, through which hot gas heated by the SN explosions can flow into the halo (Norman and Ikeuchi 1989; Heiles 1989). Before (or simultaneously with) the construction of a model of the interstellar medium, it is necessary to study the elementary processes of inflow/outflow phenomena. I summarize in this paper the results of numerical simulations of (1) the global or galactic-scale outflow and (2) the small-scale outflow phenomena, i.e. outflow through superbubbles.

2. THE GALACTIC-SCALE OUTFLOW PHENOMENA

The characteristics of galactic-scale outflow have been studied by Bregman (1980) and Habe and Ikeuchi (1980). The hot gas heated by supernova explosions in the disk flows to the halo. The flow pattern is expected to be as follows: first a hot gas flows upward; with increasing latitude, the gravity becomes weaker; the gas flows radially outward due to the centrifugal force. In this flow, there are two important criteria which determine the flow.

(1) Critical temperature: the gas temperature at which the enthalpy per mass is equal to the effective gravitational potential, i.e., $T_c(r) \equiv (\gamma-1)/\gamma \times \mu m_p/k \times \left(|\Phi(r, z=0)| - V_\phi^2/2\right)$. Gas in the disk with temperature higher than T_c flows away to infinity as a wind and forms Wind-Type Halo.

(2) Cooling time vs. dynamical time: Cooling time scale of the hot gas is estimated as $t_{cool} \simeq 3kT/(2\Lambda n) \simeq 42 \text{Myr}(T/10^6 \text{K})^{1.6}(n/10^{-3}\text{cm}^{-3})^{-1}$. If the radiative cooling does not work so effectively, the dynamical timescale is given as $t_{dyn} \simeq$ scale height/sound speed $\simeq 6.5(T/10^6\text{K})(g/10^{-8}\text{cm s}^{-2})^{-1}\text{kpc}/\{140(T/10^6\text{K})^{1/2}\text{km s}^{-1}\}$, where g represents the gravitational acceleration in z-direction. Comparing these two, $t_{cool}/t_{dyn} \sim \mathcal{O}(1)(T/10^6\text{K})^{1.1}(n/10^{-3}\text{cm}^{-3})^{-1}(g/10^{-8}\text{cm s}^{-2})$. In the case of $t_{cool} < t_{dyn}$, the outflowing gas is quickly cooled before it reaches the adiabatic scale height, and Cooled-Type Halo is realized. Contrarily, in the case of $t_{cool} > t_{dyn}$, the hot gas can not escape from the gravitational potential but can expand into wide region whose size is comparable to the adiabatic scale height. Bound (but not cooled)- Type Halo is formed. These three kinds of flow patterns (Wind-, Bound-, and Cooled-Type Halos) are actually seen in Habe and Ikeuchi (1980).

To explain High Velocity Cloud's (HVC's) by a "fountain model", the clouds must be formed from cooling gas at rather high latitudes, $z \simeq 5 - 10$ kpc. This is because in order to achieve a free-fall velocity of ~ 100 km s^{-1} sufficient distance is necessary. Thus, the most plausible values are 10^6 K and 10^{-3} cm^{-3} ($t_{cool} \simeq t_{dyn}$) (Bregman 1980), which correspond to the values that form the Bound-Type Halo (Habe and Ikeuchi 1980).

3. ELEMENTARY PROCESSES OF OUTFLOW — THE EVOLUTION OF SUPERBUBBLES

Clustered type II SNe occurring in OB associations form large-scale supernova remnants called superbubbles. The effect of clustered SNe was studied in one-dimensional hydrodynamics by Bruhweiler et al. (1980) and Tomisaka et al. (1981). It was shown that the size of the bubble exceeds the density scale height in the z-direction. More recently, the effect of the density stratification has received much attention. Many two-dimensional hydrodynamical simulations have been done to study the evolution of superbubbles in a stratified medium (Tomisaka et al. 1986; Tenorio-Tagle et al. 1987; Mac Low and McCray et al. 1988; Mac Low et al. 1989; Tenorio-Tagle et al. 1990; Igumentschev et al. 1990). When the size of the bubble exceeds several times the scale height of the disk, the bubble blows

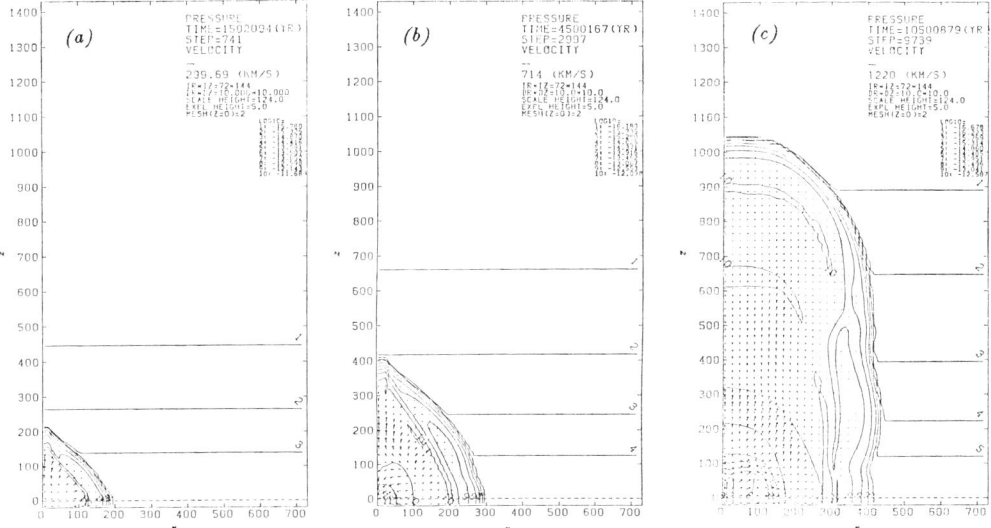

Figure 1. The evolution of a superbubble, with parameters $n_o = 0.1$ cm^{-3} and $\Delta t = 0.2$ Myr (Tomisaka and Ikeuchi 1986). The structures at (a) $t = 1.5$ Myr, (b) 4.5 Myr, and (c) 10.5 Myr are shown. It is clear that the spherical bubble is deformed into a funnel shape structure.

out. I will summarize briefly the conclusions of these simulations in the following subsections (for a review see Tenoria–Tagle and Bodenheimer 1988, and Spitzer 1990). Recently, magnetohydrodynamical simulations of magnetized superbubbles were done by Tomisaka (1990) and Shapiro et al. (1990). Magnetic fields parallel to the galactic disk lead to confinement of the bubble in the disk. I will show the recent results of three–dimensional magnetohydrodynamical simulations of a superbubble in §4.

3.1 Expansion Law

There is one characteristic time scale in the dynamics of a bubble in a density stratified medium, which is defined as

$$\tau_D \equiv H^{5/3}\left(\frac{\rho_0}{L_{SN}}\right)^{1/3} \simeq 1.2\,\mathrm{Myr}\left(\frac{H}{100\mathrm{pc}}\right)^{5/3}\left(\frac{\rho_0}{2\times10^{-24}\mathrm{g\ cm}^{-3}}\right)^{1/3}\left(\frac{L_{SN}}{10^{38}\mathrm{erg\ s}^{-1}}\right)^{-1/3}, \quad (3.1)$$

where H, ρ_o, and L_{SN} represent, respectively, the density scale height, the density in the mid-plane, and the mechanical luminosity emitted from SN explosions in an OB association. In case the bubble formed in a spherical symmetric exponential atmosphere, with $\rho = \rho_o \exp(-r/H)$, the shock front is decelerated first. The shock front and the shell begin to accelerate after $t > 2.8\tau_D$, when the size of the shock $R_s > 2.2H$ (Kompaneets 1960; Laumbach and Probstein 1969; Koo and McKee 1990).

Figure 1 shows the evolution of a superbubble with parameters $\rho_o = 2 \times 10^{-25}$ g cm^{-3} and $L_{SN} = 1.6 \times 10^{38}$ erg cm^{-3}, using two–dimensional hydrodynamics (Tomisaka and Ikeuchi 1986). Sequential SN explosions with energy $E_o = 10^{51}$ erg/s were assumed to occur every $\Delta\tau \simeq 0.2$ Myr. However, there is hardly any difference in the global structure between models of continuous energy ejection and sequential SN explosions, if L_{SN} is taken as $E_o/\Delta\tau$. It can be seen in Figure 1 that after the size exceeds the scale height H, the shell begins to elongate in the z–direction. The bubble reaches as high as $z \simeq 1$ kpc.

The results from a semi–analytical study using the Kompaneets (1960) approximation by Mac Low and McCray (1988) are useful to predict the evolution of a superbubble. In this approximation, the shell is assumed infinitely thin and the pressure distribution in the cavity is assumed uniform. They introduced a non-dimensional parameter that represents the ratio of ejected energy in the dynamical time $L_{SN}\tau_D$ to the necessary work for the bubble to expand up to the scale height, $p_e H^3$,

$$D \equiv \left(\frac{L_{SN}\tau_D}{p_e H^3}\right)^{3/2}$$

$$\simeq 940 \left(\frac{L_{SN}}{10^{38}\text{erg s}^{-1}}\right)\left(\frac{H}{100\text{pc}}\right)^{-2}\left(\frac{p_e}{10^4 k \text{ dyne cm}^{-2}}\right)^{-3/2} n_0^{-1/2}, \quad (3.2)$$

where p_e represents the pressure of external interstellar matter. Their numerical results show that, for sufficiently large D, the expansion laws of the shell are identical irrespective of D if time is measured in τ_D. Furthermore, they showed that in the case of an exponential atmosphere, $\rho(z) = \rho_o \exp(-z/H)$, the shell will eventually be accelerated upwardly due to the density stratification, if $D \gtrsim 100$.

3.2 Rayleigh-Taylor Instability and Blow-out

Transition from acceleration to deceleration occurs after $(2-3) \times \tau_D$ since the SN explosions started (Table 1). Small differences seem to come from the adopted initial density distribution. What happens in the upwardly accelerated shell? In a frame co-moving with the accelerated shell, the effective gravity (really an inertia force) works downwards. The cooled shell is pushed by tenuous hot gas against gravity. This configuration is unstable for Rayleigh–Taylor instability (Różyszka and Tenorio–Tagle 1985). The instability grows very rapidly after the transition, and finally the shell begins to break at the polar cap. This is clearly seen in numerical simulations by Tomisaka and Ikeuchi (1987), Mac Low et al. (1989), and Tenorio–Tagle et al. (1990).

TABLE 1.
Transition from Deceleration to Acceleration.

Authors	Initial Density Distribution	Transition Time τ_T	Height at τ_T	n_0
TI[a]	Fuchs & Thielheim atmosphere	$\simeq 2\tau_D$	$\simeq 2H$	0.1
MMN[b]	exponential atmosphere	$3.3\tau_D$	$2.9H$	1
	Gaussian atmosphere	$2.5\tau_D$	$1.7H$	1

[a] Tomisaka and Ikeuchi (1986).
[b] Mac Low et al. (1989).

3.3 Effect of a Thick Disk

The acceleration of the shell and the Rayleigh–Taylor instability at the polar cap may be typical for a superbubble in a *thin disk* with a scale height of $H \simeq 100$ pc. There is good observational evidence, however, for a warm HI component whose density scale height is ~ 500 pc (Lockman et al. 1986). The transition time τ_T is given by

$$\tau_T \simeq 30 \mathrm{Myr} \left(\frac{L_{SN}}{10^{38} \mathrm{erg\ s^{-1}}}\right)^{-1/3} \left(\frac{H}{500 \mathrm{pc}}\right)^{5/3} \left(\frac{n_0}{0.1\ \mathrm{cm^{-3}}}\right)^{1/3}. \quad (3.3)$$

The duration of an active phase of SN explosions in an OB association is estimated to be 30–50 Myr (from the initial mass function and stellar lifetime data). It can be concluded that the shell of a superbubble is decelerated throughout its lifetime. Actually, from the numerical results presented by Tenorio–Tagle et al. (1990), blow–out as well as acceleration is not seen in the model with an exponential atmosphere with $H = 500$ pc and $\rho_o = 1.6 \times 10^{-25}$ g cm^{-3} in 10 Myr. However, the superbubble expands and still reaches $z \simeq 1$ kpc during its lifetime.

4. MAGNETIZED SUPERBUBBLES

The flow across a magnetic field is blocked in case of a frozen–in field. Therefore, as in the case of a thick disk, the magnetic field tends to prevent expansion of the superbubble in the z–direction, if it runs parallel to the galactic disk. Moreover, there is a possibility that the magnetic field completely confines the superbubble in the galactic disk. In addition, in a magnetized interstellar medium, hydrodynamical shock waves are replaced by magnetohydrodynamical (MHD) shock waves. In this case, the fast MHD shock wave propagates and compresses the magnetic–field. This section presents recent results on the evolution of magnetized superbubbles. The strength of the magnetic field B is not well determined observationally; it is

TABLE 2.
Density Distribution.

Component	$\rho(0)^a$ $(10^{-25}\text{gcm}^{-3})$	H^b (pc)	Distribution
Cold H I	6.8	135	Gaussian[c]
Warm H I	1.8	135	Gaussian
Warm H I	2.3	400	exponential[d]
H$_2$	6.8	70	Gaussian
Warm H II	0.56	1000	exponential

[a] Density on the $z = 0$ plane.
[b] Density scale-height.
[c] $\rho = \rho(0)\exp(-z^2/2H^2)$.
[d] $\rho = \rho(0)\exp(-z/H)$.

reported to $\sim 5\mu$G in the warm ionized interstellar matter (Rand and Kulkarni 1989). Magnetic pressure, $p_{mag} \simeq 10^{-12}(B/5\mu G)^2$ dyne cm^{-2}, cannot be ignored in comparison with the ram pressure exerted on the superbubble shell, $p_{ram} \simeq 10^{-12}(\rho/2 \times 10^{-24}\text{g cm}^{-3})(V/10 \text{ km s}^{-1})^2$ dyne cm^{-2}, where ρ and V represent the ambient density and expansion speed of the shell.

4.1 Model

It is assumed that the interstellar medium in which the SN begin to explode is in hydrostatic equilibrium. A realistic hydrostatic model similar to that of Bloemen (1987) is taken, i.e., five gas components (Table 2), supported by thermal pressure against gravity $g_z(z)$ (Bahcall 1984)

$$p_{th}(z) + p_{mag}(z) = \int_0^z \rho(z)g_z(z)dz + p(0), \tag{4.1}$$

where $p(0)$ is the total (thermal + magnetic) pressure at $z = 0$, taken to be 3×10^{-12} dyne cm^{-1}. Since the scale–height of the magnetic field strength is much larger than that of the density distribution, the strength of the magnetic field is assumed uniform irrespective of z. Further, I assume that the OB association is located in the $z = 0$ plane. The mechanical luminosity emitted by SN explosions is given by $L_{SN} \simeq 1.68 \times 10^{38}(E_o/10^{51} \text{ erg})(\Delta\tau/2 \times 10^5 \text{ yr})^{-1}$ erg/s.

Cartesian coordinates are adopted: the x–axis is chosen as the direction of the initial magnetic field and the z–axis is chosen perpendicular to the galactic disk (x–y plane). Since both the density stratification and the magnetic field parallel to the galactic disk are included, the problem to be solved becomes a three–dimensional MHD flow. The "monotonic scheme" (van Leer 1977; Norman and Winkler 1986) is adopted to solve the hydrodynamical equations and the "constrained transport method" (Evans and Hawley 1988) is used to solve the induction equation of the magnetic field. The number of meshes is 81^3 or $81^2 \times 121$ in Cartesian coordinates (x, y, z). The amount of memory required is 128–192 MB. Since the code of this finite–difference scheme is totally vectorized, only 0.7 - 1 second of CPU time is needed for each time step using the HITACHI S–820/80 supercomputer at the University of Tokyo.

Sequential SN explosions were replaced by a continuous energy release, as was done by Mac Low et al. (1989), and the gas was assumed adiabatic with $\gamma = 5/3$ for simplicity. Since the amount of the energy lost by radiative cooling has not been determined definitely (Mac Low et al. 1989; Tomisaka and Ikeuchi 1986), I studied two cases with $L_{SN} = 1.68 \times 10^{37}$ erg/s and $L_{SN} = 1.68 \times 10^{38}$ erg/s to make up for the adiabatic assumption.

4.2 Numerical Results

Figure 2 shows the structure of the superbubble, 10 Myr after the SN explosions began. The structure is similar to that of the non–magnetic bubble, i.e., the matter ejected from the SNe flows outward as a free wind; ejected matter passes through the inward–facing shock; the accumulated interstellar matter and the magnetic field are spread between the contact surface and the outward–moving MHD shock. The outward–moving wave front is seen at $x \simeq 350$ pc (Fig. 2a and c), $y = 400$ pc (Fig. 2c). In the z–direction, it has already reached the upper boundary $z_{max} = 800$ pc. The hot gas is distributed over the region where the velocity vectors are plotted in Figure 2. From Figures 2a and b, it can be seen that the hot matter formed by SN explosions is confined below $z \lesssim 200$ pc. The magnetic field is swept up and compressed in the $z \gtrsim 200$ pc region by the thermal pressure of the hot matter below. Figure 2c clearly shows that the hot matter does not extend much in the y–direction ($y \lesssim 150$ pc) and that the flow is focused in the x–direction. The outer boundary of the hot matter (contact surface) expands as $(x_i, y_i, z_i) \simeq (160$ pc, 130 pc, 150 pc) [1] at 5 Myr, (200 pc, 140 pc, 180 pc) at 7.5 Myr, and (220 pc, 140 pc, 200 pc) at 10 Myr. Compared with the model of no magnetic field but the same density distribution, in which the hot gas extends up to $(x_i, y_i, z_i) \simeq (170$ pc, 170 pc, 290 pc) in $t = 10$ Myr (Figure 3), for the model I have studied it can be concluded that the magnetic field dams the flow across the field line and encourages the flow parallel to the field line (x–direction). The hot gas is confined within $z \lesssim 200$ pc by the magnetic field. Outside the hot matter, warm gas with $T \lesssim 10^4$ K, heated by the outermost wave front, is distributed in a rather thick region (250 pc $< x <$ 400 pc, 130 pc $< y <$ 450 pc, $z >$ 200 pc). The present simulation is restricted to the adiabatic process; in reality, due to radiative cooling, a cooled thin shell will replace this warm gas and surround the hot gas. Although the magnetic field is swept and compressed in this warm region, thermal pressure predominates over the magnetic pressure. Extrapolating the expansion law of x_i, the hot gas occupies the region with $z_i \lesssim 300$ pc during the active phase of the OB association (~ 30 Myr).

---

[1] This expression means that the boundary crosses the x-axis at $(x_i, 0, 0)$, y-axis at $(0, y_i, 0)$, and so on.

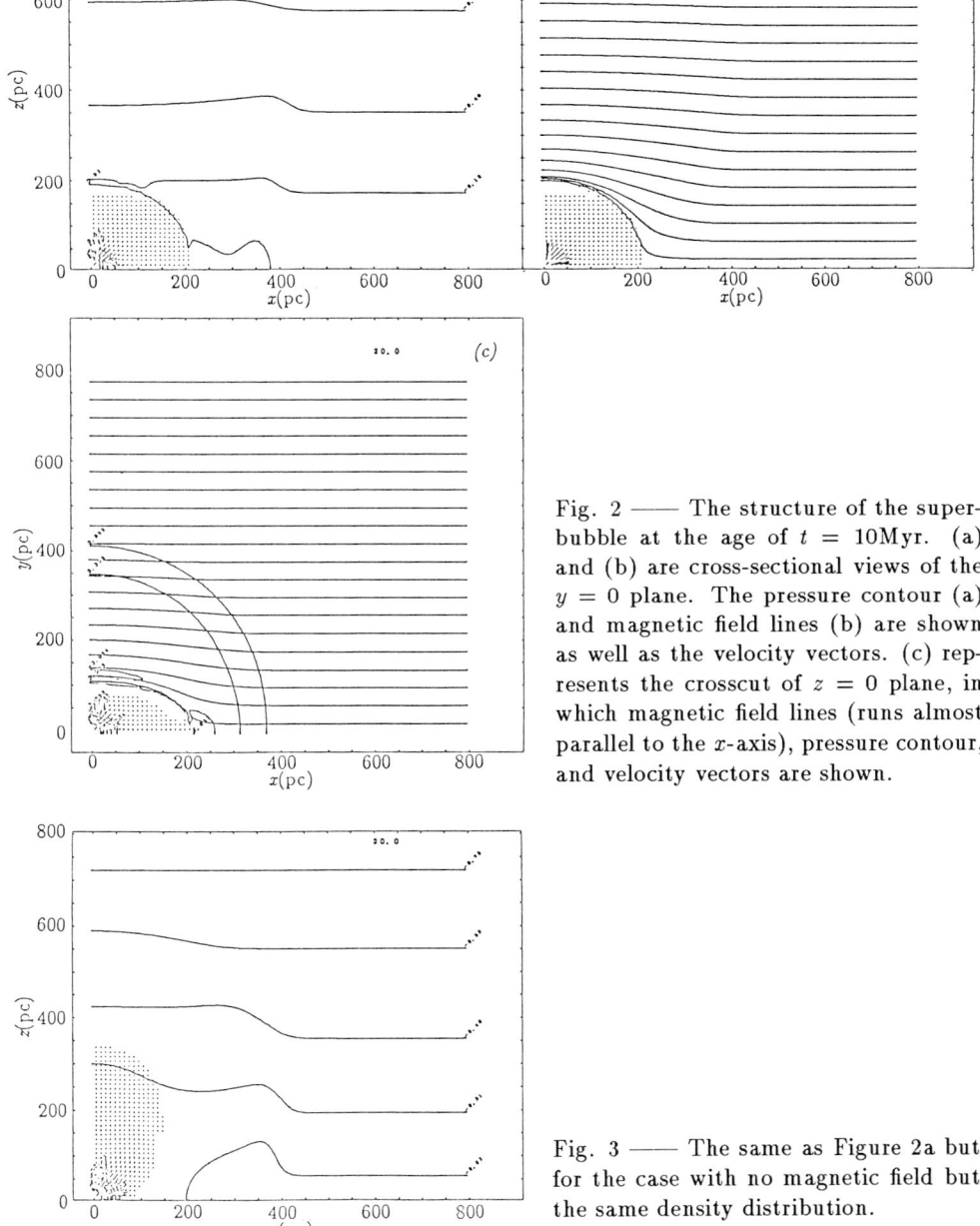

Fig. 2 —— The structure of the super-bubble at the age of $t = 10$Myr. (a) and (b) are cross-sectional views of the $y = 0$ plane. The pressure contour (a) and magnetic field lines (b) are shown as well as the velocity vectors. (c) represents the crosscut of $z = 0$ plane, in which magnetic field lines (runs almost parallel to the x-axis), pressure contour, and velocity vectors are shown.

Fig. 3 —— The same as Figure 2a but for the case with no magnetic field but the same density distribution.

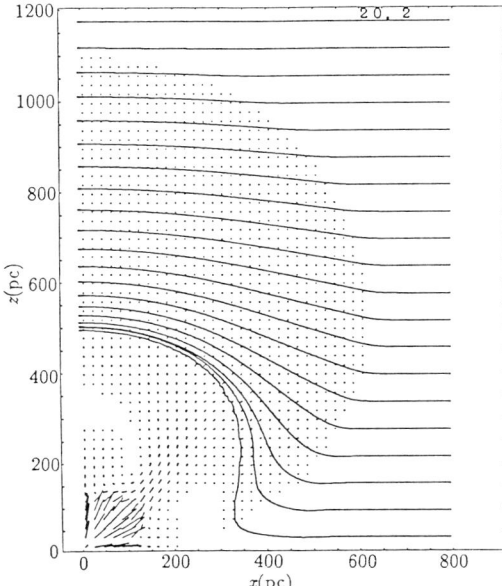

Fig. 4 —— The structure of the superbubble at the age of $t = 10$Myr for the case with the mechanical luminosity ten times larger than the previous case.

Expansion is highly anisotropic, i.e., the outermost wave front expands as $(x_s, y_s, z_s) \simeq$ (220 pc, 230 pc, 310 pc) at 5 Myr, (300 pc, 320 pc, 520 pc) at 7.5 Myr, and (370 pc, 410 pc, >800 pc) at 10 Myr. Although the height of the hot-warm interface is decelerated as $z_i \propto t^{0.4}$, that of the outermost wave front is accelerated. Acceleration of z_s is mainly driven by steep density gradient $\partial\rho/\partial z$. Perpendicularly to the magnetic field (y- and z-direction) the expansion is *encouraged by the magnetic field*, since in these directions waves propagate as the magnetosonic wave mode (Kulsrud et al. 1965). This encouragement by the magnetic field can be seen by comparing this model with that of no magnetic field but the same density distribution (Figure 3), in which the size is as large as $(x_s, y_s, z_s) \simeq$ (360 pc, 360 pc, 570 pc) at 10Myr.

In Figure 4, I plot the structure of the superbubble at the age of 10 Myr with $L_{SN} = 1.68 \times 10^{38} \text{ergs}^{-1}$, which is ten times larger than the previous case. It is interesting that the MHD shock seen in $z \lesssim 500$pc is so called "intermediate shock." Due to the large luminosity, the superbubble expands much faster: $(x_s, y_s, z_s) \simeq$ (470 pc, 500 pc, 1100 pc) at 10 Myr. Since the pressure of the of the inner hot gas is high, also the hot-warm interface expands faster than that of the previous case: $(x_i, y_i, z_i) \simeq$ (320 pc, 200 pc, 500 pc). It is clear that before the typical lifetime of massive stars in OB associations the hot gas in the bubble reaches the halo region ($z \gtrsim 1$kpc). From this result, it is concluded that if the radiative cooling is inefficient, the magnetic field assumed here $\sim 5\mu$G cannot prevent the hot matter from flowing out to the halo.

From the study of spherical symmetric nonmagnetic wind-blown bubble (Weaver et al. 1976), the pressure in the shocked wind (ejected matter) region is expressed as

$$p \simeq 0.17(L_W^2 \rho_0^3)^{1/5} t^{-4/5},$$

and the expansion of contact surface between the shocked wind and accumulated matter is

$$R_i \simeq 0.76(L_W/\rho_0)^{1/5} t^{3/5},$$

where L_W represents the mechanical luminosity of wind $= L_{SN}$. If the structure of the magnetic bubble is not far from that of the non-magnetic one, when $p \simeq p_{\text{mag}}$, the radius of the contact surface reaches

$$\overline{R}_i \simeq 2.22 L_W^{1/2} \rho_0^{1/4} B_0^{-3/2}$$
$$\simeq 230\text{pc}(L_W/1.68 \times 10^{37}\text{erg s}^{-1})^{1/2}(\rho_0/1.8 \times 10^{-24}\text{g cm}^{-3})^{1/4}(B_0/6\mu\text{G})^{-3/2}$$

If $\overline{R}_i \gg H$, before magnetic field works, due to the density stratification the superbubble breaks through the disk (the model shown in Figure 4: $L_{SN} = 1.67 \times 10^{38}\text{erg s}^{-1}$). On the other hand, if $\overline{R}_i \lesssim H$, before density stratification works, the superbubble is confined by the magnetic field (the model shown in Figure 2: $L_{SN} = 1.67 \times 10^{37}\text{erg s}^{-1}$).

This work was partially supported by Grant-in-Aid for Encouragement of Young Scientists from the Ministry of Education, Science, and Culture (02854016). I would like to express my gratitude to Yamada Science Foundation and IAU for their travel support.

	Thin Disk $H \sim 100$ pc	Thick Disk $H \gtrsim 500$ pc
$B=0$	Accelerated. Broken at the Polar Cap.	Not Accelerated. Reach $z \sim 1$kpc.
$B \sim 5\mu\text{G}$		Not Accelerated. Much Confined.

summary

REFERENCES

Bahcall, J. N. 1984, *Ap. J.*, **276**, 169.
Bloemen, J. B. G. M. 1987, *Ap. J.*, **322**, 694.
Bregman, J. N. 1980 *Ap. J.* **236**, 577.
Bruhweiler, F. C., Gull, T. R., Kafatos, M., and Sofia, S. 1980, *Ap. J. (Letters)*, **238**, L27.
Cash, W., Charles, P. Bowyer, S., Walter, F., Gamire, G., and Riegler, G. 1980, *Ap. J.* **238**, L71.
Evans, C. R., and Hawley, J. F. 1988, *Ap. J.*, **332**, 659.
Habe, A., and Ikeuchi, S. 1980, *Prog. Theor. Physics*, **64**, 1955.
Heiles, C. 1989, in *Structure and Dynamics of the Interstellar Medium*, eds. by G. Tenorio-Tagle, M. Moles, and J. Melnick (Berlin: Springer), p. 484.
Igumentshchev, I. V., Shustov, B. M., and Tutukov, A. V. 1990 *Astr. Ap.* in press.
Kompaneets, A. S. 1960, *Soviet Phys. Dokl.*, **5**, 46.
Koo, B.-C., and McKee, C. F. 1990, *Ap. J.*, **354**, 513.
Kulsrud, R. M., Berstein, I. B., Kruskal, M., Fanussi, J., and Ness, N. 1965, *Ap. J.*, **142**, 491.
Laumbach, D. D., and Probstein, R. F. 1969, *J. Fluid Mech.*, **35**, 53.
Lockman, F. J., Hobbs, L. M., and Shull, J. M. 1986, *Ap. J*, **301**, 380.
Mac Low, M.-M., and McCray, R. 1988, *Ap. J.*, **324**, 776.
Mac Low, M.-M., McCray, R. and Norman, M. L. 1989, *Ap. J.*, **337**, 141.
McKee, C. F., and Ostriker, J. P. 1977, *Ap. J*, **218**, 148.
Norman, C. A., and Ikeuchi, S. 1989, *Ap. J.*, **345**, 372.
Norman, M. L., and Winkler, K.-H. A. 1986, in *Astrophysical Radiation Hydrodynamics*, eds. by K.-H. A. Winkler and M. L. Norman (Dordrecht: Reidel), p. 187.
Rand, R. J., and Kulkarni, S. R. 1989, *Ap. J.*, submitted.
Reynolds, R. J., and Ogden, P. M. 1979, *Ap. J.*, **229**, 942.
Różyczka, M., and Tenorio-Tagle, G. 1985, *Astr. Ap.*, **147**, 209.
Shapiro, P. R., Mineshige, S., and Shibata, K. 1990, this volume.
Spitzer, L., Jr. 1990, *Ann. Rev. Astr. Ap.* in press.
Tenorio-Tagle, G., Bodenheimer, P., and Różyczka, M. 1987, *Astr. Ap.*, **182**, 120.
Tenorio-Tagle, G., and Bodenheimer, P. 1988, *Ann. Rev. Astr. Ap.*, **26**, 145.
Tenorio-Tagle, G., Różyczka, M., and Bodenheimer, P. 1990, submitted to *Astr. Ap.*.
Tomisaka, K. 1990 *Ap. J. (Letters)* in press.
Tomisaka, K., and Ikeuchi, S. 1986, *Pub. Astr. Soc. Japan*, **38**, 697.
———. 1987, *Ap. J.*, **330**, 695.
Tomisaka, K., Habe, A., and Ikeuchi, S. 1981, *Ap. Space Sci.*, **78**, 273.
van Leer, B. 1977, *J. Comput. Phys.*, **23**, 276.
Weaver, R., McCray, R., Castor, J., Shapiro, P., and Moore, R. 1977, *Ap. J.*, **218**, 377; **220**, 742.

LARGE-SCALE GAS DYNAMICAL PROCESSES AFFECTING THE ORIGIN AND EVOLUTION OF GASEOUS GALACTIC HALOS

PAUL R. SHAPIRO
Department of Astronomy
The University of Texas at Austin
Austin, Texas 78712, U.S.A.

ABSTRACT. Observations of galactic halo gas are consistent with an interpretation in terms of the galactic fountain model in which supernova heated gas in the galactic disk escapes into the halo, radiatively cools and forms clouds which fall back to the disk. The results of a new study of several large-scale gas dynamical effects which are expected to occur in such a model for the origin and evolution of galactic halo gas will be summarized, including the following: (1) nonequilibrium absorption line and emission spectrum diagnostics for radiatively cooling halo gas in our own galaxy, as well the implications of such absorption line diagnostics for the origin of quasar absorption lines in galactic halo clouds of high redshift galaxies; (2) numerical MHD simulations and analytical analysis of large-scale explosions and superbubbles in the galactic disk and halo; (3) numerical MHD simulations of halo cloud formation by thermal instability, with and without magnetic field; and (4) the effect of the galactic fountain on the galactic dynamo.

1. INTRODUCTION

In what follows, I summarize the results of several new theoretical investigations whose common goal is the elucidation of the origin and nature of galactic halo gas. This work, it is hoped, will "flesh out" some of the expectations and implications of the galactic fountain model of Shapiro and Field (1976) in order to determine how well it explains the observed properties of interstellar disk and halo gas and their interaction. The reader is referred to Spitzer (1990) for a recent review of and references to other work in this field.

2. NONEQUILIBRIUM ABSORPTION LINE AND EMISSION SPECTRUM DIAGNOSTICS FOR RADIATIVELY COOLING GALACTIC HALO GAS

Observations of ultraviolet absorption and emission lines in interstellar gas at large distance from the galactic plane have been interpreted in terms of a gaseous galactic halo in which hot gas ($\sim 10^6$ K) is radiatively cooling and recombining as in the galactic fountain model. As the first step in a new investigation of the observable properties of such halo gas, we have recalculated the nonequilibrium radiative cooling, ionization, and recombination of an optically thin gas composed of the 13 elements H, He, C, N, O, Ne, Na, Mg, Si, S, Ca, Fe, and Ni (Shapiro and Benjamin 1990). We use the abundances of Allen (1973) and the atomic data of Raymond and Smith (1977) and Raymond (1987). We have solved the ionization balance rate equations together with the equation of energy conservation for a gas cooling from 10^6 K to 10^4 K at either constant density (isochoric) or constant pressure (isobaric), including the effects of photoionization by an ambient radiation field contributed by galactic stars and supernova remnants and the metagalactic radiation background. These limiting isochoric and isobaric cases correspond, respectively, to the cases where the cooling regions are large enough for the sound crossing time to exceed the cooling time or vice versa. The UV absorption line spectrum as well as the emergent line and continuum spectrum in the soft X-ray and UV wavelengths have been calculated and compared with current observations. We have also considered other types of incident radiation and applied our results to the question of the origin of quasar absorption lines in galactic halo clouds.

2.1 The Galactic Fountain and Milky Way Halo Gas

We assume a steady-state, plane-parallel, optically thin flow of halo gas out of the galactic disk, with initial velocity v_0 and initial total hydrogen density $n_{H,0}$. The outflowing gas starts in ionization equilibrium at a temperature of 10^6 K. Column densities through the flow for an ionic species i are given by $n_{H,0} v_0 \int dt y_i = N_H(t) <y_i>_t$, where $y_i = n_i/n_H$, time t is measured from the initial time at which the fluid element is at 10^6 K, $N_H(t)$ is the total H column density between the base of the flow and the position of the fluid element at time t, and $<y_i>_t$ is the time-average of y_i between t = 0 and t. For cases without photoionization included, the product $N_H(t)<y_i>_t$ reaches an asymptotic value for times greater than a few initial cooling times. For cases with photoionization, however, the gas relaxes to thermal and ionization equilibrium, and the column densities of some ions (e.g. C IV and Si IV in the isobaric case, and all ions shown here in the isochoric case), thereafter, grow in proportion to time for t greater than this relaxation time.

A selection of our results for the column densities of C IV, N V and Si IV and the temperature versus total H column density is presented in Figures 1(a) and (b), along with the observed column density ratios for Milky Way halo gas from Savage and Massa (1987), as summarized by Savage (1988). For cases which include an incident ionizing radiation field, we adopt the spectrum of Bregman and Harrington (1986) and vary the flux level by adjusting the ionization parameter $\Gamma = n_\gamma/n_{H,0}$, where n_γ is the number density of H ionizing photons in the incident radiation field. There is a convenient scaling property for these results such that, for a fixed spectral shape and value of Γ, the results for different values of v_0 and $n_{H,0}$ are the same if plotted versus $N_H(t)/v_0 = n_{H,0} t$. The mass circulation rate for the entire galaxy in $M_\odot$ per year implied by the steady presence of the corresponding particle flux $n_{H,0} v_0$ over a fraction f of the galactic disk of area πR^2 with R = 15 kpc is $\dot{M} \approx 2.5$ $(n_{H,0}/10^{-3}$ cm$^{-3})(v_0/100$ km s$^{-1})$f. Our results *without* photoionization agree reasonably well with those of Edgar and Chevalier (1986), although we have included charge exchange reactions while their calculations did not. We find that, for $\Gamma = 0$, the asymptotic values of $N(C IV)/v_0$ are $\approx 10^7$ cm^{-3} s and $10^{6.3}$ cm^{-3} s for the isochoric and isobaric limits, respectively, implying that the observed value $N(C IV) \approx 10^{14}$ cm^{-2} requires $v_0 \sim 100$ km s^{-1} and $v_0 \sim 600$ km s^{-1} for each of these cases. The former value of v_0 is consistent with the expectations of a galactic fountain flow, while the latter is a bit too high to be reasonable. In addition, neither case reproduces the observed ratios.

With photoionization, the isochoric limit succeeds in reproducing the observed column density ratios at the same point that it achieves $N(C IV) \approx 10^{14}$ cm^{-2} as long as the flux level is in the range $10^{-1.8} \lesssim \Gamma \lesssim 10^{-2.7}$ and $v_0 \sim 100$ km s^{-1}. The isobaric limit, however, fails again even for very large flux levels. This suggests that the cooling regions are initially large enough [i.e. size $\gtrsim$ 1 kpc $(n_{H,0}/10^{-2}$ cm$^{-3})^{-1}$] that an isochoric

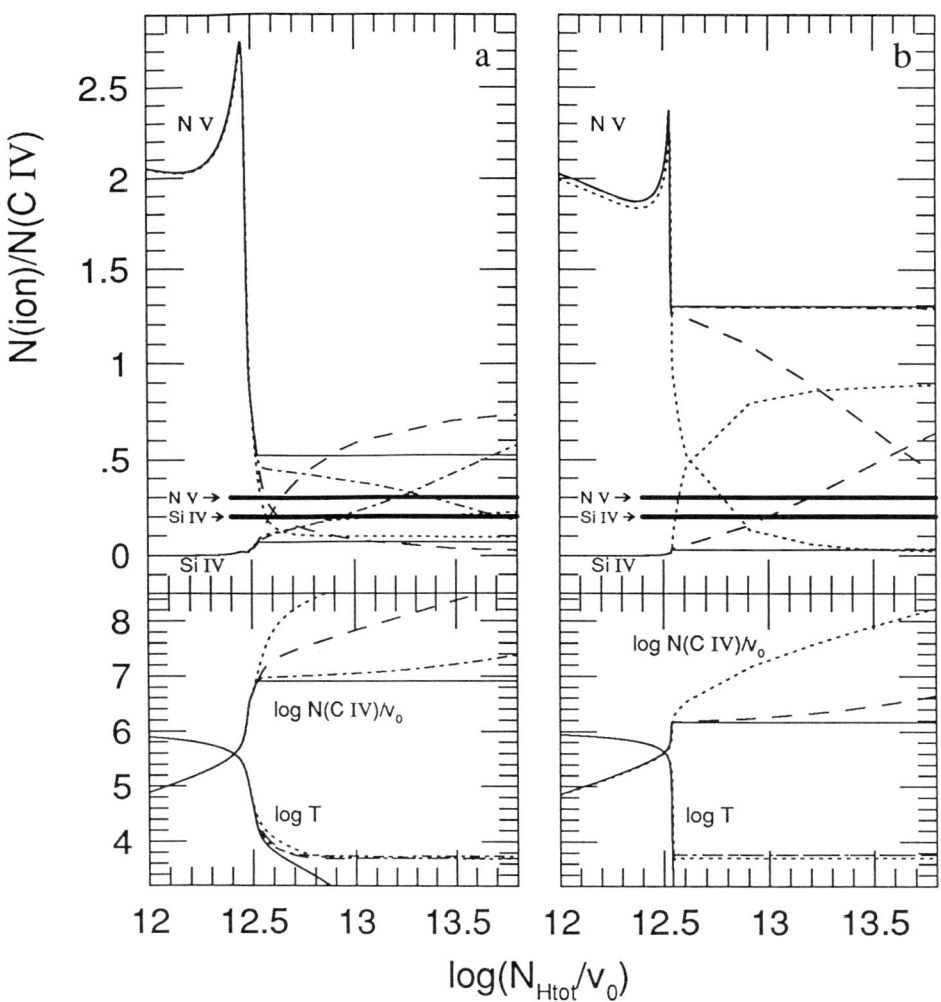

Figure 1. (a) (Upper panel) Column density ratios N(N V)/N(C IV) (descending curves) and N(Si IV)/N(C IV) (ascending curves) versus $N(H)/v_0$ for isochoric cases with no photoionization (solid line), and with an external photoionization field with $\log \Gamma = -1.2$ (dotted), $\log \Gamma = -2.2$ (dashed), and $\log \Gamma = -3.1$ (dash-dot). The observed column density ratios are indicated with bold lines. (Lower panel) Temperature and $N(C\ IV)/v_0$ versus $N(H)/v_0$. (b) Same as (a), but with isobaric cooling, for cases with no photoionization (solid line), and an external photoionization field with $\log \Gamma = -0.2$ (dotted), $\log \Gamma = -1.2$ (dashed), and $\log \Gamma = -2.2$ (dash-dot).

description is more appropriate than an isobaric one. The required flux level range given above can be compared with previous estimates of the ambient ionizing flux level to which halo gas is exposed. Bregman and Harrington (1986) estimated $\Gamma = \Gamma_{BH} = 10^{-1.2}$ $(n_{H,0}/10^{-3} \text{ cm}^{-3})^{-1}$, implying $3 \times 10^{-2} \gtrsim n_{H,0} \gtrsim 3 \times 10^{-3} \text{ cm}^{-3}$ in order to make $\Gamma_{required} = \Gamma_{BH}$. Recently, more restrictive limits have been claimed for Γ, based upon attempts to detect Hα recombination emission from halo clouds. Kutyrev and Reynolds (1989) claim $\Gamma \lesssim \Gamma_{KR} = 10^{-2.2} (n_{H,0}/10^{-3} \text{ cm}^{-3})^{-1}$, while Songaila, Bryant, and Cowie (1989) claim $\Gamma \lesssim \Gamma_{SBC} = 10^{-2.7} (n_{H,0}/10^{-3} \text{ cm}^{-3})^{-1}$. In that case, $n_{H,0} \sim 10^{-3} \text{ cm}^{-3}$ makes our results consistent with these limits.

We have also calculated the emergent radiation spectrum of the cooling fountain flow. For external $\Gamma = 0$, a convenient scaling property exists such that the quantity $\phi_\nu = I_\nu/(n_{H,0}v_0)$ is constant with respect to variations of $n_{H,0}$ and v_0, where I_ν is the emergent intensity at frequency ν. Martin and Bowyer (1990) recently reported the detection of UV emission lines of C IV and O III from the galactic halo. We find that, in order to match the observed intensity of these lines while at the same time satisfying the absorption-line data as described above, $n_{H,0} \approx 2 \times 10^{-2} \text{ cm}^{-2}$ and $v_0 \sim 100 \text{ km s}^{-1}$ are required for the isochoric case. This density makes the external photoionizing flux level required to reproduce the C IV, Si IV, and N V absorption-line data exceed the limit suggested, for example, by Kutyrev and Reynolds (1989) by a factor of several. However, such limits are subject to reinterpretation if the neutral halo gas which was observed in Hα emission was actually exposed to an attenuated or diluted version of the flux which ionizes the gas in the fountain flow.

A remarkable new result which emerges from these calculations is that the ionizing radiation emitted by the cooling fountain flow itself is sufficient to photoionize the flow so as to reproduce the observed column densities of C IV, Si IV, and N V. The flux level of this self-ionization spectrum corresponds to $\Gamma_{self} = (v_0/c)\phi$, where $\phi = \int(\phi_\nu/h\nu)d\nu$ from $\nu = 13.6$ eV/h to $\nu = \infty$. We find $\phi \approx 3.7$ for the isochoric case, and, hence, $\Gamma_{self} \cong 10^{-2.9} (v_0/100 \text{ km s}^{-1})$. The observed line ratios are matched for $v_0 \sim 100 \text{ km s}^{-1}$. This Γ_{self} is somewhat less than that required if the photoionization is attributed instead to external sources, so the self-ionization spectrum is more efficient at photoionizing the flow. In short, a cooling fountain flow can explain both the observed UV absorption and emission lines of galactic halo gas by absorbing its own photoionizing emission as it cools, if $n_{H,0} \sim 2 \times 10^{-2} \text{ cm}^{-3}$, $v_0 \sim 100 \text{ km s}^{-1}$, and the initial size of the cooling region is several hundreds of parsecs or more.

2.2 The Galactic Fountain and Quasar Absorption Lines

Can the same cooling fountain flow explain the observed quasar absorption lines of Lyman limit systems at $z \sim 3$? We have plotted in Figure 2 a selection of our results for the column densities of C II, C III, Si III, and Si IV relative to that of C IV, versus N(H I)/v_0, for the same isochoric, steady-state cooling flow as described above in § 2.1. We have also plotted the observed values for a cloud with N(H I) $\cong 10^{17.5}$ cm^{-2} at $z_{abs} = 2.9676$ in the spectrum of PKS 2126–158 ($z_{em} = 3.27$) reported by Sargent, Steidel, and Boksenberg (1990). The relative metal abundances are from Allen (1973) as before, but the overall metal abundance relative to H is adjusted in order to match the observed ratio N(C IV)/N(H I). In Figure 2(a), the incident ionizing spectrum is that for gas in our own galactic halo estimated by Bregman and Harrington (1986), while that in Figure 2(b) was a metagalactic ionizing background at $z \sim 3$ calculated by Giroux and Shapiro (1990) for AGN-type sources distributed throughout a cosmologically evolving intergalactic medium.

We find that, without the effects of photoionization included, cooling fountain gas does not match the data. With photoionization, however, a fountain flow *can* explain the data.

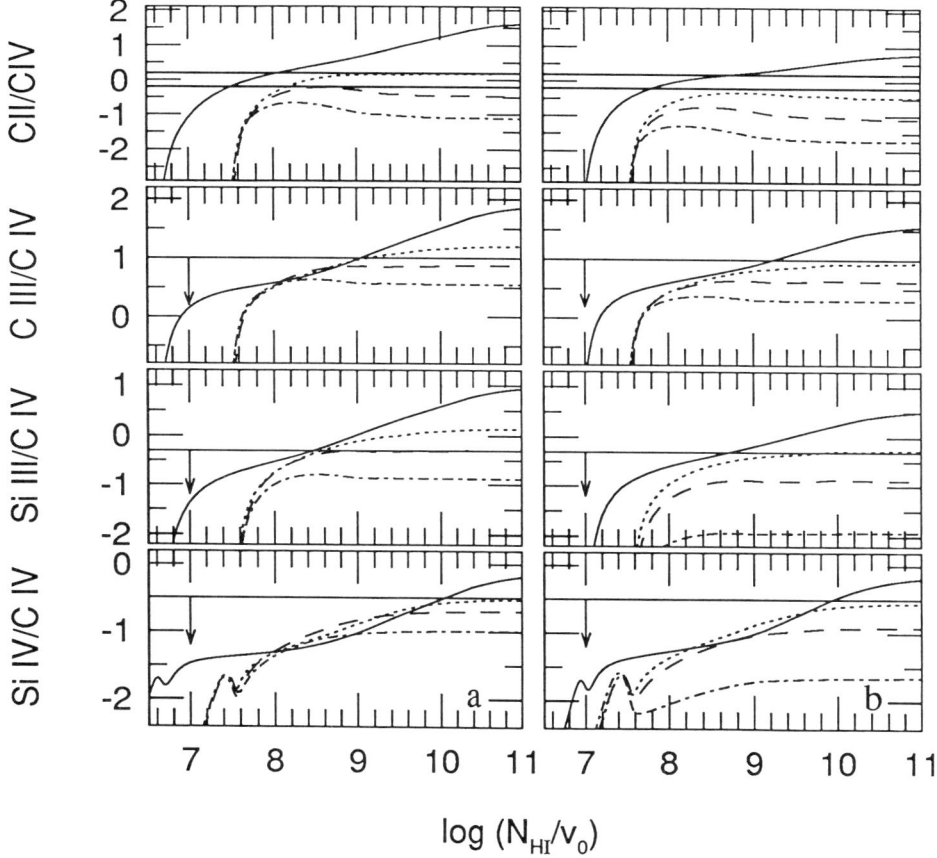

Figure 2. (a) Log (N(ion)/N(C IV)) for isochorically cooling gas with a galaxy type spectrum from Bregman and Harrington (1986), with log Γ −2.7 (solid), log Γ = −2.0 (dotted), log Γ = −1.6 (dashed), and log Γ = −1.2 (dash-dot). The observed range for C II is indicated, and upper limits for C III, Si III, and Si IV are indicated with a downward arrow. (b) Same as (a), but with an attenuated AGN-source metagalactic spectrum from Giroux and Shapiro (1990).

For the normal galaxy-type spectrum of Bregman and Harrington (1986), the data are all well matched for $v_o \sim 100$ km s^{-1}, $\Gamma_{BH} \sim 10^{-1.8}$ and metal abundances which are $10^{-2.5}$ times the solar values. For the AGN-source metagalactic spectrum of Giroux and Shapiro (1990), on the other hand, the data are matched with $v_o \sim 100$ km s^{-1}, $\Gamma_{GS} \sim 10^{-2.2}$, and metal abundances which are 10^{-2} times solar. These flux levels are significantly higher than that of the radiation emitted by the flow, itself. Hence, unlike the case of our own galactic halo gas, self-ionization of the fountain flow is not sufficient to account for the observed quasar absorption lines. Finally, we note that isobarically cooling gas with Γ even as large as 10^{-1} cannot match the observations.

3. LARGE-SCALE EXPLOSIONS AND SUPERBUBBLES IN THE GALACTIC DISK AND HALO: MHD SIMULATIONS AND ANALYTICAL APPROXIMATIONS

The evolution of the interstellar superbubbles arising from sequential supernova explosions or winds from OB associations in the galactic disk were studied numerically, using a two-dimensional magneto-hydrodynamics (MHD) code (Mineshige, Shibata, and Shapiro 1990). We find that in the presence of horizontal magnetic fields of strength comparable to that in the galactic disk, $B = 3$ µG, the vertical expansion of the superbubbles (the contact surface) can, under some conditions, be significantly inhibited by the effect of a decelerating $J \times B$ force, while, at the same time, the outermost effect of the disturbance actually propagates somewhat faster than in nonmagnetic cases, as an MHD shock or nonlinear wave. The conditions under which magnetic fields are dynamically important in the evolution of superbubbles have been discussed analytically, as well (Shapiro, Mineshige, and Shibata 1990). We have derived a new, analytical Kompaneets approximation for a two-dimensional, axisymmetric, steadily driven explosion in a nonmagnetic, exponential disk which we apply to this question.

3.1 Numerical Results

As an illustration, we show here results of a 2D, MHD, numerical simulation of a superbubble resulting from sequential supernovae. Our simulation in the (x,z) plane assumes a field $B = (3\mu G)\hat{x}$ (i.e. horizontal to the plane), a mass and energy input per explosion equivalent to 10 M$_\odot$ and 10^{51} erg deposited every 3×10^5 yr uniformly in a cylinder along the y-axis of radius 20 pc and length 500 pc, in an ambient medium at 10^4 K with density 1 cm^{-3} in the disk, which extends to height $z = 100$ pc, with a halo of density 10^{-2} cm^{-3} for $z > 100$ pc. Figures 3(a)-(c) show three time-slices for each of two simulations, one with and one without magnetic field, at roughly the same epoch. Figure 4 shows the qualitative features of the magnetized case. Our results are consistent with those described by Tomisaka elsewhere in this volume.

3.2 The Modified Kompaneets Approximation

When are magnetic fields dynamically important in the evolution of superbubbles? In order to help answer this, we have derived a new, analytical Kompaneets approximation for a steadily-driven explosion in a nonmagnetized, plane-stratified, exponential atmosphere with undisturbed ambient density $\rho \propto \exp(-z/H)$. According to this solution, the condition for "blow-out" to occur in the absence of magnetic field is given by:

$$L_{38}^{-3/2} p_{-12}^{1/2} H_2^{-2} \geq (0.23, 0.34, 0.25, 0.38) \text{ for (AO, AS, IO, IS)}, \qquad (3.2.1)$$

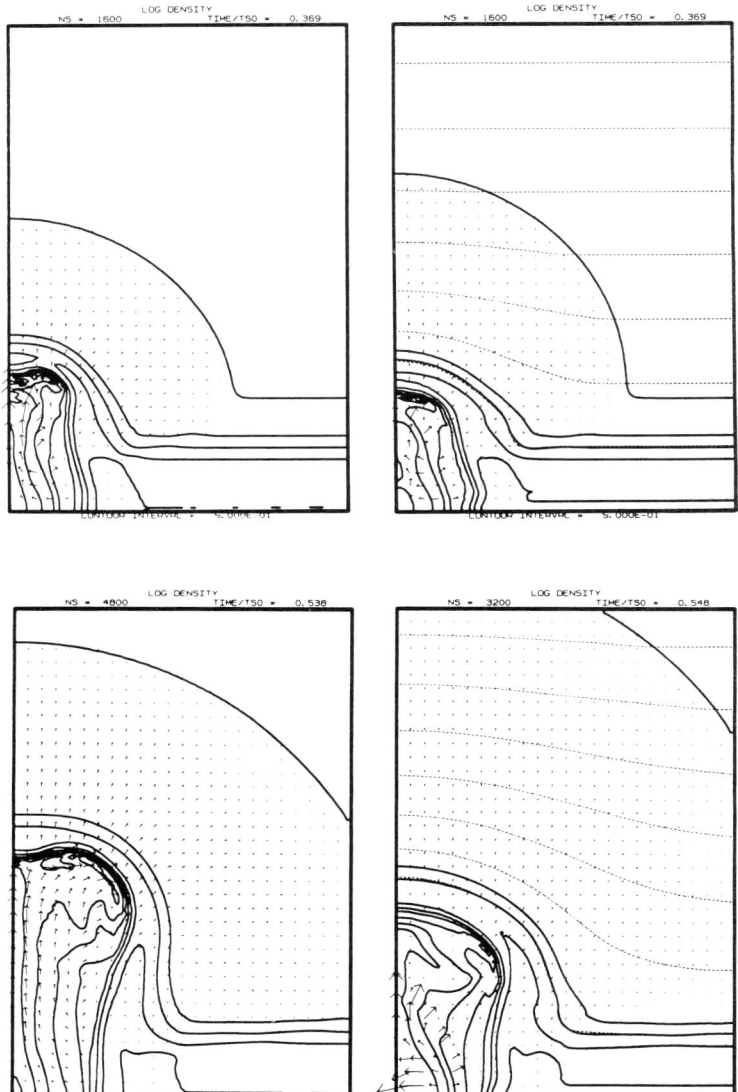

Figure 3. (a) (top) and (b) (bottom). Superbubble from sequential SN in disk-halo, with and without magnetic field. Left (B = 0). Right (B = 3μG). Density contour (thick lines) are logarithmically spaced by 0.5 dex from $\rho = 10^{-27} - 10^{-24}$ g cm^{-3}. Velocity vectors (arrows) and magnetic field lines (dashed lines) are also shown. Elapsed times in Myr = 10^6 yr are (a) 3.0 and (b) 4.4.

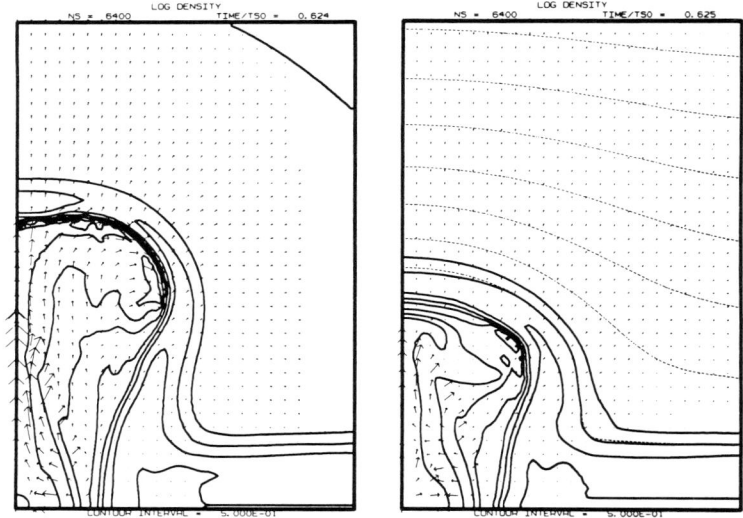

Figure 3. (c) Same as Figures 3(a), (b) except elapsed time is 5.1 Myr.

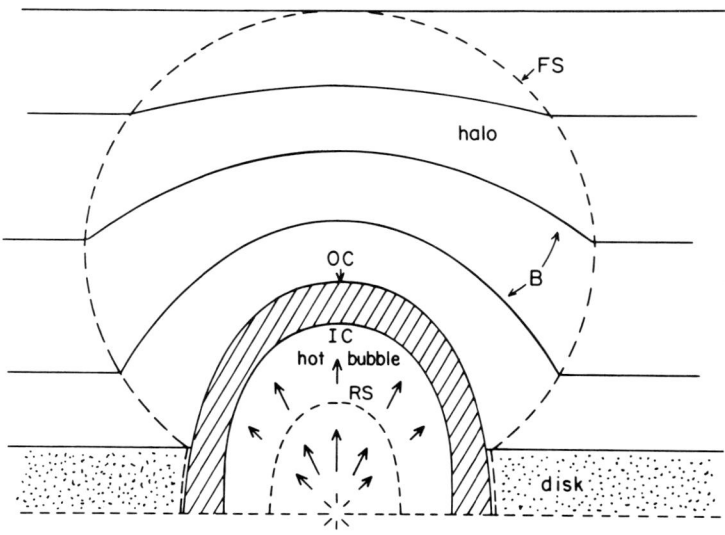

Figure 4. Schematic diagram of magnetic superbubble. Outer contact discontinuity ("OC"), inner contact discontinuity ("IC"), reverse shock ("RS"), and forward shock ("FS") are labeled.

where the central explosion energy injection rate at $z = 0$ has the constant luminosity $L_{38} \equiv L/(10^{38} \text{erg s}^{-1})$, the ambient undisturbed gas pressure at $z = 0$ is $p_{-12} = p(z=0)/(10^{-12} \text{erg cm}^{-3})$, $\rho_{-24} = \rho(z=0)/(10^{-24} \text{g cm}^{-3})$, and $H_2 = H/(100 \text{ pc})$, "AO" = adiabatic shock, one-sided case [$\rho \propto \exp(-z/H)$], "AS" = adiabatic shock, symmetric case [$\rho \propto \exp(-|z|/H)$], "IO" = isothermal shock, one-sided, and "IS" = isothermal shock, symmetric. This analytical expression is roughly consistent with the approximate numerical solutions of Mac Low and McCray (1988).

In order for a magnetic field to influence this explosion, it must be roughly true that the original magnetic energy in the volume swept up by the explosion exceeds the explosion energy. Suppose the undisturbed gas has $\beta = \beta_0 = p_{gas}/p_{mag}$ = constant, independent of z, and condition (3.2.1) is already met. In that case, we find that magnetic fields are dynamically important and might inhibit "blow-out" if the following condition is met:

$$\beta_0 \lesssim (2.0, 2.1, 2.1, 2.3) \times \left[L_{38} p_{-12}^{-3/2} \rho_{-24}^{1/2} H_2^{-2} \right]^{-1} \text{ for (AO, AS, IO, IS).} \quad (3.2.2)$$

4. MHD SIMULATION OF HALO CLOUD FORMATION BY THERMAL INSTABILITY

The nonlinear evolution of thermal instability in a density perturbation was studied as a model for the formation of clouds in the galactic halo (Mineshige, Shibata, Shapiro, and Tajima 1990). We have followed the evolution of a cloud analytically, using the isobaric approximation. The temperature of the cloud decreases explosively and reaches the minimum value in a time shorter than the initial cooling time scale. These features are confirmed by two-dimensional numerical simulations for nonmagnetic cases: the cloud formation takes place symmetrically in space and explosively in time. We have further included magnetic fields and gravity in our simulations. In contrast to the nonmagnetic cases, the shape of a cloud is vertically elongated by the presence of horizontal magnetic fields because the ambient gas cannot accrete onto the cloud across the field lines. In the case of strong magnetic fields with the initial ratio of the gas pressure to the magnetic pressure $\beta = 1$, the cloud is nearly suspended by the horizontal magnetic fields.

4.1 Numerical Results

As an illustration, we show results from our 2D, MHD simulations (in the x, z-plane) for a cloud forming by thermal instability in an initially, slight over-dense region in a radiatively cooling gas which starts at 10^6 K in magneto-hydrostatic equilibrium at constant $\beta = p_{gas}/p_{mag}$ in a uniform Galactic gravitational field, with mean density 10^{-27} g cm^{-3} at height $z = 0$. Comparison in Figure 5 of the magnetized case ($\beta = 1$ and B|| $\hat{x}$) with the unmagnetized case shows that in the former, magnetic fields decelerate the vertical precipitation motion of the cloud perpendicular to B (notice the much smaller velocity vectors in the bottom time-slices for $\beta = 1$ compared to the top time-slices for B = 0).

5. THE GALACTIC DYNAMO IN THE PRESENCE OF A GALACTIC FOUNTAIN

The conventional picture of the galactic dynamo mechanism for the generation and maintenance of the large-scale magnetic field in disk galaxies (e.g. Parker 1971) neglects effects due to the presence of a gaseous halo and of the disk-halo interaction. Possible

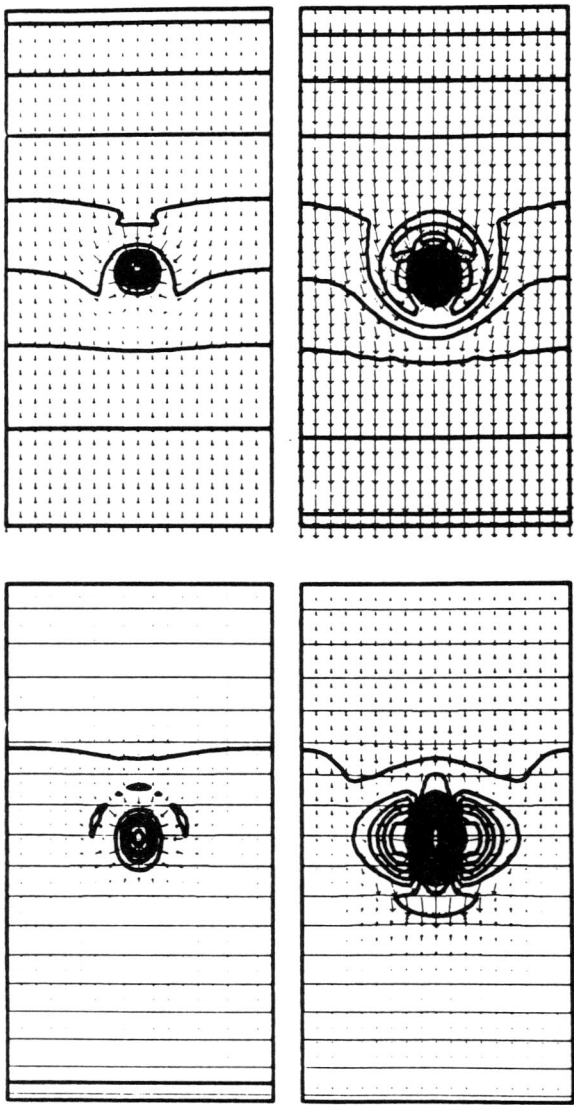

Figure 5. Time-slices of simulation, with (bottom panels) and without (top panels) magnetic field. Density contours are thick lines with logarithmic spacing of 0.5 dex in density, magnetic field lines are thin lines, velocity vectors are arrows. Times (in units of 10^8 yr) are 0.15 (left panels) and 0.298 (right panels).

such effects include buoyant and explosion-driven escape of magnetic flux from the disk into the halo, helicity associated with galactic fountain flows which circulate gas from the disk to the halo and back to the disk, velocity shear associated with a gradient in galactic rotation velocity in the direction perpendicular to the galactic plane, resistive diffusion of the field in the z-direction or the possibility that turbulent resistivity is, instead, confined to the disk and absent from the halo, and the presence of z-motions and of gradients in the z-direction in general.

The standard equation of $\alpha\omega$-dynamo theory is the modified induction equation

$$\frac{\partial \mathbf{B}}{\partial t} = \nabla \times (\mathbf{V} \times \mathbf{B}) + \eta \nabla^2 \mathbf{B} + \nabla \times (\alpha \mathbf{B}), \qquad (5.1)$$

where $\mathbf{B}$ = magnetic field, $\mathbf{V}$ = fluid velocity, both averaged locally over space, η = turbulent resistivity and α represents the effect of helical turbulence, or cyclonic convection. This helical convection is characterized by the occurrence of vertical flows in a differentially rotating gaseous disk which diverge (expand) or converge (compress) for upward or downward motion, respectively, as expected if the disk gas density decreases with distance from the central plane. As estimated by Vainshtein and Ruzmaikin (1972), for example, this leads to an α-effect of strength $\alpha \approx \ell^2 \omega / H$, where ℓ is the length scale of an individual turbulent eddy, H is the disk vertical density scale height, and ω is the angular rotation velocity of the disk. For the conventional picture of the galactic dynamo confined to the galactic disk, $\ell_{disk} \sim H_{disk} \sim 100$ pc are assumed.

Galactic fountain/superbubble flows, however, provide even larger-scale, cyclonic convection suitable for generating its own α-effect. We estimate that such flows can produce a "galactic fountain" α-effect characterized by $\alpha_{GF} \approx \ell^2_{GF} \omega / H_{GF}$, where $\ell_{GF} \gtrsim 1$ kpc (e.g. ℓ_{GF} is at least as large as the size of a superbubble that breaks out of the disk) and $H_{GF} \gtrsim$ kpc's (e.g. H_{GF} ranges from the size of a superbubble at break-out to the density scale-height of the galactic halo gas). Hence, $\alpha_{GF}/\alpha_{disk} \sim (\ell_{GF}/\ell_{disk})(\ell_{GF}/H_{GF}) \sim 10$ is possible. Apparently, a galactic-fountain-dominated galactic dynamo may contribute significantly to the origin and evolution of the large-scale magnetic fields in galaxies.

ACKNOWLEDGEMENTS

I am grateful to Robert Benjamin, Shin Mineshige, Kazunari Shibata, Toshiki Tajima, and Robert Bishop for their collaboration in the research which I have previewed here. Robert Benjamin and I are grateful to John Raymond for making his hot gas atomic parameter data set available to us. We also thank Mark Giroux for his advice and assistance. This work was supported in part by Robert A. Welch Foundation Grant No. F-1115, Texas Advanced Research Program Grant No. 4132, NASA Training Grant No. NGT-50519, and an Alfred P. Sloan Foundation Fellowship in Physics. All of our numerical calculations were performed on the University of Texas Center for High Performance Computing Cray X/MP.

REFERENCES

Allen, C. W. (1973) *Astrophysical Quantities* (London: The Athlone Press).
Bregman, J. N., Harrington, J. P. (1986) *Ap. J.*, **309**, 833.
Edgar, R. J., Chevalier, R. A. (1986) *Ap. J. (Lett.)*, **310**, L27.

Giroux, M. L., Shapiro, P. R. (1990) in *Physical Processes in Fragmentation and Star Formation*, eds. R. Capuzzo-Dolcetta, C. Chiosi, and A. De Fazio (Boston: Kluwer Academic), 71.
Kutyrev, A. S., Reynolds, R. J. (1989) *Ap. J. (Lett.)*, **344**, L9.
Mac Low, M-M., McCray, R. (1988) *Ap. J.*, **324**, 776.
Martin, C., Bowyer, S. (1990) *Ap. J.*, **350**, 242.
Mineshige, S., Shibata, K., Shapiro, P. R. (1990) *Ap. J.*, submitted.
Mineshige, S., Shibata, K., Shapiro, P. R., Tajima, T. (1990) in preparation.
Parker, E. N. (1971) *Ap. J.*, **163**, 255.
Raymond, J. C. (1987) private communcation.
Raymond, J. C., Smith, B. W. (1977) *Ap. J. Suppl.*, **35**, 419.
Sargent, W.L.W., Steidel, C. C., Boksenberg, A. (1990) *Ap. J.*, **351**, 364.
Savage, B. D. (1988) in *QSO Absorption Lines: Probing the Universe*, eds. J. C. Blades, D. Turnshek, C. A. Norman (New York: Cambridge U. Press), 195.
Savage, B. D., Massa, D. (1987) *Ap. J.*, **314**, 380.
Shapiro, P. R., Benjamin, R. A. (1990) in preparation.
Shapiro, P. R., Field, G. B. (1976) *Ap. J.*, **205**, 762.
Shapiro, P. R., Mineshige, S., Shibata, K. (1990) *Ap. J.*, submitted.
Songaila, A., Bryant, W., Cowie, L. L. (1989) *Ap. J. (Lett.)*, **345**, L71.
Spitzer, L. (1990) *Ann. Rev. Astron. Astrophys.*, **28**, in press.
Vainshtein, S. I., Ruzmaikin, A. A. (1972) *Sov. Astr.*, **15**, 714.

NONLINEAR EVOLUTION OF THE PARKER INSTABILITY

R. MATSUMOTO
College of Arts and Sciences, Chiba University,
Yayoi-cho, Chiba 260, Japan

K. SHIBATA
Department of Earth Sciences, Aichi University of Education,
Kariya, Aichi 448, Japan

ABSTRACT. Two-dimensional MHD simulations are performed to study the nonlinear evolution of the Parker instability in galactic gas disks. When the most unstable mode grows, magnetic field lines kink across the equatorial plane of the disk and thin spur-like structures are formed above dense regions in magnetic pockets. In low β ($= p_{gas}/p_{mag} < 3$) disks, shock waves are produced at the footpoint of magnetic loops, while in high β (> 3) disks, nonlinear oscillations are excited and the loop length increases with time up to $\lambda_c \simeq (3.5\beta + 6)H$, where H is the half-thickness of the disk.

1. INTRODUCTION

It is suggested that the Parker instability is related to the formation of interstellar cloud complexes (Parker 1966; Mouschovias et al. 1974) and the Radio spurs and HI spurs observed in our galaxy (Sofue 1973,1976). Although the linear stage of the Parker instability has been studied extensively (e.g., Parker 1966), the nonlinear stage of this instability is not well understood. Thus we performed two-dimensional MHD simulations to study its nonlinear evolution (Matsumoto et al. 1988,1990; Shibata et al. 1990).

2. NUMERICAL MODEL

We consider a local part of galactic gas disks in a Cartesian geometry (x, z), where x-axis is parallel to the azimuthal direction and z-axis is perpendicular to the disk. For simplicity, we assume a plane parallel gas layer located at a distance R from a point mass M which is the origin of the gravity, and neglect effects of cosmic ray pressure, rotation, and self-gravity of the gas disk. Thus the gravitational acceleration in z-direction is $g(z) = GMz/(R^2 + z^2)^{3/2}$. Magnetic fields are assumed to be initially horizontal in magnetostatic equilibrium. Two kinds of initial magnetized disk (and halo) are studied: (1) the isothermal gas layer with constant Alfvén speed (i.e., the magnetic flux is not localized), (2) the magnetized disk with a hot halo (magnetic flux is localized around the equatorial plane). Parameters describing the equilibrium model are $\beta = p_{gas}/p_{mag}$ and $\epsilon = GM/[(1 + 1/\beta)C_s^2 R]$, where the plasma β and the sound speed C_s are evaluated at the equatorial plane. Note that $p(z = \infty)/p(z = 0) = \exp(-\epsilon)$ for isothermal disk with constant β. Thus the external (or halo) pressure is large for small ϵ disk.

3. RESULTS OF NONLINEAR NUMERICAL SIMULATIONS

3.1 Spurs and Shock Waves

Figure 1 shows the fully nonlinear stage of the Parker instability in the isothermal gas disk with initially constant Alfvén speed ($\beta = 1$ and $\epsilon = 6$). As the instability develops for the most unstable mode, magnetic field lines kink across the equatorial plane of the disk and dense regions are created in the pockets of magnetic loops. Above these dense regions, spur-like structures perpendicular to the disk plane are formed. These spurs explain thermal spurs observed in our galaxy (e.g., Müller et al. 1987). At the footpoint of magnetic loops, shock waves are formed because the speed of downflow along magnetic loops is comparable to the initial Alfvén speed ($= 1.4C_s$ in this model) and is larger than the sound speed. The configuration of these shock wave fronts is similar to the distribution of molecular clouds in the Orion region (e.g., Maddalena et al. 1986). We propose that shock waves produced by the Parker instability trigger the formation of molecular clouds. After the stage shown in Figure 1, expansion of magnetic loops is almost stopped owing to the high external pressure. The region behind shock waves settles into a quasi-static state.

3.2 The Length of Magnetic Loops

Figure 2 shows the results of parameter survey for the isothermal disk. Here, the wavelength of perturbation (λ) is normalized by the local density scale height H. In low β (< 3) disks, shock waves are formed when the most unstable perturbation grows. Owing to the energy dissipation due to the shock, the system quickly settles into the quasi-static state where the loop length is equal to the most unstable wavelength. On the other hand, in high β (> 3) disks, nonlinear oscillations are excited and the loop length gradually increases with time owing to the mode coupling. The final length of the loop is determined by the condition of shock wave formation (the upper hatched region in Figure 2), and is $\lambda_c \simeq (3.5\beta + 6)H$. Therefore, the loop length in the high β disk is much longer than the most unstable wavelength. A long (~ 4.5kpc) arc-like structure of the magnetic field observed in M31 (Beck and Berkhuijsen 1990) may correspond to our long loop (~ 4kpc) for $\beta \simeq 10$.

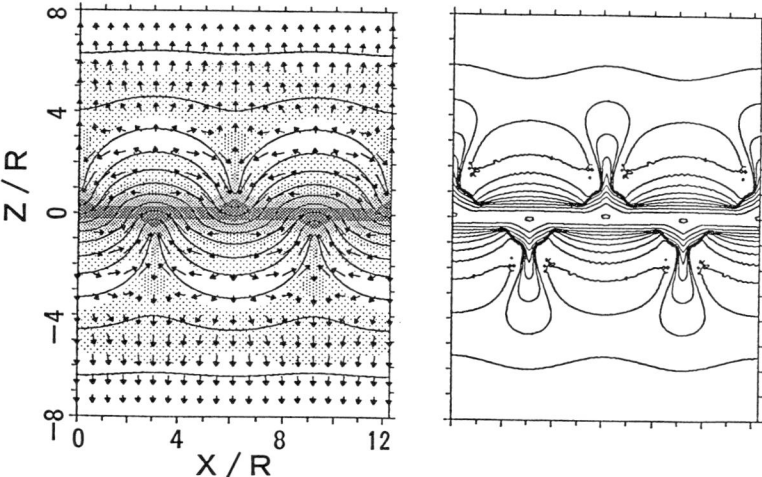

Figure 1. Nonlinear stage of the Parker instability in the isothermal disk with $\beta = 1$ and $\epsilon = 6$. The left panel shows the density distribution (grey scale), velocity field (arrows), and magnetic field lines (solid curves) at $t = 8.55 R/C_s$. The right panel shows the density contours. The contour level step width is $\Delta \log \rho = 0.25$. The region where the contour lines are close each other represent shock wave fronts.

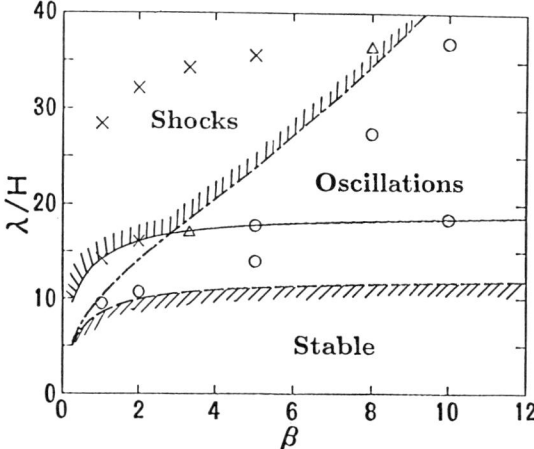

Figure 2. Criterion of shock wave formation and nonlinear oscillation for the isothermal disk with $\epsilon = 6$. Crosses show the cases where shock waves are formed. Circles denote the cases where the system shows nonlinear oscillations. Triangles are intermediate cases. Solid curve shows the linearly most unstable wavelength. Upper hatched area denotes the expected size of quasi-static magnetic loops.

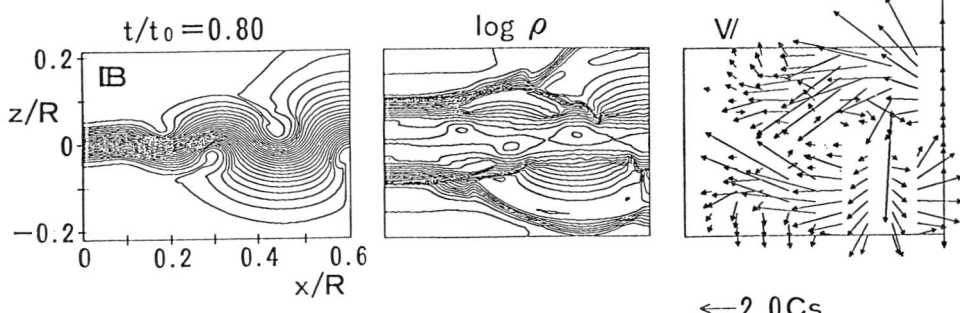

Figure 3. Numerical results of the Parker instability of a localized magnetic flux sheet.

3.3 Self-Similar Expansion of Magnetic Loops

Figure 3 shows the results when the magnetic flux is initially localized around the equatorial plane. In the initial equilibrium model, hot halo ($T_{halo}/T_{disk} = 100$) exists in $|z| > 0.1R$. Other parameters are $\beta = 1$ and $\epsilon = 1000$. Symmetric boundary condition is used at $x = 0$ and $x = 0.6R$, and the imposed perturbation is localized in $0.45R < x$. In this low halo pressure ($p_{halo}/p_{disk} < 10^{-3}$) model, magnetic loops continue to expand into the halo as long as the magnetic pressure in the loop is larger than the external halo pressure. The expansion of the magnetic loop is self-similar and the expansion speed increases linearly with height. These results are similar to the results of two-dimensional MHD simulations of emerging magnetic flux from the solar photosphere (Shibata et al. 1989). Figure 3 also shows that the Parker instability produces various structures such as dense blobs, magnetic loops, shock waves, neutral sheets, and spurs.

REFERENCES

Beck, R., Berkhuijsen, E.M. (1990) *Proc. IAU Symp. No. 140, "Galactic and Intergalactic Magnetic Fields"*, eds. R. Beck, P. Kronberg, R. Wielebinski, Reidel, p. 198
Maddalena, R.J., Morris, M., Moscowitz J., Thaddeus, P. (1986) *Ap. J.* **303**, 375
Matsumoto, R., Horiuchi, T., Shibata, K., Hanawa, T. (1988) *Pub. Astr. Soc. Japan* **40**, 171
Matsumoto, R., Horiuchi, T., Hanawa, T., Shibata, K. (1990) *Ap. J.* **356**, 259
Mouschovias, T.Ch., Shu, F.H., and Woodward, P.R. (1974) *Astr. Ap.* **33**, 73
Müller, P., Reif, K., Reich, W. (1987) *Astr. Ap.* **183**, 327
Parker, E.N., (1966) *Ap. J..* **145**, 811
Shibata, K., Tajima, T., Matsumoto, R., Horiuchi, T., Hanawa, T., Rosner, R., Uchida, Y. (1989) *Ap. J.* **338**, 471
Shibata, K., Tajima, T., Matsumoto, R. (1990) *Ap. J.* **350**, 295
Sofue, Y. (1973) *Pub. Astr. Soc. Japan* **25**, 207
Sofue, Y. (1976) *Astr. Ap.* **48**, 1

THE INTERSTELLAR DISK-HALO CONNECTION IN GALAXIES: REVIEW OF OBSERVATIONAL ASPECTS

CARL HEILES
1989-1990 Visiting Fellow, Joint Institute for Laboratory Astrophysics
University of Colorado, Boulder CO 80309-0440 USA

An impressively large amount of data was presented at this meeting and a review by an ordinary human cannot hope to do full justice to either the authors or the research. Accordingly, I shall concentrate on those aspects that particularly piqued my interest and apologize to those authors whom I overlook. I shall give references in the usual style except for references to papers presented at this meeting, for which I shall simply mention the author(s) with no date.

I had considerable difficulty deciding how to organize this review. First I discuss the various gas components in rough order of increasing scale height. Section 1 discusses neutral gas, section 2 the 'warm' and 'not-so-warm' ionized gas, section 3 the $T \sim 10^5$ K component at higher z that is detected in UV absorption and emission, section 4 the high-velocity neutral gas, section 5 the cosmic-ray halo as revealed by synchrotron emission, and section 6 the magnetic field. Next, section 7 covers the interaction between the low-z gas and the halo, which is the main topic of this symposium; and finally, section 8 discusses some aspects of the interstellar medium that are relevant to this interaction, with emphasis on the uncertainties.

1. NEUTRAL GAS

1.1. H I

Lockman reviewed the distribution of Galactic H I. It is concentrated toward the plane, but has a high-z low-density 'tail' originally discovered by Shane (1971). This tail has a scale height $h_{HI} \sim 500$ pc for Galactocentric radius $3 \lesssim R_G \lesssim 9$ kpc. For larger R_G the thickness increases dramatically. For $R_G \lesssim 3$ kpc the high-z component of the H I does not exist.

The velocity dispersion of the H I is gravitationally commensurate with its scale height at the Solar Galactocentric radius $R_G \approx 9$ kpc. However, as emphasized by H. de Boer, the constancy of the scale height over the range $3 \lesssim R_G \lesssim 9$ kpc is puzzling because the z-component of the gravitational field increases strongly towards smaller R_G. This might be understandable if the velocity dispersion of the H I increased together with the gravity, but the dispersion appears to be independent of R_G (Kulkarni and Heiles 1987; KH). Why, then, should h_{HI} remain constant over this range of R_G? This is a long-standing puzzle.

1.2. CO

In the Galaxy, most of the CO resides in molecular clouds and, locally, has z scale height ~ 100 pc (Scoville and Sanders 1987). However, there is a distinguishable minority component ($\sim 15\%$ of the molecular cloud component near the Sun) seen at high latitudes (Blitz). Much of this component is not located in clouds, but instead is associated with H I filaments and sheets (Blitz 1988), and it seems reasonable to infer that the CO is formed in shocks associated with expanding H I shells. The local scale height of this component is about the same as that of the cloud component, which is somewhat smaller than that of the H I (KH), which may imply that the CO is only formed on the low-z side of the shells. The velocities of the extended-component CO clouds are comparable to those of the H I, which are somewhat larger than the velocities of molecular clouds.

Three of the high-latitude clouds have anomalously large negative velocities, ranging as high as -45 km s^{-1}. They appear to be somewhat unusual objects from the standpoint of morphology. One is part of the Draco complex, which has been modelled as an interaction between high-velocity (section 4) and low-velocity gas, and one is colliding with a low-velocity H I cloud. The distance to the Draco complex is not absolutely certain, but Goerigk and Mebold (1986) have derived a distance of ~ 800 pc. In this one case, then, high velocity probably implies high z. Does high velocity generally imply large z? This question seems important, because if so there is a population of molecular clouds at z-distances far beyond those we ordinarily associate with molecular clouds.

In the edge-on galaxy NGC891, Garcia-Burillo (using the 30-m IRAM telescope) and Handa et al. (using the Nobeyama array) presented CO data indicating that most of the CO has $h_{CO} \approx 140$ pc, which is larger than that of the Blitz clouds. This is larger than the mean Galactic h_{CO}, which is ~ 45 pc in the Galactic interior. Apart from the quantitative difference in scale height, it would be nice to be able to conclude that the molecules in NGC891 are reasonably well-confined to the galactic plane, as they are in the Galaxy. However, the IRAM results also show a 'plateau' component with $h_{CO} \approx 840$ pc. This component is not seen at Nobeyama. It is important to determine whether this component is indeed real: the Nobeyama observations may not be sensitive enough, and the IRAM observations may be affected by effects that can plague single-dish observations such as sidelobes. If this extended component is real, it drastically departs from our standard notions of molecular clouds as being confined, in the main, to very small z-heights. However, it might have the same scale height as the H I in NGC891, for which only an upper limit of 1 kpc has been established (Sancisi and Allen 1979). Or if the 'plateau' component is real, it might be completely different, for example if it has high velocities.

2. THE WARM IONIZED MEDIUM (THE WIM; ALSO CALLED THE DIFFUSE IONIZED GAS, THE DIG)

2.1. The $\sim 10^4$ K component

Walterbos reviewed the WIM in both our Galaxy and external galaxies. It is a diffusely-distributed, $T \sim 10^4$ K gas, distinguished from ordinary H II region gas by its several-times higher [S II]/Hα line ratio, which is characteristic of gas that is photoionized by a very weak radiation field from distant O stars (Mathis 1986). In external galaxies, where in some senses it is easier to observe, it contributes a significant fraction $\sim 30\%$ of the total Hα luminosity and much of the emission is in sheet or shell structures. The properties of this 'Reynolds component' in our Galaxy were reviewed by Reynolds and are rather well-determined by pulsar and Hα observations. The total column density from $z = 0$ to ∞ is $\approx 10^{20}$ cm^{-2} and the scale height $h_e \approx 1$ kpc; the volume filling factor is $\sim 10\%$ in the Galactic plane and increases with z (KH). Energetically this component may be somewhat less important than it is in the external galaxies.

The WIM is also seen in NGC891. Dettmar and Dahlem find $h_e \approx 600$ to 1000 pc, depending on position; if this is correct, it implies that NGC891 is comparable to our own Galaxy as regards the WIM. However, Hester *et al.* (also Rand, Kulkarni, and Hester 1990) find a completely different result, $h_e \approx 4$ kpc. It is important to resolve this discrepancy. Both groups see lots of structure in NGC891 that resembles Galactic worms and supershells. In another galaxy, NGC3079, Hester *et al.* see many such structures in Hα, and many are remarkably well correlated with structure seen in the nonthermal radio continuum by Irwin and Seaquist.

The source of ionization of the WIM has long been a mystery. The total energy requirement is comparable to the total power output of supernovae in our Galaxy (Reynolds 1990), and analyses of the problem have shown that only the young, massive O stars produce enough ionizing photons to produce the WIM. The problem lies in getting the photons from the stars to the gas. The neutral gas is so opaque to ionizing photons, and the neutral gas is itself so pervasive, that the ionizing photons cannot get very far. However, the cylindrical cavities indicated by the presence of worms and by the H I 'holes' observed in our own and in external galaxies (section 7) provide unobstructed pathways for the ionizing photons. The photons can escape the stars in straight lines, providing cones of ionizing radiation with the apex located at the stars and the cone angle defined by the diameter of the cylindrical cavity. In addition, Norman pointed out that the photons can also scatter off of the sides of the cavities; however, the importance of this mechanism depends on the reflection efficiency, which remains to be worked out. This mechanism would probably change the photon energy distribution in such a way as to reproduce the observed [S II/Hα] ratio and to produce a broad, lower-z region. It would be very nice if these ionization mechanism were to work well enough to solve the 'ionization source problem' for the WIM.

2.2. The 'Not-so-Warm Ionized Gas' (the 'NSWIM')

Israel presented observational evidence for a new component of the ISM, a cool ionized phase with (for the clumpy model) $n_e \approx 1.0$ cm^{-3}, $T \lesssim 1000$ K (possibly $\ll 1000$ K), scale height $h_{NSWIM} \sim 2$ kpc, and a filling factor $\sim 10\%$ (Israel and Mahoney 1990). As this component is in some sense similar to the WIM but has much lower temperature, we temporarily adopt the somewhat awkward name 'not-so-warm ionized medium', and anticipate the day when a better name is invented by somebody more clever than we. Cox reminded us that such a component was predicted back in the early 1970's when time-dependent models of the ISM were popular (Gerola, Kafatos, and McCray 1974); it can exist because the cooling time scale is shorter than the recombination time scale. Thus from the *physical* standpoint, the NSWIM is likely to be different from the WIM, because we regard the WIM to be in ionization equilibrium and, in contrast, the NSWIM is likely not to be.

My knee-jerk reaction is to question the reliability of the observational evidence. However, the observed effect is a correlation with the inclination angle of a deficiency the low-frequency nonthermal radiation (relative to the power-law extrapolation from higher frequencies). Such a correlation is difficult to ascribe to selection effects or measurement errors, because an external galaxy has no knowledge of our location. The correlation strongly implies an opacity effect.

Our Galaxy exhibits no obvious NSWIM component. The low-frequency absorption of our Galaxy is easily produced by the same WIM that emits the Hα radiation (KH). However, the properties of the Galactic ionized gas are derived from the latitude dependence of the absorption and Hα emission, so are restricted to the Solar vicinity's 'Local Bubble' (Cox and Reynolds 1987); the local properties may not be representative of those in the Galaxy as a whole. Perhaps our Galaxy is unusual in not having the NSWIM; alternatively, perhaps it does, or perhaps the low-frequency observations or their interpretation might be incorrect.

The existence of this component is an important issue that should be confirmed on a larger sample of galaxies. The low-frequency observations were performed at Clark Lake Observatory in the U.S.A.; unfortunately, the U.S. National Science Foundation used its well-known quality of wisdom to decide that the relatively small operating costs for this observatory, which had only recently been made fully functional under NSF funding, were too costly. Thus further observations will have to be done elsewhere.

3. HOT IONIZED GAS

Savage reviewed the high-z ionized gas. Observations of UV absorption lines are best matched by a gas having $N_e \approx 2 \times 10^{18}$ cm^{-2} and $T \approx 2 \times 10^5$ K. Thus the mass of this gas is negligible compared to the mass of the neutral and WIM components. The scale height is determined from distances of the background stars and is $h_e \approx 3$ kpc. UV emission lines have recently been observed at high Galactic latitudes (Martin and Bowyer 1990), and if it is the same gas as seen in absorption then we have $n_e \approx 10^{-2}$ cm^{-3} and $T \approx 10^5$ K, yielding a pressure $P/k \approx 1300$ cm^{-3} K. This is probably close to the pressure expected at $z = 3$ kpc.

A major question is what keeps this gas warm. Its cooling time is $\sim 2 \times 10^5$ yr. Looked at in another way, the locally-observed gas has a cooling rate of nearly *half the local supernova power!* The cooling time is much shorter than the infall time.

Martin and Bowyer (1990) argue (from detections at only 6 positions) that the observed line intensity increases towards the Galactic pole, that the gas resides not just locally, and that the gas is part of the Galactic Fountain. Personally, I am not completely convinced by their arguments. In my opinion, we need more data to establish statistical reliability; at that point a definitive interpretation can ensue. If this gas is truly globally distributed within the Galaxy, it will be one of the most important components of the ISM from the standpoint of energetics and, hence, theoretical significance (section 7). Clearly, more extensive observations are urgently required.

From the relatively short cooling time one might infer that the gas is not in thermal equilibrium. One possibility is that the gas represents the turbulent mixing layer between a cool ($T \sim 10^4$ K?) 'cloud' component moving within a much hotter ($T \sim 10^7$ K?) diffuse component; the mixing layer tends to take on a temperature which is roughly the geometric mean between the two components (Begelman and Fabian 1990).

4. HIGH-VELOCITY CLOUDS (HVC'S)

Galactic HVC's were long ago discovered by the Dutch astronomers, led by Oort (1966), whose presence at this meeting we are privileged to have. They are prominent in the 21-cm line and cover a non-trivial $\sim 7\%$ of the sky (Wakker). Braun reported that 'HVC's' also exist in some other galaxies, although it is my impression that any extragalactic HVC that is in fact observable is a much bigger entity than a Galactic HVC.

In her review, Danly reported that the HVC's are seen in UV absorption lines against extragalactic objects. This shows that the HVC heavy-element abundances are consistent with those of ordinary interstellar gas, although the uncertainties leave a wide margin for differences between the abundances. Nevertheless, the significant heavy-element abundances make it unlikely that the HVC's are primordial gas.

K. de Boer and Kuntz and Danly showed convincingly that the previous distance limits of Songaila, Cowie, and Weaver (1988), derived from optical observations of Ca II lines against background stars, are based on incorrect interpretation of the data. Thus we must revise our thinking concerning the cloud distances: 'Complex A', which we assume to be representative of the classical HVC's located at positive Galactic latitudes, is more distant than 4 kpc (Schwarz and van Woerden).

Where do HVC's come from? Mirabel analyzed the velocity distribution of HVC's, restricting himself to sectors toward the Galactic center and anticenter, locations chosen to eliminate the complications of Galactic rotation. He concludes that the clouds have little angular momentum, so are presumably extragalactic, and are falling towards the Galactic center. In contrast, Wakker examined the entire sample at all longitudes. He finds that the velocity distribution is consistent with Galactic rotation plus a large random component, and concludes that the clouds may be the returning 'fountain' gas.

It is curious that two such completely different models fit the data. Mirabel's model is based on a restricted sample of HVC's, which perhaps argues against it, but I find the correspondence between the data and his model quite impressive. Could there be two (or more?) populations of HVC's?

I would like to make some possibly extraneous comments on the HVC's. First, why are they neutral? They are located in an environment comparable to that of the other halo gas, all of which is much hotter and quite highly ionized. The volume density inside an HVC is much larger than that of the ambient halo gas, and this must be in part responsible for the difference in ionization state. Nevertheless, an HVC should have an ionized edge, because of either photoionization or evaporation. One such edge has probably been detected (Kutyrev and Reynolds 1989), and such work is worth further effort.

Second, HVC's must be confined by the ambient halo gas. Thus there is an interface between the cool, neutral HVC gas and the ambient halo gas. It seems to me that this interface should depend on at least two things: one, the physical conditions of the ambient halo gas; and two, whether the interface is on the front or the back of the HVC (as defined by its direction of motion). We might learn something by studying these interfaces.

Third, HVC's have reasonably large column densities and might be detectable in gamma rays produced by interaction of the high-z cosmic rays with the gas. Because of their location in the halo, they are unique probes of the cosmic rays in the halo, and perhaps this information would be useful in understanding the role of cosmic rays in halo structure.

5. THE COSMIC RAY HALO

Direct observations of the cosmic rays come only from the intensity of synchrotron emission. However, the synchrotron emission traces only the electron component, which is a poor substitute for the far more dominant proton component. When cosmic rays are produced, the energy of the electron component is usually taken to be the canonical 1% of the total component. Even if this fraction is universally valid, the electrons are subject to loss mechanisms that hardly affect the protons. Energy losses for electrons are observationally demonstrated by the steepening of the spectral index in regions far from where the electrons are produced and as reviewed by Hummel their lifetimes are inferred to be of order 4×10^7 yr. Thus the *absence* of synchrotron emission cannot be taken to be a reliable indication of the absence of cosmic rays. On the other hand, the *presence* of synchrotron emission does definitely indicate the existence of cosmic rays.

Cosmic rays were reviewed by Dogiel. Indirect evidence for a cosmic ray halo comes from theoretical arguments based on the roughly constant cosmic ray density within the Galaxy, lifetimes, and grammage. These arguments are compelling. Thus, whether or not our Galaxy has a synchrotron-emitting halo, we must conclude that it does have a cosmic-ray halo—and, correspondingly, a magnetic-field halo.

6. MAGNETIC FIELDS

Information on the magnetic field comes from both the intensity and the polarization of synchrotron emission. As with cosmic rays, the intensity is not a perfect tracer of magnetic field, because relativistic electrons are also required.

Hummel reviewed the morphology of synchrotron emission observed in edge-on galaxies. Galaxies exhibit a thin disk, a thick disk, and in some cases a halo. Often non-axisymmetric structures such as jets and plumes are seen. The thin disk scale height is typically ~ 1 kpc, and occasionally as large as 3.5 kpc. The morphology of the synchrotron emissivity of our own Galaxy is not directly observable because we are immersed within it, but it can be obtained by modelling the observed angular distribution of intensity. These models indicate that our Galaxy has both a thick disk and a halo.

The strength of the local Galactic field can be inferred both from observations and from theory. Observationally, there are two independent results. One is Faraday rotation: the recent study of pulsars by Rand and Kulkarni (1989) derives a field strength of ≈ 5 μG, most of which is in the 'random' component. The other is synchrotron emissivity: as reviewed by K. de Boer, consistency with both the angular distribution of intensity and the measured spectrum of relativistic electrons is also obtained with a field strength of ≈ 5 μG, although Phillips et al. (1981) derive a somewhat smaller value, ≈ 4 μG. Theoretically, Cox presented a straightforward argument favoring a high magnetic field for the Galaxy: the weight of the interstellar gas must be supported by pressure, but the pressure of gas and cosmic rays appears to be inadequate. A field strength of ≈ 5 μG is required. However, this estimate is uncertain, both because the total weight is uncertain and because H I line widths always exceed the thermal width (KH) so that a significant portion of the gas pressure arises from 'turbulence'. Relying on the observational data alone, it seems that the canonical value of the field strength in the Galactic plane near the Sun should be taken as 4-5 μG.

Beck reviewed observations of the magnetic field in external galaxies. The direction of the field is revealed by linear polarization of synchrotron radiation. Polarization observations of edge-on galaxies reveal the direction of the field relative to the plane of the disk. At low z, one galaxy has $\vec{B}$ primarily perpendicular to the disk, 4 have $\vec{B}$ parallel to the disk, and 2 show polarization in limited, bubble-like regions. If we include our own Galaxy in these statistics, which also has $\vec{B}$ parallel to the disk, the 4 become 5. In the 4 external galaxies, $\vec{B}$ is not everywhere parallel but is sometimes perpendicular; this tends to happen in regions that are 'active' in some way, characterized by morphological features in the disk such as perturbed nonthermal emission, star formation, or holes in the disk. It also tends to happen in the outer parts of disks and higher in the halo. Above $z \sim 3$ kpc, the halo fields tend to become tangled.

The fact that $\vec{B}$ is sometimes perpendicular to the disk in active regions implies a direct connection to the halo, one that was probably produced by the activity. This is most important for the topic of this conference.

The fact that $\vec{B}$ is parallel to the disk says nothing about its direction within the disk. We consider the field to be a spiral, which may be so tightly wound in some cases that the field is essentially circular. The direction of the field in this spiral can be revealed only by Faraday rotation. For a spiral galaxy tilted with respect to the line of sight, the Faraday rotation indicates whether the plane-of-the-galaxy field points towards or away from the observer. If it points towards the observer on one side of the galaxy and away on the other, then the field winds around the galaxy in one direction and is referred to as an Axially Symmetric Spiral (A.S.S.); otherwise, it winds into the center on one side and out on the other and it is called a BiSymmetric Spiral, or B.S.S., field. Of those galaxies for which reliable measurements exist, two are A.S.S., two (plus possibly one more) are B.S.S., and three (including our Galaxy) are neither.

For those galaxies which the configuration can be reliably determined to be either A.S.S. or B.S.S., the data—which consist of the variation of rotation measure along the major axis—are usually quite unambiguous, in the sense that the systematic variation is obviously larger than the uncertainties and is statistically significant. In some cases the degree of statistical significance varies with R_G. Also, sometimes there occur very large departures from the pattern, which are most reasonably interpreted as isolated large perturbations instead of a poor fit to the model. These departures tend to be associated with other morphological oddities, which reinforces the perturbation idea.

In the verbal version of this paper I suggested that the fact that there are more 'neither' than A.S.S. or B.S.S. galaxies suggests that perhaps the 'neither' category is the basic one and that A.S.S. or a B.S.S. configurations might be only the first term in a Fourier-series representation of the randomness that characterizes the actual field distribution in cases that are, fundamentally, 'neither'. However, after some reflection I now believe this suggestion is incorrect.

Instead, it is my impression that at least one of the representatives of each class (IC342 [Krause, Hummel, and Beck 1989] and M31 [Beck 1982] for A.S.S.; M81 [Krause, Beck, and Hummel 1989] for B.S.S.) seems qualitatively different from the 'neither' galaxies. In these representatives, the intrinsic polarization of the synchrotron emission is higher than for the 'neither' galaxies, which means that the uniform component of the large-scale field is more important, relative to the random component. Also the rotation measures are larger for these representatives, again an indication that the uniform field component is larger. For these three galaxies, the ratio of uniform to random component ≈ 0.8, while for our Galaxy (a representative of the 'neither' case) it is ≈ 0.3. These numbers are subject to error from various depolarization effects; Beck and his colleagues are addressing this issue with multiwavelength observations. The larger uniform component should imply that the dynamo, which is responsible for the uniform component, is better established in these cases. Beck provides additional arguments against the 'neither' hypothesis.

The large-scale field distribution is a fascinating topic, and of course should be a direct probe of the dynamo processes in a galaxy. Existing data show that the distribution can take on any of the simplest forms with roughly equal probability. We would like to know how the field configuration, and thus the dynamo, is related to other properties of a galaxy. Obtaining reliable field configurations is a difficult observational task, but to address these questions we need a larger sample—more objects observed!

7. SUPERSHELLS VS. WORMS VS. CHIMNEYS ...

We now come to the most important part of this summary, at least in terms of the topic of this meeting. The connection between the gaseous disk and halo almost certainly arises in the chimneys.

On Wednesday night I asked for a vote on the *existence* of worms and related structures. The response was almost unanimously positive. Based on evidence presented at this meeting, this is hardly surprising!

Sofue showed that vertical dust lanes are prominently visible in some spiral Galaxies. Beck showed that the magnetic field, which tends to lie in the plane of a galaxy, sometimes runs vertically to high z in active regions. Braun showed that both M31 and M33 contain > 100 H I holes, and IC10 and the Magellanic Clouds also contain prominent holes. Some are also seen in Hα and are associated with peculiar velocities. Koo cataloged > 100 worms, supershells, and H I holes in our Galaxy in H I, IRAS emission, and radio continuum; also, the impressive dm-wavelength radio continuum maps in both the southern (Jonas and Baart) and northern (reviewed by Reich) hemispheres seem to exhibit many such structures.

These structures are seen in magnetic fields, Hα, dust absorption, and non-thermal radio emission in external galaxies; and in H I, IR emission, and radio continuum emission in our own Galaxy. The appearance in these different observables corresponds extremely well in many cases. The structures tend to be oriented perpendicular to the disk. Theoretically, such vertical structures are expected as a result of large explosions in a sufficiently thin disk, and the interpretation of their having been produced by multiple supernova explosions and injection of stellar winds seems unassailable. The existence of these structures in the Galaxy and in all external galaxies so far observed implies that these are widespread and common phenomena.

Clearly, I asked the wrong question on Wednesday night. As emphasized by Walterbos in his review, the *real* question is whether these structures do, in fact, connect the disk to the halo. To phrase it another way, the question is whether or not these structures are really *chimneys*. Many observers call these structures chimneys as a descriptive term, but this nomenclature implies more than the observations actually provide. We observers must adhere to Cox's first moral principle: we must never give an empirically-defined object a name that connotes a physical effect suggested by theory. The theory may be incorrect, or it may change, but the empirically-defined object remains itself, to be modified only by the evolution of observational technique and accumulated data.

There are two excellent reasons for believing that most worms are *not* chimneys. Theoretically, the thick disk of low-density WIM electrons makes it difficult for shells to break out of the disk and connect to the halo. This is seen in numerical treatments (e.g. Mac Low, McCray, and Norman 1989; Palouš; Shapiro; Tomisaka) which show that adding just the H I z-extended component, which is only about half the thickness of the WIM layer, considerably reduces the chance for breakout. Observationally, a very important fact (Cox) is that in those galaxies that have been studied, no more than 1% of the total supernova power is emitted as diffuse X-rays from $T \sim 10^6$ K gas. A qualification for the Galaxy: this estimate rests on assuming that the observed X-rays, which are sampled only locally because of absorption by intervening neutral matter, are representative of the whole Galaxy.

If gas flows up into the halo from the disk through chimneys, driven by correlated supernovae, it must be hot. If it is hotter than $\sim 7 \times 10^6$ K, it will escape as a wind (Heiles 1987) unless it cools rapidly enough either by expansion or radiation, in which case it will eventually reach the 10^6 K at which it would be easily observable in X-rays. If the gas is injected at $T < 10^6$ K, it must have lost thermal energy either during the explosion process or on its way out to the halo; current theory does not suggest that this occurs.

Alternatively, worms may be chimneys. Suppose that the halo gas lies between $\sim 2 \times 10^6$ K and 7×10^6 K so that it is neither observable in X-rays nor escapes the Galaxy, and that the energy is emitted by the $T \sim 10^5$ K halo gas observed in UV absorption and emission (section 3). If this gas is distributed over the whole Galaxy, then the total luminosity in these lines really amounts to half the total supernova power. If the gas really lies above the WIM at $z \gtrsim 1$ kpc, then the supernova power permeates the halo and we are almost forced to conclude that most of the worms *are* chimneys.

Thus, while the absence of observable X-ray emission, particularly at this level of 1% of the supernova power, is a powerful constraint, it may not be relevant. It is intriguing that these UV emission lines, which are so very difficult to observe, might highlight the most energetically important phase of the diffuse ISM! We desperately need more observations of the UV emission lines to definitively establish their pervasiveness.

Given these uncertainties, observers *must not* call these objects chimneys. We do not know whether they connect to the halo or not. We can imagine that a worm does either: when it dies it may go to heaven (up into the halo) or not. Thus the term 'worm', or some other suitable term defined on a purely *empirical* basis, is better for these entities that are empirically defined by their sharp, well-defined typically vertical structure.

Observationally, how can we determine whether a worm is indeed a chimney? We cannot use the mere existence of an H I hole, because only if the hole continues all the way through the higher-lying WIM can we be sure that there is a direct connection. We cannot use the cones of ionization caused by O stars located within H I holes because the WIM does not absorb those photons; these cones should exist whether or not there is a hole in the WIM.

I can think of just two observables, neither being very promising. One is to observe the upward-moving gas itself. It should be very hot; it might be detectable in X-ray emission or in absorption lines of specific highly-ionized species. Another is to observe the hole in the WIM itself. This is difficult, because the WIM has a very low emission measure and its *presence* is barely detectable; detecting its *absence* is even more difficult. But if holes in either the WIM or in the higher-z, $T \sim 10^5$ K gas can be detected, they would be indications of breakout into the halo.

We regard it as essentially certain that supershells and worms are produced by clusters of supernovae and stellar winds. However, there is another mechanism that operates in certain specific cases. Observationally, evidence for interaction of HVC's and disk gas to produce the Galactic 'anticenter shell' was reviewed by Mirabel, and in external galaxies evidence for spectacular interactions of HVC and ambient gas was reviewed by van der Hulst. Theoretically, the very largest supershells cannot be produced by correlated supernovae because the energy gets transferred to vertical instead of horizontal motion in the disk, unless the disk is very thick. It would be nice to economize by invoking the minimum number of mechanisms to produce the observed effects and assume that *all* supershells and worms are produced by HVC interaction. However, the total energy in HVC's is only $\sim 1\%$ that in supernovae and is insufficient for the task.

There are a number of fundamental, currently unanswered questions concerning superbubbles and related structures. The holes they produce in a galactic disk should be round, except as modified by differential rotation with age (Palouš, Franco, and Tenorio-Tagle 1990); thus, when observed in external galaxies, H I hole shapes should follow a well-defined distribution which depends on the inclination angle and the rotation curve of the galaxy. The holes are supposed to have been produced by shocks, which sweep up the matter in the hole and, after becoming radiative, deposit it in a dense shell on the outside of the hole; these dense shells have never been observed in either atomic or molecular gas. Why? Supershells are almost never observed as complete spheres, but only as hemispheres or less. Is this because the supernovae that produce them blow up next to dense molecular clouds, so that the explosion energy is free to drive a fast shock in only one direction? If so, what happens to the molecular cloud, and from the theoretical standpoint what are the shell dynamics in such a macroscopically inhomogeneous medium? What is the effect of the partly ordered, mainly random ambient interstellar magnetic field on the shell dynamics?

Braun estimated that the observing time required to attack these problems on some of the world's great telescope arrays runs into several months. Similarly, I suspect, the computing time required on the world's greatest computers also seems prohibitive. However, I suggest that both we and the directors of such facilities alter our attitudes. Most observatories parcel out time in small chunks in order to satisfy a large group of users, and never award very large amounts of time to individual, important projects. However, there are some projects that are so important to our understanding of fundamental issues that the expenditure of significant resources—be they observing time or money—is justified. Some of the small satellites, such as IRAS and COBE, are prime examples. Similarly, I believe, a few well-selected projects that will elucidate the fundamentals of the disk-halo interaction have enough merit to justify altering our traditional criteria for awarding telescope time.

8. BASIC PROPERTIES OF THE ISM

Theorists cannot concoct applicable theories for conditions that differ from those they assume. As observers, we have the responsibility to provide this information. After decades of work and the expenditure of much telescope time and taxpayers' money, I'm afraid we have failed. This is not entirely our fault. The ISM is a complicated multiphase medium, and whenever we make a new type of observation that highlights any temperature we in fact see gas at that temperature. Interpretational difficulties are compounded by the facts that the optical, UV, X-ray, and most high-latitude observations can sample only nearby material, and the local region is not very representative (Cox and Reynolds 1987).

What component of the ISM occupies most of the volume? Back in the 1960's, the two-phase model was popular and predicted that the warm neutral medium (WNM) would do so. This is quite consistent with the observations: H I is distributed rather smoothly over the sky, from which we infer that it is rather smoothly distributed in 3-d space.

The theoretical picture changed in the 1970's with the realization, mainly by Cox and Smith (1974) and McKee and Ostriker (1977), that supernovae are more than just perturbers of the ISM: instead, they dominate it. The interior of a supernova remnant is filled with hot gas (the hot ionized medium, or HIM), and the remnants grow so big that this gas should fill most of space. This picture seemed to agree with the observation of soft X-ray emission from thermal gas located in the Solar vicinity, but seems to disagree with the H I data. Recently, Cox has changed his mind for reasons he has explained in this meeting.

Meanwhile, H I observers have been trying, in spirit if not in fact, to accommodate the theoretical picture of HIM filling most of space. Braun, who I believe was talking primarily of M31, expressed this possibility in discussing the idea of having a large *2-d*, or *area*, filling factor together with a small *3-d*, or *volume*, filling factor. In spirit, I personally have come to realize that a large 2-d filling factor of H I, which is what we really mean by saying that the H I distribution as observed from the Earth looks 'smooth', does not necessarily imply a large 3-d filling factor. Much of the H I is distributed in sheets or shells. A bedsheet covers a bed but does not occupy very much volume, and if the interstellar H I covers the Galactic plane in the same way a sheet covers a bed we might reproduce the H I observations—with a large 2-d but a small 3-d filling factor, as might be produced in a HIM-dominated ISM. Observationally, the point is this: we tend to assume that gas at different velocities lies at different distances. This is certainly not always the case. To what degree is it 'not always the case'? This question needs to be answered, but the answer will not come easily.

We don't even know whether H I 'clouds' are primarily filaments or sheets. We often observe real shells, and these are certainly best described as sheets. The *Copernicus* satellite definitively established the existence of sheets, for example in front of ζ Oph (Morton 1975). However, H I maps also exhibit objects that look more like filaments. They are curved, and reminiscent of shells, but their insides have very low column densities; we discuss the specific case of the 'NCP' shell below. However, the fact that something *looks* like an isolated filament does not mean that it *is* one. The reason is that most shells are not complete. An incomplete shell, if approaching us, is recognizable as a portion of a shell. But an incomplete shell that moves across our line of sight looks more like a filament because only the 'tangentially viewed' portion of the shell exists. In principle we should be able to distinguish a filament from a partial shell from the velocity distribution, because the radial velocity of the tangentially-moving portion of a shell varies rapidly with line-of-sight distance. Unfortunately, detailed studies of a reasonable sample have never been done.

Not only do we not know the geometry of the ISM, we do not know its topology. Many theoretical models predict a 'Swiss cheese' structure for the ISM; others predict a 'spaghetti' type structure, with or without 'meatballs'. Some of these models predict that the cold H I clouds fill the holes in the cheese, and others that the HIM bubbles, immersed in a much cooler medium, fill the holes. Other models (now out of fashion) have predicted that the HIM fills tunnels—a spaghetti structure, which has never been either observed or ruled out. We observe cold H I and molecular filaments, which are either true filaments or the caustics of nearly edge-on sheets. Topologically, what is the connectedness of any particular ISM component?

Filling factors of the various components of the ISM are uncertain. So is the magnetic field. And what little knowledge we do have of all these matters is generally restricted to the Solar circle, or more specifically the Solar neighborhood.

The well-defined arching structure centered near $(l, b) \approx (130°, 28°)$ is an excellent example of some of these points. I call this the 'North Celestial Pole', or NCP shell, because it passes right through the pole. Studies of this object have been the subject of several excellent poster papers presented at this meeting. This object does not appear to be a complete shell because its interior area is almost completely empty—it has one of the lowest H I column densities anywhere in the sky. Meyerdierks and Heithausen model it as a cylindrical cavity formed by collision of nearby HVC's with the ordinary disk gas. Alternatively, such a cylindrical cavity might be a chimney viewed end-on! However, the cylindrical geometry is not the only interpretation. Grenier models it as gas ejected from a nearby well-defined expanding shell. And we at Berkeley believe that it may be a portion of a shell, oriented such that it expands primarily in the plane of the sky; we see the shell portion tangentially so it looks like a filament. The NCP shell has a strong magnetic field (Heiles 1989), which presumably should be incorporated in a successful model.

Another point about the NCP shell is that it seems to cause spectacular scintillation of background radio sources. There is a new class of scintillating radio sources whose intensity varies by very large factors (Fiedler et al. 1987); their 'light curves' make it appear as if the sources are being occulted by interstellar structures. These 'extreme scattering events' are probably a result of 'refractive scintillation' (Coles et al. 1987). The archetype is the source 0954+658, located at $(l, b) = (146°, 43°)$, which lies behind the NCP shell. The scintillation is caused by very small scale high-density fluctuations in electron density that have scale lengths across the line of sight of the order of 1 a.u. (1.5×10^{13} cm) and electron column densities along the line of sight of order 10^{18} cm (Clegg, Chernoff, and Cordes 1988). If they are produced by a thin sheet such as a shock with a line-of-sight length 100 times the sheet thickness—an assumption designed to minimize the inferred electron density n_e—then $n_e \sim 10^3$ cm^{-3}. This corresponds to an enormous gas pressure! It strikes me that we might learn a lot about the dynamics of shells, supershells, worms, and the ISM in general by having excellent statistical studies of these events. A related observation is the study of time variability of pulsar dispersion measures.

I believe that one way to learn much about the ISM, and particularly the influence of various forces on it such as supernovae and gravity, is to compare its properties at different Galactocentric radii R_G and in different galaxies. Supernova rates vary from galaxy to galaxy, and with R_G within a galaxy. So does the z-component of the gravitational field. In the extreme cases of starburst galaxies, the Heckman 'superwinds' reviewed by Norman give us relieved confidence that extreme supernova rates do, in fact, produce the expected effects. Walterbos presented some of the first comparative results concerning the WIM in external galaxies; a much larger sample of galaxies is needed!

With regard to H I, Braun's study of temperatures in M31, using background continuum sources to measure the absorption, is an admirable first step. He finds that the H I in M31 is warmer than that in the Galaxy. Another admirable step, not presented at this meeting, is the determination of H I temperatures as a function of R_G within our Galaxy by Garwood and Dickey (1989); they found that the Galactic H I gets warmer toward the center of our Galaxy.

These are contradictory results in at least one sense. The supernova rate *increases* toward the center of our Galaxy, and the rate is thought to be *smaller* in M31 than in the Galaxy. But the H I temperatures are higher in *both* regions. This implies that the H I temperature is not affected strongly by the supernova rate. This is a disappointing result because theoretically the gas temperature and the supernova rate should be linked. Wang and Cowie (1988) argued that the ISM pressure should increase with SN rate, and Cioffi (1985) found that the ISM pressure should increase with the fraction of SN that are correlated. For both reasons, the gas pressure should increase towards the Galactic interior. Increased pressure should result in higher-density, cooler clouds.

ACKNOWLEDGMENTS

It is a great pleasure to thank the Joint Institute for Laboratory Astrophysics, where I have spent much of this year, for the ideal research environment. My colleagues there commented in detail on an earlier version of this paper, which crystallized my thoughts and resulted in significant improvements; these people include Mitch Begelman, Denis Cioffi, Andrew Hamilton, Marthijn de Kool, Dick McCray, Mike Shull, and Ellen Zweibel. I received similar help from Rainier Beck, Leo Blitz, Don Backer, and Vladimir Dogiel. My travel was supported by the Research Committee of the University of California, Berkeley, and grant number 443836-21705 from the National Science Foundation.

REFERENCES

Beck, R. 1982, *Astron. Ap.*, **106**, 121.
Begelman, M.C. and Fabian, A.C. 1990, *Mon. Not. Roy. Astr. Soc.*, **244**, 26P.
Blitz, L. 1988, in *The Evolution of Galaxies*, ed. Jan Palouš, Czechoslovak Academy of Sciences.
Cioffi, D.F. 1985, Ph. D. thesis, University of Colorado, chapter 4.
Clegg, A.W., Chernoff, D.F. and Cordes, J.M. 1988, in *Radio Wave Scattering in the Interstellar Medium*, ed. J.M. Cordes, B.J. Rickett, and D.C. Backer, p. 174.
Coles, W.A., Frehlich, R.G., Rickett, B.J., and Codona, J.L. 1987, *Ap. J.*, **315**, 666.
Cox, D.P. and Reynolds, R.J. 1987, *Ann. Rev. Astr. Ap.*, **25**, 303.
Cox, D.P. and Smith, B.W. 1974, *Ap. J.*, **189**, L105.
Fiedler, R.L., Dennison, B., Johnston, K.J., and Hewish, A. 1987, *Nature*, **326**, 675.
Garwood, R.W. and Dickey, J.M. 1989, *Ap. J.*, **338**, 841.
Gerola, H., Kafatos, M., and McCray, R. 1974, *Ap. J.*, **189**, 55.
Goerigk, W., and Mebold, U. 1986, *Astron. Ap.*, **162**, 279.
Heiles, C. 1987, *Ap. J.*, **315**, 555.
Heiles, C. 1989, *Ap. J.*, **336**, 808.
Israel, F.P. and Mahoney, M.J. 1990, *Ap. J.*, **352**, 30.
Kulkarni, S. and Heiles, C. 1987, in *Interstellar Processes*, D.J. Hollenbach and H.A. Thronson, Jr. (eds.), 87.
Krause, M., Beck, R., and Hummel, E. 1989, *Astron. Ap.*, **217**, 17.
Krause, M., Hummel, E., and Beck, R. 1989, *Astron. Ap.*, **217**, 4.
Kutyrev, A.S. and Reynolds, R.J. 1989, *Ap. J.*, **344**, L9.
Mac Low, M, McCray, R., and Norman, M.L. 1989, *Ap. J.*, **337**, 141.
Martin, C. and Bowyer, S. 1990, *Ap J.*, **350**, 242.
Mathis, J.S. 1986, *Ap. J.*, **301**, 423.
McKee, C.F. and Ostriker, J.P. 1977, *Ap J.*, **218**, 148.
Morton, D.C. 1975, *Ap. J.*, **197**, 85.
Oort, J.H. 1966, *Bull. Astron. Inst. Netherlands*, **18**, 421.
Palouš, J., Franco, J., and Tenorio-Tagle, G. 1990, *Astron. Ap.*, **227**, 175.
Phillips, S., Kearsey, S., Osborne, J.L., Haslam, C.G.T., and Stoffel, H. 1981, *Astron. Ap.*, **98**. 286.
Rand, R.J. and Kulkarni, S.R. 1989, *Ap. J.*, **343**, 760.
Rand, R.J. and Kulkarni, S.R., and Hester, J.J. 1990, *Ap. J.*, **352**, L1.

Reynolds, R.J. 1990, *Ap. J.*, **349**, L17.
Sancisi, R. and Allen, R.J. 1979, *Astron. Ap.*, **74**, 73.
Scoville, N.A. and Sanders, D.B. 1987, in *Interstellar Processes*, D.J. Hollenbach and H.A. Thronson, Jr. (eds.), 21.
Shane, W.W. 1971, *Astron. Ap. Suppl.*, 4, 315.
Songaila, A., Bryant, W., and Cowie, L.L. 1989, *Ap. J.*, **345**, L71.
Songaila, A., Cowie, L.L., and Weaver, H.F. 1988, *Ap. J.*, **329**, 580.
Wang, Z. and Cowie, L.L. 1988, *Ap. J.*, **335**, 168.

CONTENTS POSTER BOOK *

I. THE DISK-HALO INTERFACE IN OUR GALAXY

P. Abrahám
  The "Cepheus Bubble", Birth of a Galactic Chimney

O. Bienamé, A. Robin, M. Crezé, and V. Mohan
  The Vertical Distribution of Stars

J.H. Grobbelaar and D.J. van der Walt
  Star Formation, Supernovae and the Longitude Distribution of Galactic Gamma Rays

U. Haud
  Dynamics of the Outer Arm HVC

F. Jansen, I. Halm, and D. de Niem
  Cosmic-ray Antiprotons from the Central Region of the Galaxy?

P. Jenniskens and F.-X. Désert
  Emission and Extinction of Very Small Grains

J.L. Jonas and E.E. Baart
  The Rhodes 2300 MHz Survey

K.D. Kuntz and L. Danly
  The HVC/IVC Complex CII – Its Distance and Ionization

S.E. Labov and B.H.W. Yan.
  The Anticorrelation between the Soft X-ray Background and Neutral Interstellar Material on Small Angular Scales

H. Meyerdierks and A. Heithausen
  A Galactic-disk Loop Caused by Infalling Halo Clouds?

J. Milogradov-Turin
  Temperature Parameter as a Proof for the Absence of an Elliptical Galactic Radio Halo

---

* "The Interstellar Disk-Halo Connection in Galaxies – Poster Proceedings" (ISBN 90-72523-02-4 / available from Hans Bloemen, Space Research Leiden, P.O. Box 9504, 2300 RA Leiden, The Netherlands)

S. Sakamoto, T. Hasegawa, M. Hayashi, T. Handa, K. Sunada, N. Kaifu
   The University of Tokyo-Nobeyama CO(J=2–1) Galactic Plane Survey Project

V.R. Shutenkov
   Study of the Magnetic Field in the Galactic Halo using the Correlation Function of the Intensity of the Synchrotron Background Radiation

L.V. Tóth and L.G. Balázs
   Strip-like Structures in Cepheus on the IRAS Sky Flux Maps

II. THE DISK-HALO INTERFACE IN OTHER GALAXIES

A. Barteldrees and R.-J. Dettmar
   The Large-scale Structure of Stellar Disks

D.V. Bowen
   The Disk-Halo Interface of NGC 4319

C.L. Carilli, J.H. van Gorkom, and E.M. Haxthausen
   Radio Observations of Quasar-Galaxy Pairs: A Study of Extended Gas in Disturbed Systems

M. Dahlem, B. Koribalski, U. Mebold, and R. Wielebinski
   Mass Outflow from the Nuclear Region into the Halo of NGC 1808

R.-J. Dettmar and M. Dahlem
   A Comparison of the Diffuse Ionized Gas and Radio Continuum Emission in Edge-on Galaxies

R.-J. Dettmar, J.W. Keppel, M.S. Roberts, and J.S. Gallagher
   Observations of the Diffuse Ionized Gas Perpendicular to the Plane of NGC 891

G. Golla and R. Beck
   Radio-continuum Emission of the Edge-on Galaxy NGC 5775

T. Handa, R. Kawabe, S. Ishizuki, S. Ikeuchi, and Y. Sofue
   Thickness of Molecular Gas in the Edge-on Galaxy NGC 891

J.J. Hester, S.R. Kulkarni, R.J. Rand, and W.T. Deich
   Warm Ionized Gas in the Edge-on Spirals NGC 891 and NGC 3079

J.A. Irwin and E.R. Seaquist
   Heiles Shells in NGC 3079

J. Kamphuis
   High-velocity Gas and Superbubbles in the HI of M 101

D.-W. Kim, G. Fabbiano, and G. Trinchieri
   X-ray Spectra of Normal Galaxies and their Emission Mechanisms

K.M. Lanzetta
   The Evolution of High-redshift Lyman-limit Absorption Systems

G.B. Lima-Neto, D. Gerbal, F. Durret, and M. Lachièze-Rey
   The X-ray Analysis of Cluster A85 Revisited

M. Prieto, E. Battaner, E. Florido, E. Mediavilla, and M.L. Saavedra
   Star Formation at Large Galactic z and R of NGC 4013 and NGC 6504

M. Prieto, A.M. Varela, C. Muñoz-Tuñon, J.E. Beckman, D.P. Longley
   Photometric Studies of the Disk-Halo Interface in Sbc Galaxies

H.-P. Reuter, U. Klein, and R. Wielebinski
   Star Forming Regions and Relativistic Particles in M82

D. J. Saikia, A. Pedlar, S.W. Unger, D.J. Axon, and G.J. Yates
   The Large- and Small-scale Radio Structures of Nearby Galaxies

H. Schulz
   An HII Supershell in NGC 253

III. THEORY AND MODELLING

R.A. Benjamin and P.R. Shapiro
   Nonequilibrium Absorption Line and Emission Spectrum Diagnostics for Radiatively Cooling Galactic Halo Gas

R.C. Bishop and P.R. Shapiro
   The Galactic Dynamo in the Presence of a Gaseous Halo

A. Boulares
   Galactic Stability by the Warm Intercloud Gas

L. Brett and F.D. Kahn
   Magnetic Fields and Magnetic Reconnection in the Disk and Halo

S.P. Chakrabarti
    Properties of Self-similar Spiral Shocks in Disks

M.K. Dougherty, C.M. Ko, and J.F. McKenzie
    A Static One-dimensional Galactic Halo Model, with Cosmic-ray Diffusion and Alfvén Wave Effects

A. Ferrara and G. Einaudi
    The Role of Instabilities in the Hydrodynamics of the Galactic Halo

F. Ferrini, F. Matteucci, C. Pardi, and U. Penco
    A Model for the Chemical Evolution of Halo and Disk

M. Hattori, A. Habe, and T. Yoshida
    Non-linear Thermal Instability in X-ray Halos of Galaxies and Cluster Cooling Flows

M. Hernanz, E. Garcia-Berro, J. Isern, and R. Mochkovitch
    Theoretical Halo and Disk White Dwarf Luminosity Functions

C.M. Ko and M.K. Dougherty
    Flux-tube Formulation of a Static Galactic Halo

F. Li and S. Ikeuchi
    Model Reaction Equations of the Connected Disk-Halo System

S. Mineshige, P.R. Shapiro, K. Shibata, and T. Tajima
    MHD Simulation of Halo Cloud Formation by Thermal Instability

J. Palouš
    Expanding Bubbles in 3D

Peng Qiu-he
    An Exponential Disk Model of 3D Galaxies and a New Method of Determining the Thickness of Galaxies

P.R. Shapiro, S. Mineshige, and K. Shibata
    Large-scale Explosions and Superbubbles in the Galactic Disk and Halo: MHD Simulations and Analytical Approximations

I. Souvatzis
    The Velocity Structure of Intermediate Negative Velocity Clouds Created by High-velocity-cloud – Disk Interaction

M. Urbanik
    The Parker Instability and Halo Polarization in Spiral Galaxies

AUTHOR INDEX

Abrahm, P.	*3*	Einaudi, G.	*87*
Allen, R.J.	287	Fabbiano, G.	*57*
Ashe, G.A.	387	Fabian, A.	237
Axon, D.J.	*69*	Ferrara, A.	*87*
Baart, E.E.	*15*	Ferrini, F.	397, *89*
Balzs, L.G.	*29*	Florido, E.	*65*
Barteldrees, A.	*33*	Gallagher, J.S.	295, *43*
Battaner, E.	*65*	Garcia-Berro, E.	*93*
Beck, R.	233, 267, *47*	Garcia-Burillo, S.	299
Beckman. J.E.	*66*	Gerbal, D.	*63*
Benjamin, R.A.	*75*	Gilmore, G.	97
Berkhuijsen, E.M.	233	Golla, G.	233, 307, *47*
Bienamé, O.	*5*	van Gorkom, J.H.	*37*
Bishop, R.C.	*77*	Grobbelaar, J.H.	*7*
Blitz, L.	41	Guélin, M.	299
de Boer, H.	333	Habe, A.	*91*
de Boer, K.S.	161	Halm, I.	*11*
Boulares, A.	*79*	Handa, T.	307, *25, 49*
Bowen, D.V.	*35*	Hasegawa, T.	*25*
Brand, J.	121	Hattori, M.	*91*
Braun, R.	213	Haud, U.	*9*
Bregman, J.N.	387	Haxthausen, E.M.	*37*
Breitschwerdt, D.	373	Hayashi, M.	*25*
Brett, L.	*81*	Heiles, C.	165, 433
Carilli, C.L.	*37*	Heithausen, A.	*21*
Chakrabarti, S.P.	*83*	Herbstmeier, U.	161
Chi, X.	197, 198	Hernanz, M.	*93*
Cox, D.P.	143	Hester, J.J.	*51*
Crézé, M.	313, *5*	van der Hulst, J.M.	201
Dahlem, M.	299, *39, 41*	Hummel, E.	257
Danly, L.	53, *17*	Ikeuchi, S.	*49, 97*
Deich, W.T.	*51*	Irwin, J.A.	*53*
Désert, F.X.	149, *13*	Isern, J.	*93*
Dettmar, R.-J.	295, *33, 41, 43*	Ishizuki, S.	*49*
Dogiel, V.A.	175	Israel, F.P.	303
Dougherty, M.K.	*85, 95*	Jansen, F.	*11*
Durret, F.	*63*	Jenniskens, P.	*13*

Pagenumbers in italics refer to the poster book

Jonas, J.L.	15	Reach, W.T.	163
Jones, F.C.	359	Reich, W.	187
Kahn, F.D.	1, *81*	Reuter, H.-P.	*67*
Kaifu, N.	*25*	Reynolds, R.J.	67
Kamphuis, J.	201, *55*	Roberts, M.S.	295, *43*
Kawabe, R.	*49*	Robin, A.	*5*
Keppel, J.W.	295, *43*	Saavedra, M.L.	*65*
Kim, D.-W.	*57*	Saikia, D.J.	*69*
Klein, U.	*67*	Sakamoto, S.	*25*
Ko, C.M.	*85, 95*	Savage, B.D.	129
Koo, B.-C.	165	Schlickeiser, R.	377
Koribalski, B.	*39*	Schulz, H.	*71*
Kulkarni, S.R.	51	Sciama, D.W.	77
Kuntz, K.D.	*17*	Seaquist, E.R.	281, *53*
Labov, S.E.	*19*	Shapiro, P.R.	417, *75, 77, 99, 107*
Lachièze-Rey, M.	*63*	Shibata, K.	429, *99, 107*
Lanzetta, K.M.	*59*	Shutenkov, V.R.	*27*
Li, F.	*97*	Sofue, Y.	169, 307, 309, *49*
Lockman, F.J.	15	Souvatzis, I.	*109*
Longley, D.P.	*66*	Stephens, S.A.	323
Malin, D.F.	309	Sukumar, S.	287
Matsumoto, R.	429	Sunada, K.	*25*
Matteucci, F.	*89*	Tajima, T.	*99*
McKenzie, J.F.	373, *85*	Tobin, W.	109
Mebold, U.	161, *39*	Tomisaka, K.	407
Mediavilla, E	*65*	Tth, L.V.	*29*
Meyerdierks, H.	*21*	Trinchieri, G.	*57*
Milogradov-Turin, J.	*23*	Unger, S.W.	*69*
Mineshige, S.	*99, 107*	Urbanik, M.	*111*
Mirabel, I.F.	89	Valentijn, E.A.	245
Mochkovitch, R.	*93*	Varela, A.M.	*66*
Mohan, V.	*5*	Verschuur, G.L.	93
Muñoz-Tuñon, C.	*66*	Völk, H.J.	345, 373
Nakai, N.	307	Wakamatsu, K.	309
de Niem, D.	*11*	Wakker, B.P.	27
Norman, C.A.	337	van der Walt, D.J.	7
Odegard, N.	281	Walterbos, R.A.M.	223
Palous, J.	*101*	Wielebinski, R.	307, *39, 67*
Pardi, C.	*89*	Wolfendale, A.W.	197, 198
Pedlar, A.	*69*	Wouterloot, J.G.A.	121
Penco, U.	*89*	Wyse R.F.G.	*97*
Peng Qui-He	*103*	Yates, G.J.	*69*
Pohl, M.	369	Yoshida, T.	*91*
Prieto, M.	*65, 66*		
Rand., R.J.	*51*